SUPERNOVA REMNANTS AND THE INTERSTELLAR MEDIUM

INTERNATIONAL ASTRONOMICAL UNION
UNION ASTRONOMIQUE INTERNATIONALE

SUPERNOVA REMNANTS AND THE INTERSTELLAR MEDIUM

Proceedings of the 101st Colloquium of the
International Astronomical Union
held in Penticton, British Columbia, June 8-12 1987

Co-sponsored by the Herzberg Institute of Astrophysics
and the Canadian Institute for Theoretical Astrophysics

Edited by

R. S. ROGER AND T. L. LANDECKER

Dominion Radio Astrophysical Observatory
Herzburg Institute of Astrophysics
National Research Council of Canada

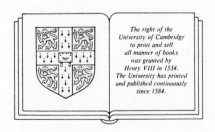

*The right of the
University of Cambridge
to print and sell
all manner of books
was granted by
Henry VIII in 1534.
The University has printed
and published continuously
since 1584.*

CAMBRIDGE UNIVERSITY PRESS
Cambridge
New York   New Rochelle   Melbourne   Sydney

Published by the Press Syndicate of the University of Cambridge
The Pitt Building, Trumpington Street, Cambridge CB2 1RP
32 East 57th Street, New York, NY 10022, USA
10 Stamford Road, Oakleigh, Melbourne 3166, Australia

© Cambridge University Press 1988

First Published 1988

Printed in Great Britain at the University Press, Cambridge

British Library cataloguing in publication data available

Library of Congress cataloguing in publication data available

ISBN 0 521 35062 X

TABLE OF CONTENTS

Contents

PREFACE

Supernovae and their remnants play a vital role in the evolution of a galactic disk, both by injecting heavy elements and cosmic rays and by energizing the interstellar material through radiative and magneto-hydrodynamic processes. At the same time, the detailed state of the circumstellar and interstellar medium surrounding a supernova, sometimes modified by the progenitor star, determines the subsequent evolution of an individual remnant nebula throughout its lifetime. This mutual interaction was the theme for the International Astronomical Union Colloquium 101 held in Penticton, British Columbia from June 8 to 12, 1987.

One hundred and seventy thousand years ago a B3 supergiant exploded in the Large Magellanic Cloud. The light from SN1987a reached the Earth a mere three months before the Colloquium, with considerable effects on astronomy as a whole and on our meeting in particular. The results are plain to see in the proceedings that follow.

The Colloquium attracted 132 participants from 17 countries. Twelve invited review papers and 86 contributed papers were presented. All but five are published in this volume.

We are indebted to the members of the scientific organizing committee

Dr. S. van den Bergh - Canada   Dr. V.N. Fedorenko - U.S.S.R.
Dr. R.D. Blandford - U.S.A.   Dr. D.J. Helfand - U.S.A.
Dr. J.L. Caswell - Australia   Dr. H. Itoh - Japan
Dr. R.A. Chevalier - U.S.A.   Dr. S. D'Odorico - Germany
Dr. B.T. Draine - U.S.A.   Dr. I.R. Tuohy - Australia
Dr. R.A. Fesen - U.S.A.   Dr. H.W. Yorke - Germany

for their help in planning the programme. We are also grateful for the encouragement given us by the Presidents of Commissions 33, 34, 40 and 48 of the I.A.U. and by its Assistant General Secretary, Dr. Derek McNally.

Financial assistance was received from the co-sponsors, the International Astronomical Union, the Herzberg Institute of Astrophysics of the National Research Council of Canada and the Canadian Institute for Theoretical Astrophysics. We are also indebted to the Natural Sciences and Engineering Research Council for a financial grant.

Our organizing efforts were aided by the entire staff of the host institution, the Dominion Radio Astrophysical Observatory. We are particularly grateful to Lloyd Higgs, Erika Rohner, Cindy Furtado, David Lacey and Bette Jones for making everything happen. Serge Pineault, David Routledge and Fred Vaneldik came from far away to help run the show. Sidney van den Bergh organized and chaired the concluding panel discussion.

Tom Landecker and Rob Roger

SUPERNOVA REMNANTS AND THEIR SUPERNOVAE

Robert P. Kirshner
Harvard-Smithsonian Center for Astrophysics
Cambridge, MA 02138, USA

Abstract: Observing supernova remnants provides important clues to the nature of supernova explosions. Conversely, the late stages of stellar evolution and the mechanism of supernova explosions affect supernova remnants through circumstellar matter, stellar remnants, and nucleosynthesis. The elements of supernova classification and the connection between super- nova type and remnant properties are explored. A special emphasis is placed on SN 1987a which provides a unique opportunity to learn the connection between the star that exploded (whose name we know) and the remnant that will develop in our lifetimes.

Introduction:

The bright supernova 1987a offers the unique oppor- tunity to watch the development of a supernova into a supernova remnant, given moderate living and good health. It illuminates the intimate connection between supernova explosions and the remnants they create. The link to the young supernova remnants is clear: the composition depends on stellar interiors, the environment on stellar mass loss, and the energetics and the stellar remnant depend on the details of the collapse or explosion. The link to older remnants is less obvious, and we often act as if there is no shard of stellar history left to affect the remnant's behavior. But this may be misleading. Some remnants which are old enough to have swept up a large mass of interstellar matter may yet bear the signature of their stellar origin in unmixed debris, or in the structure of the surrounding interstellar medium.

Classifying Supernovae:

Supernovae are classified into two bins, based on their spectra near maximum light (Zwicky 1965). Type I supernovae(SN I) have no hydrogen lines in their optical spectra, and Type II supernovae (SN II) do. An unknown supernova can be classified by comparison to the prototypes for each class (Kirshner et al 1973, Oke and Searle 1974, Branch et al 1981, Branch et al 1983). This is a good empirical approach, although it does not guarantee that the types correspond to the stellar origins or to the physics of the explosion. The ideal classification scheme would describe whether a supernova results from a nuclear explosion or from a core collapse, and whether the star was of low mass or of high mass. The presence or absence of hydrogen on the surface may not be the ideal indicator for these more basic properties. Although the SN I probably do correspond to violent nuclear burning in low mass stars and SN II probably do correspond to high mass stars with core collapse, the other morphological boxes may be populated. For example, supernovae may result from core collapse in massive stars which have lost their surface hydrogen through stellar winds. The recent isolation of a subclass of SN I, the SN Ib, illustrates this point (Porter and Filippenko 1987). These explosions are widely (though not universally) thought to arise from core collapse in massive Wolf-Rayet stars, which could have a small amount of surface hydrogen, but the heart of a SN II.

SN I:

A successful picture of the SN I consists of exploding white dwarf stars, nudged over the Chandrasekhar limit by mass transfer from a binary companion. This fits the circumstantial evidence that they are the only supernovae seen in elliptical galaxies since they need not have very massive progenitors. It fits the interpretation of the spectrum at maximum light, which can be synthesized from the expected composition of a carbon/oxygen white dwarf and the observed colors at maximum. It fits the energetics of the late time photometry, too, with the long exponential decline (Doggett and Branch 1985), which is known to persist for at least two years (Kirshner and Oke 1975) powered by the radioactive decay of a few tenths of a solar mass of nickel which beta decays to cobalt and then to iron. The deflagration (subsonic burning) of a carbon/oxygen white dwarf is expected to produce this material as a result of the fusion reactions that disrupt the star. A model for the emission spectrum seen at late times in SN I is consistent with the excitation of this iron-peak material by the radioactive decay chain (Kirshner and Oke 1975, Axelrod 1980).

A serious problem with this picture is posed by the failure of X-ray observations to find large iron abundances in the remnants of supernovae that are widely thought to be from SN I (Hamilton et al. 1985). Of course, the spectroscopic evidence provided by Tycho on his 1572 event is no better than for SN 1006, so the classification is not really comparable to that for contemporary supernovae, but these two, along with Kepler's SN (SN 1604) are widely thought to

be remnants of SN I. A solution to this riddle comes from
the suggestion that the iron is too cold. (Hamilton,
Sarazin, and Szymkowiak 1986). If the iron, slowly ejected
from near the core, lies in the interior of the remnant, the
reverse shock may not yet have reached that material, and
may not yet have heated and ionized it to produce X-ray
emission.   Although such a picture may appear contrived,
there is some direct evidence from UV observations that this
may actually be the case.

IUE observations of the SN 1006 remnant, originally
carried out by Wu et al. (1983) suggested the presence of
cold iron seen in absorption against the ultraviolet
continuum of a background star. A painstaking analysis of
the old and some new IUE spectra by Fesen et al (1987) makes
this original suggestion convincing, and reveals several Fe
II lines, presumably due to the cold iron in the interior of
the remnant.

This combination of theoretical and observational work
makes it quite plausible that the deflagrating white dwarf
model for SN I is consistent with the observed remnants.
The SN I light curves, especially their peak luminosities,
are expected to have a narrow dispersion in this case, since
the explosion takes place in a well-defined stellar setting
by a sharply constrained physical mechanism (Arnett, Branch,
and Wheeler 1984). The empirical evidence is that SN I do
have very similar properties, and might make good standard
candles for cosmology (Sandage 1985).

A second puzzle associated with SN I has been the optical spectra of the young remnants. Although they result
from the violent disruption at 12 000 km s^{-1} of a star with no
hydrogen, the spectra of SN 1006 and of SN 1572 show only
hydrogen lines at zero velocity. The spectrum of H alpha
for SN 1006 actually has two components, a narrow feature
and a broad one with a FWHM of about 2600 km s^{-1} (Kirshner,
Winkler, and Chevalier 1987). This type of structure was
predicted by Chevalier, Kirshner and Raymond (1980) based on
similar observations of SN 1572 and pursued, but not
detected, by Lasker (1981).

The model for this emission is that the "non-radiative" supernova shock is overrunning neutral material in the
neighborhood. Otherwise, the expected post-shock temperature would be so high that no optical emission would be
seen. Of course, the very presence of neutral material
close to a supernova puts a limit on the uv flash produced
at the surface of the star when the supernova shock wave
arrives. Whether the gas itself results from mass loss from
the binary system is not known. Since the shock in the
interstellar medium generated by the disrupted star is

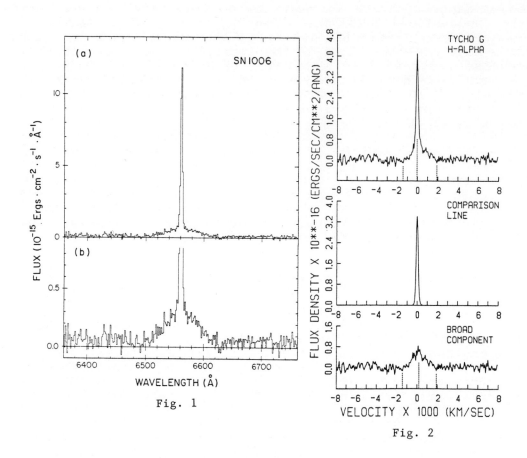

Fig. 1

Fig. 2

Fig. 1-- The H alpha line of SN 1006. (a) Illustrates the strength of the narrow component relative to the broad. (b) Shows the width of the broad component.

Fig. 2-- The H alpha line of SN 1572. (a) The observations. (b) A comparison line profile observed with the identical equipment. (c) The observations with a suitably scaled instrumental profile subtracted.

mediated by charged particles, the neutral hydrogen atoms suddenly find themselves surrounded by fast moving electrons and protons. These particles might excite the H atoms: the subsequent radiative decay would give rise to a narrow emission line at zero velocity. An alternative possibility is charge exchange, producing a fast-moving atom in an excited state. These decays create a broad line whose width reflects the velocity distribution. Since the ratio of the direct excitation to the charge exchange is a function of velocity, the relative strength of the two components, as well as the width of the broad line, reflect the shock velocity.

The interesting fact is that the velocity derived from the width and the velocity derived from the ratio of narrow lines to broad lines give consistent answers: 2300 km s^{-1} for SN 1572 and 3300 km s^{-1} for SN 1006. These velocities can be compared with the proper motions measured by Kamper and van den Bergh (1978) and by Hesser and van den Bergh (1981) to give distances to these two SN I: 1.4-2.1 kpc for SN 1006 and 2.0-2.8 kpc for SN 1572. The evidence is consistent with SN 1006 expanding into a lower density medium than SN 1572, just as required by the detailed models for the X-ray emission (Hamilton et al. 1986)

In principle it would be interesting to use this analysis of the remnant together with the contemporary accounts of the supernova explosion to establish the absolute magnitude of SN I explosions and calibrate the extragalactic distance scale. In practice, the uncertainties in the early records make this a precarious enterprise.

The follow-up to the Einstein X-ray survey of the LMC by Tuohy et al. (1982) revealed four LMC remnants with narrow-band images that showed H alpha, but not [S II]. Whether or not these are the remnants of SN I is not shown by the images or by spectra, but the resemblance to SN 1572 and SN 1006 is suggestive. Winkler and Kirshner have observed these four remnants at the CTIO 4m: we find that one has H alpha only, but no broad component, one has H alpha only with a broad component, one has H alpha with a broad component and weak [S II] emission, and one has H alpha with no broad component, [N II], and [S II]. It is possible that these represent the stages of evolution from the high velocity "non-radiative" shock down through the beginning of an ordinary cooling shock in ionized interstellar material. In any event, it may be possible to measure the shock velocity for some of these, for comparison with X-ray models. As a class, these Balmer-dominated remnants provide a strong warning that searches for extragalactic SNR's (such as that reported in this volume by Long et al) which depend on the [S II]/H alpha ratio will miss some remnants.

The other young SNR which is widely thought to result from a SN I is Kepler's SN 1604. There, the late stellar evolution and mass loss may be decisive in shaping the observed remnant. Spectroscopically, it shows strong [N II] lines, suggesting a high nitrogen abundance, such as might arise from the CNO-cycle hydrogen burning in a massive star. One way to account for this, the large distance from the galactic plane and the X-ray morphology has been suggested by Bandiera (1987) who considered extensive mass loss from a runaway star as a possible origin for Kepler's SNR. If that is correct, the usual classification as a SN I will have to be reconsidered.

SN Ib :

Another complication in the SN I picture has arisen from the recent evidence that there is a distinct subclass of SN I, the SN Ib, which show no hydrogen (and are there- fore SN I) but which do not have the strongest absorption feature (at about 6150 Å) that is seen in classical SN I (now called SN Ia). The observational situation has been summarized by Porter and Filippenko (1987). Several indirect lines of evidence converge to indicate that this spectroscopic difference is not just a detail, but that the SN Ib may come from a different type of star and may have distinct remnants. Those hints are (Uomoto and Kirshner 1985, Harkness et al 1987) that the SN Ib are fainter at maximum light, redder at maximum light, associated with H II regions, in Sc galaxies, and have some radio emission. These are generally taken to indicate that the progen- itors of SN Ib are massive stars, but not very extended ones, with some circumstellar matter: a good possibility would be Wolf-Rayet stars. This view is strengthened by the analysis of late time spectra by Begelman and Sarazin (1986), which indicates a mass of oxygen in excess of 5 solar masses, and the models of Schaeffer, Casse, and Cahen (1987) which show that the light curve of an exploding W-R star would conform well with the observed properties of SN Ib.

A recent set of observations of Cas A by Fesen, Becker and Blair (1987) shows that the connection between this well-known remnant and W-R stars may be important. Cas A has fast moving knots which are oxygen and oxygen-burning products from the interior of a massive star (Kirshner and Chevalier 1977, Chevalier and Kirshner 1978, 1979) and quasi-stationary flocculi (QSF), which have hydrogen and strong nitrogen lines and low velocities, and are presumably the relics of mass loss from the star. What Fesen et al have found is fast moving material with composition like the

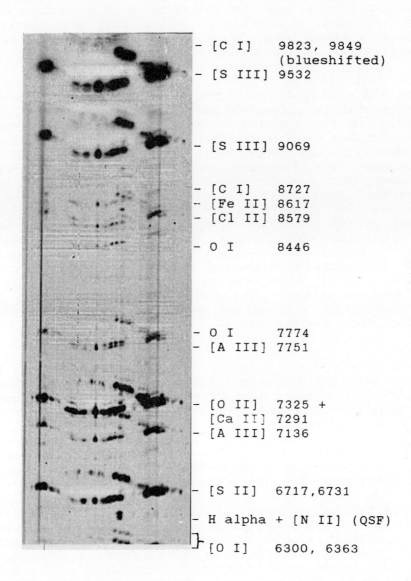

— [C I]   9823, 9849
            (blueshifted)
— [S III] 9532

— [S III] 9069

— [C I]   8727
— [Fe II] 8617
— [Cl II] 8579

— O I   8446

— O I   7774
— [A III] 7751

— [O II]   7325 +
  [Ca II] 7291
— [A III] 7136

— [S II]   6717,6731

— H alpha + [N II] (QSF)

— [O I]   6300, 6363

Fig. 3-- The near IR spectrum of Cas A. Note the large
velocity range of the approaching and receding shells. The
lines of [C I], [Cl II] and permitted O I are of special
interest.

QSFs.   They suggest that this nitrogen-rich, hydrogen rich material must have been on the surface of the massive star when it exploded, following an episode of mass loss, and that the star would have resembled a WN 7, a Wolf-Rayet star with substantial nitrogen on its surface.

The detailed link between the mass of the star that explodes, its evolutionary state at the moment of explosion, and the remnant it leaves behind can be investigated in part by looking at the chemical composition of the fast-moving debris.   For Cas A, the chemical analysis of the fast-moving knots leaves little doubt that this is the interior of a massive star, when compared with the models for massive stars (see for example Woosley and Weaver 1986).   New observational techniques can push this investigation into the UV (see for example Blair et al in this volume), and CCD's on ground-based telescopes make the near IR accessible.   Winkler and Kirshner have examined the near IR spectra of Cas A and we find it to be a rich source of chemical information, with O I recombination lines at 7774 and 8446Å, [Cl II] 8579, and [C I] 9823,9849 among the most novel features seen.   The abundance of carbon in these oxygen-rich knots will be an especially interesting number to extract from the spectra.

SN 1987a:

The external phenomena associated with the explosion of massive stars may hinge on whether or not the star has undergone extensive mass loss.   In that case, SN Ib might have the same underlying physical mechanism as SN II, core collapse, but the luminosity and spectral evolution would appear quite different because of the absence or presence of an extensive hydrogen atmosphere.   The same bomb can have different effects when placed in different suitcases.

This point has been illustrated vividly by SN 1987a, where we know the properties of the star that exploded, based on optical observations of the field before it exploded and UV observations afterwards.   The optical measurements show that the star, Sanduleak -69° 202, is coincident with the supernova (West et al 1987). Careful image analysis by Walborn et al (1987) shows that the Sanduleak star has two close neighbors, imaginatively dubbed star 2 and star 3.   Because the supernova faded rapidly in the UV, it has been possible to observe the neighborhood using IUE (Kirshner et al 1987).   Careful deconvolution of the IUE spectra by Sonneborn, Altner, and Kirshner (1987), Gilmozzi et al (1987) shows that the separation, magnitude, and spectra of the two stars seen is consistent with the identification of the survivors as stars 2 and 3 of the Sanduleak -69° 202 trio.   The clear implication is that Star

1, the 12th magnitude B3 I star, has disappeared, and it may be identified as the progenitor of SN 1987a.

The fact that SN 1987a brightened exceptionally rapidly, was intrinsically fainter than other SN II, and changed in color so quickly (Blanco et al 1987) can all be attributed to the fact that the star, plausibly a 15 - 20 solar mass object when on the main sequence, was blue, and hence more compact than a red supergiant, when it exploded.

An important question is whether the star was a red giant before it became a blue one. Recent evidence from ultraviolet spectra show that SK -69° 202 probably did have an episode of substantial mass loss before the explosion, plausibly as a red supergiant wind. This evidence comes from short wavelength IUE spectra (1100-1900Å). This is the region where the supernova itself has little detectible flux. However, narrow emission lines, principally of nitrogen have been detected in this region (Wamsteker et al 1987, Kirshner et al 1987) and are growing in strength. These lines would be absorbed if formed in the debris of the supernova itself, and are probably coming from a circumstellar shell at a distance of order a light year. This shell has been excited by UV emission from the supernova emitted in the first few hours as the shock wave from the interior hit the surface of the blue supergiant. The material is nitrogen rich, with $N/C > 10$, and N/O also high. This is the type of composition expected in the atmosphere of a red giant which has converted its carbon and oxygen to nitrogen as a by-product of hydrogen burning in the CNO cycle. The quantitative question associated with the mass loss: how much hydrogen was on the star when it exploded? probably will be answered by modelling the supernova's spectral evolution and the supernova's light curve.

These issues are relevant to studying supernova explosions, supernova remnants, and the interaction with the interstellar gas because the state of the atmosphere and the circumstellar environment can have a substantial effect on the supernova event and on the early evolution of the remnant. This is clearly important for the SN II, which are thought to arise from red supergiants, which are stars with considerable mass outflow. The radio emission for these is thought to arise in the circumstellar matter (Chevalier 1982), but there is also direct evidence for circumstellar matter in the spectra of SN II. For example, the spectra taken before maximum light by Niemela et al. (1985) of SN 1983k showed strong emission lines of highly ionized C and N, similar to the lines seen in Wolf-Rayet stars. Dopita et al (1984) observed narrow emission lines of H and He which they attributed to a wind surrounding this SN II. The interpretation of the UV spectra of SN II by Fransson et al

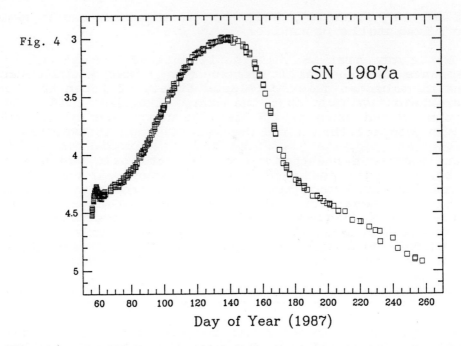

Fig. 4

SN 1987a

Day of Year (1987)

Fig. 4-- A V-like magnitude versus time for SN 1987a as measured with the IUE satellite fine error sensor. The exponential decline at late times is a strong clue that radioactive decay is important. The slope corresponds well with the 114 day decay of ^{56}Co, and the magnitude corresponds to about 0.07 solar masses of radioactive nickel.

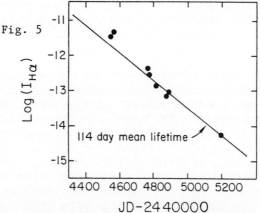

Fig. 5

JD-2440000

Fig. 5-- The H alpha line flux for the SN II 1980k. The straight line is not a fit to the data, but is a line with the 114 day e-folding time of ^{56}Co.

(1984) also requires the treatment of radiation reprocessed by the circumstellar gas.

Winds may have an important effect on the properties of remnants. One example is Puppis A, where the radio remnant is embedded in nitrogen-rich knots of material with low expansion velocities. This material is presumably analogous to that seen in SN 1987a, Kepler, and Cas A: the result of mass loss that includes the products of CNO cycling. At the same time, the chemical abundances of the stellar interior for Puppis A's progenitor are accessible to observation both through X-ray spectroscopy (Canizares and Winkler 1981) and optical spectroscopy (Kirshner and Winkler 1985). Both these lines of evidence suggest that Puppis A came from a star that produced large amounts of oxygen and neon: the stellar interior of a star around 15 solar masses. It is surprising, but true, that unmixed stellar debris is still present in Puppis A, even though the kinematic age is about 3700 years (see Winkler et al in this volume). Whether this is related to the interaction between the stellar wind and the more general interstellar medium is not yet clear.

There is a handful of cases where unmixed debris from deep within the star can be observed in the remnant: we ordinarily see high velocity gas with strong oxygen lines and no hydrogen emission. Those cases are: 1E0120 in the SMC, 0540-69 and N132D in the LMC, G292, Cas A, and Puppis A in our Galaxy, and the SNR in NGC 4449 (see Blair, Kirshner, and Winkler 1983). While this is strong evidence for massive progenitors, in most cases more detailed study of the remnants will be required to see whether an inference about the mass loss prior to the explosion can be supported. Those that exploded with a large hydrogen mass (presumably as red supergiants) would be the result of SN II, those which lost so much mass that they were compact at the moment of core collapse might correspond better to low luminosity SN II like SN 1987a or to exploding Wolf-Rayet stars such as Cas A might have been, and SN Ib might be.

SN II:

In discussing the evolution of massive stars with mass loss to become SN Ib, or SN 1987a, we have already alluded to the picture of SN II. In general, the core collapse to a neutron star in the red supergiant envelope of a massive star accounts for many of the observed features of the light curve near maximum light, the spectrum, and the circumstantial evidence that links SN II to regions of recent star formation. The observations of neutrinos from SN 1987a lends a satisfying note of reality to a picture in which the gravitational binding energy of a neutron star (of order

10^{53} erg) is converted 99% to previously undetected neutrinos, and 1% to kinetic energy that we see in the disrupting star, and eventually in the supernova remnant.

While we are accustomed to considering the shock
generated by the explosion as the source for energy in
supernova remnants, other sources of energy are present and
they may be of importance at early times. For example, the
formation of a neutron star could have significant effects
on the supernova remnant. The photoionization from the
pulsar in the Crab Nebula is the principal source of
excitation for the filaments we see there. In the past, the
relativistic pressure may have played a role in the dynamics
of the Crab. The pulsar in 0540-69 (Seward et al 1984) may
play an important role there, while the pulsar in Vela
probably is not an important factor in that much older
remnant.

Other features of the explosion may be significant in
the excitation of the very young remnant as it makes the
transition from a star to a nebula. In particular, the
inner zones of the star may be subjected to sufficient
heating that nuclear burning takes place. Even though the
principal energy source is the neutrino mediated shock, the
fusion to the iron peak (especially the doubly magic nucleus
^{56}Ni) can be significant. This decays to ^{56}Co which decays
to ^{56}Fe with an e-folding time of 114 days. Over the first
months of an ordinary SN II, the diffusion of shock energy
dominates the luminosity, but the radioactive energy input
from 0.1 solar mass of Ni can become significant at later
times. This effect can be seen in the light curve of SN
1987a, and in other SN II, where Uomoto and Kirshner(1986)
have shown that the H alpha flux tracks the radioactive
decay rate with surprising fidelity. Very young remnants
are an especially interesting subject for study because they
provide the links to stellar interiors and to the circumstellar surround that we seek to understand. The
presence of an energy source to keep this material warm and
visible in the first years after the explosion is a very
useful circumstance that deserves further exploration.

Acknowledgments:

Research on supernovae and their remnants at Harvard is
supported by the National Science Foundation through grant
AST85-16537 and by NASA through grants NAG5-841 and NAG5-
87. Many thanks to the organizers of IAU Colloquium 101 for
their indulgence after dinner and their patience after the
meeting.

REFERENCES

Arnett, W.D., Branch, D., and Wheeler, J.C. 1984, Nature 314, 337.

Axelrod, T.S. 1980, Ph. D. thesis UCSC, available as preprint 52994 from Lawrence Livermore National Lab

Bandiera, R. 1987, Ap.J. 319, 885.

Begelman, M.C. and Sarazin, C.L. 1986, Ap.J.(Lett.) 302, L59.

Blair, W.P., Kirshner, R.P., and Winkler, P.F. 1983, Ap.J. 272, 84.

Blanco et al 1987, Ap.J. 15 Sept 87

Branch et al 1983, Ap.J. 270, 123.

Branch et al 1981, Ap.J. 244, 780.

Canizares, C.R. and Winkler, P.F. 1981, Ap.J. (Lett.) 246, L33.

Chevalier, R.A. and Kirshner, R.P 1978, Ap.J. 219, 931.

Chevalier, R.A. 1982, Ap.J. 259, 302.

Chevalier, R.A., Kirshner, R.P. and Raymond, J.C. 1980, Ap.J. 235, 186.

Chevalier, R.A. and Kirshner, R.P. 1979, Ap.J. 233, 154.

Doggett, J.B. and Branch, D. 1985, A.J. 90, 2303.

Dopita, M.A., Evans, R., Cohen, M, and Schwartz, R.D. 1984, Ap.J.(Lett.) 287, L69.

Fesen, R.A., Wu, C.C., and Leventhal, M. 1987 Ap.J. in press.

Fesen, R.A., Becker, R.H., and Blair, W.P. 1987 Ap.J. 313, 375.

Fransson, C. et al. 1984, A & A 132, 1.

Gilmozzi,R. et al. 1987, Nature 328, 318.

Hamilton, A.J.S., Sarazin, C.L., Szymkowiak, A.E., and Vartanian, M.H. 1985, Ap.J. (Lett.) 297, L5.

Hamilton, A.J.S., Sarazin, C.L. and Szymkowiak, A.E. 1986, Ap.J. 300, 698.

Harkness, R.P et al 1987, Ap.J. 317, 355.

Hesser, J.E., and van den Bergh, S. 1981, Ap.J. 251, 549.

Kamper, K., and van den Bergh, S. 1978, Ap.J. 244, 851.

Kirshner, R.P. et al. 1987, IAUC 4435.

Kirshner, R.P., Sonneborn, G., Crenshaw, D.M., and Nassiopoulis, G.E. 1987, Ap.J. 15 September 1987.

Kirshner, R.P. and Oke, J.B. 1975, Ap.J. 200, 574.

Kirshner, R.P. and Winkler, P.F. 1985, Ap.J. 299, 981.

Kirshner, R.P., Oke, J.B., Penston, M.J., and Searle, L. 1973, Ap.J. 185, 303.

Kirshner, R.P. and Chevalier, R.A. 1977, Ap.J. 218, 142.

Kirshner,R.P., Winkler, P.F. and Chevalier, R.A. 1987, Ap.J.(Lett.) 315, L135.

Lasker, B.M. 1981, Ap.J. 244, 517.

Niemela, V., Ruiz, M.T., and Phillips, M.M. 1985, Ap.J. 289, 52.

Oke, J.B. and Searle, L. 1974, Ann. Rev. Astron. Astophys.
    12, 315.
Porter, A.C. and Filippenko, A.V. 1987, A.J. 93, 1372.
Sandage, A.R. 1985 in Lecture Notes in Physics 224, <u>Super-
    novae as Distance Indicators</u>, N.Bartel, ed. Springer
    Verlag.
Schaeffer, R., Casse, M., and Cahen, S. 1987, Ap.J. 316, 31.
Seward, F.D., Harnden, F.R., and Helfand, D.J. 1984, Ap.J.
    (Lett.) 287, L19
Sonneborn, G., Altner, B.A., and Kirshner, R.P. 1987, Ap.J.
    (Lett.) in press.
Tuohy, I.R., Dopita, M.A., Mathewson, D.S., Long, K.L. and
    Helfand, D.J. 1982, Ap.J. 261, 473.
Uomoto, A.K. and Kirshner, R.P. 1986, Ap.J. 308, 685.
Uomoto, A.K. and Kirshner, R.P. 1985, A & A 149, L7.
Walborn, N., Lasker, B.M., Laidler, V. and Chu, Y.-H. 1987,
    Ap.J.(Lett), in press.
Wamsteker, W. et al. 1987, IAUC 4410.
West, R.M., Lauberts, A., Jorgensen, H.E. and Schuster,
    H.-E. 1987, A & A (Letters) 177, L1.
Woosley, S.E. and Weaver, T.A. 1986, Ann. Rev. Astron.
    Astrophys. 24, 205.
Wu, C.C., Leventhal, M., Sarazin, C.L., and Gull, T.R. 1983,
    Ap.J. (Lett.) 269, L5.
Zwicky, F. 1965, in <u>Stars and Stellar Systems</u>, Vol VIII:
    Chap.7, ed. L.H. Aller, D.B. McLaughlin. U of Chicago
    Press.

THE CIRCUMSTELLAR STRUCTURE AROUND SUPERNOVAE

P. Lundqvist, Lund Observatory, Sweden
C. Fransson, Stockholm Observatory, Sweden

Abstract: The time dependent ionization and temperature structure of the circumstellar medium around supernovae has been calculated, in order to interpret recent supernova radio observations. For a stellar wind origin of the circumstellar medium, we relate the time of radio turn-on to the progenitor mass loss rate. We also show that large column densities for the UV resonance lines are expected. The results are applied to SN 1979c, SN 1980k and SN 1987A.

1. Introduction

Observations of recent Type Ib and Type II supernovae (cf. reviews by Chevalier, 1984, Fransson, 1986 and Panagia, 1987) can be explained by a model where the supernova ejecta expands into a dense circumstellar medium (Chevalier, 1982). The wavelength dependent delay between the optical and radio outbursts (Weiler et al., 1986) is well explained in terms of free-free radio absorption in the circumstellar medium outside the expanding blast wave. If we parametrize the pre-supernova mass loss with a steady rate $\dot{M}$ and expansion velocity u, the free-free optical depth between the radio emitting region and the observer scales as $\tau_{ff} \propto (\dot{M}/u)^2 \, T_e^{-3/2} \, x_e^2 \, \lambda^2 \, R_s^{-3}$, where T_e the wind temperature, x_e the fraction of free electrons in the wind, λ the wavelength and R_s the shock radius. The delay between optical and radio outbursts, and hence also the estimated mass loss rate, is thus, in addition to the mass loss rate, sensitive to both the temperature and the degree of ionization. To calculate these parameters, we have made time dependent photoionization calculations, including the ionizing effects of the supernova. A detailed discussion may be found in Lundqvist and Fransson (1987), here we only summarize the main results. In addition, we report similar calculations for SN 1987A.

2. Results and Discussion

The progenitors of 'normal' Type II SNe are thought to be red supergiants with masses $>10 \, M_\odot$ and extended atmospheres. Stars of this type are observed to ~ have strong winds with mass loss rates 10^{-6}–$10^{-4} \, M_\odot$/yr and wind velocities of ~10 km/s. For this type of supernovae, the wind structure is mainly influenced by the ionizing photons produced by Compton scattering. Here, photospheric photons are scattered by hot (~10^9 K) electrons in the shocked stellar wind close to the blast wave, thereby roughly doubling their energy in each scattering (eg. Fransson, 1986). For SN 1979c we find that the total number of ionizing photons is ~10^{60}. The photons will ionize both the wind and create an H II-region of radius ~20 $n_H^{-1/3}$ pc in the interstellar gas. This is roughly consistent with the observations of Cas A (~5.7 pc for n_H~15 cm^{-3}, Peimbert and van den Bergh, 1971). For low

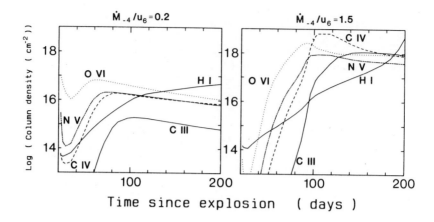

Fig. 1. The temporal dependence on the column densities of the most interesting ions for models representing SN 1980k (left) and SN 1979c. $\dot{M}_{-4}$ is the pre-supernova mass loss rate in 10^{-4} $M_\odot$/yr and u_6 the wind speed in 10^6 cm/s.

density stellar winds, $\dot{M}_{-4}/u_6 \lesssim 0.01$, where $\dot{M}_{-4}$ is the pre-supernova mass loss rate in 10^{-4} $M_\odot$/yr, and u_6 the wind speed in 10^6 cm/s, the initial EUV/soft X-ray outburst dominates the ionization of the circumstellar medium, creating roughly 10^{58}-10^{59} ionizing photons (Chevalier and Klein, 1979). Another important ionizing source for low density winds is the reverse shock propagating into the supernova ejecta. For dense winds, this contribution is severely absorbed in the thin interaction region between shocked supernova ejecta and shocked stellar wind, but increases in importance with time and dominates the total ionization after 50-300 days.

For parameters representing SN 1979c, we find that the wind temperature close to the blast wave rises to ~2×10^5 K during the first month, and decreases to ~1.3×10^4 K after ~400 days (Fig. 2). At the same time, the initially fully ionized wind recombines to ions like C II, N II and O I close to the blast wave. Outside ~0.1 pc, the density is too low for the gas to recombine and it remains almost fully ionized. This temporal behaviour is reflected in the column densities of the UV absorbing ions, shown in Fig. 1. For normal Type II supernovae, the optical depth in both C IV(λ1548-51) and N V(λ1239-43) should increase during the first months, due to recombination in the innermost part of the wind. They then remain roughly constant due to the increasing recombination time scale close to the blast wave. Considering the large column densities, these lines should be easily observable with the resolution of the Space Telescope.

The temperature and ionization structure close to the blast wave determines the radio absorption, and in Fig. 2 we show these parameters together with the resulting radio light curves for a model with $\dot{M}_{-4}/u_6=1.0$, roughly representing SN 1979c. An interesting point is that the dip at ~350 days in the 6 cm light curve is due to rapid cooling of

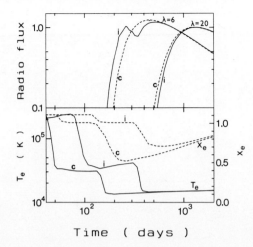

Time (days)

Fig. 2. Radio light curves for $\dot{M}_{-4}/u_6$=1.0. Lines marked 'i' are for equal ion and electron temperatures in the shocked wind, whereas those marked 'c' are for Coulomb heating of the electrons only. Note that the dip in the 6 cm light curve coincides with the temperature drop in the gas close to the blast wave, shown in the lower panel. This should be compared with a similar drop seen for SN 1979c.

the gas, temporarily increasing τ_{ff}. A similar dip is present in the observations of SN 1979c (Weiler et al., 1986) and thus reflects the cooling of the gas.

In Fig. 3 we relate the time of radio onset to the precursor mass loss rate. The full lines are for equal ion and electron temperatures in the shocked wind, dashed lines for Coulomb heating of the electrons only. For comparison the dotted lines show the $\dot{M}/u$ versus t relation for a fully ionized wind with temperature 10^4 K. For high mass loss rates with a turn-on of more than ~200 days at 20 cm, the optical depth is determined mainly by the recombination and the cooling of the wind. It is thus insensitive to the properties of the early UV and EUV flux. For low mass loss rates, the turn-on is set by the temperature of the wind, which depends on the flux and spectral shape of the flux from the shock and early EUV burst. Comparing with the observations by Weiler et al. (1986), we find mass loss rates of 1.2×10^{-4} $M_\odot$/yr and 3×10^{-5} $M_\odot$/yr for the pre-supernovae of SN 1979c and SN 1980k, respectively, for a wind speed of 10 km/s.

The early radio turn-on of SN 1987A, at 20 cm only ~2.1 days after the core collapse (Turtle et al., 1987), indicates a much less dense wind than around SN 1979c and SN 1980k. Using the circumstellar absorption model, Chevalier and Fransson (1987) find a mass loss rate of 8.8×10^{-6} $M_\odot$/yr for a wind velocity of 550 km/s, assuming a wind temperature of ~10^5 K. With the results for the photosphere calculated by Shigeyama et al. (1987), and the ionizing radiation from the reverse shock estimated by Chevalier and Fransson (1987), we have made a calculation for this wind density, using the same method as described above. In this case all the elements in the wind attain their helium-

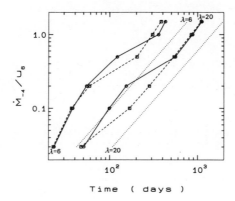

<u>Fig. 3</u>. The relation between radio turn-on at 6 cm and 20 cm, and pre-supernova mass loss rate. Solid and dashed lines have the same meaning as in Fig. 2. For a comparison, dotted lines are for a fully ionized wind at temperature 10^4 K. The radio turn-on is defined to occur when the radio flux is a factor e^{-1} times that extrapolated from the power law part of the light curve, corresponding to a free-free optical depth, $\tau_{ff}=1$.

like stages, except close to the blast wave where the wind becomes almost completely ionized. The temperature of the circumstellar medium is after ~1 day ~7.0×10^4 K throughout the wind, not very different from that used by Chevalier and Fransson. A more detailed discussion on the results for the structure around SN 1987A is presented elsewhere.

References:

Chevalier, R.A.: 1982, Ap. J. <u>259</u>, 302.
Chevalier, R.A.: 1984, Ann. N.Y. Acad. Sci. <u>422</u>, 215.
Chevalier, R.A., Fransson, C.: 1987, Nature <u>328</u>, 44.
Chevalier, R.A., Klein, R.I.: 1979, Ap. J. <u>234</u>, 597.
Fransson, C.: 1986, in <u>Radiation Hydrodynamics in Stars and Compact Objects</u>, eds. D. Mihalas and K.H.A. Winkler, Springer, p. 141.
Lundqvist, P., Fransson, C.: 1987, Astr. Ap., in press.
Panagia, N.: 1987, in <u>Cargèse Advanced Study Institute on High Energy Phenomena Around Collapsed Objects</u>, ed. F. Pacini, Reidel, in press.
Peimbert, M., van den Bergh, S.: 1971, Ap. J. <u>167</u>, 223.
Shigeyama, T., Nomoto, K., Hashimoto, M., Sugimoto, D.: 1987, preprint.
Turtle, A.J., Campbell-Wilson, D., Bunton, J.D., Jauncey, D.L., Kesteven, M.J., Manchester, R.N., Norris, R.P., Storey, M.C., Reynolds, J.E.: 1987, Nature <u>327</u>, 38.
Weiler, K.W., Sramek, R.A., Panagia, N., van der Hulst, J.M., Salvati, M.: 1986, Ap. J. <u>301</u>, 790.

ANALYSIS OF THE SPECTRUM OF THE TYPE V SUPERNOVA SN1986J

A. J. S. Hamilton, W. D. Vacca, A. K. Pradhan (JILA, Univ. of Colorado, Boulder, CO 80309), M. P. Rupen, J. E. Gunn (Princeton University Observatory) and D. P. Schneider (IAS).

Abstract: The supernova SN1986j resembles the prototypical Type V supernova SN1961v in the relatively slow ~1000 km/s expansion velocity, the slow light curve, and also in the $H\alpha$ dominated spectrum. The optical spectrum is similar to the spectra of some novae, and some OB stars with massive winds, being characteristic of a nebular plasma at about $10^{10} cm^{-3}$ and 10^4 K. What makes SN1986j exceptional is its tremendous radio luminosity, the brightest radio supernova observed to date. The radio emission indicates the presence of a massive circumstellar wind, with which the SN ejecta are now colliding. Since the cooling time of the optically emitting gas is about an hour, a heat source is required to power the light curve. Shocks moving back into the ejecta offer a natural heat source, and account quantitatively for the observed luminosity and spectral character of SN1986j. The large $H\alpha/H\beta$ ratio is attributed to trapping of $Ly\alpha$, which pumps the $n=2$ level of hydrogen, causing a finite optical depth in Balmer lines, and converting $H\beta$ to $P\alpha$ and $H\alpha$. The ratio of the derived $H(n=2)$ density and column density yields a size for the $H\alpha$ emitting region consistent with the thickness of a cooling shock, but less than 10^{-7} of the $10^{17} cm$ VLBI size. An important discriminant between shock models and photoionization models of the spectrum is that shocks predict Lyman 2-photon emission. The mass of the optically emitting material in SN1986j is about $1 M_\odot$, substantially less than the $2000 M_\odot$ argued in the case of SN1961v by Utrobin. However, there may be, and probably is, considerably more unobserved ejecta. This material should reveal itself as the remnant of SN1986j continues to evolve.

Introduction: SN1986j was discovered at the VLA (Rupen *et al.* 1987) as a radio source in the edge-on spiral galaxy NGC 891. The distance to NGC 891 is $\approx 7.6 h^{-1}$Mpc, making SN1986j the most luminous radio supernova to date. Prior epoch radio observations at the VLA and Westerbork show that SN1986j was present as a 6 cm radio source at ~1/3 the present luminosity in 1984, before which only upper limits are available. Thus SN1986j is at least 2 years old. The radio spectral index indicates that the source is optically thick at 20 cm. The VLBI FWHM radius of the object is about 7×10^{16}cm at 7.6 Mpc. Rupen *et al.* reported an optical spectrum, taken in September 1986, which reveals SN1986j as a low ionization emission line object, showing emission lines of H I, He I, O I and Fe II, with FWHM line widths of ~1000 km/s. The Balmer decrement is unusually large, $H\alpha/H\beta \approx 60$. The visual magnitude was ≈ 19.5 in September 1986, about 1 magnitude fainter than a prior epoch observation taken in January 1984. Between 1977 and 1984 upper limits exist at ~18th magnitude. Based on the similarities between this object and the Type V supernova SN1961v (Zwicky 1965), including an unusually low expansion velocity, a slow (years near maximum) light curve, a large $H\alpha/H\beta$ ratio, and prominent He I recombination lines, Rupen *et al.* classified the supernova as Type V.

Plasma Diagnostics: An optical spectrum of SN1986j is reported by Rupen *et al.* (1987). We obtained a second spectrum in October 1986; the new spectrum is similar to the old, but covers a slightly wider range 4000-9800 Å.

The weakness of Paschen lines (H 3-10,11,12) indicates that the large $H\alpha/H\beta$ ratio is most likely not due entirely to interstellar absorption; from the Paschen to Balmer line ratios we derive an extinction $A_V \approx 2$ in agreement with Rupen *et al.*

If it is assumed that hydrogen is collisionally ionized, and that the ionization of oxygen is tied to hydrogen by charge exchange, then the observed ratio [OII] $\lambda7320$/[O I] $\lambda6300 \approx 2$ implies a temperature $\approx 13,000$ K. The temperature is lower if hydrogen is photoionized, but this paper does not pursue that possibility. Since the O I temperature is probably lower than the O II temperature (see below), 13,000 K is probably a lower limit on the temperature of the region where O II and $H\alpha$ emit. If the O/H abundance is at least cosmic, then the large ratio $H\alpha$/[O II] $\lambda7320 \approx 50$ requires either a low temperature or a high density to suppress the [O II] emission; for a temperature $\gtrsim 13,000$ K, the required electron density is $n_e \gtrsim 3\times10^9$ cm^{-3}. On the other hand, the weakness of the OI$\lambda7774$ recombination line relative to [O II]$\lambda7320$ implies an upper limit of $n_e \lesssim 3\times10^{10}$ cm^{-3}. We conclude that the density and temperature of the optically emitting region are $n_e \approx 10^{10}$ cm^{-3} and $T \approx 10^4$ K. The high density is consistent with the presence (absence) of (un)observed forbidden lines in the spectrum.

There are a number of indications that the plasma is inhomogeneous, and that the optically emitting region may be only a fraction of the total. The O I temperature deduced from [O I]$\lambda5577$/[O I]$\lambda6300$ is only $T \approx 5,000$ K for any electron density higher than 10^8 cm^{-3}. At the same time, the presence of He I recombination lines requires temperatures $T \gtrsim 20,000$ K (while the absence of [O III]$\lambda4363$ constrains $T \lesssim 20,000$ K). The optical spectrum shows no evidence of any continuum; this implies an electron scattering optical depth less than unity, which in turn implies a mean electron density $\bar{n}_e \lesssim 2\times10^7$ cm^{-3} over a VLBI radius of 7×10^{16} cm. The inference is that the bulk of the plasma may be relatively cool, neutral, and unobserved.

What Powers the Light Curve?:   At $n_e \approx 10^{10}$ cm^{-3} and $T \approx 10^4$ K, the cooling time of the emitting plasma is about the recombination time which is about 1 hour. Consequently, a heat source is required to power the light curve. Possible heat sources include:
(1) Photons still diffusing out of the SN fireball? No - there is no evidence of any optical continuum.
(2) Radioactivity? No - none left at ~3 years old; also, the supernova should have been brighter in the past.
(3) Photoionization? This possibility appears difficult to rule out; it requires a hot (nonthermal?), nonextended source with a luminosity $\approx 10^7 L_\odot$, such as, perhaps, a Very Massive Object, or a mini-quasar. Chevalier (1987) has proposed a central pulsar to power the optical display.
(4) Shocks? As we now argue, shocks are expected, and account quantitatively for the observed luminosity and spectral character of SN1986j.

Shocks: The picture we are proposing of SN1986j is illustrated schematically in the Figure below.

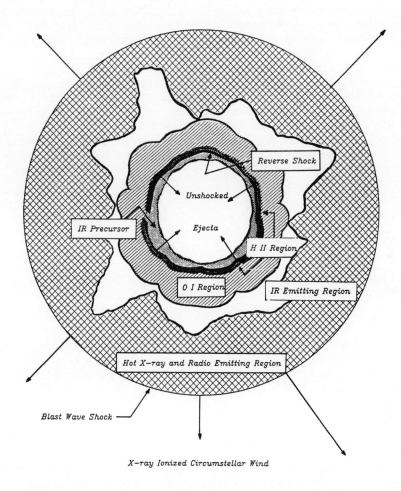

The luminous radio emission implies the presence of a massive circumstellar wind, with which the supernova ejecta are presumably colliding (Chevalier 1982). A free-free optical depth greater than unity at 20 cm implies an electron density in the wind of $n_e(\text{wind}) \gtrsim 5\times10^4 \text{cm}^{-3}$ at a radius of $r = 7\times10^{16}$cm, for a wind density profile going as r^{-2}. The collision of ejecta with such a wind will drive shocks into the ejecta at ~10 km/s if the expansion velocity of the ejecta is a few 1000 km/s and the ejecta density ~10^{10}cm^{-3}. Note that the inferred mass of the wind is $\gtrsim 0.3\,\text{M}_\odot$ outside 7×10^{16}cm; the original wind mass would be a factor $7\times10^{16}\text{cm}/R_{\text{star}}$ greater.

The expected Hα luminosity, which is about 1 photon per H atom entering the shock, $4\pi r_{\text{shock}}^2 n_{\text{H}} v_{\text{shock}}$ photons per unit time, equals the observed Hα luminosity if the shock radius is $r_{\text{shock}} \approx 5\times10^{16}$cm, close to that observed.

At an electron density $n_e \approx 10^{10} \mathrm{cm}^{-3}$, and given the observed size and expansion velocity of SN1986j, $\mathrm{Ly}\alpha$ is trapped, escaping mainly by 2-photon emission, provided that no alternative escape route is available, for example absorption by dust grains. If $\mathrm{Ly}\alpha$ escapes mainly by 2-photon emission, then the density of excited $n = 2$ hydrogen is $n_{\mathrm{H}(n=2)} \approx 10^6 \mathrm{cm}^{-3}$ for an electron density of $n_e \approx 10^{10} \mathrm{cm}^{-3}$. The excited hydrogen density is lower if other escape routes exist. The high density of excited hydrogen suggests the possibility that Balmer lines may have an appreciable optical depth; this would then account for both the large $\mathrm{H}\alpha/\mathrm{H}\beta \approx 60$ ratio, since $\mathrm{H}\beta$ is converted into $\mathrm{P}\alpha$ and $\mathrm{H}\alpha$, and also for the strength of $\mathrm{OI}\lambda8446$, which is pumped by $\mathrm{Ly}\beta$ converted from $\mathrm{H}\alpha$. Note that consistency between the optical depths of $\mathrm{H}\alpha$ and $\mathrm{H}\beta$ implies that the O/H abundance is near cosmic. The required Balmer optical depths imply a column density in excited hydrogen of $N_{\mathrm{H}(n=2)} \approx 10^{14} \mathrm{cm}^{-2}$, whence a total $\mathrm{H\,II}$ column density of $N_{\mathrm{H\,II}} \approx 10^{18} \mathrm{cm}^{-2}$, much smaller than the total hydrogen column density of perhaps $10^{26} \mathrm{cm}^{-2}$ through the ejecta, but roughly consistent with the cooling column density in a shock.

In a shock, [O I] is emitted in a cooler quasi-neutral region downstream of the $\mathrm{H}\alpha$ and [O II] emitting region, consistent with the lower temperature deduced from the spectrum. The [O I] emitting zone is ~ 5000 (ratio of cooling column densities) times thicker than the $\mathrm{H}\alpha$ zone. The observed [O I] luminosity implies an O I mass $\approx 3 \times 10^{-4} M_\odot$, corresponding to a total mass in the O I zone of $\approx 1 M_\odot$ for cosmic O/H. The mass of the $\mathrm{H}\alpha$ emitting zone is 1/5000 of this.

One curious feature of the optical spectrum of SN1986j is the presence of several lines, notably $\mathrm{O\,I}\lambda8227$, arising from recombination of $\mathrm{O\,II}$ in an excited ground state configuration. The only viable mechanism for exciting these lines appears to be charge exchange between $\mathrm{H}(n=2)$ and excited $\mathrm{O\,II}$.

Conclusions: We argue that the emission line spectrum of SN1986j is powered by a reverse shock moving at $\sim 10\,\mathrm{km/s}$ into SN ejecta at a density of $\sim 10^{10} \mathrm{cm}^{-3}$. The shock is driven by the collision of the ejecta with a massive circumstellar wind. The wind itself is being shocked at $\sim 1000\,\mathrm{km/s}$, producing the observed bright nonthermal radio emission.

The large observed $\mathrm{H}\alpha/\mathrm{H}\beta$ ratio is attributed to $\mathrm{Ly}\alpha$ trapping, which pumps the $n = 2$ level of hydrogen, causing a finite optical depth in Balmer lines, and converting $\mathrm{H}\beta$ to $\mathrm{P}\alpha$ and $\mathrm{H}\alpha$.

An important prediction of the shock model of the spectrum is of Lyman 2-photon continuum. Photoionization models predict no such continuum.

The inferred mass of the optically emitting regions is only $\sim 1\,M_\odot$, well short of the $\sim 2000\,M_\odot$ deduced by Utrobin (1984) for the prototypical Type V supernova SN1961v. However, it is likely that the emitting material in SN1986j represents only a small fraction of the total.

Oxygen and nitrogen abundances appear cosmic.

References
Chevalier, R. A. 1982, *Ap. J.*, **259**, 302.
Chevalier, R. A. 1987, preprint.
Rupen, M. P., Van Gorkom, J. H., Knapp, G. R., Gunn, J. E., and Schneider, D. P. 1987, preprint.
Utrobin, V. P. 1984, *Ap. Sp. Sci.*, **98**, 115.
Zwicky, F. 1965, in *Stellar Structure, Stars and Stellar Systems Vol. VIII*, ed. L. H. Aller and D. B. McLaughlin (Chicago: University of Chicago Press), p. 367.

RADIO SUPERNOVAE: THE DETECTION OF SN 1961V IN NGC 1058

John J. Cowan, Richard B. C. Henry, and David Branch
University of Oklahoma
Norman, OK

Abstract :   We report the radio detection of supernova 1961v in NGC 1058. SN 1961v has a spectral index of -0.4 $\pm$ 0.1. At the distance of NGC 1058, the absolute monochromatic luminosity of this source is comparable to that of Cas A. A second nonthermal source with a spectral index of -0.3 $\pm$ 0.1 was also detected in NGC 1058 and is likely to be a remnant of a supernova that was not optically detected. The two radio sources, and two optically faint H II regions that coincide with the radio sources, are separated by only 3.5".

Introduction :   Until recently radio emission had only been observed from supernovae within a few years of optical maximum (see *e.g.* Gottesman *et al.* 1972; Weiler *et al.* 1981; Sramek, Panagia, and Weiler 1984; Weiler *et al.* 1986) or from supernova remnants several hundred years old. Cowan and Branch (1985) detected radio emission from supernova 1957d in M 83 and probably from supernova 1950b in the same galaxy. The detection of radio emission from supernovae decades after optical maximum indicates that at least some supernovae remain radio bright at "intermediate- age".

In an attempt to detect more of these objects, Branch and Cowan (1985) used the Very Large Array (VLA) to observe NGC 1058. This galaxy contains the unusual supernova 1961v, as well as SN 1969l. SN 1961v is the primary example of a Type V supernova (Zwicky 1965). While the spectra of this supernova contain hydrogen lines (like the spectra of ordinary Type II supernovae), the expansion velocity of SN 1961v, determined from spectral line widths, was very low compared to normal Type II supernovae (Bertola 1964; Branch and Greenstein 1971). Until the discovery of SN 1987a in the Large Magellanic Cloud, SN 1961v was the only supernova with a known progenitor - a star seen near 18th magnitude in NGC 1058 for decades before the explosion (Zwicky 1964). The optical light curve of SN 1961v was also unusual and declined very slowly over more than six years (Bertola and Arp 1970; Doggett and Branch 1985). No other supernova has been followed optically for more than two years after optical maximum. The observed properties of SN 1961v suggest the explosion of a very massive star (Utrobin 1984).

Results :   Branch and Cowan (1985) had previously reported the detection of radio emission at 20 cm from two sources in NGC 1058, with one of the sources very near the reported position of supernova 1961v. Recently, we used the VLA to observe NGC 1058 at 6 cm. Using the program IMEAN, the peak flux at 20 cm

from the eastern source in the field of view was determined to be 0.18 mJy, while
the western source had a measured peak flux of 0.14 mJy. At 6 cm the peak flux
of the eastern source was 0.14 mJy and the western source had a peak flux of 0.10
mJy. These values lead to spectral indices of -0.4 ± 0.1 for the eastern source and
-0.3 ± 0.1 for the western source. At an assumed distance of 12 Mpc for NGC 1058
(Kirshner and Kwan 1974; Schurmann, Arnett, and Falk 1979), the measured flux
values are comparable in absolute radio luminosity to Cas A.

The position, determined at both 20 cm and 6 cm, for the eastern source was
$\alpha(1950) = 02^h40^m29.^s70 \pm 0.3''$ and $\delta(1950) = 37° 08' 01.''6 \pm 0.3''$ and for the
western source was $\alpha(1950) = 02^h40^m29.^s51 \pm 0.3''$ and $\delta(1950) = 37° 08' 01.''89$
$\pm 0.3''$. Since the two sources are faint the positional accuracy is no better than
approximately 0.3''. Klemola (1986) has measured with astrometric precision the
position of SN 1961v to be $\alpha(1950) = 02^h40^m29.^s694 \pm0.1''$ and $\delta(1950) = 37°$
$08' 01.''64 \pm 0.1''$. It is therefore clear that the peak radio emission from the
eastern source comes from the remnant of the supernova 1961v. The western
source is most likely a remnant from a supernova that was not optically identified.
We did not detect the Type II supernova 1969l in NGC 1058 at a 3 σ limit of 0.09
mJy at 6 cm. This upper limit corresponds to an absolute luminosity that is less
than that of Cas A.

The size of the radio emitting region at the positions of SN 1961v and the
western source is slightly larger than the beam and thus is partially resolved.
Optical H_α observations, using the 2.1 meter telescope at Kitt Peak National
Observatory, indicate faint H II regions, first found by Fesen (1985), at the exact
positions of the two radio sources. These H II regions are most likely responsible
for the extended radio emission around the two supernova remnants. In both the
H_α and [S II] images, these two H II regions are relatively faint, but otherwise
normal, in comparison to a number of other H II regions in NGC 1058.

Discussion and Summary :   The radio observations at 20 cm and 6 cm confirm
the nonthermal nature of the two sources in NGC 1058, and the accurate position
of Klemola (1986) confirms the identification of the eastern source with SN 1961v.
This detection marks the second definite (and most probably third) detection of a
supernova decades after optical maximum. The first of these "intermediate-age"
supernovae detected was supernova 1957d in M 83 (Cowan and Branch 1985).
We also detected a nonthermal radio source very close to the reported position of
supernova 1950b, but no accurate position is available for this supernova. It is,
therefore probable, but not definite, that SN 1950b was detected.

It is not clear why the two radio sources, the supernova remnants, are so
close together in NGC 1058. At a distance of 12 Mpc, the 3.5'' separation between
the two sources corresponds to a linear separation of approximately 200 parsecs.
Whether or not this small separation is coincidental is not obvious.

No spectra were obtained for supernova 1957d in M 83, but its position on the
inner edge of the spiral arm suggests that its progenitor star was massive (Cowan
and Branch 1985). SN 1961v was the prototypical Type V supernova (Zwicky

1965). Utrobin (1984) argued, on the basis of the very slow decay of the light curve, that the progenitor of SN 1961v lost 2000 $M_\odot$ prior to supernova outburst. Thus, both "intermediate-age" supernova so far detected may have resulted from the explosion of a massive star, with extensive mass loss prior to outburst. This extensive mass loss might be necessary to produce detectable radio emission from supernovae decades after optical maximum.

Radio emission has been detected from the Type II supernovae 1970g in M 101 (Gottesman *et al.* 1972), SN 1979c in M 100 (Weiler *et al.* 1981) and SN 1980k in NGC 6946 (Weiler *et al.* 1982). Chevalier (1984a, b) has explained radio emission from supernovae as coming from the interaction of the supernova ejecta with the surrounding circumstellar material, which was lost from the supernova progenitor as a result of mass loss. Both this "mini-shell" model of Chevalier (1984a, b) and pulsar driven models (see *e.g.* Pacini and Salvati 1973), also known as "mini-plerion" models, provide reasonable fits to the observations of the supernovae 1979c and 1980k (Weiler *et al.* 1986). Radio observations have also been made of SN 1981k in NGC 4258 by van der Hulst *et al.* (1983). On the basis of its radio properties (no optical spectra were obtained), Weiler *et al.* (1986) argue that SN 1981k was also a Type II supernova. Upper limits on radio emission from several other Type II supernovae have been reported by Branch and Cowan (1985) and Cowan and Branch (1985).

No classical Type I (*i.e.*, Type Ia) has yet been seen at radio wavelengths (Weiler *et al.* 1986), possibly owing to a lack of circumstellar material. Radio emission has been detected from the peculiar Type I (*i.e.* Ib) supernovae 1983n in M83 (Sramek, Panagia, and Weiler 1984) and 1984l in NGC 991 (Weiler *et al.* 1986).

It is not clear whether the "intermediate-age" supernovae that have been detected are the result of fading radio supernovae. Extrapolation of models of Chevalier (1984a, b) for the linear Type II radio supernovae SN 1979c and SN 1980k bracket the radio emission of SN 1957d and SN 1961v at their present ages. On the other hand, SN 1957d or SN 1961v (or possibly both) could be very young supernova remnants that are just starting to brighten. Models by Gull (1973) [see also Cowsik and Sarkar (1984)] might also be able to account for the radio emission from SN 1957d and SN 1961v. These models predict radio emission from young remnants as a result of synchrotron emission when the mass of swept-up interstellar material exceeds the mass of ejected material from the supernova explosion. We will have to study SN 1957d and SN 1961v over a long time period to determine whether their radio emission is decreasing or increasing with time. We will also have to detect additional such "intermediate-age" supernovae to learn more about this class of objects.

In summary, we have observed NGC 1058 at 6 cm and 20 cm and detected two nonthermal radio sources. One of the sources is coincident with the position of supernova 1961v. The second source, located 3.5" to the west of SN 1961v, probably is a remnant of a supernova that was not seen optically. The detection of SN 1961v represents the second (and probably the third) detection of a supernova of intermediate age. The observed spectral index of SN 1961v is -0.4 ± 0.1, while the western source has a spectral index of -0.3 ± 0.1. The absolute radio luminosity at 20 cm of SN 1961v, at the distance of NGC 1058, is 4.7×10^{34} ergs sec^{-1} - comparable to Cas A. The absolute radio luminosity of the western source is only slightly less. Optical images of NGC 1058, taken using the 2.1 meter telescope at KPNO, indicate the presence of two faint H II regions at the positions of SN 1961v and the western source. Both radio sources are partially resolved, probably indicating additional thermal radio emission from the underlying H II regions. While the two supernovae apparently exploded inside the H II regions, brightness comparisons among several bands indicate that these two H II regions are normal with respect to the many optically brighter H II regions in NGC 1058. Radio emission was not detected from these optically bright H II regions.

This work has been supported by NSF grants AST-8521705 and AST-8419734.

REFERENCES

Bertola, F. 1964, *Ann. Ap.*, **27**, 319.

Bertola, F., and Arp., H. 1970, *P.A.S.P.*, **82**, 894.

Branch, D. and Cowan, J.J. 1985, *Ap. J. (Letters)*, **297**, L33.

Branch, D., and Greenstein, J. L. 1971, *Ap. J.*, **167**, 89.

Chevalier, R. A. 1984a, *Ann. N. Y. Acad. Sci.*, **422**, 215.

Chevalier, R. A. 1984b, *Ap. J. (Letters)*, **285**, L63.

Cowan, J. J., and Branch, D. 1985, *Ap. J.*, **293**, 400.

Cowsik, K. R., and Sarkar, S. 1984, *M.N.R.A.S.*, **207**, 745.

Doggett, J. B., and Branch, D. 1985, *A. J.*, **90**, 11.

Fesen, R. A. 1985, *Ap. J. (Letters)*, **297**, L29.

Gottesman, S.T., Broderick, J.J., Brown, R.L., Balick, B., and Palmer, P. 1972, *Ap.J.*, **174**, 383.

Gull, S.F. 1973, *M.N.R.A.S.*, **161**, 47.

Kirshner, R. P., and Kwan, J. 1974, *Ap. J.*, **193**, 27.

Klemola, A. 1986, *P.A.S.P.*, **98**, 464.

Pacini, F. and Salvati, M., 1973, *Ap.J.*, **186**, 249.

Schurmann, S. R., Arnett, W. D., and Falk, S. W. 1979, *Ap. J.*, **230**, 11.

Sramek, R., A., Panagia, N., and Weiler, K. W. 1984, *Ap. J. (Letters)*, **285**, L59.

Utrobin, V. P. 1984, *Ap. Sp. Sci.*, **98**, 115.

van der Hulst, J. M., Hummel, F., Davies, R. D., Pedlar, A., and van Albada, G. D. 1983, *Nature*, **306**, 566.

Weiler, K.W., van der Hulst, J.M., Sramek, R.A., and Panagia, N. 1981, *Ap.J.(Letters)*, **243**, L151.

Weiler, K.W., Sramek, R.A., van der Hulst, J.M., and Panagia, N. 1982, in *Supernovae: A Survey of Current Research*, eds. M.J. Rees and R.J. Stoneham (Dordrecht:Reidel), p. 281.

Weiler, K. W., Sramek, R. A., Panagia, N., van der Hulst, J. M., and Salvati, M. 1986, *Ap. J.*, **301**, 790.

Zwicky, F. 1964, *Ap. J.*, **139**, 514.

Zwicky, F. 1965, in *Stellar Structure*, ed. L. H. Aller and D. B. McLaughlin (University of Chicago Press: Chicago), p. 367.

ASYMMETRICAL EJECTION OF MATTER IN A THERMONUCLEAR MODEL OF A SUPERNOVA EXPLOSION

V.M. Chechetkin, A.A. Denisov, A.V. Koldoba,
Yu.A. Poveschenko, and Yu.P. Popov
Institute of Applied Mathematics, USSR Academy of
Sciences, Moscow, 117259

With the recent Supernova 1987a in the LMC, new and interesting possibilities have arisen for the solution of a problem relating to the explosion mechanism of supernovae. The presupernova was probably a B3Ia, the blue supergiant, and not a red giant as it was earlier thought likely[1][2]. Calculations of the evolution, which have been made recently, show that the loss of hydrodynamical stability may be connected with carbon burning in the stellar core[2] during the blue giant stage. This loss of stability of the CO core is the main factor in our explanation of the recent event of SN 1987a.

Before we consider the thermonuclear model of a supernova based on a thermal flash in the degenerate CO core, let us dwell on the present situation in the theory of supernovae. The main problem in supernova theory is the simultaneity of births of a compact remnant and an expelled envelope. The compact remnant later becomes a neutron star. The expelled envelope determines the curve of brightness. Now, it is clear that a supernova explosion may result in either the total disruption of a star or in the simultaneous production of a compact remnant and an expelled envelope.

What are the sources of the energy in a supernova explosion which is equal, within an order of magnitude, to 10^{51} ergs? In early studies[3][4][5] the gravitational energy, which is released during the birth of the compact remnant, was taken as the source. However, such a mechanism failed[6]. Later the so-called bounce models of supernovae also failed. This gave rise to the conclusion that a model of supernovae based on thermonuclear burning in a degenerate CO or O-Ne-Mg core is the most likely in terms of stellar evolution as well as for supernova simulation.

Including rotation into this model is a further step towards the real physical situation. The first paper on the simulation of a thermonuclear explosion in a rotating stellar CO core resulted in the conclusion that the runaway velocity along the axis of rotation somewhat exceeds that on the equator[8]. However, it was shown[9] that the shock wave from a supernova explosion becomes spherical when it propagates through an interstellar medium with uniform density.

In our study two-dimensional calculations of the thermonuclear burning of a rotating CO core were carried out[10]. As an initial model, a degenerate CO core with a mass of 1.4 $M_\odot$ and a central density of $2.33 \cdot 10^9$ g cm^{-3} was assumed. In this model, for rotational parameters $\Omega=1.95$ s^{-1} = 0.51 Ω_{cr}, $\Omega_{cr} = (GM/v^3)^{1/3}$ (where Ω is the angular velocity of uniform rotation, G the gravitation constant, M the stellar core mass and v the equatorial radius), the equilibrium configuration of the star should be an ellipsoid of rotation. In Figure 1(a) the initial configuration with a computational grid is shown. The grid

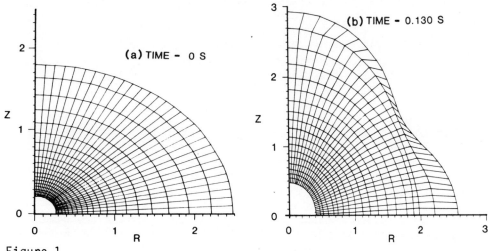

Figure 1

lines, which are parallel to the surface of the core, correspond to levels of equal density. At the initial moment roughly 20% of the total core mass is instantly burned off and 10^{50} erg of nuclear energy is released, generating the detonation wave.

Figure 2 shows the propagation of the burning front across the stellar core. In external layers of the core the detonation burning occurs in the over-driven wave mode. This can be seen from Table I, where the burning front radius (r_f) and velocity (v_f) and the respective Chapman-Juge velocities (v_{CJ}) are given. The overcompressed detonation (over-driven wave) has a tendency to be damped if the supporting pressure profile beyond the burning front is decreased[11]. As can be seen from Figure 2d the burning front reaches the surface of the star in the region of the poles. It leads to the appearance of the rarefraction wave going from the region of the shock wave exit to the surface. Since, in the overcompression detonation mode, the propagation velocity of the burning front is less than the sound velocity, the rarefraction wave may overtake the burning front and damp the detonation wave. Such modes can be observed in tests.

Table I.

t ms	$r_f/10^8$ cm	$v_f/10^9$ cm s^{-1}	v_{CJ}
29	0.5	1.5	1.5
48	0.7	1.6	1.4
65	0.95	1.5	1.36
80	1.1	1.7	1.26
102	1.4	1.7	1.2
122	1.8	2.0	1.13
132	2.1	2.1	1.1

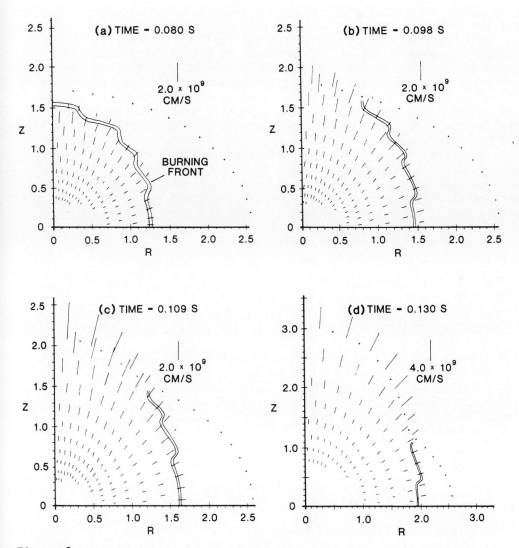

Figure 2

Due to the coarseness of the grid we could not finish the
calculations in our model. As recent studies have shown, in the case
of the burning instabilities, special computational algorithms are
needed to resolve this problem. A similar problem arises in
calculations of the deflagrational burning regime[12]. Here a
contradiction arises between the small size of the burning zone and the
large size of the object under investigation. Therefore there is no
self-consistent solution for the transition from spontaneous burning
into either the detonational or deflagrational regimes[12].

Calculations were also carried out to investigate burning wave
propagation in the spherically symmetric core and at lesser values of
the angular velocity. It was shown that for small angular velocities

the results obtained for the burning front propagation agree with those for the spherically symmetric case.

What astrophysical consequences follow from the results which include the possibility of the formation of a toroidal remnant containing unburnt carbon and oxygen? With further evolution such a remnant will resemble a colder region against a hotter region with an increased content of primary matter from the degenerate stellar core. The relationship between carbon and oxygen is determined by the rate of the reaction $^{12}C + \alpha \rightarrow {}^{16}O + \gamma$ and the star's evolutionary track in the presupernova stage.

Recently, in young supernova remnants, similar structures have been observed[13]. Furthermore, supernovae are known which have an enhanced content of oxygen[14]. These structures and the increased content of oxygen can be explained easily in the framework of this model. It is interesting to note the fact that, depending on the initial rotation of the stellar core, different chemical elements are produced in the thermonuclear explosion.

References:
(1) Woosley, S.E., Pinto, P.A., Ensman, L. "Supernova 1987a: six weeks later", 1987, submitted to Astrophys. J.
(2) Hillebrant, W., Höflich, P., Truran, J.W., Weiss, A. "Explosion of Blue Supergiant: A model for Supernova 1987a", 1987, Preprint MPA, 286.
(3) Ivanova, L.N., Imshennik, V.S., Nadyozhin, D.K. 1967, Nauchnii Information Astrosovjet of AN USSR.
(4) Colgate, S.A. 1968, Astrophys. J., 153, 335.
(5) Nadyouzhin, D.K. 1978, Astrophys. Space Sci., 53, 131.
(6) Trimble, V. Supernova, Rev. of Modern Phys., 1982, 54, 1183.
(7) Chechetkin, V.M., Gershtein, S.S., Imshennik, V.S., Ivanova, L.N., Khlopov, M.Yu. 1980, Astrophys. Space Sci., 67, 61.
(8) Mahaffy, J.H., Hansen, C.J. 1975, Astrophys. J., 201, 695.
(9) Bysnovaty-Kogan, G.S., Blinnikov, S.I. 1982, Astr. Zh., 59, 876.
(10)Denisov, A.A., Koldoba, A.V., Poveschenko, Yu.A., Popov, Yu.P., Chechetkin, V.M. 1986, Preprint No. 99, Inst. Applied Math., Moscow.
(11)Williams, E.A. Combustion Theory
(12)Blinnikov, S.I., Khokhlov, A.M. 1986, Pis'ma Astr. Zh., 12, 318.
(13)Losinskaya, T.A. 1986, Stellar Winds and Supernova Remnants, (Nauka, Moscow).
(14)Chugai, N.N. 1986, Astr. Circular No. 1469.

THE EARLY EVOLUTION OF SUPERNOVA REMNANTS

R. A. Chevalier
Department of Astronomy, University of Virginia
Charlottesville, VA U.S.A.

Abstract: The density distribution of the supernova ejecta
and that of the surrounding medium are the most important
parameters for the early evolution of supernova remnants.
The distribution of the ejecta depends on the detailed hydrodynamics of the explosion, but the outer parts of a
supernova can probably be represented by a steep power law
density distribution with radius. Self-similar solutions
are especially useful for modeling the interaction of a
supernova with its surroundings. The supernova first interacts with mass loss from the progenitor star. Evidence for
circumstellar interaction is present in a number of extragalactic supernovae, including SN1987a. The explosions of
massive stars probably interact with circumstellar gas for
a considerable time while Type Ia supernovae interact more
directly with the interstellar medium. X-ray spectroscopy
is a good diagnostic for the physical conditions in young
supernova remnants and for the composition of the supernova
gas.

1. Introduction: The early evolution of supernova remnants is determined by the nature of the supernova explosion and the properties of
the ambient medium. The interaction between them involves shock waves
and gas dynamical processes. This review will concentrate on developments since 1982, the time of I.A.U. Symposium No. 101 (Danziger and
Gorenstein 1983). Energy input by a central pulsar will not be considered here. Section 2 discusses the expansion of supernovae, with an
emphasis on the density and composition structure. Hydrodynamic features of the interaction are discussed in section 3; self-similar solutions are especially useful in this area. Interactions with circumstellar gas and with interstellar gas are presented in sections 4 and
5 respectively. The conclusions are in section 6.

2. Supernova Expansion: Supernovae are observationally divided into
two major classes depending on whether hydrogen lines are absent (Type
I) or present (Type II) in their spectra. It has recently been realized
that there are two categories of Type I events. The Type Ia events are
associated with an old stellar population and have strong Fe line
emission in their late spectra while the Type Ib supernovae are associated with a young stellar population and have lines of O and other
intermediate elements in their late spectra (Kirshner and Oke 1975;
Gaskell et al. 1986; Filippenko and Sargent 1986).

The most successful theoretical model for Type Ia supernovae is the
carbon deflagration of a white dwarf (Chevalier 1981; Sutherland and

Wheeler 1984; Nomoto, Theilemann, and Yakoi 1984). This model can generally reproduce the light curves, early time spectra, and late time
spectra (Axelrod 1980) of the supernovae. The models of Nomoto et al.
(1984) are particularly detailed with regard to the composition and
density structure and their model W7 has been used to model the observed
spectra of Type Ia supernovae (Branch et al. 1985). The result of this
work is that mixing of the intermediate element layers with velocities
above 8000 km s^{-1} significantly improves the spectral fit. The density
structure of the gas in the free expansion phase is complex because of
the partial incineration of the gas. The process that mixes the gas
may also smooth some of the dense features in density profile.

The mechanism that is responsible for the probable mixing is not
known. However, the two-dimensional carbon burning calculations of
Muller and Arnett (1986) are suggestive of the complex motions that can
accompany the propagation of a burning front. They find that the burning creates hot bubbles that are Rayleigh-Taylor unstable and result
in a corrugated burning front. It is not clear to what extent the numerical resolution plays a role in the calculated structure or whether
the flow becomes fully turbulent.

The class of Type Ia supernovae is quite homogeneous and it has
been difficult to definitively identify differences between specific
events. Spectra of SN1986g have settled this question. The $\lambda6355$ line,
identified with Si II, shows more rapid evolution in wavelength than
did the same feature in spectra of SN1981b (Phillips et al. 1987). This
is suggestive of a flatter density profile in SN1986g. Thus, Type
Ia supernovae exploding in identical media are not expected to produce
identical remnants.

The observed characteristics of Type Ib supernovae make an interpretation as the explosions of massive stars that have lost their hydrogen envelope attractive (Wheeler and Levreault 1985; Chevalier 1986).
These are Wolf-Rayet stars. The deduction that SN1985f ejected at
least 5 M$_\odot$ of oxygen (Begelman and Sarazin 1986) is particularly
suggestive of this interpretation. Schaeffer, Casse, and Cahen (1987)
have shown that the light curves of exploding Wolf-Rayet stars are
roughly consistent with those of Type Ib supernovae. The hydrogen envelope can be lost either through massive single star evolution or
through mass transfer in a close binary system. Blaauw (1985) has estimated that 18% of early B type stars are in close binaries. One-
dimensional models of Wolf-Rayet star explosions show the ejection of
layers of heavy elements (Ensman and Woosley 1987). However, the line
profile of [OI]$\lambda6300$ in the spectrum of SN1985f implies the ejection of
oxygen over a broad velocity range of 0 to > 3000 km s^{-1} (Filippenko
and Sargent 1986; Fransson 1986a). The origin of this mixing is unknown, but it may be related to the inhomogeneous ejecta observed in the
Cassiopeia A supernova remnant. There, some fast moving knots of heavy
element gas are observed with heavier elements having higher velocities
than lighter elements (Chevalier and Kirshner 1979). This gradient is
opposite to that expected in the models.

The explosions of Wolf-Rayet stars are expected to have a range of progenitor masses. The uniformity of Type Ib supernovae, including the radio regime (Panagia, Sramek, and Weiler 1986), is thus a possible problem for the massive star model. The possibility that mass loss from massive stars drives the core toward a common structure is not anticipated in stellar evolution theory.

If a massive star explodes with its hydrogen envelope, the result is a Type II supernova. This is the least controversial of the various supernova models. The explosion is generally expected to occur in a red supergiant envelope, although the recent supernova 1987a shows the possibility of a blue supergiant progenitor. The available evidence points to the explosion of the B3 Ia star Sk-69 202 in this case (Kirshner et al. 1987; Woosley, Pinto, and Ensman 1987).

If the progenitor star has not undergone extensive mass loss, the heavy element mantle gas is effectively decelerated by the envelope. This process is expected to be Rayleigh-Taylor unstable and can lead to mixing of mantle and envelope gas. The spectroscopic study of Type II supernovae in their late phases should lead to information on the density and composition structure of the ejecta.

Numerical simulations of supernova explosions typically show that the outer part of the density profile can be approximated by a power law in radius. A recent example is Arnett's (1987) model for SN1987a which shows an outer profile of the form $\rho \propto r^{-n}$, where n is 8 to 9. Some insight into the production of power law profiles can be gained from self-similar flow theory. Sakurai (1960) found self-similar solutions for the propagation of a shock wave in a medium with $\rho = Ax^{\beta}$ where A and β are constants and x is the distance from the boundary of the star. The solutions are planar, but they should apply to the thin layers at the surface of a star. The solution can be continued into the regime where the gas expands into the vacuum. In the limit $t \to \infty$, the density profile asymptotically approaches a homologous free expansion. It is of the power law form $\rho \propto (-x)^{-(1+\lambda+\beta)/\lambda}$, where λ is an eigenvalue from the self-similar solution. For the adiabatic index $\gamma = 4/3$ and $\beta = \infty$, which is the limit of an initially exponential medium, we have $\rho \propto (-x)^{-6.67}$. Raizer (1964) had already noted that the planar expansion of a shock wave in an exponential medium asymptotically approaches a power law form. For $\beta < \infty$, the power law index is larger in magnitude (Chevalier and Jones 1987).

This solution can be expected to describe the propagation of a strong shock wave through the exponential atmosphere of a star if the density scale height is much less than the stellar radius and radiative losses are negligible. These assumptions should hold for the explosion of a star like Sk-69 202. The initially planar free expansion develops into a spherically symmetric expansion. Conservation of mass shows that this steepens the power law by two powers of radius, to $\rho \propto r^{-8.7}$. This result appears to agree with numerical computations and generally applies for the expansion of an initially exponential atmosphere.

3. Hydrodynamic Evolution: The result of the supernova explosion is a
radial flow in free expansion. The velocity field is given by $v = r/t$
where t is the age of the explosion and the density by $\rho = Bt^{-3}f(v)$
where B is a constant and $f(v)$ is a function that depends on the initial
hydrodynamic evolution. The pressure in the expanding gas is negligible
because of adiabatic expansion. The pressure and velocity of the
ambient medium can generally be neglected because of the high initial
supernova velocities and only the density distribution is relevant.
Because of these simplifications, self-similar solutions can be very
useful in delineating the major features of the interaction. These
solutions are calculated for one-dimensional flows. Although two-
dimensional flows are expected to be self-similar when the ambient or
supernova density can be separated into radial and angular functions,
they are not easily calculated because partial differential, not
ordinary differential, equations are involved. For spherically symmetric flows, the ambient medium is generally taken to be of the form
$\rho \propto r^{-s}$, where s = 0 (interstellar medium) or 2 (circumstellar medium).
This circumstellar medium results from a constant velocity wind with
a constant mass loss rate.

The first case is a point explosion in a power law medium. Sedov
(1959) obtained an analytic solution for this case and noted that the
shock radius increases as $t^{2/(5-s)}$. For s=0, the shocked gas is concentrated at the shock front. For s = 2, the solution is particularly
simple, with velocity $v \propto r$, density $\rho \propto r$, and pressure $p \propto r^3$. These
solutions assume adiabatic postshock flow. In a young supernova
remnant, heat conduction will tend to flatten the temperature profile
if conduction is not impeded by magnetic fields. Korobeinikov (1956)
and Solinger, Rappaport, and Buff (1975) discussed isothermal blast
waves. The transport of heat into the shock reduces the shock compression ratio to 2.38 from 4. During the early phases, there is unlikely to be energy equipartition between ions and electrons. Electron
heat conduction occurs on a faster timescale than proton conduction so
that the electrons may be isothermal with the ions adiabatic. Cox and
Edgar (1983) have investigated self-similar solutions in this case and
for s = 0 find a shock compression of 3.24. They assume that ion-
electron energy equilibrium is achieved in the collisionless shock
front. Even if the magnetic field geometry is favorable for the action
of heat conduction, it is quite possible that plasma instabilities
reduce the mean free paths of the ions and electrons below the Coulomb
interaction values. This has the effect of reducing heat conduction.

The other class of self-similar solutions for young remnants includes the shocked supernova ejecta gas as well as the shocked ambient
medium. Chevalier (1982) and Nadyozhin (1985) found solutions for the
interaction of a power law ejecta profile ($\rho \propto t^{-3}v^{-n}$) with a power law
ambient medium. Dimensional analysis shows that the shock waves and
contact discontinuity expand as $t^{(n-3)/(n-s)}$. For n = 5, the expansion
law is the same as that for the Sedov solution, so for $n \leq 5$ the flow
should approach that for a point explosion. For s=0, the density at
the contact discontinuity drops to 0 for both the shocked supernova
ejecta and the ambient medium while for s=2, the density at the contact

discontinuity becomes infinite for both media. The density ratio between the shocked supernova gas and the shocked ambient medium increases for larger values of n.

Self-similar solutions with reverse shock waves and isothermal gas have not been found. Bedogni and d'Ercole (1987) have carried out numerical computations of the above case with heat conduction included and two-fluid flow. They find that the flow becomes complex with reverse shocks forming close to the contact discontinuity and thermal conduction driving a broad inner shocked region. Smooth self-similar flows may not exist in this case.

The reverse shock solutions do assume that the expanding supernova gas is smoothly distributed. Some remnants, like Cas A, indicate that the ejecta may be clumpy. Hamilton (1985) has investigated the inter-action of clumpy ejecta with an ambient gas for cases similar to the reverse shock solutions discussed above. In order to preserve self-similar flow, certain assumptions, such as undecelerated clump motion, were necessary. Ablation of the clumps was allowed. If the clumps interacted strongly with the ambient medium, the solution for smooth flow was recovered. For weaker interaction, the clumps moved out ahead of the shock front in the ambient medium. This type of behavior is qualitatively expected.

The above reverse shock solutions assume a relatively steep power law density profile. Hamilton and Sarazin (1984a) have found self-similar solutions for the initial phases of a reverse shock in a medium with a flat density profile (n < 1). The solutions apply to the time when the distance between the reverse shock wave and the edge of the "freely expanding" ejecta is much less than the radius so that the flow is approximately planar. For uniform ejecta, the reverse shock propagates as $z \propto t^{(5-s)/2}$, where z is the distance to the edge of the ejecta if it continued in free expansion. As opposed to the steep power law case, the shocked supernova ejecta have a density peak at the contact discontinuity for both s = 0 and s = 2. The solutions for the shocked ambient medium resemble those for the steep power law case, although the flow is not exactly self-similar.

For the transition from the early reverse shock flow and for more general density distributions than power-laws in radius, numerical hydrodynamic calculations are needed. Because of the steep density profiles present in the early flows, many computational zones are needed in the one-dimensional calculations to reproduce the self-similar solu-tions (Jones and Smith 1983; Hamilton and Sarazin 1984a). A number of computations have been carried out on the ineraction with circumstellar matter (Fabian, Brinkmann, and Stewart 1983; Itoh and Fabian 1984; Dickel and Jones 1985). The general expectations for the expansion of a massive star are that it will first interact with the dense wind ejected in the red supergiant phase, it will then approach free expan-sion in the bubble created by the fast wind lost in the main sequence phase, and will finally interact with the swept up wind bubble shell.

The interaction with the interstellar medium takes place in a late evolutionary phase. If Type Ib supernovae have Wolf-Rayet star progenitors, the dense wind does not occur close to the stellar surface, but may be present further out from an earlier evolutionary phase. Type Ia supernovae may interact more directly with the interstellar medium.

4. Circumstellar Interaction: There is excellent evidence for interaction with a dense circumstellar wind for the Type II supernovae SN1979c and SN1980k (see reviews by Chevalier 1984a and Fransson 1986b). The evidence includes radio emission from the interaction region for both supernovae (Weiler et al. 1986), infrared dust echoes for both supernovae (Dwek 1983), thermal X-ray emission from the interaction region in SN1980k (Canizares et al. 1982), and the ultraviolet line emission from highly ionized atoms in SN1979c (Fransson et al. 1984). The radio emission is a particularly good diagnostic because the early absorption of the radio emission can be interpreted as free-free absorption by the preshock gas and an estimate of the circumstellar density is obtained. Lundqvist and Fransson (1987) have made a detailed study of the temperature and ionization of the circumstellar gas in the radiation field of the supernova and have been able to reproduce detailed features of the radio light curves. They derive mass loss rates of 12×10^{-5} $M_\odot$ yr^{-1} and 3×10^{-5} $M_\odot$ yr^{-1} for a wind velocity $v_w = 10$ km s^{-1} for SN1979c and SN1980k respectively.

There are presently 5 radio supernovae with fairly extensive data, including the rising part of the radio light curve. They are SN1979c, SN1980k, SN1983n (Sramek, Panagia, and Weiler 1984), SN1986j (Rupen et al. 1987), and SN1987a (Turtle et al. 1987). Table 1 lists the supernova type, the time of optical depth at 20 cm, t_{20}, and the circumstellar density given in terms of the presupernova mass loss rate divided by the wind velocity. SN1986j was probably not observed near maximum light and the Type II designation given here is based solely on the presence of hydrogen line emission. Rupen et al. (1987) have suggested a Type V designation. Of the 5 supernovae, only SN1979c and SN1980k show clear evidence for circumstellar interaction outside of

Table 1
Radio Supernovae

Supernova	Type	t_{20} (days)	$\dot{M}/v_w$ ($M_\odot$ yr^{-1})/(km s^{-1})	
1979c	II	950	1×10^{-5}	a
1980k	II	190	3×10^{-6}	a
1983n	Ib	30	5×10^{-7}	b
1986j	II	1600	2×10^{-5}	c
1987a	II	2	1×10^{-8}	d

a.   Lundqvist and Fransson, 1987
b.   Chevalier 1984b; Sramek, Panagia, and Weiler 1984
c.   Chevalier 1987
d.   Chevalier and Fransson 1987

radio wavelengths. For SN1983n and SN1987a this is attributable to the
low circumstellar density and for SN1986j to the late discovery.

The results show that the winds around SN1979c, 1980k, and 1986j
are consistent with the dense slow winds expected around red supergiant
stars. The density around the Type Ib event SN1983n is considerably
lower, but it roughly consistent with the value expected around a Wolf-
Rayet star. A wind velocity of 1000 km s^{-1} and $\dot{M} = 10^{-4}$ $M_\odot$ yr^{-1} leads
to a value of $\dot{M}/v_w$ that is a factor of 5 below the estimated value.
SN1987a had an even earlier turn-on and was a faint radio supernova,
but the estimated value of $\dot{M}/v_w$ is roughly consistent with the density
expected around a B3 Ia star like the Sk-69 202 progenitor star
(Chevalier and Fransson 1987). The observational estimate is again a
factor of a few larger than the expected value. If there is clumping
in the circumstellar wind, the observational estimates are reduced.

In the circumstellar interaction model for the radio emission,
the radio luminosity at a given age should be correlated with the
density of circumstellar material. This is observed. It appears
that the circumstellar interaction does give information on the properties of the supernova progenitor.

An exciting development is the possibility of resolving radio
supernovae with very long baseline interferometry (VLBI) techniques.
The expansion of SN1979c has been measured (Bartel et al. 1985; Bartel
1986) and Bartel (private communication) has estimated that if the
expansion follows $R \propto t^m$, then m = 0.9 $\pm$ 0.1. The radius and the
expansion law are consistent with circumstellar interaction. SN1986j
which is currently the brightest radio supernova, has also been resolved by VLBI observations (Bartel, Rupen, and Shapiro 1987). Measurements of the expansion of the radio source should allow the age of
the supernova to be estimated.

During the next phase of evolution the supernova may approach free
expansion in a low density wind bubble. When SN1979c enters this phase,
an accelerated rate of decline of the radio emission is expected. One
way to identify supernova remnants in this evolutionary stage would be
to search for pulsar nebulae which show little or no evidence for interaction with a surrounding medium. This may be the explanation for
"Crabs without shells" which are about equal in number to the Crabs
with shells (Becker and Helfand 1987). Of course the Crab Nebula itself is lacking a shell.

Of the young supernova remnants, Cassiopeia A is the most likely
to be related to Type Ib supernovae. It has fast moving oxygen-rich
gas and is interacting with dense nitrogen-rich circumstellar gas;
Fesen, Becker, and Blair (1987) have suggested that the progenitor was
a Wolf-Rayet WN star. The problems with the Type Ib identification are
that Cas A has recently been found to have fast-moving hydrogen-rich
gas (Fesen et al. 1987) and the supernova was probably too faint to be
a typical Type Ib event, even if it was observed by Flamsteed in 1680
(Ashworth 1980). Although the presence of hydrogen would appear to rule

out a Type I supernova, it is perhaps possible that the hydrogen would not have been detectable spectroscopically near maximum light.

An interesting recent suggestion is that Kepler's supernova remnant is the result of a Type Ib supernova. An analysis of the X-ray emission indicates that the initial stellar mass was > $7M_\odot$ (Hughes and Helfand 1985), although this is rather uncertain. Bandiera (1987) has argued that the progenitor star was a runaway Wolf-Rayet star from the galactic plane. Proper motion studies of the dense optical knots do imply a high space velocity (van den Bergh and Kamper 1977) and the asymmetry of the supernova remnant may be due to the interaction of presupernova mass loss with an ambient medium (Bandiera 1987).

5. <u>Interstellar Interaction</u>: The physical properties of young supernova remnants are probably best studied by X-ray spectroscopy and there have been a number of recent theoretical studies in this area. The first case to be examined in detail is the X-ray emission from a self-similar Sedov blast wave (Gronenschild and Mewe 1982; Hamilton, Sarazin, and Chevalier 1983). The properties of the emission are determined by two parameters, i.e. $n_o^2 E$ and t where n_o is the ambient density, E is the total energy, and t is the age. The most important property of the flow is that the gas is underionized compared to equilibrium values because ionization timescales can be longer than the hydrodynamic timescales. Since underionization can favor line emission, the X-ray luminosity from a nonequilibrium flow may be a factor of 10 higher than that from an equivalent flow assumed to be in ionization equilibrium.

It is unknown to what extent electrons are heated in collisionless shock fronts so that the amount of heating is often a parameter in theoretical studies (e.g. Hamilton, Sarazin, and Chevalier 1983). Even if collisionless heating does take place in the shock, electrons may be released by ionization in the postshock flow which have not been subject to this heating. Itoh (1984) noted that some fast shocks appear to be moving into a partially neutral medium and that the electrons released from the neutrals in the postshock flow are only subject to Coulomb heating. Hamilton and Sarazin (1984c) noted a similar process for the postshock ionization and heating of a heavy element gas. In either case, it is necessary to take into account two populations of electrons.

Hamilton and Sarazin (1984b) found that the X-ray emission from a variety of self-similar flows can be estimated without carrying out detailed calculations for each case. Two important parameters are an ionization time, τ, which is weighted by a Boltzmann factor and an emissivity parameter, ε, which is a function of radius. Two supernova remnants that have similar values of τ versus ε through the remnant belong to the same structural type. The two basic types are the Sedov type, which approaches a hot, low density medium in the postshock flow, and a type which approaches a cold, high density medium in the postshock flow. The interaction of a steep power law density profile with an s = 0 medium is of the first type. The interaction of a steep profile with an s = 2 medium and the reverse shock wave for uniform ejecta are of the second type. Two remnants of the same type that have similar values of average τ, average temperature, and total emissivity are expected to produce similar X-ray spectra. The fact that detailed spectra have been calculated for Sedov blast waves makes this method quite useful.

A more general way to calculate X-ray spectra is to solve the time dependent ionization equations along with a numerical hydrodynamic computation (Itoh and Fabian 1984; Nugent et al. 1984; Hughes and Helfand 1985). These calculations can be time consuming, but Hughes and Helfand (1985) have developed a useful matrix method that speeds up the calculation of the ionization equations.

As discussed in the previous section, the explosions of massive stars are likely to interact with circumstellar matter during their early phases. Direct interaction with the interstellar medium is most likely for Type Ia supernovae and Tycho's supernova (SN 1572) and SN1006 may belong to this class. Detailed modeling of the X-ray spectra of these remnants has been carried out by Hamilton, Sarazin, and Szymkowiak (1986a,b). They find that in both cases, the spectra are best fit by models of the second type discussed above, i.e. models with cool dense ejecta. A range of temperatures is needed to give an approximate power law continuum and to produce emission that approximates ionization equilibrium. Hamilton et al. concentrate on models in which constant density ejecta with a sharp edge expand into the interstellar medium. For both supernova remnants, the models can accommodate $\gtrsim 0.5\ M_\odot$ of Fe as expected in a Type Ia supernova because the Fe is either unshocked or is at low density. The presence of cold Fe in SN1006 appears to be confirmed by the presence of broad ultraviolet Fe absorption in the direction of the Schweizer-Middleditch star (Wu et al. 1983; Fesen et al. 1987).

The X-ray spectra of Tycho and of SN1006 are very different in that Tycho shows strong line emission while SN1006 does not. Hamilton et al. attribute this to a low density surrounding SN1006 so that its "ionization age" is less than that of Tycho and the ionization has not yet proceeded to the stage which gives X-ray line emission. Kirshner, Winkler, and Chevalier (1987) have recently confirmed this hypothesis by measuring the Balmer line emission from two remnants and using it to estimate shock velocities. SN1006 has a higher shock velocity even though it is an older remnant, which implies it is expanding into a low density medium.

The models of Hamilton et al. appear to be very promising, but they do not allow for the presence of a steep outer power law component to the density profile that is expected from supernova modeling (see section 2). A related problem may be that the expansion rates of the supernova remnants in the models are larger than is indicated by optical and radio observations. A power law region with n somewhat greater than 5 would lead to greater deceleration of the outer shock front. Possible resolutions of these problems are that there is an outer power law region but with a relatively small amount of mass or that clumping of the ejecta plays an important role.

6. Conclusions: Although it often seems as if a separate physical picture is needed for each supernova and supernova remnant, some general trends in the interpretation of these objects are becoming clear. For massive stars, mass loss plays a crucial role both for the supernova explosions and their remnants. Type Ib events may be closely related to Type II supernovae, but have lost their hydrogen envelopes in pre-supernova evolution. If the explosion of Sk-69 202 as a blue supergiant is related to presupernova mass loss, it may be an intermediate case (Woosley, Pinto, and Ensman 1987). Radio supernovae give good evidence for the expansion of massive explosions into the nearby circumstellar wind. The further evolution may be related to a wind bubble and its associated shell. Current observations of Type Ia supernovae are consistent with direct interaction with the interstellar medium. They are not observed as radio supernovae (Weiler et al. 1986) and the remnants of SN1572 and SN1006 appear to be interacting with the interstellar medium.

With regard to hydrodynamical modeling, spherically symmetric models for the interaction of a supernova with an ambient medium have now been calculated in considerable detail. However, there is the expectation of hydrodynamic instabilities and the observational data show evidence for clumpiness and mixing. It will eventually be important to clarify the three-dimensional evolution of supernovae and their remnants.

This work was supported in part by NSF grant AST-8615555 and NASA grant NAGW-764.

REFERENCES

Arnett, W. D. 1987, preprint.
Ashworth, W. B. 1980, J. Hist. Astr., 11, 1.
Axelrod, T. S. 1980, Ph.D. thesis, University of California, Santa Cruz.
Bandiera, R. 1987, Ap. J., in press.
Bartel, N. 1986, Highlights of Astr., 7, 655.
Bartel, N., Rogers, A. E. E., Shapiro, I. I., Gorenstein, M. V., Gwinn,
    C. R., Marcaide, J. M., and Weiler, K. W. 1985, Nature, 318, 25.
Bartel, N., Rupen, M. and Shapiro, I. I. 1987, IAU Circ. 4292.
Becker, R. H. and Helfand, D. J. 1987, these proceedings.
Bedogni, R. and d'Ercole, A. 1987, Astr. Ap., submitted.
Begelman, M. C. and Sarazin, C. L. 1986, Ap. J. (Letters), 302, L59.
Blaauw, A. 1985, in Birth and Evolution of Massive Stars and Stellar
    Groups, W. Boland and H. van Woerden, eds. (Dordrecht: Reidel),
    p. 211.
Branch, D., Doggett, J. B., Nomoto, K., and Thielemann , F. K. 1985,
    Ap. J., 294, 619.
Canizares, C. R., Kriss, G. A, and Feigelson, E. D. 1982, Ap. J.
    (Letters), 253, L17.

Chevalier, R. A. 1981, Ap. J., 246, 267.
Chevalier, R. A. 1982, Ap. J., 258, 790.
Chevalier, R. A. 1984a, Ann. N.Y. Acad. Sci., 422, 215.
Chevalier, R. A. 1984b, Ap. J. (Letters), 285, L63.
Chevalier, R. A. 1986, Highlights of Astr., 7, 599.
Chevalier, R. A. 1987, Nature, submitted.
Chevalier, R. A. and Fransson, C. 1987, Nature, in press.
Chevalier, R. A. and Jones, E. M. 1987, in preparation.
Chevalier, R. A. and Kirshner, R. P. 1979, Ap. J., 233, 154.
Cox, D. P. and Edgar, R. J. 1983, Ap. J., 265, 443.
Danziger, I. J. and Gorenstein, P. eds. 1983, IAU Symposium 101,
     Supernova Remnants and their X-Ray Emission (Dordrecht: Reidel).
Dickel, J. R. and Jones, E. M. 1985, Ap. J., 288, 707.
Dwek, E. 1983, Ap. J., 274, 175.
Ensman, L. and Woosley, S. E. 1987, in preparation.
Fabian, A. C., Brinkmann, W., and Stewart, G. C. 1983, in IAU Symposium
     101, Supernova Remnants and their X-Ray Emission, I. J. Danziger
     and P. Gorenstein, eds. (Dordrecht: Reidel), p. 83.
Fesen, R. A., Becker, R. H., and Blair, W. P. 1987, Ap. J., 313, 378.
Fesen, R. A., Wu, C. -C., Leventhal, M., and Hamilton, A. J. S. 1987,
     preprint.
Filippenko, A. V. and Sargent, W. L. W. 1986, A. J., 91, 691.
Fransson, C., et al. 1984, Astr. Ap., 132, 1.
Fransson, C. 1986a, Highlights of Astr., 7, 611.
Fransson, C. 1986b, in Radiation Hydrodynamics in Stars and Compact
     Objects, D. Mihalas and K. H. A. Winkler, eds. (Berlin: Springer),
     p. 141.
Gaskell, C. M., Cappellaro, E., Dinerstein, H. L., Garnett, D. R.,
     Harkness, R. P. and Wheeler, J. C. 1986, Ap. J. (Letters), 306, L77.
Gronenschild, E. H. B. M. and Mewe, R. 1982, Astr. Ap. Suppl., 48, 305.
Hamilton, A. J. S. 1985, Ap. J., 291, 523.
Hamilton, A. J. S. and Sarazin, C. L. 1984a, Ap. J., 281, 682.
Hamilton, A. J. S. and Sarazin, C. L. 1984b, Ap. J., 284, 601.
Hamilton, A. J. S. and Sarazin, C. L. 1984c, Ap. J., 287, 282.
Hamilton, A. J. S., Sarazin, C. L., and Chevalier, R. A. 1983, Ap. J.
     Suppl., 51, 115.
Hamilton, A. J. S., Sarazin, C. L., and Symkowiak, A. E. 1986a, Ap. J.
     300, 698.
Hamilton, A. J. S., Sarazin, C. L., and Symkowiak, A. E. 1986b, Ap. J.,
     300, 713.
Hughes, J. P. and Helfand, D. J. 1985, Ap. J., 291, 544.
Itoh, H. 1984, Ap. J., 285, 601.
Itoh, H., and Fabian, A. C. 1984, M.N.R.A.S., 208, 645.
Jones, E. M. and Smith, B. W. 1983, in IAU Symposium 101, Supernova
     Remnants and their X-Ray Emission, I. J. Danziger and P. Gorenstein,
     eds. (Dordrecht: Reidel), p. 83.
Kirshner, R., Nassiopoulas, G. E., Sonneborn, G., and Crenshaw, D. M.
     1987, Ap. J., in press.
Kirshner, R. P. and Oke, J. B. 1975, Ap. J., 200, 574.
Kirshner, R. P., Winkler, P. F. and Chevalier, R. A. 1987, Ap. J.
     (Letters), 315, L135.

Korobeinikov, B. P. 1956, J. Acad. Sci. Soviet Union, 109, 271.

Lunqvist, P. and Fransson, C. 1987, Astr. Ap., in press.

Muller, E. and Arnett, W. D. 1986, Ap. J., 307, 619.

Nadyozhin, D. K. 1985, Ap. and Sp. Sci., 112, 225.

Nomoto, K., Thielemann, F. -K., and Yokoi, K. 1984, Ap. J., 286, 644.

Nugent, J. J., Pravdo, S. H., Garmire, G. P., Becker, R. H., Tuohy, I. R., and Winkler, P. F. 1984, Ap. J., 284, 612.

Panagia, N., Sramek, R. A., and Weiler, K. W. 1986, Ap. J., (Letters), 300, L55.

Phillips, M. M. et al. 1987, preprint.

Raizer, Yu. P. 1964, Zh. Prikl. Mat. Tekh. Fiz., No. 4, 49.

Rupen, M. P., van Gorkom, J. H., Knapp, G. R., Gunn, J. E., and Schneider, D. P. 1987, preprint.

Sakurai, A. 1960, Comm. Pure Appl. Math., 13, 353.

Schaeffer, R., Casse, M., and Cahen, S. 1987, Ap. J. (Letters), 316, L31.

Sedov, L. 1959, Similarity and Dimensional Methods in Mechanics, Academic Press, New York.

Solinger, A., Rappaport, S., and Buff, J. 1975, Ap. J., 201, 381.

Sramek, R. A., Panagia, N., and Weiler, K. W. 1984, Ap. J. (Letters), 285, L59.

Sutherland, P. G. and Wheeler, J. C. 1984, Ap. J., 280, 282.

Turtle, A. J. et al. 1987, Nature, 327, 38.

van den Bergh, S. and Kamper, K. W. 1977, Ap. J., 218, 617.

Weiler, K. W., Sramek, R. A., Panagia, N., van der Hulst, J. M., and Salvati, M. 1986, Ap. J. (Letters), 301, 790.

Wheeler, J. C. and Levreault, R. 1985, Ap. J., 294, L17.

Woosley, S. E., Pinto, P. A., and Ensman, L. 1987, Ap. J., in press.

Wu, C. -C., Leventhal, M., Sarazin, C. L., and Gull, T. R. 1983, Ap. J. (Letters), 269, L5.

INSTABILITIES DRIVEN IN YOUNG SUPERNOVA REMNANTS BY ELECTRON HEAT CONDUCTION

R. Bedogni*, A. D'Ercole
Osservatorio Astronomico di Bologna, Italy
*Present address: Minnesota Supercomputer Institute,
                   Minneapolis, USA

1. INTRODUCTION

Adiabatic models of supernova remnants (SNRs) show very large temperature gradients. The effect of thermal conduction on the Sedov solution was studied by a number of authors (Solinger et al., 1979; Cox and Edgar, 1983; Cowie, 1977).

The observations show, however, that young SNRs such as SN 1006 (Hesser and van den Bergh, 1981) and SN 1572 (Strom, Goss and Shaver, 1982) are in an intermediate state between free expansion and the Sedov phase. In these cases stellar matter cannot be neglected. Following Chevalier (1982), the freely expanding ejecta of a Type I SNR can be modeled in such a way that the inner 4/7 of the mass have constant density and the outer 3/7 have a $\rho \propto r^{-7}$ profile (the "ramp"). The interaction of the ejecta with the uniform circumstellar medium (CSM) gives rise to a pair of shocks. As long as the reverse shock is within the r^{-7} part of the density profile, the interaction region is described by a self-similar solution (Chevalier, 1982). Such a solution holds for an adiabatic single fluid; the temperature gradient, however, is so large that it may give rise to a quite high heat flux.

2. ASSUMPTIONS AND RESULTS

In order to investigate the effect of such a flux on the Chevalier solution, we performed numerical computations based on a number of assumptions:
i)   We adopted an expression for the electron heat conduction which takes into account saturation effects (cf. Cowie and McKee, 1977), the plasma being collisionless.
ii)   For the same reason, the electron (T_e) and ion (T_i) temperatures are expected to be different inside the remnant, and we calculated the ion and electron fluids separately.
iii) Observational data from young SNRs such as Cas A and Tycho (Pravdo and Smith, 1979) suggest that non-Coulomb electron-ion coupling occurs on the collisionless shock fronts; we therefore assumed that T_e and T_i are equal on the shocks.
iv)   We suppressed heat conduction through the shocks, as suggested by X-ray observations of SNRs.

Our numerical results show that the solution remains self-similar until the inner shock is reached by the top of the ramp. The remnant expansion follows the same temporal power law as in the adiabatic case.

Fig. 1 shows the density distribution of our solution at different times, as long as the inner shock propagates into the ramp. The most striking feature is represented by two reverse shocks in the ejecta. Much of the thermal energy generated at the external shock is transported inward, raising the temperature of the inner shock. The stellar material, once having entered this latter shock, at first slows down but then quickly starts to expand almost freely because of the high temperature; the flow eventually becomes supersonic, and a second reverse shock is thus formed as the shocked ejecta impinge on the contact surface.

The non-adiabatic solution presented here is clearly not consistent with our initial assumption that thermal conduction is suppressed through shock fronts. If the thermal flux is quenched at the intermediate shock, the latter is pushed by the heat and accelerates. On the contrary, the inner shock is no longer sustained by this flux and slows down (in a reference frame co-moving with the contact surface). It is eventually overtaken by the other reverse shock while new intermediate shocks develop. Fig. 2 shows the density distribution after six years. After ten years several shocks have developed, interacting and giving rise to additional sub-shocks. The motion ceases to be tractable; we may however conclude that the flow becomes chaotic in a very short time.

3.   DISCUSSION

It is quite difficult to assess the influence of heat flux on the flow of young SNRs, since plasma turbulences are still an open field. On one hand we may assume that heat flux suppression and electron-ion coupling happen only in well developed shock fronts; on the other hand it is possible that these effects happen as soon as the plasma perturbations grow out of the linear regime.

In this latter case thermal conduction is regulated by a very efficient feed-back mechanism. The heat flux is reduced just when its dynamical effect starts to become effective, while T_e and T_i tend to be equal because of the energy equipartition. The remnant, therefore, can be described by the adiabatic, one-fluid Chevalier model; heat conduction produces plasma turbulence on small scales only.

Radio observations effectively show a great deal of small scale turbulence (Dickel, 1983). The feed-back mechanism set up by thermal conduction could explain such turbulence.

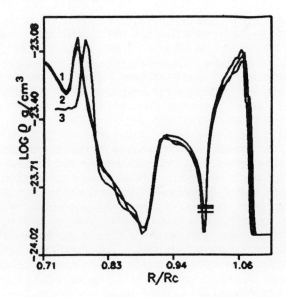

Fig. 1.   Density distribution at three different times.   1) t=4.3 yr;
2) t=73.2 yr; 3) t=309.6 yr.   This latter curve refers to the time when
the inner shock is reached by the top of the ramp.   Distances are
normalized to the value of the radius of the contact surface.

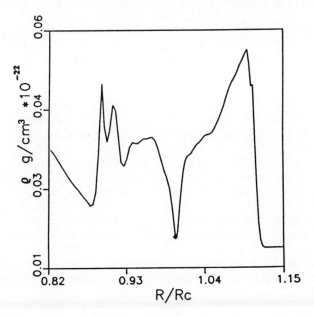

Fig. 2.   Density distribution at t=6 yr when the heat flux is
suppressed through the intermediate shocks.

Consider now the possibility that the heat flux is inhibited only in well developed shock fronts. Then the fluid motion must be chaotic. The remnant is likely to assume a clumpy appearance, losing the well defined two-shell structure. Such a structure is however confirmed for Tycho's remnant by X-ray observations (Seward, Gorenstein and Tucker, 1983). Clumps are definitely present, but they are distributed in the inner shell and their presence may easily be explained in terms of Rayleigh-Taylor instability.

In contrast to Tycho, Cas A shows a quite different aspect. Tuffs (1983) reported predominantly chaotic proper motions of some 342 radio features. They present a poorly defined mean expansion age of 949 yr. Several structures have components of proper motion directed towards the optical expansion centre. This slow, chaotic expansion remains puzzling; heat-flux driven instabilities provide a likely explanation. As a possibility, we may assume that plasma turbulences are such that the feed-back mechanism does not work. In this case heat conduction in Tycho is impeded by some other process which could not be active in Cas A; in this latter remnant additional shocks would be free to form, driving chaotic motions.

Alternatively, we may assume that the CSM surrounding Cas A is cloudy, as testified by the quasi-stationary flocculi (Peimbert and van den Bergh, 1971). The ambient clouds may be able to penetrate into the inner shell and partially convert the kinetic energy of ejecta into thermal energy. In this case the heat flux would be greater than that allowed by the feed-back mechanism, giving rise to a chaotic motion.

References

Chevalier, R.A.: 1982, Astrophys. J. 258, 790
Cox, D.P., Edgar, R.J.: 1983, Astrophys. J. 265, 443
Cowie, L.L.: 1977, Astrophys. J. 215, 226
Cowie, L.L., McKee, C.F.: 1977, Astrophys. J. 211, 135
Dickel, J.R.: 1983, in Proc. IAU Symp. 101 'Supernova Remnants and their X-ray emission', ed. J. Danziger and P. Gorenstein, Reidel, Dordrecht
Hesser, J.E., van den Bergh, S.: 1981, Astrophys. J. 251, 549
Peimbert, M., van den Bergh, S.: 1971, Astrophys. J. 167, 223
Pravdo, S.H., Smith, B.W.: 1979, Astrophys. J. (Letters) 234, L195
Seward, F., Gorenstein, P., Tucker, W.: 1983, Astrophys. J. 266, 287
Solinger, A., Rappaport, S., Buff, J.: 1975, Astrophys. J. 201, 381
Strom, R.G., Goss, W.M., Shaver, P.A.: 1982, Monthly Notices Roy. Astron. Soc. 200, 473
Tuffs, R.J.: 1983, in Proc. IAU Symp. 101 'Supernova Remnants and their X-ray emission', ed. J. Danziger and P. Gorenstein, Reidel, Dordrecht

NOVA GK PERSEI – A MINIATURE SUPERNOVA REMNANT?

E. R. Seaquist and D. A. Frail, University of Toronto, Canada
M. F. Bode, J. A. Roberts, and D. C. B. Whittet, Lancashire Polytechnic, U. K.
A. Evans and J. S. Albinson, University of Keele, U. K.

Abstract: We present radio and optical images of the shell-like remnant of the 1901 outburst of Nova GK Persei. The behaviour of this object is remarkably similar to supernova remnants. The synchrotron radiation-emitting shell is polarized with the magnetic field oriented radially, as in young SNR's. This similarity plus extensive data we have acquired on the expansion and the interstellar environment of GK Per indicate that the nova shell is colliding with ambient gas whose density is substantially higher than the ISM.

Furthermore, there is strong evidence that the ambient gas is circumstellar rather than interstellar, and that this material is the shell of an ancient planetary nebula associated with the white dwarf companion of GK Per.

The production of synchrotron radiation-emitting electrons in a supernova remnant (SNR) occurs when the ejecta shell sweeps up interstellar or circumstellar gas with mass comparable to that in the shell. This effect is evidently occurring in the nova remnant GK Persei 1901, as originally discovered by Reynolds and Chevalier (1984).

We have made extensive observations of the nonthermal radio emission from GK Per using the VLA in D configuration at 6 cm and C configuration at 20 cm. Figure (1) shows the 6 cm map superposed on a [NII] CCD image kindly provided to us by H.W. Duerbeck. The bright [NII] emission, together with the coincident radio emission, suggests an interaction between a shell of expanding nova ejecta and an ambient medium located to the southwest. Note, however, that there is no detailed correspondence between radio emission and optical knots. The phenomenon is similar to that occuring in SNR's.

Figure (2) shows the radio spectrum of the integrated emission from GK Per demonstrating that the mechanism is synchrotron radiation, and that a flattening occurs in the spectrum at low frequencies. Our studies show that this flattening must be caused by curvature in the energy distribution of the emitting relativistic electrons, rather than by any absorption process or the Razin Effect.

Figure (3) shows the distribution of polarization E-vectors superimposed on the total intensity contours at 6 cm and 20 cm, as well as the corresponding distributions of the percentage polarization. The key features are first that the E-vectors at 6 cm are tangent to the shell signifying a magnetic field directed radially. The 20-cm E-vectors are rotated by galactic Faraday rotation occurring over the intervening 500 pc. These characteristics are again reminiscent of a young SNR such as Cas A, except that the entire phenomenon is scaled down by five orders of magnitude in energy!

Note that the 20-cm emission is depolarized relative to that at 6 cm by a factor of three in the outer part of the intense ridge. This effect is produced by Faraday effects occurring within the ridge itself due to the magnetic field (equipartition value 7×10^{-5} gauss) and a plasma with mean electron density of $10 - 100$ cm^{-3}. This plasma probably originates from the [NII]-emitting ejecta, since the depolarization occurs where the [NII] emission is generally the brightest.

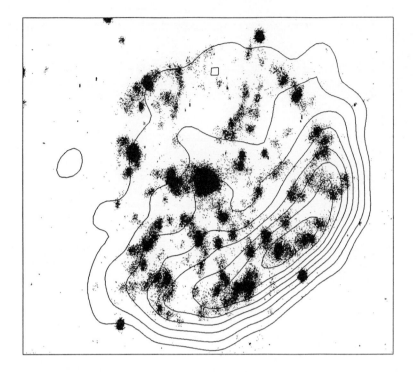

Figure (1) : Radio emission at 6 cm superimposed on an [NII] CCD image.

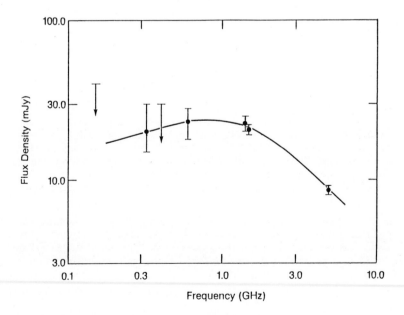

Figure (2) : Integrated radio spectrum of GK Per.

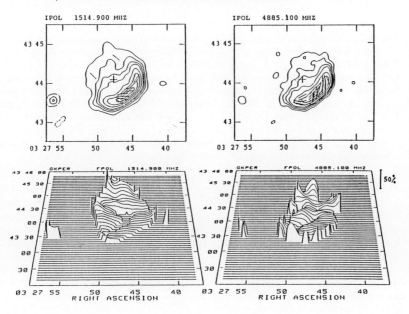

Figure (3) - Top : Polarization vectors at 20 cm and 6 cm superimposed on total intensity contours.
Bottom : Percentage polarization at 20 cm and 6 cm.

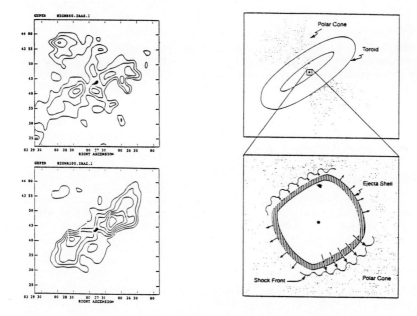

Figure (4) - Left : An IRAS $60\,\mu$m map (top) and a $100\,\mu$m map (bottom), showing the IR emission surrounding GK Per. The radio remnant is shown to scale by the blackened spot at the center.

Figure (5) - Right : A schematic showing our interpretation, which involves the interaction between the nova ejecta and the remnant of an ancient planetary nebula associated with GK Per.

Figure (4) shows IRAS maps at 60 μm and 100 μm of an extended region surrounding GK Per. The position of the GK Per is shown to scale as the black dot in the middle of the IR contours. Evidently GK Per is at the centre of a cloud of dust with maximum dimension 4 pc and with GK Per situated in a saddle of the IR brightness distribution. The cloud appears elongated in the same direction as the radio emitting ridge of the remnant. The colour temperature of the dust is about 23 K, similar to that of the IR cirrus. HI 21-cm emission from this region (not shown) suggests an elongated feature parallel to the axis of the IR emission. Analyses of these data suggest that GK Per is embedded in a toroidal ring of gas and dust whose radius and total mass are respectively about 1 pc and 1 $M_\odot$, and that gas associated with this cloud is responsible for the interaction which produces the synchrotron radiation.

We are led to hypothesize that the cloud surrounding GK Per is an ancient planetary nebula, associated with the evolution of one member of the binary to the white dwarf stage (Bode *et al.* 1987a). The morphology of the nebula is shaped by the presence of the binary. The age of this planetary would be between 50,000 and 200,000 years. Scattering of light by the dust associated with the planetary nebula may have been responsible for the apparent "superluminal" expansion of the nebulosity in this nova seen at the time of the 1901 outburst (Couderc 1939). Figure (5) shows how the nova ejecta may be interacting with the remnant of a bipolar outflow associated with the planetary nebula phase whose axis is coincident with that of the torus. The association of such bipolar structures with planetary nebulae is fairly common (Balick 1987). If this picture is correct, then the observed interaction would be possible only if the 1901 nova outburst in this system is the first. Otherwise the medium interacting with the ejecta would have been swept away by a previous outburst.

A search for nonthermal emission from other old fast novae similar to GK Per failed to turn up similar phenomena in other cases (Bode 1987b). Thus the phenomenon appears so far to be unique to GK Per.

Conclusion:

The radio and optical features associated with GK Per, coupled with the morphology of extended IR emission surrounding the nova, suggest to us that the nova ejecta are presently interacting with an ancient planetary nebula associated with GK Per . A detailed study shows that the interaction is very similar to that produced in young SNR's. The uniqueness of this phenomenon among novae supports the case for an interaction of a nova with its own planetary nebula.

If this picture is correct, then the study of this object may provide some unusual insights into relativistic particle production in shock waves, and into the timescales involved in the production of white dwarfs and in the onset of nova outbursts.

Acknowledgements:

ERS acknowledges the support of an operating grant from the National Sciences and Engineering Research Council of Canada (NSERC). DAF was supported by an NSERC scholarship. MFB and JSA are supported by the UK Science and Engineering Research Council. JAR is supported by the by the National Advisory Body to the Polytechnics.

References:

Balick, B., in *Proceedings of the Calgary Workshop on Late Stages of Stellar Evolution, Calgary, Alberta, Canada, June 2-5, 1986*, in press, 1987.

Bode, M. F., Seaquist, E. R., Frail, D. A., Roberts, J. A., Whittet, D. C. B., Evans, A., and Albinson, J. S., submitted to *Nature*, 1987a.

Bode, M. F., *et al.* , *Mon. Not. R. Astron. Soc.*, in press, 1987b.

Couderc, P., *Ann. d'Astrophys.*, **2**, 271, 1939.

Reynolds, S. P. and Chevalier, R. A., *Ap. J. (Letters)*, **281**, L33, 1984.

THE BULK RADIO EXPANSION OF CASSIOPEIA A

D.A.Green,
Mullard Radio Astronomy Observatory, Cavendish Laboratory,
Madingley Road, Cambridge CB3 0HE, United Kingdom.

Abstract: Comparison, in the visibility plane, of radio observations of Cassiopeia A made at 151 MHz over a 2.3 yr interval indicate that the bulk of the radio emitting material has not been decelerated strongly.

Introduction: Cassiopeia A (hereafter simply Cas A) is thought to be the remnant of a relatively recent SN. Optical observations reveal, among other things, a population of fast-moving knots which show a *low* expansion timescale (i.e. current angular size/current angular expansion rate) of ~300 yr (Kamper & van den Bergh 1976). Interpreting these knots as undecelerated ejecta from the SN explosion (e.g. Kamper & van den Bergh) gives an explosion date around AD1671 — indeed it has been suggested that Flamsteed may have catalogued Cas A's parent SN in AD1670 (Ashworth 1980; Kamper 1980). On the other hand, observations of compact radio knots indicate a *high* expansion timescale of ~950 yr (Bell 1977; Tuffs 1986), implying material that has been decelerated appreciably. This is, at first sight, difficult to reconcile with the complex structure of the radio emission from Cas A (Bell, Gull & Kenderdine 1975), which is quite different from that expected (Gull 1975) from the blast-wave which would result if it had indeed decelerated because of sweeping up a large amount of circumstellar/interstellar material. (cf. Tycho's SN, which has decelerated considerably to near Sedov expansion (Strom, Goss & Shaver 1982; Tan & Gull 1985), and has a sharply-defined shell of radio emission (e.g. Duin & Strom 1975; Green & Gull 1983).) However, the radio knots represent only a small proportion of the total radio emission, so that the expansion timescale deduced from them is not representative of the current dynamics of the bulk of the radio emitting material.

Here I present preliminary results from a comparison of observations made of Cas A at 151 MHz over an interval of 2.3 yr. These indicate an expansion timescale for the bulk of the radio emission considerably smaller than that deduced from the compact knots.

Observations: The Cambridge Low-Frequency Synthesis Telescope (CLFST) is an east-west synthesis telescope which, when operating at 151 MHz, consists of 60 multi-yagi aerials. The 28 eastern aerials (which are divided into four huts: A, B, C and D) are correlated against the 32 western aerials (huts E, F, G, and H) to give an almost uniform coverage of the visibility plane with baselines from 6λ to 2352λ in steps of 3λ. (One each of the small and large baselines are missing, and eight baselines are duplicated. Also, the aerials in hut H lie some way from the ideal east-west line, so that coverage of the visibility plane in regions involving baselines from these aerials is not as uniform as elsewhere.)

Cas A was observed with the CLFST at 151 MHz for ~12 hr on 1984 July 18 and 1986 November 1. Expansion timescales of 300–900 yr would result in expansions of 0.77–0.26 per cent between these observations, and since Cas A is ~5 arcmin in diameter, this corresponds to an expansion of the remnant of 2.3–0.8 arcsec. Such a change is only a small fraction of the beam of the CLFST at 151 MHz (~70 arcsec), so that comparison in the map plane would require very high dynamic range maps. In practice the quality of the synthesized maps is limited by residual hut- and aerial-based calibration errors. These produce circular errors in the map plane that would contaminate, in a difficult to appreciate way, any misfit statistic used to determine the expansion of Cas A. Instead, the data have been compared in the visibility plane, where residual calibration errors can be better appreciated and the contaminated data avoided.

If there are no changes in the shape of the radio emission from Cas A between the two observations, but only an overall expansion, then the contraction of *any* feature in the visibility plane corresponds to the expansion of the *whole* of the emission from Cas A. There are, however, reports of relatively large changes ('flares') in the total radio flux of Cas A at 38 MHz (Read 1977a, 1977b; Walczowski & Smith 1985), although the evidence for them is marginal. These flares, which apparently occur on the timescale of a few years, must have steep spectra as they were not evident at frequencies above 38 MHz, and might be due to the 'switching-on' of the radio emission from compact features in the remnant. The comparison at 151 MHz will be complicated if such a flare occurs during the observations.

Day-to-day variations in the gains of the aerials of the CLFST, plus the effect of varying ionospheric refraction, are usually eliminated by phase-rotating the observed visibilities to a relatively bright compact source in the field. In the case of the observations of Cas A this is not possible, but, by comparing adjacent baselines, the effects of changes in amplitude sensitivity and instrumental phase of aerials and huts relative to each other can be largely eliminated. (The relative scaling between huts A, B, C and D remains the most uncertain, as there are few baselines which correspond to overlap in the aperture plane between these huts.) However, due to ionospheric refraction, and poor initial phase calibration of the 1984 observations, the absolute phases, particularly for the larger baselines, are still uncertain. For these reasons, comparison of the observations has so far been restricted to the *amplitude* of the visibility function.

Results, Conclusion (and Future Prospects)

Results: Figure 1 shows the amplitude of the visibility function for Cas A from the 1984 observations (several minutes' data were lost near HAs -6^h and 5^h due to interference). The visibility function of an ideal shell source is centrally peaked, with circles of zero amplitude. Cas A is basically a distorted shell, and its visibility function shows a central peak, with approximately elliptical minima, the first of which is much more sharply defined than the second. The first minimum falls to less than 1 per cent of the total flux density of Cas A between HA~1.5^h and HA~3.5^h. Between the two observations, this well-defined portion of the first minimum shows a clear contraction corresponding to an expansion timescale of ~400 yr. This result is, however, only from regions with very low amplitude, and from a limited range of HAs, so it may be biased by the effects of flares or any shape changes.

In order to extend the comparison to other HAs, and to regions that do not show very deep minima, a series of visibility planes were made from the 1986 data, on grids 0.1, 0.2... per cent smaller than that used for the 1984 data. These were then compared, and a simple statistic (the relative misfit of the 1986 and 1984 amplitudes: the sum of $|1-(amp_{1986}/amp_{1984})|$ over valid pixels within the region) was computed for various regions to determine the scaling change that best represents the contraction of the visibility function of Cas A between the two observations.

Fortunately the baselines near first minimum all involve aerials in hut C, which avoids the difficulties due to calibration uncertainties between huts A, B, C and D. The minimum misfit for a region containing the whole of the first minimum corresponds to a scale change between 0.5 and 0.6 per cent (an expansion timescale between 460 and 380 yr), similar to that deduced from only the very deep part of the first minimum. This result does *not*, however, just depend on the lowest amplitude regions or a limited HA range.

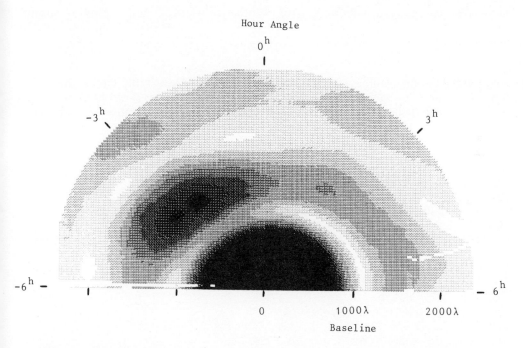

Figure 1. *Visibility plane of Cas A at 151 MHz from the 1984 observations. The central peak is ~8500 Jy, and the greyscale used is (white to black in ten steps) from 0 to 2500 Jy.*

The second minimum is poorly defined, and in places spans the boundaries between huts A and B, where relative amplitude scaling errors are still a problem. Restricting the comparison to the best defined portions of the second minimum gives a range of best-fit scale changes ranging from 0.2 per cent to 0.7 per cent depending on hour angle (i.e. 1150 to 330 yr for the expansion timescale). Interpretation of these results is confused not only by residual calibration problems, but also be the range of values deduced. It is notable, however, that only one of the deeper features of the second minimum requires an expansion timescale larger than that deduced from compact radio knots.

Conclusion: These results show that, although there are variations depending on HA, the expansion of the bulk radio emission from Cassiopeia A is on a timescale nearer that deduced from the fast optical knots than that deduced from the compact radio knots, implying little deceleration of the bulk of the radio emitting material. One consequence of this *low* expansion timescale is that continuing particle acceleration within the remnant may not be needed (Green 1987) to explain the observed rate of decrease of its radio emission.

Future Prospects: It is hoped that the calibration of these observations (and future observations) can be be improved to a level where the effect of overall expansion of the remnant can be quantified more accurately by separating it from the HA-dependant effects due to any compact features or shape-changes. Also, with improved calibration, subtraction of synthesized maps may provide useful constraints on, or positions for any radio flares (at 151 MHz)occurring between the observations.

Acknowledgments: I am very grateful to Dr J.A. Baldwin, with whom this work has been done in collaboration, and Mr P.J. Warner, for their help with the reduction and calibration of these observations. I am also grateful to Churchill College, Cambridge for a Junior Research Fellowship.

References:
Ashworth, W.B., 1980. *J. Hist. Astr.*, **11**, 1.
Bell, A.R. , 1977. *Mon. Not. R. astr. Soc.*, **210**, 642.
Bell, A.R., Gull & Kenderdine, S.K., 1975. *Nature*, **257**, 463.
Duin, R.M. & Strom, R.G., R.G., 1975. *Astr. Astrophys.*, **39**, 33.
Green, D.A., 1987. In *Genesis and Propagation of Cosmic Rays*, eds Shapiro, M.M. & Wefel, J. Reidel, Dordrecht, in press.
Green, D.A. & Gull, S.F., 1983. In: *Supernova Remnants and their X-ray emission*, eds Danziger, I.J. & Gorenstein, P., Reidel, Dordrecht, p329.
Gull, S.F., 1975. *Mon. Not. R. astr. Soc.*, **171**, 263.
Kamper, K.W., 1980. *Observatory*, **100**, 3.
Kamper, K.W. & van den Bergh, S., 1976. *Astrophys. J. Suppl.*, **32**, 351.
Read, P.L., 1977a. *Mon. Not. R. astr. Soc.*, **178**, 259.
Read, P.L., 1977b. *Mon. Not. R. astr. Soc.*, **181**, 63P.
Strom, R.G., Goss, W.M. & Shaver, P.A., 1982. *Mon. Not. R. astr. Soc.*, **200**, 473.
Tan, S.M. & Gull, S.F., 1985. *Mon. Not. R. astr. Soc.*, **216**, 579.
Tuffs, R.J., 1986. *Mon. Not. R. astr. Soc.*, **219**, 13.
Walczowski, L.T. & Smith, K.L., 1985. *Mon. Not. R. astr. Soc.*, **212**, 27P.

3C58'S FILAMENTARY RADIAL VELOCITIES, LINE INTENSITIES, AND PROPER MOTIONS

R. A. Fesen[1], R. P. Kirshner[2], and R. H. Becker[3]

[1] CASA, University of Colorado, Boulder CO
[2] Center for Astrophysics, Cambridge, MA
[3] Dept. of Physics, UC Davis, Davis, CA

ABSTRACT: Optical spectroscopy on nearly 50 filaments of 3C58 indicates a maximum expansion velocity of 1100 ± 100 km s^{-1}. A considerable range in radial velocity with projected distance from remnant center is found suggesting that the remnant's optical emission is not confined to a thin shell. Typical filament electron densities are between 200 - 500 cm^{-3} with Hα/[N II] ratios in the range 0.2 - 0.5. Optical extinction to 3C58 is modest with E[B-V] = 0.68 ± 0.08. Preliminary radial proper motion measurements for a few outlying filaments suggest values of order 0.05" - 0.07" yr^{-1}.

INTRODUCTION

The galactic SNR 3C58 is perhaps the best studied member of the subclass of remnants which have properties resembling those of the Crab Nebula. In the radio, 3C58 exhibits a filled center morphology, high degree of linear polarization, and a relatively flat spectral index (Weiler 1980; Reynolds and Aller 1985; Green 1986). It also possesses a centrally peaked X-ray emission structure with hints of an X-ray point source, suggesting the presence of a central neutron star (Becker et al. 1982). If 3C58 is indeed the remnant of SN 1181 as suggested by Stephenson (1971) and Clark and Stephenson (1977), it then also has about the same age as the Crab Nebula. However, with angular dimensions of 5' x 9' and a distance of 2.6 ± 0.2 kpc estimated from 21 cm absorption studies (Green and Gull 1982), 3C58 appears nearly 50% larger than the Crab despite being slightly younger. Since optical filament radial velocities only as large as 900 km s^{-1} have been observed (Fesen 1983), a firm connection between 3C58 and SN 1181 has not yet been established and such an association has been recently questioned based on new interpretations of the historical observations (Huang 1987).

OBSERVATIONS

In order to further investigate 3C58's optical properties and kinematics, we obtained several long slit CCD spectra covering portions of the remnant's central 5' diameter region using the Cryogenic CCD Camera attached to the KPNO 4 m telescope. These data provided radial velocity (±75 km s^{-1}) and relative emission line intensity information on nearly 50 individual filaments. In addition, we recently obtained (Nov. 1986) deep CCD Hα interference-filter images of four of the brightest outlying emission knots. The positions of these knots on these images were compared to their locations on van den

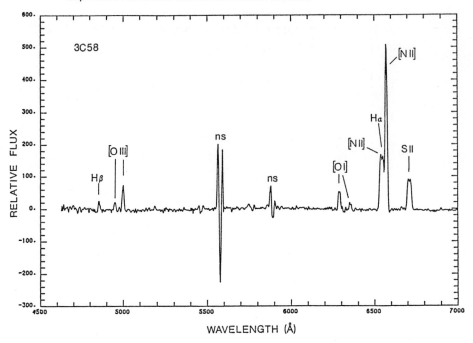

RELATIVE FLUX

WAVELENGTH (Å)

Figure 1: Optical spectrum of one of 3C58's brighter emission filaments covering the wavelength region from 4500 - 7000 Å at 13 Å resolution. Note the strong [N II] 6548,6583 and [O I] 6300,6364 lines relative to the strength of Hα. The features at 5577 and 5890 are due to imperfect subtraction of the night sky emission lines.

Bergh's (1978) red photograph in order to estimate the remnant's proper motion.

RESULTS

A representative spectrum for one of 3C58's brighter filaments is shown in Figure 1. The presence of strong [S II] 6717,6731, [O I] 6300,6364, and [O III] 4959,5007 line emissions relative to the strength of Hα is strong evidence that the observed optical emission represents shock-heated gas associated with the remnant. The electron density-sensitive [S II] 6717/6731 line ratio was found to be typically between 1.0 - 1.2 with a total range of 0.85 to 1.4. This implies a range of electron densities for 3C58's filaments of $\leq$ 100 to 1000 cm^{-3} with average values of between 200 - 500 cm^{-3}. These values in turn suggest preshock densities of order 2 - 5 cm^{-3}.

The majority of 3C58's filaments exhibit strong [N II] 6548,6583 line emission with Hα/[N II] ratios typically in the range 0.2 - 0.5, but as small as 0.15. Such strong [N II] emission is seen in Kepler's SNR and Cas A's QSF's and is believed to signify a nitrogen enrichment several times over the solar abundance. Filaments showing considerably weaker [N II] emission (Hα/[N II] = 1.0 - 1.5) like that observed in shocked interstellar gas appear limited to the remnant's northern edge (cf. Kirshner and Fesen 1978).

Measured radial velocities for nearly 50 of 3C58's filaments were found to range from +1000 to −1075 km s^{-1} (see Fig. 2). The highest velocities observed are for filaments

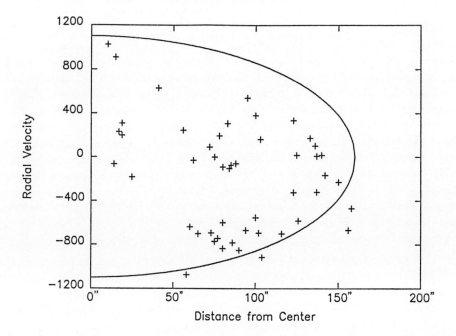

Figure 2: Observed filament radial velocity (km s^{-1}) plotted versus projected radial distance from the 3C58's X-ray point source in arcseconds. Curve represents an expansion velocity of 1100 km s^{-1} at a radius of 160".

located near the remnant's projected center suggesting a maximum line of sight expansion velocity of about 1100 ± 100 km s^{-1}. However, as can be seen from Figure 2, considerable low-velocity gas ($|V_r| \leq 300$ km s^{-1}) is also observed throughout the remnant including 3C58's center and positions immediately adjacent to filaments with V_r in excess of ± 500 km s^{-1}. This indicates that the remnant's optical emission is not confined to a thin shell as is the case in the Crab Nebula. This lower radial velocity gas does not appear significantly different from the remnant's higher velocity filaments with respect to morphology, electron density, or [N II] line emission strength.

An estimate of the reddening to 3C58 was made using the observed Hα/Hβ ratio for the brighter filaments where Hβ could be accurately measured. Assuming an intrinic Hα/Hβ ratio of 3.0, the observed range of 5.6 - 6.6 suggests only a modest amount of optical extinction; i.e., E[B-V] $= 0.68 \pm 0.08$ ($A_v = 2.0 \pm 0.25$ mag). While larger than the $A_v = 1.3 \pm 0.2$ value estimated by Green and Gull (1982) from H I column density observations but less than the $A_v = 3.1$ suggested by Panagia and Weiler (1980), this value is in good accord with reddening estimates for distances between 1 and 4 kpc in this general direction (Neckel and Klare 1980).

Proper motion estimates for four relatively bright outlying filaments (northern knots "A" and "B" observed by Kirshner and Fesen 1978; WNW knots at r $= 170$"; and a NE knot at r $= 158$") indicate values on the order of 0.05" to 0.07" yr^{-1} and appear to exclude values as high as 0.15" yr^{-1} for filaments having radial distances less than 160".

DISCUSSION

Filament radial velocities of up to 1100 km s^{-1}, the presence of strong nitrogen emission indicating a probable N/H enhancement, and filament electron densities as large as 10^3 cm^{-3} suggest that 3C58's optical emission is that of a young SNR. However, if 3C58 is the remnant of SN 1181 and therefore only 800 yrs old, our measured proper motions of 0.05" - 0.07" yr^{-1} are considerably less than its average value (radius/age) of 0.2" yr^{-1}. Similarly, the observed 1100 km s^{-1} radial expansion velocity is less than half 3C58's average expansion velocity if at a distance of 2.6 kpc. A kinematic distance estimate for 3C58 from our data is complicated by the uncertainity of the remnant's tangential expansion velocity in the E-W or N-S directions. Assuming $V_T = V_R$, then our preliminary proper motion estimates suggest a distance of between 3.0 - 4.5 kpc. A 2.6 kpc distance is still quite possible, however, requiring only $\mu = 0.09$" yr^{-1}.

A possible way to reconcile 3C58's dimensions with an 800 year age despite its observed low proper motions and radial velocities would be for its ejecta to have undergone a rapid deceleration. Such an explanation would require a substantial ambient interstellar gas density of order 5 - 10 cm^{-3} in the remnant's immediate vicinity. Interestingly, 3C58's low-velocity emission filaments might indicate the existence of such a high-density medium which might be due in part to a pre-supernova mass loss episode. In terms of radial velocity and [N II] emission strength, 3C58's low-velocity gas appears similar to the nitrogen-rich, pre-SN mass loss material found in Cas A and Kepler. The relatively bright filaments along 3C58's northern rim which exhibit weaker [N II] line emission also suggest that the remnant lies is a region containing at least some high-density interstellar gas. In any case, the presence of low-velocity gas within 3C58 represents an important difference between its optical emission and that of the Crab Nebula. Therefore, despite lower than expected radial velocities and proper motions considering 3C58's size and age relative to the Crab Nebula, 3C58 appears the likely remnant of SN 1181.

REFERENCES

Becker, R. H., Helfand, D. J., and Szymkowiak, A. E. 1982, *Ap. J.*, **255**, 557.

Clark, D. H., and Stephenson, F. R. 1977, *The Historical Supernovae*, (Oxford: Pergamon).

Fesen, R. A. 1983, *Ap. J. Letters*, **270**, L53.

Green, D. A., and Gull, S. F. 1982, *Nature*, **299**, 606.

Green, D. A. 1986, *M.N.R.A.S.*, **218**, 533.

Huang, Y.-L. 1987, *Nature*, submitted.

Kirshner, R. P., and Fesen, R. A. 1978, *Ap. J. Letters*, **224**, L59.

Neckel, Th., and Klare, G. 1980, *Astr. Ap. Suppl Ser.*, **42**, 251.

Panagia, N. and Weiler, K. W. 1980, *Astr. Ap*, **82**, 389.

Reynolds, S.P., and Aller, H. D. 1985, *A. J.*, **90**, 213.

Stephenson, F. R. 1971, *Quart. J. R. A. S.*, **12**, 10.

van den Bergh, S. 1978, *Ap. J. Letters*, **220**, L9.

Weiler, K. W. 1980, *Astr. Ap.*, **84**, 271.

THE REIONIZATION OF UNSHOCKED EJECTA IN SN1006

A. J. S. Hamilton and R. A. Fesen
Joint Institute for Laboratory Astrophysics
University of Colorado and National Bureau of Standards
and Center for Astrophysics and Space Astronomy,
University of Colorado, Boulder, Colorado 80309

ABSTRACT: The fortuitous positioning of the Schweizer and
Middleditch OB subdwarf behind SN1006 has permitted the
detection and subsequent confirmation by IUE of broad
(±5000 km/s) Fe II absorption features which probably arise
from unshocked iron ejecta in the center of SN1006. The
mass of detected Fe II, ~0.012 $M_\odot$, is however only 1/25 of
the ~0.3 $M_\odot$ of Fe within ±5000 km/s predicted by carbon
deflagration models. IR and optical observations exclude
any appreciable iron in grains or Fe I, but high ion stages,
Fe III and up, could be present. Promising mechanisms for
ionizing the unshocked iron in SN1006 include the radioactive decay of ^{44}Ti, and photoionization by UV and X-ray
emission from the reverse shock. Although the photoionization model works, insofar as it permits as much as
0.2 $M_\odot$ of unshocked iron in the center of SN1006, agreement
with the IUE data requires that the ejecta density profile
be flatter, less centrally concentrated, than the W7
deflagration model of Nomoto, Thielemann, and Yokoi.

I. The Problem: Figure 1 shows the problem: Nomoto, Thielemann and
Yokoi's (1984) W7 carbon deflagration model fits nicely to the IUE
observations of Fe II absorption in SN1006 -- except that the predicted model density of Fe (all ion stages) is 25 times the observed
density of Fe II.

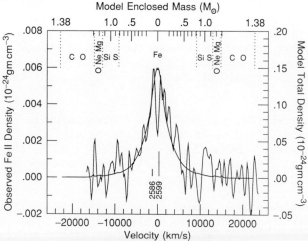

Figure 1. The Fe II λ2586, 2599 absorption line profile observed in
SN1006 (Fesen et al. 1987), converted to density assuming
no saturation, compared to Nomoto et al.'s (1984) W7 model
evolved in free expansion to 980 years old.

Could the Fe be hidden in Fe grains? No. Any appreciable
depletion into grains in the Fe ejecta would make SN1006 a strong IRAS
source, not observed. Fe I? No. No Fe I absorption is present in the
Schweizer and Middleditch (1980) optical spectrum. Fe II? No. The
relative equivalent widths of strong and weak Fe II absorption lines in
the IUE data indicate that the Fe II lines are not saturated. Fe III,
IV? Could be.

Mechanisms for ionizing unshocked ejecta:
(1) Radioactivity - the decay of ^{56}Ni leaves the plasma mainly neutral
     at 4 years old, but ^{44}Ti can have some effect later on -- see § II.
(2) Photoionization by ambient UV starlight - ionizes Fe I to Fe II in
     100 years, but no higher.
(3) Cosmic rays - zilch.
(4) The ejecta has been reverse shocked after all - no - the reverse
     shock kills Fe II dead, contrary to the IUE data.
(5) Photoionization by UV and x-rays from reverse-shocked ejecta - the
     most probable answer.

II. The Radioactive Decay of ^{44}Ti

The familiar $^{56}\text{Ni} \xrightarrow{6d} {}^{56}\text{Co} \xrightarrow{77d} {}^{56}\text{Fe}$
decay scheme is no good for causing persistent ionization in SNRs: it
dumps its energy too early, while the density is high, and recombination
and cooling times are shorter than the age of the remnant.

The radioactive decay $^{44}\text{Ti} \xrightarrow{47yr} {}^{44}\text{Sc} \xrightarrow{4h} {}^{44}\text{Ca}$, with a 47 year half-
life, is more effective. The principal decay scheme, with a branching
ratio of 0.932 for positron emission, is:

(1) $^{44}\text{Ti} \xrightarrow{47yr} {}^{44}\text{Sc}^+ + \nu$ (orbital electron capture)
(2) $^{44}\text{Sc} \xrightarrow{4h} {}^{44}\text{Ca}^* + e^+ + \nu$ (positron emission, mean energy .767 MeV)
(3) $^{44}\text{Ca}^* \longrightarrow {}^{44}\text{Ca} + \gamma$ (excited Ca emits 1.159 MeV gamma-ray).

The neutrinos and gamma-rays escape, but the positrons Coulomb scatter
off electrons in the plasma before annihilating. The heated electrons
then collisionally excite and ionize the plasma. Since at 47 years old
the recombination time exceeds the age of the remnant, any ionization
which occurs as the result of the decay of ^{44}Ti persists to the present
time.

It is believed that the explosive nucleosynthesis of radioactive
^{44}Ti is the dominant source of ^{44}Ca in the galaxy. If ^{44}Ti is synthesized in the cosmic ratio of $^{44}\text{Ca}/^{56}\text{Fe} = 1.41 \times 10^{-3}$, then ^{44}Ti decays
produce $1.41 \times 10^{-3} \times 0.767$ MeV $\times 0.932 = 1.01$ keV per Fe ion, sufficient in principle to ionize Fe several times.

Detailed calculations including adiabatic and collisional cooling
losses show that the amount of Fe ionization is a sensitive function of
ejecta density and ^{44}Ti abundance, with appreciable ionization occurring
only for at least a cosmic fraction of ^{44}Ti.

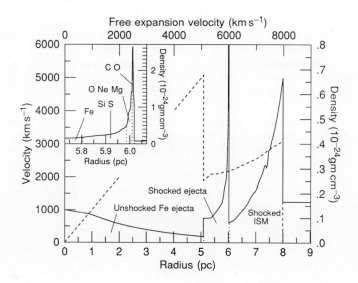

Figure 2. Velocity (dashed line) and density (solid line) structure in
          Nomoto et al.'s W7 model, evolved by spherically symmetric
          hydrodynamic simulation into a uniform ambient medium, to the
          point where the reverse shock has penetrated as far as the
          5000 km/s free expansion radius, as appropriate for SN1006 at
          the present time, according to the IUE Fe II line widths. At
          the present 980 year age of SN1006, the inferred ambient inter-
          stellar density is 0.07 H atoms cm^{-3}, the swept up inter-
          stellar mass is 5.3 $M_\odot$, the distance is 1.8 kpc, and the blast
          wave velocity is 4100 km/s, corresponding to an expansion rate
          $r \propto t^{0.52}$.

Problems:In current carbon deflagration models,
(a) ^{44}Ti is synthesized in only 1/10 the cosmic ratio of $^{44}Ca/^{56}Fe$;
(b) ^{44}Ti is synthesized in a shell outside the iron, not mixed in it.

III. Reverse Shock Photoionization Model:
(1) Use Nomoto, Thielemann and Yokoi (1984) W7 model as starting point
     for spherically symmetric hydrodynamic simulation.
(2) Evolve W7 model into uniform ambient medium until the reverse shock
     has reached the 5000 km/s free expansion radius, as indicated by
     IUE, and illustrated in Fig. 2.
(3) Adopt a two-layer approximation to ejecta composition, 0.8 $M_\odot$
     of Fe on the inside, 0.6 $M_\odot$ of Si on the outside.
(4) Compute time-dependent photoionizing UV and x-ray emission from the
     reverse shock in the "instantaneous" approximation, where material
     entering the reverse shock is collisionally excited and ionized to
     a high ionization state immediately it is shocked.
(5) Calculate collisional excitation and ionization of shocked gas in
     the high-temperature Bethe approximation.

(6) Include processes of collisional excitation, autoionization,
     fluorescence, multiple ionization. Approximately 70 photoionizing
     "lines" of Fe II to Fe XVI, and 40 "lines" of Si II to Si XII, each
     "line" standing for a complex of several individual lines. Use
     oscillator strength sum rules to check that no important source of
     photoionizing emission has been missed.
(7) Include ambient photoionizing starlight.
(8) Follow detailed self-consistent time-dependent radiative transfer
     and photoionization of expanding unshocked Fe and Si ejecta.
(9) Ionization state of material entering the reverse shock determined
     self-consistently.

     Figures 3 and 4 show the results of the phiotoionization calcula-
tions just described.

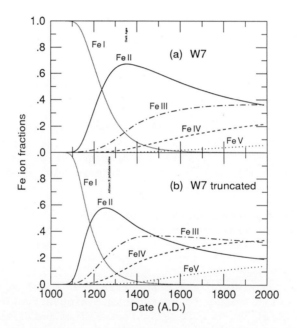

Figure 3. The evolution of Fe ion fractions at the center of SN1006,
           photoionized by ambient starlight and UV and x-ray emission
           from the reverse shock in (a) Nomoto et al.'s W7 model (top
           panel), and (b) Nomoto et al.'s W7 model with a truncated
           central ejecta density (bottom panel). Ionization is faster
           in the truncated model because the optical depth is smaller.

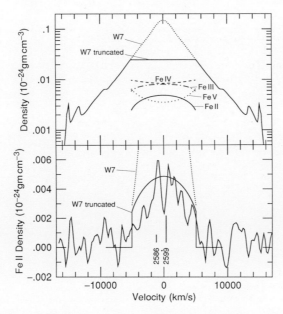

Figure 4. Fe II density (lower panel) computed in the truncated W7
          model, compared to the IUE observations of the Fe II $\lambda 2600$
          line. The unmodified W7 model predicts too much Fe II (dotted
          line). The upper panel shows the total and Fe ionic densities
          predicted by the truncated W7 model, along with the total
          density in the unmodified W7 model. The truncated model
          contains 0.20 $M_\odot$ of Fe within ± 5000 km/s, as opposed to the
          0.36 $M_\odot$ of Fe in the plain W7 model.

IV. Conclusions: The radioactive decay of ^{44}Ti should have an appreciable effect in ionizing the layer of ejecta where the ^{44}Ti is.
However, it probably has a negligible effect on inner layers of iron
ejecta.

    UV and x-ray emission from the reverse shock is able to cause
appreciable photoionization of unshocked iron ejecta. However,
agreement with the observed IUE Fe II absorption line profiles requires
the ejecta density profile to be flatter, less centrally concentrated,
than the W7 model of Nomoto, Thielemann and Yokoi (1984).

Acknowledgments: We are grateful to Ken Nomoto for providing the W7
model evolved into the free expansion phase.

References
Fesen, R. A., Wu, C.-C., Leventhal, M., and Hamilton, A. J. S. 1987
     Ap. J., submitted.
Nomoto, K., Thielemann, F.-K., and Yokoi, K. 1984 Ap. J., 286, 644.
Schweizer, F., and Middleditch, J. 1980 Ap. J., 241, 1039.

KINEMATICS OF OXYGEN-RICH FILAMENTS IN PUPPIS A

P. Frank Winkler, John H. Tuttle
Middlebury College, Middlebury, VT, U.S.A.

Robert P. Kirshner
Harvard-Smithsonian Center for Astrophysics, Cambridge, MA, U.S.A.

Michael J. Irwin
Institute of Astronomy, Cambridge University, Cambridge, U.K.

Abstract: We have measured proper motions for fast, oxygen-rich knots in Puppis A, which we demonstrate are probably uncontaminated ejecta from the progenitor star's core. Typical fast knots show motions of 0.1-0.2 arcsec yr^{-1} diverging from a point 4' northeast of the center of the radio shell. A model assuming constant expansion fits the data well and gives an age of 3700 ± 300 yr for Puppis A. We also present new spectra which indicate the presence of neon along with oxygen in the fast knots.

I. Introduction

Puppis A was one of the first supernova remnants (SNRs) in which optical nebulosity was identified (Baade and Minkowski 1954), and the chaotic morphology of its optical filaments has long been puzzling. The spectra of the brightest filaments indicate a high abundance of nitrogen (Dopita, Mathewson, and Ford 1977) and modest radial velocities: < 300 km s^{-1} (Elliott 1979; Shull 1983). In striking contrast are the faint filaments discovered by Winkler and Kirshner (1985) which are extremely rich in oxygen and have radial velocities up to 1500 km s^{-1}. Winkler and Kirshner interpreted these fast filaments as vestiges of ejecta from the core of the supernova star, similar to the fast-moving knots in Cas A. This picture meshes with the X-ray spectra of Puppis A, which indicate a hot plasma heavily enriched in oxygen and neon (Canizares and Winkler 1981). A few solar masses of ejecta similar in composition to the fast-moving optical filaments could have provided the enrichment inferred from the X-ray plasma.

The near absence of hydrogen in the fast filaments of Puppis A suggests that they have interacted rather little with interstellar or circumstellar material, and thus that they may remain relatively undecelerated since the supernova event. Our proper-motion study has enabled us to measure the expansion of the fast filament system and to determine the kinematic age of Puppis A to be 3700 ± 300 yr. The evolution of SNRs and their interaction with the interstellar medium is now a sufficiently mature subject to merit its own IAU Colloquium, yet only a handful of the youngest remnants have well-established ages. The measurement based on kinematics inducts Puppis A as the senior member into this heretofore exclusively youthful society.

II. Astrometry and the Expansion Model

Our measurements are based on 4-m prime-focus plates taken from CTIO at three epochs: 1978.2, 1984.1, and 1986.2. Plates were taken in two broad wavelength bands, IIIaJ + GG495 ("green", ~ 490-530 nm) and IIIaF + RG610 ("red", ~ 610-700 nm), of two overlapping fields covering virtually the entire Puppis A remnant, which has a diameter of about 55'. The ejecta-rich filaments are easily seen on the green plates due to their strong [O III] emission, but are virtually invisible on the red plates. For this study we have used five green plates, which include three pairs with identical centers and baselines of 5.9 years or more.

We used a digital technique to carry out the astrometry. The plates were scanned and digitized using the Automatic Plate Measuring System (APM) at the Institute of Astronomy, Cambridge (Kibblewhite *et al*.1984). The pixel size of 30 μm, equivalent to 0".565 at the plate scale of 18.6 arcsec mm^{-1}, resulted in 30 million pixels per plate.

Measurement of the displacement of an individual knot between a reference (R) and a comparison (C) plate from different epochs is accomplished by first selecting a region about 2' x 2' in size surrounding the knot. Centroids are determined for all the features with stellar profiles; the optimum translation and rotation of the C plate to bring its stars into alignment with those on the R plate is calculated; and the C region is then rebinned coincident with the R one.

A knot to be measured is then selected interactively with a cursor; ideally the knot should be "bright," well defined, and have no stars within the cursor-defined area. The displacement of the filament on plate C relative to plate R is determined using a statistic that minimizes the absolute value of the difference between pixel values on the two plates, summed over the knot area. This statistic is more robust than a least-squares one, and leads to uncertainties of 0".1 - 0".25 in each component of displacement from a single plate pair.

We measured displacements for 11 knots on each of the three pairs of plates, one pair separated in epoch by 5.9 yr, the other two pairs by 8.0 yr. The 11 knots have proper motions in the range 0".09 - 0".22 yr^{-1}. All the measured knots are shown with vectors indicating their proper motions in Figure 1.

We have fit the data with a model which assumes the undecelerated expansion of all the filaments from a common origin in space and time. This model has only three parameters: the x and y coordinates of the expansion center and the age of the remnant. The model gives a good fit to the data, with a χ^2_v value of 1.4. Although the ejecta-rich filaments are distributed over a sector of only 70°, the best-fit parameters are nevertheless well determined:

Age = 3700 ± 300 yr;   Expansion Center: R.A. = $8^h20^m44^s.3$ (1950)
Dec. = -42° 47' 48"

The expansion center is located 4' northeast of the center of the radio shell. The error ellipse is shown in Figure 1, and all the proper motion data are plotted in Figure 2, along with the best-fit model.

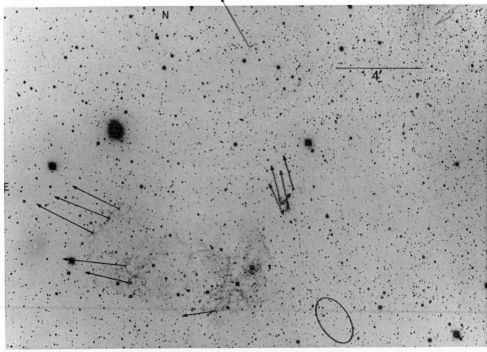

Figure 1. Section of CTIO green plate shows primarily [O III] and Hβ emission. The arrows indicate proper-motion vectors for approximately 1000 yrs. The 90-percent-confidence contour for the expansion center is shown by the ellipse.

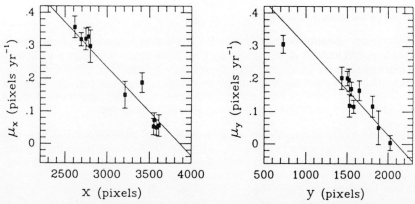

Figure 2. X and Y components of proper motion vs. position; the line is the best-fit free expansion model.

II. Spectra and Radial Velocities

Through spectroscopy with the CTIO 4-m telescope, we have extended the spectra shown in Winkler and Kirshner (1985) to shorter wavelengths and to additional filaments. The spectrum of the omega filament shown in Figure 3 is typical; it may be compared with the longer wavelength spectrum shown in Figure 2 of Winkler and Kirshner. It is particularly interesting to note the presence of neon, indicated by [Ne III] $\lambda\lambda$ 3869, 3967. These lines provide further evidence that the fast-moving filaments are similar in composition to the material which enriched the Puppis A X-ray plasma, where strong lines of Ne IX and Ne X indicate an overabundance of neon as well as oxygen (Canizares and Winkler 1981).

All of the filaments for which oxygen lines dominate the spectra show radial velocities 400 < | vr | < 1600 km s^{-1}, most of them blue-shifted. There is little systematic relationship between radial velocity and proper motion. This is not an inconsistency; it merely means that the knots we are seeing have a range of space velocities, and presumably lie at different distances from the center. While we have not been able to use kinematics to improve the distance measurement to Puppis A, at a distance of 2 kpc the transverse and radial velocities of the fast filaments are comparable. Further details of the spectroscopy and the proper motions will appear in Winkler, Kirshner, and Irwin (1988).

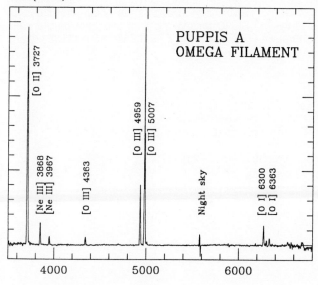

Figure 3. Spectra of the fast-moving filaments indicate a composition rich in oxygen and neon, and almost devoid of hydrogen.

IV. Origin and Survival of Ejecta Knots

In view of their unusual composition, high velocities, and common center of expansion, the case that the fast knots are supernova ejecta seems secure. Their space velocities are 1500-3000 km s^{-1}, lower than those typically associated with Type II supernovae. But since today's fast knots represent material from the inner layers of the progenitor star, they were probably laggards at the time of the explosion. Indeed, they have managed to survive only because they are moving into the low-density shell swept out by the faster-moving supernova blast wave.

We may consider the knots as high-density bullets moving out through a low-density surround. One obvious criterion for knot survival is that the bullet can encounter no more than its own mass in ambient material without being ablated into oblivion, but Nittmann, Falle, and Gaskell (1982) have determined a more stringent criterion by considering dynamic instabilities of the bullet. As shown in Winkler, Kirshner, and Irwin (1988), knots larger than 0.1 pc with density ~200 O-atoms cm^{-3} (typical of the minimum dimension and density of the observed knots) could reach their present distances through a surround of density < 3 cm^{-3}.

V. The Age of Puppis A

Previous estimates of Puppis A's age have used the X-ray spectrum to determine the shock velocity and then used a Sedov model to obtain the age. Values in the range 4000-8000 yr have typically been obtained (Culhane 1977). Uncertainties involving lack of resolution in the X-ray spectra, non-equilibrium ionization effects, the applicability of the Sedov model, and the distance to Puppis A all attend the interpretation of the X-ray data. Thus the direct kinematic measurement obtained here is probably much more reliable.

It is interesting to note that an age of only 3700 years puts Puppis A's supernova within the range of recorded human history: 2000-1400 BC based on constant expansion, and possibly somewhat younger if some deceleration has occurred. The Sumerians and Babylonians made note of astronomical events, so it is conceivable that some mention of a bright southern star may be found on cuneiform tablets. If we assume a typical peak luminosity for Type II supernovae of M_V = -18, a distance of 2 kpc, and A_V = 1.5, we may estimate the peak brightness of the Puppis A event at $V \approx$ -5, surely bright enough to attract notice.

We congratulate the organizers of IAU Colloquium 101 for an exceedingly well-organized meeting in which it was a pleasure to participate. We would like to thank the CTIO staff for their unflagging and good-natured support during the several observing runs necessary to obtain the plates for this project, and the Institute of Astronomy and SERC for making time available on the APM system. Financial support has come through NSF grants AST-8520557, AST-8516537, and NASA grant NAG-8389.

References

Baade, W., and Minkowski, R. 1954, *Ap. J.*, **119**, 206.
Canizares, C.R., and Winkler, P.F. 1981, *Ap. J. (Letters)*, **246**, L33.
Culhane, J.L. 1977, in *Supernovae*, D.N. Schramm, ed. (Dordrecht: D. Reidel), pp. 29-
          51.
Dopita, M.A., Mathewson, D.S., and Ford, V.L. 1977 *Ap. J.*, **214**, 179.
Elliott, K.H. 1979, *Mem. della Soc. Astr. Italiana*, **49**, 477.
Kibblewhite, E., Bridgeland, M., Bunclark, P., Cawson, M., and Irwin, M. 1984, in
          Astronomy with Schmidt-Type Telescopes, M. Capaccioli, ed. (Dordrecht: D. Reidel), 89.
Nittmann, J., Falle, S.A.E.G., and Gaskell, P.H. 1982, *M.N.R.A.S.*, **201**, 833.
Shull, P., Jr. 1983, *Ap. J.*, **269**, 218.
Winkler, P.F., and Kirshner, R.P. 1985, *Ap. J.*, **299**, 981.
Winkler, P.F., Kirshner, R.P., and Irwin, M.J. 1988, in preparation for *Ap. J.*

ADIABATIC SUPERNOVA EXPANSION INTO THE CIRCUMSTELLAR MEDIUM

DAVID L. BAND AND EDISON P. LIANG
Institute of Geophysics and Planetary Physics and Physics Department
Lawrence Livermore National Laboratory, P.O. Box 808, Livermore, CA 94550

ABSTRACT. We perform one dimensional numerical simulations with a Lagrangian hydrodynamics code of the adiabatic expansion of a supernova into the surrounding medium. The early expansion follows Chevalier's analytic self-similar solution until the reverse shock reaches the ejecta core. We follow the expansion as it evolves towards the adiabatic blast wave phase. Some memory of the earlier phases of expansion is retained in the interior even when the outer regions expand as a blast wave. We find the results are sensitive to the initial configuration of the ejecta and to the placement of gridpoints.

The surface of a young supernova (SN) does not expand freely until it sweeps up a sufficient mass of the surrounding material to form a blast wave. Due to the steep density gradient in the SN envelope, and a circumstellar medium (CSM) denser than the average interstellar medium (ISM), the density jump between the expanding ejecta and the surrounding medium may be relatively small, resulting in a decelerating interaction region long before a blast wave forms. The CSM may be the product of a slow, high density wind emitted by the progenitor for a short period before the explosion. An observable interaction region bounded by a forward and reverse shock can then form at the contact surface between the SN envelope and the surrounding medium. Most of the historical supernova remnants (SNRs) are in this pre-blast wave phase, and indeed some extragalactic SNs show radio emission from this early interaction (Weiler *et al.* 1986). Because this phase has observable consequences, and the relevant physical processes are not clear, we are undertaking a series of numerical studies that will trace the early evolution of a SN until a blast wave is formed. In the first study reported here we assume adiabatic hydrodynamic expansion.

The homologously expanding ejecta can be approximated as a constant density core surrounded by an envelope with a steep density gradient $\rho \propto r^{-n}$. The surrounding medium is either a constant density ISM, or a CSM with a $\rho \propto r^{-2}$ profile due to a constant velocity wind. Chevalier (1982a) developed a self-similar solution (SSS) for the expansion of an envelope with $\rho \propto r^{-n}$, $n > 5$, into a stationary region $\rho \propto r^{-s}$, assuming ideal hydrodynamics. This solution should be a valid approximation for the early SNR as long as the reverse shock remains within the envelope, and the forward shock remains within the CSM (if there is a CSM). After the SSS breaks down, the remnant eventually evolves into a Sedov-Taylor blast wave (which is also self-similar, but here will be referred to as the blast wave). The SSS reproduces observed radio light curves reasonably well with simple prescriptions for modeling the radio emission and absorption (Chevalier 1982b; Weiler *et al.* 1986). The transition between the SSS and the blast wave phases is of interest in modeling observed supernovae and understanding the evolution of the interstellar medium (Cioffi 1986, private communication). The transition phase cannot be studied analytically, and numerical calculations are required. In addition, the numerical calculations of Itoh and Fabian (1984) cast doubt on whether the SSS is reached before the conditions for its validity break down. Also, physical effects such as heat conduction, magnetic fields and inefficient coupling between electrons and ions may be relevant within the interaction region (Liang and Chevalier 1985).

The applicable physical processes are not clear. As a result of the low densities and the high temperatures, the Coulomb mean-free-path is comparable to the length scales of the interaction region. Thus, it would appear that hydrodynamics is a questionable assumption within the interaction region; thermal conduction is likely to be flux limited; and the shocks are collision-less. However, the electron gyroradius in the magnetic fields (whose presence is indicated by the radio synchrotron emission) is shorter than the scale of the SN, justifying the use of hydrodynam-ics (Chew, Goldberger, and Low 1956). Similarly, the magnetic fields should suppress thermal conduction across field lines, particularly since shock compression and spherical expansion will in-crease the ratio of the tangential to radial magnetic fields. Finally, observations of interplanetary shocks and numerical plasma simulations indicate that plasma processes nearly equilibrate the electron and ion temperatures in a collisionless shock. Consequently, even though the Coulomb mean-free-path is very long, ideal hydrodynamics is probably a good approximation of the appli-cable physics. In the first stage of our numerical studies we therefore assume ideal hydrodynam-ics; in future studies we will include the effects of flux-limited thermal conduction and different kinds of electron-ion thermal coupling.

The interaction region appears to be Rayleigh-Taylor unstable since the pressure and density gradients are opposite either in the shocked ejecta or the shocked CSM. The instability will be reduced if thermal conduction smooths out the temperature gradient. However, the mag-netic fields will suppress thermal conduction; moreover, the magnetic fields should be amplified in the turbulence produced by the instabilities. This turbulence is beyond the capabilities of our current numerical methods, and consequently will not be included at this time.

The numerical calculations were performed on a Cray computer using an explicit Lagran-gian finite difference scheme in which shocks are smoothed by von Neumann artificial viscosity. Typically energy is conserved to 1%. We found that the spacing of the Lagrangian gridpoints is crucial to the success of the numerical calculations. The density varies by large factors over the various regions and the mass per zone must vary to provide the desired resolution. However, abrupt changes in the mass per zone introduce interfaces from which spurious shocks can be re-flected. For example, the reflected shock found by Itoh and Fabian (1985) when the reverse shock reached the core-envelope interface is probably an artifact of the zoning; we found similar spuri-ous reflections in test runs.

As an initial condition in our calculations, the power law envelope was extended to a ra-dius close to the radius predicted by the SSS, and consequently the runs rapidly reached the SSS. In the calculations of Itoh and Fabian (1984) the SN did not expand with the predicted power law behavior because they truncated the envelopes at a much smaller radius than the predicted postion of the SSS contact surface. The density jump at their initial contact surface was very large, and thus the ejecta expanded at a nearly constant velocity. Indeed, in their runs with a CSM the contact surface expanding at its initial velocity would have intersected the predicted SSS contact surface only near the outer edge of the CSM. Thus the relevance of the SSS to ac-tual SN expansion depends upon the radius to which the envelope extends.

We considered three different models. In all cases the explosion released an energy of 10^{51} ergs and the star was surrounded by an ISM density of 10^{-24} gm-cm^{-3}. When there was a CSM interposed between the ejecta and the ISM, it had a mass of 1 $M_\odot$ and had been emitted for 10^4 years at a wind velocity of 10 km-sec^{-1}. Models of Type I SNs, with an ejecta mass of 1.4 $M_\odot$ and a $\rho \propto r^{-7}$ envelope, were allowed to expand into two types of media: a CSM followed by an ISM (the CSM case); and into an ISM alone (the ISM case). The model of a Type II SN, with a 10 $M_\odot$ ejecta with a $\rho \propto r^{-12}$ envelope, was only expanded into a CSM followed by an ISM. The Type II SN calculation was qualitatively similar to the Type I SN calcu-lation with a CSM, hence we will concentrate here on the two Type I calculations.

The position of the forward shock for the two Type I SN cases is shown on Figure 1. The initial SSS phase and final blast wave phase are both self-similar; for self-similar expansion the position of the forward shock is $r_{f_s} \propto t^\eta$ (note $v_{f_s} = \eta r_{f_s}/t$). The effective power law index

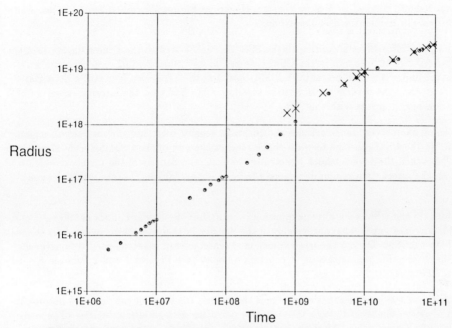

Figure 1—Position of Forward Shock. The expansion of a 1.4 $M_\odot$ star with a $\rho \propto r^{-7}$ envelope and $E_k = 10^{51}$ ergs-sec^{-1} expanding into a $\rho = 10^{-24}$ gm-cm^{-3} ISM (x's) is compared with the same star expanding first into a CSM that ends at 3.2×10^{17} cm (dots).

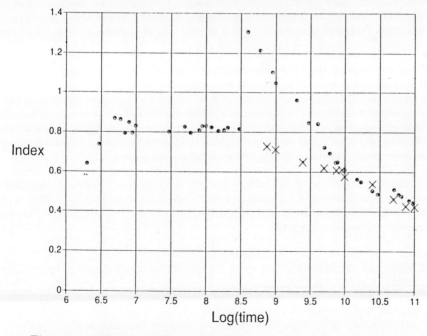

Figure 2—Effective Forward Shock Power Law Indices. The index $\eta = v_{fs}t/r_{fs}$ for the two cases in Figure 1. For a blast wave $\eta = .4$.

of the forward shock, defined as $\eta = v_{fs}t/r_{fs}$, is shown on Figure 2 for both Type I cases. Four different phases can be identified in the expansion:

- First, the SN evolves according to the SSS. As can be seen in the figures, in the ISM case $r_{fs} \propto t^{4/7}$ until 8×10^9 sec and in the CSM case $r_{fs} \propto t^{4/5}$ until 2×10^8 sec. The run was begun near the end of the SSS phase in the ISM case, and the expansion was still a little faster than predicted when the reverse shock left the envelope. We find that the internal structure of the interaction region agrees with the SSS.

- Second, the reverse shock reaches the constant density core, and the interaction region slows down. In the CSM case the forward shock reaches the end of the CSM before it slows down substantially. When the reverse shock penetrates the core, the density of the unshocked ejecta impinging on the interaction region decreases, the force of the "piston" lessens, and the expansion slows.

- Third, in the CSM case a free expansion occurs after the forward shock reaches the end of the CSM. The free expansion continues until the density in the interaction region falls to levels comparable to the ISM. During the free expansion thermal energy is reconverted to kinetic energy, and the interaction region travels at constant velocity (see Figures 1 and 2 between 8×10^8 sec and 2×10^{10} sec).

- Fourth, as the reverse shock approaches the center, the drive of the "piston" become inconsequential as the unshocked ejecta falls to lower densities with smaller velocities. The remnant evolves into a blast wave. After the reverse shock hits the core, the pressure gradient causes material to rush back to the core, increasing the pressure there, and eventually reversing the inward flow. Bertschinger (1986, private communication) found in a related calculation that weak shocks bounce back and forth between the core and the contact surface, reheating the ejecta; in our calculation the ejecta are numerically noisy at this time, perhaps due to waves bouncing back and forth. While the inner region of the ejecta is complicated as the reverse shock approaches the center, and some of the resulting structure may remain long afterwards, the amount of mass involved is an insignificant fraction of the total SNR mass.

The free expansion in the CSM ends with the forward shock at approximately the same radius as the forward shock in the ISM case. At that point the ejecta have swept up approximately their own mass. Note, however, that the SNR is still expanding more rapidly than predicted by the blast wave solution ($\eta = .4$), and even when the SN has swept up several times its own mass there remains structure inside the SNR not found in the Sedov-Taylor blast wave solution. Our results suggest that the accuracy with which the blast wave solution approximates the SNR is of the order of the ratio of swept up ISM mass to ejecta mass.

We thank E. Bertschinger, R. Chevalier, D. Cioffi, C. Max and C. McKee for insightful discussions, and George Zimmerman for assistance in the use of the computer code. This work was supported by the U.S. Department of Energy, under contract No. W-7405-ENG-48 to the Lawrence Livermore National Laboratory.

References

Chevalier, R. 1982a, *Ap. J.*, **258**, 790.
Chevalier, R. 1982b, *Ap. J.*, **259**, 302.
Chew, G. F., Goldberger, M. L., and Low, F. E. 1956, *Proc. Roy. Soc. (London)*, **A236**, 112.
Itoh, H. and Fabian, A. C. 1984, *M.N.R.A.S.*, **208**, 645.
Liang, E. P. and Chevalier, R. A. 1984, *Ann. New York Acad. of Sci.*, **422**, 233.
Weiler, K. W., Sramek, R. A., Panagia, N., van der Hulst, J. M., and Salvati, M. 1986, *Ap. J.*, **301**, 790.

OVERVIEW OF THE INTERSTELLAR MEDIUM: SUPERNOVA RELATED ISSUES

Donald P. Cox
Department of Physics
University of Wisconsin-Madison

Abstract: Establishing that the interstellar pressure is higher than usually realized and that the magnetic component is probably dominant, I propose a drastic revision for our understanding of the interstellar landscape.

I have spent much of the past year thinking about and collaborating on a review paper (with Ron Reynolds) on the Local Interstellar Medium (LISM) (1). Since I have come to regard the LISM as an anomalous region, atypical of the interstellar medium at large (ISM), this activity has done little to prepare me for giving this talk. Fortunately, other recent reviews (2,3,4,5) have satisfied me that a general presentation of the diverse properties of the ISM would be superfluous. At this time an honorable man would probably return to his seat. What I plan instead is to discuss a wide range of ISM issues relevant to both the nature of the medium and the interaction of supernovae with it.

In order to make a detailed predictive model of the evolution and appearance of a remnant, one needs to know a great deal about its interstellar environment. Roughly speaking, the requirements are the spatial distributions of material (including density, temperature, elemental abundances, dust grains, and ionization stages) as well as those of the magnetic field, nonthermal particles, background sources of ionization, and mass motions.

Super Bubbles, Etc.: As you know, many of the relevant environmental parameters can be seriously perturbed by the presence of a massive presupernova star, or worse, by an association of such stars, some of which exploded in the medium prior to t=0 for the supernova remnant about to be modeled.

I have no intention of amusing you with my ignorance of the physics of sequential supernovae and the possibilities of generating superbubbles and worms thereby. Superbubbles and worms, like the LISM, fall in the category of anomalous regions. A paper (6) by Tenorio-Tagle, Bodenheimer, and Rozyczka that I recently refereed, however, added an interesting twist to this subject, showing that shell formation could be Rayleigh–Taylor unstable when driven by sequential explosions.

The Cygnus Loop Environment: Not surprisingly, the way one actually learns the sort of detailed ISM information needed for remnant modeling is by observing remnants. For example, studies (7) of the Cygnus Loop have indicated that there are at least four density regimes with which it is currently interacting. Away from regions of bright optical emission, the shock speed is thought to be about 400 km s^{-1}, into an effective density $n_0 \sim 0.16$ cm^{-3} (8,9), generating diffuse x-ray emitting gas with $T \simeq 2.4 \times 10^6$ K. In many places around the circumference, however, the Hα signature of "non-radiative" shocks is evident in deep exposures (10,11,12,13). In some of these areas, at least, the shock velocity is more like 150 to 200 km s^{-1}, into $n_0 \sim 1$ cm^{-3}. Interestingly enough, this preshock gas must be neutral in order to generate the Hα signature. With a recombination timescale of about 10^5 years, this places an interesting lower limit on the time since the explosion, or an upper limit on the combined ionizing UV of the preexplosion and exploding star.

In regions of the bright optical emission of radiative shocks, the UV and optical spectra suggest shock velocities in the neighborhood of 100 km s^{-1} and preshock densities $n_0 \sim 8$ cm^{-3} (7). Abundances seem normal, but with evidence of depletion of Si and Fe relative to other elements. The weak 2 photon continuum shows that this higher density medium has been preionized by the recent UV emission of the shock itself.

Finally, there are a few small dense knots of material (e.g. Miller position 2, XA of Hester and Cox (14,7)).

Some years ago when it was first realized that the Cygnus Loop had the gross properties of shocks with speeds of both 100 and 400 km s^{-1}, it was suggested (15) that one should envision a "blast wave" traveling in a low density (0.16 cm^{-3} for example) within which there are clouds (8 cm^{-3} for example). The high pressure behind the blast wave was imagined to drive the slower radiative shock waves into the clouds. In a sense, this is the picture I just described. But I want to encourage a bit of caution. The cloudlet/intercloud blast wave picture was originally directed toward understanding how the x-rays can be brightest in precisely those areas where the radiative shock waves are found. The cloudlet scale was presumed infinitesimal ($\lesssim 3 \times 10^{-3}$ pc) and their numbers huge so that seemingly smooth continuous filaments could be regarded as loci of the radiative shocks of a population of recently encountered cloudlets.

I wish to state categorically that this microcloudlet picture has absolutely nothing to do with the reality of the Cygnus Loop (14). The X-rays are bright in the regions of radiative shock waves because these regions are large and recently encountered. The reflected shock front in the lower density material just interior is sufficient for the observed x-ray production. A cloud/intercloud picture is appropriate only on much larger scales.

My overall impression is that the density varies between the "0.16 cm^{-3}" and "1 cm^{-3}" values on fairly large scales, since it is possible to follow the "non-radiative" structures for very large distances along the Loop perimeter (16). Perhaps the preshock density of these is frequently close to the lower value above, with the higher value more characteristic of the brighter areas chosen for detailed study. The "8 cm^{-3}" density regions, however, are found in several discrete patches over the surface of the Loop (in my thesis I estimated they covered 12% of the surface) with scales of 10 to 20 pc. Within these large structures, the density is > 10 cm^{-3} in a few dense knots but is more typically in the range $\lesssim 5$ to 10 cm^{-3}. Gradual density variations of factor of 2 are found along the 0.3 pc length of one filament near the smaller Miller 2 knot (7). This gradient may have caused much of the apparent rotation of the filament from tangential alignment. In most regions away from knots, the density gradients are probably smaller. Much of the caustic surface structure of the filament pattern would follow from much smaller density variations (17).

We shall shortly find that the ISM should have a gap in its density distribution, between roughly 0.5 cm^{-3} and 15 cm^{-3}, hence disallowing the densities which are so common in the large clouds around the Cygnus Loop. This density gap, however, is appropriate for optically thin material bathed in a full complement of starlight. Since the regions around the Loop show noticeable obscuration of background stars, perhaps the heating rate and equilibrium densities are lower. In any case, either by the action of the precursor star or by chance, the Cygnus Loop environment too appears to be in the noticeably atypical category.

As a small aside, I would like to point out that all of the above analyses have ignored the possible role of strong cosmic ray acceleration by the shock fronts. One of my students, Ahmed Boulares, is currently of the opinion that starting from the observed post shock temperatures and densities and using reasonable values for the preshock B and p_{CR}, that much the same observational picture could derive from shocks of 3 times higher velocity, putting 90% of their energy into cosmic rays. The Loop age is thereby reduced by a factor of 3 and its energy increased by a factor of 10. An indication of the possible dominance of non-thermal pressure in the Spur filament was recently found observationally. Although these inferences may sound outlandish, we will have to continue to be careful with the foundations of our house of cards.

The Interstellar Pressure and Its Scale Height: The interstellar pressure can be estimated from the vertical distributions of density and gravity (4,18,19). The resulting weight of the interstellar material significantly exceeds estimates made of the midplane pressure by other means (19,1). The size of this discrepancy has increased as we have gained appreciation of the magnitude of the densities of HI and H$^+$ at high z.

As a rough approximation, material distributed exponentially with scale height z; and midplane density ρ_{0i} in the galactic gravity of the solar neighborhood contributes a midplane pressure

$$\Delta p_i \simeq \rho_{0i} z_i (10^{-8} \text{ cm/s}^2)/(1+500\text{pc}/z_i)$$

or

$$\Delta p_i/k \simeq 8500 \text{ cm}^{-3} \text{ K}\{\sigma_i (M_\odot/\text{pc}^2)\}/(1+500\text{pc}/z_i)$$

where $\sigma_i = 2\rho_{0i}z_i$ is the full disk surface density of the component. With $\sigma_i \sim 2M_\odot \text{pc}^{-2}$ each for HI with $z_i \sim 400$ pc and H$^+$ with $z_i \gtrsim 1000$ pc (2,3), the combined p/k contribution is at least 1.9×10^4 cm^{-3} K. The cold HI (and associated warm HI) closer to the plane contribute much less pressure per gram because of their reduced weight at lower z. In addition, that contribution is to first order balanced by the velocity dispersion and can be disregarded in our quest for the magnitude of the general diffuse interstellar pressure (19). Subtracting 3000 cm^{-3} K cosmic ray pressure from the above estimate we have the contemporary estimate of the combined thermal, magnetic and wave (or turbulence) pressures in the diffuse interstellar material at z=0:

$$p/k|_{1987} \simeq 16,000 \text{ cm}^{-3} \text{ K}.$$

This neglects a potentially significant contribution from halo material (4,18). Even so, it is a factor of four higher than many typical estimates of the thermal component. In those regions for which the thermal and wave component sums to 6000 cm^{-3}K, the residual magnetic field must have a value of 5 µG. This is probably more typical of the interstellar field than the roughly factor of 2 lower value commonly estimated from various measurements (20,4).

As an aside, the thermal pressure within the hot gas of the Local Bubble is probably more nearly 10^4 cm^{-3} K, consistent with a reduced B field in the very low density cavity (1,19). The fact that at higher interstellar densities there is virtually no dependence of B on density (20) is a direct consequence of the dominant contribution of the magnetic field to the overall pressure.

Pressure scale height information has been somewhat confusing. Certainly the observed mass scale heights indicate the scale of significant pressure gradients, namely 400 pc to 1 kpc. The latter is said to be a lower limit to the electron scale height and since the corresponding ions seem now to provide the largest weight contribution, the pressure gradient may extend significantly beyond 1 kpc. Information from γ-rays indicates that cosmic rays have a thicker distribution than the interstellar material of the lower disk (4). The ^{10}Be studies have suggested that cosmic rays sample a mean density of only 0.1 cm^{-3} (21), implying a probable scale height somewhat in excess of 1 kpc. The ratio of pole brightness to plane emissivity in nonthermal radio at 10 MHz corresponds roughly to an emission path of 1.4 kpc (4,22). Since that emissivity derives from a product of magnetic field and energetic electron density, its distribution may drop off slightly faster than B^2 or the cosmic ray pressure. Collectively these evidences seem to push for effective scale heights for cosmic ray and magnetic pressures of about 2 kpc. The H$^+$ scale height may be similar.

The confusing part is that we have effectively assumed that the nonthermal pressures are the major form of support for the material, yet unless the distributions are pushed somewhat beyond reasonability, it appears that a significant portion of the weight to be supported is lower (e.g., around z ~ 400 pc) than much of the gradient in the nonthermal pressure (e.g., z ~ 1 to 2 kpc).

One way around this difficulty has been to introduce a halo component to the density distribution (4,18). At low z the thermal pressure gradient in the halo component helps to support the H I distribution while at high z the weight of the halo material helps hold down the magnetic field and cosmic rays. For best results, halos with pressure minima around 1 to 3 kpc have been invoked.

It seems to me that the high z distribution of electrons and ions probably extends to the low density halo of the coronal ion (23) population. I have no problem with the idea that material may be present at high z. This does, however, increase the midplane pressure. In theory this can be compensated by the separate inclusion of the pressure of a hot interstellar component of significant filling factor in the plane. This is just a bit tricky however. The rms value of B must be kept high (~ 5 µG) while the thermal pressure in low density regions is very significantly enhanced. This can be done in a steady state fashion only by having B within the H I even higher (say 7 µG) with a lower value in the very hot low density material. This pushes a bit beyond credence for my taste.

A simple alternative is that we have been fooled into the use of a laminar B field. By consideration of magnetic tension it is straightforward to tie high z fields and cosmic rays to the weight of lower lying material. A direct consequence is that field lines will have upward curvature below 400 pc and downward curvature above 1 kpc. This picture has been present for some time, but I and others have occasionally forgotten its significance. High z B field is "anchored" by lower z weight. It is the natural picture espoused by students of the instability of the laminar configuration (See ref 4 for a survey of some of the literature on the Parker instability).

We can illustrate the mechanism of tension while at the same time deriving an interesting limit on its effectiveness. Consider the configuration in figure 1 with a flux tube of cross sectional area A, length L, interior mass density ρ, in gravity g.

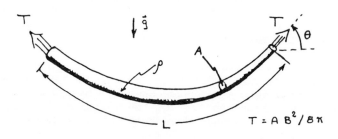

Figure 1.   Geometry of a tension supported cloud.

The force balance requires $\rho ALg = 2AB^2 \sin\theta/8\pi$. Since the dominant pressure form is magnetic, the tube cross section is nearly constant and B is essentially the ambient field strength. As a consequence, there is an upper limit to the magnitude of g (and hence the height z) to which a tube of given ρL can be supported:

$$g_{max} = \frac{2}{\rho L} \cdot \frac{B^2}{8\pi} \simeq \frac{p}{(\rho L)} .$$

For $z \lesssim 200$ pc, we have $p \sim 2 \times 10^{-12}$ dyn cm^{-2} and $g \simeq 2 \times 10^{-9}$ cm s^{-2} (z/100 pc). In this range, the absolute maximum (sin θ = 1) height of suspension is

$$\frac{z}{100 \text{ pc}}\Big|_{\text{max}} = \frac{160 \text{ pc cm}^{-3}}{L \text{ n}} \ .$$

Stable configurations will likely avoid heights greater than about half the maximum ($\sin \theta \sim 1/2$). As a consequence, the denser diffuse clouds (say $n \sim 40 \text{ cm}^{-3}$ and $L \sim 2$ pc) should never be supported above 200 pc and rarely above 100 pc. Flux tubes with this mass density on them but greater linear extent (e.g., 20 pc) should in equilibrium be found only very near the plane. Conversely, low density (e.g., 0.16 cm^{-3}) flux tubes shorter than 100 pc can be tension supported at any height.

There are two further interesting relationships following directly from this result. One is that for a spherical cloud, the maximum downward force on the field is essentially A $B^2/8\pi$ and hence the maximum downward force of a cloud population is (Σ A) $B^2/8\pi$.* As a result, the maximum downward force per unit area is roughly the "sky coverage factor" of the clouds times the pressure.

For stringy clouds, the maximum downward force is further reduced by the aspect ratio. On the whole, the observed cloud population is not likely to anchor more than roughly one fourth of the total pressure. The remainder must be provided by the diffuse intercloud material.

The second aspect involves consideration of the net effective vertical magnetic pressure as a function of z, including both the non verticality of the "pressure" perpendicular to $\vec{B}$ and the tension along B. The result is that

$$P_B^{(eff)} = \langle \frac{B^2}{8\pi} (\cos^2\theta - \sin^2\theta) \rangle = \frac{1}{8\pi} \langle B^2(z) - 2 B_z^2(z) \rangle$$

Here $B_z = B \sin\theta$ is the vertical component of B, asymptotically zero at $z = 0$ and z large. The resulting possible effective magnetic pressure distributions are shown schematically in Figure 2, for three levels of tension induced distortion of the field. Clearly the introduction of magnetic tension through a vertical component of B provides the same sort of effect achieved in the halo models. There is possibility of a strong gradient in the effective pressure close to

---

* I thank Charlie Goebel for a useful discussion leading to this point.

the plane, while having a much weaker gradient in B^2 and the cosmic ray density, allowing their large scale heights.

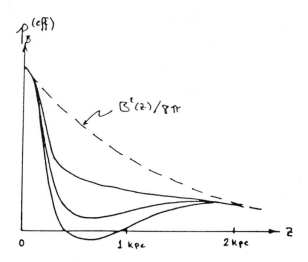

Figure 2.   Effective magnetic pressure distributions.

We will return to further discussion of the interstellar pressure, but for the moment we have the result that the nominal interstellar pressure exclusive of cosmic rays is 16,000 cm^{-3} K, much of which is magnetic, and that a comparable pressure extends to high z (scale height 1 to 2 kpc). This then is the typical pressure with which isolated supernova remnants must contend. (In the absence of appreciable magnetic tension, the midplane pressure is probably even higher, but I then have trouble understanding the low measured values of B.)

Further Aspects of Cold HI Regions: Statistics of H I column density measurement indicate that the number of clouds with column density > 10^{20} N cm^{-2} is 5.7 $N^{-0.8}$ per kiloparsec, for 0.32 < N < 2.2 (2,3). A remarkably similar result in both slope and normalization was found by Hobbs using K I absorption lines (24). His quoted range of validity is 1.7 < N < 17, consistent with the bias of the K I line toward high H I column densities. The T-τ relation, as discussed by Kulkarni and Heiles, augmented with the assumption that thermal pressures have nT ~ 3000 cm^{-3} K nominally implies cloud temperature, line of sight depth, and local density given by

$$T \sim 96 \text{ K } N^{-2/3}$$

$$L \sim 1 \text{ pc } N^{1/3}$$

$$n_{H\ I} \sim 31 \text{ cm}^{-3} N^{2/3}$$

although there is great scatter in the relation and much of it may be due to complications of radiative transfer in inhomogeneous clouds rather than systematic parameter variation between clouds (Kulkarni, private communication). With these results, however, one concludes that the space filling factor of all clouds in this range is

$$f = 0.02 \ (nT/3000 \text{ cm}^{-3} \text{ K}).$$

If these clouds were spheres, then L = 4 r/3, the number of clouds of radius exceeding r is

$$\# \ (> r) = 5.5 \times 10^{-4} \text{ pc}^{-3} \ (1 \text{ pc}/r)^{4.4}$$

for r >0.53 pc, at which point $\# = 9 \times 10^{-3} \text{ pc}^{-3}$. Hence a remnant the size of the Cygnus Loop (R~ 20 pc) could be expected to have about 300 such clouds of radius 0.5 to 1 pc and typical separation of 4.8 pc – or one fourth the Loop radius. Quite clearly a filling factor of only 0.02 is not equivalent to implying that large remnants will not have a significant cloud presence.

On the other hand, I can think of no reason whatever to suppose that clouds should be spheres, while many processes lead reasonably to linear or planar structures. We could, for example consider 10 pc square cloud sheets of effective depth 1 pc, equivalent to roughly 100 of the spherical clouds. We would then find the Loop interacting with only a few such cloud sheets, roughly as observed. Furthermore, as Kulkarni has shown (private communication), the cloud number per column density interval is consistent with all clouds being sheet like, with the same normal column density, viewed at random angles.

Pressure measurements in the cold H I clouds, as summarized by Kulkarni and Heiles (2,3) seem to reinforce the notion that the usual thermal pressure is of order 3000 cm^{-3}K, but with factor of 3 variations possible in either direction. Even with a fairly nonviolent ISM, such thermal pressure variations are possible because of the dominance of the magnetic contribution. In a truly static case, however, such variation would not be present at a given z because the absence of magnetic force along B leads to the usual

hydrostatic conditions involving only the thermal pressure. My sense
is that the interstellar dynamical timescale is too short for us to
have to worry about pressure equilibration along flux tubes, except
over short distances.

An interesting aspect of the diffuse clouds is that cold H I
($40\ K \lesssim T \lesssim 100\ K$) seems to be closely associated with warm H I
($100\ K \lesssim T \lesssim 400\ K$). It has been suggested that the latter envelops
the former, as though there were a core mantle relationship, perhaps
due to the attrition of some heating mechanism important on the
outside. Let us, however, consider the simple heating - cooling
balance for optically thin clouds heated by "starlight" at a rate C_*
per atom. A region with density n is heated per unit volume at a rate
$C_*\ n$ and cooled at a rate $L(C\ II,T)n^2$. Hence the density temperature
relationship in equilibrium is

$$n\ (T)\ =\ C_*\ /\ L\ (T)$$

$$nT\ =\ C_*\ T\ /\ L\ (T)$$

from which we can construct the nT versus n relationship shown in
Figure 3. (The values of C_* and L(T) were inferred from reference 2
and 3 but the magnitude of C_* was slightly reduced to make the
pressure minimum consistent with other measures.) A secondary segment
of the graph indicating the warm neutral medium with T ~ 8000 K is
sketched in as well.

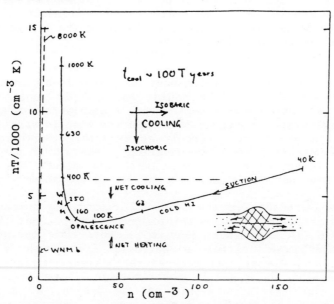

Figure 3.   Phase diagram for equilibrium between
             starlight and C^+ cooling.

So far this figure seems to provide us with a mild variant of the phase diagram upon which the Field, Goldsmith, and Habing ISM model (25) was based. In their view, at high z the total pressure was lower than the lowest cloud pressure, allowing only the intercloud component. But at lower z the overlying material weight was great enough to allow both clouds and intercloud gas as stable phases. If the scale heights were fixed in such a way that the pressure was provided mainly by the weight of the intercloud component, then the total pressure at midplane (or actually at cloud top) would hover closely around the minimum pressure allowing clouds, independent of the total H I surface density. The intercloud surface density would (for the same g, etc) always would be just the constant amount needed to stabilize most of the mass in clouds.

The above description makes a lovely picture but doesn't apply directly to our system where the pressure is largely magnetic and heating and cooling occur mainly at constant density rather than constant pressure. For isochoric cooling, material at too high (or low) nT for given n cools (heats) and moves downward (upward) to the equilibrium line. This differs drastically from isobaric cooling where the motion is left-right, and the warm leg of equilibria (100 K $\lesssim$ T < 1000 K shown) unstable. In our case such equilibria are in fact stable, although they occur over a very narrow density range (and fairly rapidly varying thermal pressure above ~ 400 K). The total temperature range with nT $\lesssim 10^4$ cm^{-3}K is ~ 34 to 700 K, although stability is greater lower in the pressure well where 45 K $\lesssim$ T $\lesssim$ 400 K. Within this range there is still a slow instability due to the thermal pressure gradients along field lines. At the two extremes the thermal pressure is high and the field is bloated. In each case the system wants to move toward relaxation by adjusting the density to settle into the thermal pressure minimum. Since this minimum is broad and shallow, ranging say from (n ~ 25 cm^{-3}, T ~ 160 K) to (n ~ 50 cm^{-3}, T ~ 80 K), the fluid has zero bulk modulus (the pressure is independent of density over times larger than the thermal time ~ 100 T years). As a result it is similar to a fluid such as CO_2 at its critical point. Such fluids exhibit critical point opalescence, a notion which may have some interstellar relevance.

In any case, I have hoped to show in this section that cold H I is perfectly capable of being in close association with warm H I ($\lesssim$ 400 K say) with exactly the same heating rate per atom; that in a magnetically dominated system, clouds seek (via suction) the local pressure minimum; and that as a consequence, thermal pressures significantly lower than the total pressure are the expected norm. Notice, however, that the "warm neutral medium" component associated with cold material in this way (WNMa) should not have a density lower than about 15 cm^{-3} (in an optically thin environment, subject to details of the heating rate). As a consequence it cannot contribute appreciably to filling the intercloud spaces. In addition, there is no advantage to any sort of core halo or onion-like structure for clouds.

Volume Filling Versus the Porosity Imperative: A recurrent topic in ISM discussions involves the fraction of the interstellar volume occupied by warm HI (or H^+) versus a very hot (or coronal) phase. In 1977 we were impressively convinced by McKee and Ostriker (26) that Coxswain's Myth* (27) had underestimated the interstellar porosity. The then popular supernova rate acting on the then popular intercloud medium (0.1 cm^{-3}) would cause total disruption into a mix of hot gas and cold dense sheets within one generation (about 2×10^6 years).

Yet it is ever more likely that an 8000 K diffuse HI component of mean density 0.1 cm^{-3} is an observational reality, along with an ionized one of average density about 0.03 cm^{-3} (2,3). By comparing the dispersion and emission measures of the latter, the volume filling factor of warm H^+ is estimated to be about 0.11, making the local density about 0.25 cm^{-3} and $p/k \sim 4000$ cm^{-3} K (28). If the neutral HI has similar thermal pressure and temperature (as various indicators suggest) its local density should be 0.5 cm^{-3} and its filling factor about 0.2. Including the small contributions of the denser phases, we are lead to believe that the total noncoronal filling factor is about 0.35. Some pundits may be uncomfortable with a value this high, others quite satisfied with the 0.65 left for the hot component.

Certainly however, the arguments above are not airtight. In some scenarios the HI is more likely to have the same density as the H^+ rather than the same pressure (3). That could raise the noncoronal occupation to 0.55. If a larger fraction of the warm HI is in the diffuse form (WNMb, 8000 K) versus cloud form (WNMa, 100–400 K), or if observational uncertainties are included, I would guess that nearly 100% occupation is possible.

One stumbling block to accepting a high spatial filling factor is that it requires a rather low intercloud thermal pressure. Full occupation would probably need $p/k \sim 1500$ to 2000 cm^{-3} K. Actually, the diffuse HI could typically have a thermal pressure this low. The one place we have a point measurement of such a diffuse phase, the VLISM (local fluff flowing through the solar system), we find that the neutral   particles   actually   measured   provide   only $nT \sim 300–800$ cm^{-3} (1). The observed HI density is less than ten times the helium density. If this were due to ionization then we infer additional electron and proton densities and $p/k \sim 1300–3000$ cm^{-3} K. Hence normal interstellar thermal pressures are found only at the upper end of the range of possibility, and only then after a factor of at least 4 extrapolation from the measurements. We do not at first mind such an extrapolation because the ionization seems needed to make sense of the He/H ratio. But recently there have been problems with the ionization explanation (MgI studies, discussed in ref. 1; Voyager data on anomalous CR component – E. C. Stone private communication). All prejudices aside the measurements are more comfortable with

---

*Barry Smith's 1973 perfect pun, pronounced Cox'nsmyth

$p/k \sim 10^3$ cm^{-3} K, a value consistent with full occupation of space by the warm HI intercloud component. Of course we cannot presently deny the existense of large discrete regions of hot material – the Local Bubble, Loop I, etc., but their general volume fraction may be small.

The theoretical reason that $p/k \leq 3000$ cm^{-3} K is perfectly acceptable for the intercloud component is that the ISM does not in fact have thermal pressure balance between phases. The cold and lukewarm HI is caught in the local pressure minimum of the phase diagram of figure 3, but the intercloud phase is not. Its stable equilibria probably include much lower pressures. Given the suction mechanism described earlier, we even expect the phase to tend toward low p, except that the dynamical and thermal timescales are both of order 10^6 years and the above tendency is frequently disrupted.

We are still faced with the porosity imperative of McKee and Ostriker (26). Supernovae would seem to disrupt the system totally on a short timescale. A careful review of the argument leading to this conclusion, however, shows that it is incorrect (29). Let me summarize. The early evolution of a remnant of energy 10^{51} E_{51} ergs in a medium of density n_0 is evaluated for the radius, mass M_c, and shell velocity v_c at the cooling epoch. Assuming negligible ambient pressure, the expanding shell is followed thereafter until the remnants overlap with their neighbors. From that, the characteristic pressure at the time of overlap is evaluated and found to be

$$P_{ov} \sim \frac{M_c v_c}{8 \pi R_g^2 \tau_{sn}} \sim 10^{-12} \text{ dyn cm}^{-2} E_{51} (E_{51} n_0^2)^{-1/14}$$

where the numerical value assumes $\tau_{sn} = 30$ years in a galactic disk of radius, $R_g = 15$ kpc. Characteristic values of the radius, velocity, and time at overlap are 104 pc, 13 km s^{-1}, and 2 x 10^6 years in the simplest model with $E_{51} = 0.5$ and $n_0 = 0.1$ cm^{-3}. We thus refind the McKee and Ostriker result that the volume occupation is of order 1 when the shell dynamical pressure is comparable to the general pressure of the ISM.

Have we thus found that the SN would immediately generate a foam of holes and shells, meaning that we calculated the SN evolution in a medium unlike the one left by SNe? Well, no we don't. We must reevaluate the mental image. These very large remnants are not sweeping the ICM into thin dense shells. Owing to magnetic field and cosmic ray pressure, the compression possible even in a radiative shock of this strength is at most a factor of 2. The kinetic energy of the shell and the energy radiated are both very small compared to the energy in the compressed field and compressed or accelerated cosmic

rays: The late expansion is very nearly adiabatic, and a comparatively
weak disturbance.

Rather than a foam of thin dense shells surrounding large volumes
of hot gas, we find large radii volumes of slightly compressed and
slightly accelerated material with bubbles of hot material buried deep
inside. The medium is certainly no longer homogeneous, but a dominant
fraction of the volume contains material with a density not very
different from that assumed initially. Rather than saying that the
remnants have destroyed the initial condition, it is more accurate to
say that they have rearranged it slightly. High volume occupation by
weak remnants does not imply high porosity.

The medium in which supernovae evolve would seem to be one in
which there are density variations of roughly a factor of 2 about
0.15 cm^{-3} over scales of order 40 pc or so, with those components in
motion at characteristic speeds ≤ 15 km s^{-1}. In addition there are
included bubbles of gas at coronal temperatures created by the intense
heating of material close to the explosion sites.

Although I am convinced from the above that supernovae would not
immediately sweep the interstellar medium into a cloud/coronal
configuration, I can't prove that some sort of cloud/coronal state
would not persist if formed, or even that a diffuse warm intercloud
state is stable against gradual accumulation of long lived hot bubbles
as originally advocated by Barry and me (27). One's understanding of
stability depends very much on assumptions about how supernova
remnants "acquire" interstellar matter. Do remnants propagate
primarily in a coronal component but augment its density by thermal
evaporation or hydrodynamical stripping? Or does the pressure wave
simply interact with all material it engulfs, everywhere heating but
at any moment to very different temperatures? My prejudice has leaned
to the latter picture for many years, and at times I have been able to
describe a feedback mechanism that guaranteed a noncoronal filling
factor of about 50%. In its absence the remnant would be unable to
acquire sufficient mass for radiation of its energy. The medium goes
thermally unstable, starting a strong wind. The pressure drops,
destabilizing the clouds and voila, the warm intercloud component
seizes the space. Something like that.

Hot bubbles, generated by supernovae in a warm intercloud ISM
dominated by magnetic pressure, seem to provide a very good model for
the origin of the OVI measurements (29). The bubbles are essentially
in pressure equilibrium with the surroundings, cooling mainly by
thermal conduction to their surfaces where the energy is radiated.
The OVI is found mainly in these boundary layers, condensation rather
than evaporative boundaries. The mean interstellar OVI density is,
like the supernova contribution to the pressure, directly proportional
to the supernova power. The observed mean density is consistent with
the observed pressure, both consistent with the estimated supernova
power. The typical OVI feature column density, mean free path, low
speed, and narrow width are all compatible with observations.

We have seen that the largest reasonable scale for an individual remnant is about 100 pc. (For various reasons, McKee and Ostriker found a somewhat larger value, about 180 pc.) For years this lead me to view the system as two dimensional, remnants occurring at some rate per unit area, essentially in the plane, and then expanding until they broke out above the HI layer, dumping their hot interiors at the base of the corona, thereby causing a galactic fountain. Since it has become clear that the dominant volume filling HI phase extends 400 pc or more above the plane, and possibly that the relevant scale height for Type I SNe does also (5), the above picture cannot be correct. In my view now, the "base of the corona" is at $z \sim 2$ kpc. Remnants occur in an essentially three dimensional space through which there may be an appreciable transport of energy to the corona. Individual remnants do not expand to the point that they deposit their hot interior material directly into a fountain or other coronal form.

System Configuration: I would like one day to be able to say that I understand how a certain total gas surface density σ_g when placed in a stellar disk σ_*, scale height h_*, will distribute itself among the various phases, and in space. The fact that I cannot now do that makes its discussion all the more important. As a consequence, let us consider some plausible attributes of the interstellar system.

As we have seen, supernovae cause the interstellar system to have at least a certain total pressure, which in some simple models is proportional to the supernova power per unit area of the disk. Let us assume that this is the dominant contribution to the total pressure requirement, that the supernova rate is proportional to the stellar density, and the scale height of this total pressure is thus approximately that of the stars, h_*.

The thermal pressure of clouds is the minimum pressure of the nT versus n diagram (always near $T \sim 100$ K) and is directly proportional to the heating rate per atom. Given that the latter is proportional to the density of starlight, the ratio of thermal pressure in clouds to total pressure provided by remnants is independent of the stellar density. It is a mass function parameter. We have also seen that a cloud population with small sky coverage factor is incapable of confining the pressure provided by the remnants. The job thus falls to an intercloud component.

Presumably there is some organization to the mass motions in the disk, leading to a dynamo generating the magnetic field. The action saturates when the field strength is sufficiently large that it interferes with the mass motions (the crank on the dynamo becomes stiff). Cosmic ray acceleration also consumes SN power at a relatively constant rate, the density building up until it can distort the magnetic field, allowing escape at a rate equal to production (30). (These are the equipartition excuses that I've found most reassuring over the years e.g. ref. 31).

The intercloud component is not constrained to have the same thermal pressure as the clouds. It is furthermore not easily disrupted into a coronal/cloud configuration because the supernovae are perturbations on an environment they have themselves created. As individuals of large scale, they necessarily constitute rather weak disturbances in a highly elastic medium. The intercloud component can apparently be supported by the B field to high z, and as a consequence we expect its scale height to be a significant fraction of that of the pressure (h_* in the current discussion.) As a result, the intercloud weight which balances the supernova induced pressure requires an intercloud surface density σ_{IC} which is independent of the density and scale height of the stars. (Both the pressure and the effective gravity have been made proportional to σ_* by our prior assumptions, leaving σ_{IC} independent of it). In addition, the signal velocity in the intercloud medium $v_{IC} \propto \sqrt{p/\rho}$ is some constant fraction of the stellar dispersion velocity.

In this simplistic picture, we have in essence supposed that clouds, intercloud gas, interstellar pressure, the supernova rate density, and stars have the same relative z distributions, so long as stars dominate the mass density. The total and thermal pressures depend on supernovae and starlight, respectively, such that their ratio is mass function rather than density dependent. Finally, the surface density of intercloud gas needed to confine the interstellar constituents to the disk is nearly constant.

If pushed to the extreme, I would say that flaring of the galactic HI disk at large radii probably follows from breakdown of the assumption that stars dominate gravity. In addition, the HI hole at small r may follow from there having been too little gas there to provide even the intercloud component, in which case the system is probably unstable to wind evacuation. But enough of this speculation. It's time to return to survey the carnage and construction in our recent path.

Summary: The total interstellar pressure at midplane can be read directly from the weight of the ISM. The resulting value is roughly a factor 4 higher than estimates of the diffuse cloud thermal pressure, and is consistent with $B_{rms} \simeq 5\mu G$, largely independent of density.

Nonthermal indicators suggest a pressure scale height of 1 to 2 kpc while much of the mass is located below 400 pc. This is inconsistent with the simple hydrostatic condition that the pressure gradient should be located with the weight. Magnetic tension, however, is almost certainly sufficient to anchor the high z pressure in the lower z intercloud ISM weight. The tension can support intercloud material at great z, but clouds have a size dependent upper limit on their support height. The maximum force per unit area of the cloud population downward on the magnetic field is $(B^2/8\pi)(D/L)$ x (sky coverage factor) where D/L is the diameter to length ratio of the clouds. This contribution is certainly small compared to the total interstellar pressure. (Cloud support tends to be largely dynamic rather than hydrostatic.)

Cloud and intercloud components are not constrained to have similar thermal pressures. Interphase pressure equilibrium is dominated by the magnetic field. Clouds are suctioned to the local minimum of the p, n diagram, with a broad temperature range around 100 K. The actual value of that pressure is proportional to the heating rate, presumably to the local starlight density. The thermal timescale is roughly 100 T years at 10^{-25} erg s^{-1} per atom. The minimum density of warm HI associated with clouds appears theoretically to be about 15 cm^{-3}. (The Cygnus Loop's involvement with such a component, however, seems to show material down to at least 5 cm^{-3} in the cloud regions, and even lower density in the surroundings. The lower density material should be overheated and in transit to the intercloud component.)

The intercloud HI (at 8000 K) is not subject to the cloud pressure minimum. Its particular distribution is dominated by dynamics. It probably has a filling factor ≥ 0.3 (with the H$^+$) and could approach 0.8. The low local densities needed for the high filling factor are seen in the VLISM and the preshock density of the intercloud component of the Cygnus Loop as well as other locations.

Supernova disruption of the intercloud medium is much less severe than previous estimates have suggested; the medium is highly elastic to the disturbances. Modest scale bubbles of hot gas are necessarily created by remnants, and their late evolution easily accounts for the interstellar OVI observations. Some large regions of hot gas exist (Local Bubble, Loop I, etc.); but it is not clear whether SN bubbles will collect to generate a hot interstellar phase. There is no clear observational need for such a phase, but if it exists, it occupies less than 0.65 of the volume and comingles with a very smoothly distributed intercloud component.

Our understanding of the control mechanisms for pressure, phase segregation, equipartition, and scale heights is exceedingly limited. Simple considerations suggest that scales should all be proportional to the SN source distribution, that the ratio of cloud thermal pressure to supernova dominated total pressure should be fixed by the population rather than density of stars, and that the total intercloud component surface density should be fairly constant.

I would like to thank Ron Reynolds and Charlie Goebel for useful discussions, and Ron for a helpful reading of the manuscript in early form. This material is based on work supported in part by the National Science Foundation under grant No. AST - 8643609. It was also supported in part by the National Aeronautics and Space Administration under grant NAG5 - 629.

REFERENCES

1. Cox, D.P. and Reynolds, R. J., <u>Ann. Rev. Astron.</u> and Astrophys.,25, 303 (1987)
2. Kulkarni, S. R. and Heiles, C. in <u>Galactic</u> and <u>Extragalactic Radio Astronomy</u>, ed. Kellerman, K. I.,Verschuur, G. L., CH. 3, 2nd ed. (in press) (1987)
3. Kulkarni, S. R., and Heiles, C., in <u>Interstellar Processes</u>, ed. Hollenbach, D. J. and Thronson, H. A. Jr., (Dordrecht:Reidel), 87 (1987)
4. Bloemen, J. B. G. M., <u>Ap. J.</u>, (in press) (Nov. 15, 1987)
5. Heiles, C., <u>Ap. J.</u>, 315, 555 (1987)
6. Tenorio-Tagle, G., Bodenheimer, P., and Rozyczka, M., <u>Astron. and Astrophys.</u>, (in press)
7. Raymond, J. C., Hester, J. J., Cox, D. P., Blair, W. P., Fesen, R. A. and Gull, T. R., <u>Ap. J.</u>, (in press) (1987)
8. Tucker, W., <u>Science</u>, 172, 372 (1971)
9. Ku, W. H.-M., Kahn, S. M., Pisarski, R. and Long K. S., <u>Ap. J.</u>, 278, 615 (1984)
10. Chevalier, R. A. and Raymond, J. C., <u>Ap. J. (Lett)</u>, 225, L27 (1978)
11. Raymond, J. C., Blair W. P., Fesen, R. A., and Gull, T. R., <u>Ap. J.</u>, 275, 636 (1983)
12. Fesen, R. A., and Itoh, H., <u>Ap. J.</u>, 295, 43 (1985)
13. Hester, J. J., Raymond, J. C., and Danielson, G. E., <u>Ap. J. (Lett)</u>, 303, L17 (1986)
14. Hester, J. J. and Cox, D. P., <u>Ap. J.</u>, 300, 675 (1986)
15. McKee, C. F., and Cowie, L. L., <u>Ap. J.</u>, 195, 715 (1975)
16. Hester, J. J., this volume (1987)
17. Hester, J. J., <u>Ap. J.</u>, 314, 187 (1987)
18. Badhwar, G. D. and Stevens, S. A., <u>Ap. J.</u>, 212, 494 (1977)
19. Cox, D. P. and Snowden, S. L., in <u>Adv. Space Res.</u>, 6, 97 (1986)
20. Troland, T. H. and Heiles, C., <u>Ap. J.</u>, 301, 339 (1986)
21. Garcia-Munoz, M., Mason, G. M., and Simpson, J. A., <u>Ap. J.</u>, 217, 859 (1977)
22. Rockstroh, J. M. and Webber, W. R., <u>Ap. J.</u>, 224, 677 (1978)
23. Savage, B. D., in <u>Interstellar Processes</u>, ed. Hollenbach D. J., and Thronson, H. A. Jr., (Reidel: Dordrecht), 123 (1987)
24. Hobbs, L. M., <u>Ap. J.</u>, 191, 395 (1974)
25. Field, G. B., Goldsmith, D. W., and Habing, H. J., <u>Ap. J. (Lett)</u>, 155, L49 (1969)
26. McKee, C. F. and Ostriker, J. P., <u>Ap. J.</u>, 218, 148 (1977)
27. Cox, D. P. and Smith B. W., <u>Ap. J. (Lett)</u>, 189, L105 (1974)
28. Reynolds, R. J., <u>Ap. J.</u>, 216, 433 (1977)
29. Cox, D.P., in <u>Proceedings</u> of <u>Meudon Conference</u> on <u>Model Nebulae</u>, ed. Pequinot, D., (1985)
30. Kraushaar, W. L., <u>Proc. Int. Conf.</u> on <u>Cosmic Rays</u> <u>at Jaipur</u>, 3, 379 (1963)
31. Cox, D. P., <u>Ap. J.</u>, 245, 534 (1981)

A MODEL FOR THE INTERACTION BETWEEN STARS AND GAS IN THE INTERSTELLAR MEDIUM

Wei-Hwan Chiang
Department of Astronomy, University of Texas
Austin, Texas

Joel N. Bregman
NRAO, Charlottesville, Virginia

Abstract: A global model for the ISM is described in which separate fluids of stars and gas interact through star formation, mass loss, stellar heating, and gaseous cooling processes. In all simulations, a steady state develops in which rarefied ionized regions separate ridges of neutral gas.

Global models for the interstellar medium have been discussed by Cox and Smith (1974), McKee and Ostriker (1977), and Cox (1981) under the assumption that the primary source of heating is from supernovae. We have extended these models by studying the fluid dynamics of gas that is heated by young stars. In our model, which is based on the work of Chiang and Prendergast (1985), gas can form into stars while stars can lose mass and heat the gas through stellar winds and supernovae. In addition, gas can cool radiatively. Because the stars as well as the gas are approximated as fluids, only one kind of star is considered, and this is an average Pop I star. The two fluids are coupled by a star formation rate proportional to the gas density, and a mass loss and heating rate proportional to the stellar density. In addition, the gas cools radiatively.

When only wind heating is considered, the heating process is continuous in time and the stability of the system can be examined. For conditions appropriate to interstellar gas, the equilibrium state, in which heating balances cooling, star formation balances mass loss, and both fluids have the same mean velocity, is subject to a star formation instability (Chiang and Prendergast 1985; Chiang and Bregman 1987). The nonlinear behavior of the system was studied with one and two dimensional numerical hydrodynamic models. After a characteristic time (the star formation and stellar mass loss time, taken to be 10^8 yr), the gaseous component forms dense connected ridges of neutral gas separated by large volumes of rarefied ionized gas (Fig. 1a). Most of the mass (70%) is in the neutral ridges, which have a typical length of 100-200 pc, yet occupy about 20% of the volume. The ridges have a lifetime of about the star formation time and move with a velocity of a few km/sec. Enhancements in the stellar density lie adjacent to the ridges (rather than at the bubble centers) and drive the motion of the gas. The ionized gas is never hotter than 20,000 K, which is a consequence of the smoothness of the heating process and the steepness of the cooling curve above 10^4 K. If the calculation had been performed in three spatial dimensions, the ridges would become shells or sheets of HI. The most prominent HI features would occur at the interface of two shells (HI sheets), of three shells (HI filaments), or at the vertex of four shells (dense HI clouds).

The two dimensional models were extended to include the effects of supernovae, which lead to the creation of million degree gas that cools slowly. Three models were calculated in which wind 1.4×10^{49} erg/pc, and the supernova rates were taken to be 0.142, 0.0472, and 0.0157 SN/kpc^2/10^8 yr for models s1, s2, and s3 respectively (Fig. 1 b,c,d). The local supernova rate is assumed to be proportional to the Pop I star density, although the actual occurrence of an event at a particular time is random. For model s1, the amount of heating provided by the supernovae is equivalent to that provided in the above wind model.

The porosity of the gas changes dramatically as a function of the supernova heating rate. In model s1, the overwhelming volume of gas is occupied by dilute ionized gas, with neutral HI appearing as islands. In model s3, the dilute ionized gas appears as isolated bubbles, and little of this gas is hotter than 10E6 K. The intermediate case, s2 has several well defined bubbles, some of which have merged to form larger structures. The typical size of a bubble, 200-300 pc is reminiscent of holes in the HI distribution seen in M31 (Brinks and Bajaja 1986)) and of the superbubbles seen in our Galaxy (Heiles 1979, 1984).

We have calculated line profiles through the grid for both the wind and supernova models (Fig. 2) and find that the line width increases with the supernova heating rate, as expected. It is not straightforward to interpret peaks in the line profile as distinct HI ridges since velocity crowding sometimes leads to such features in the profile. In the wind model, and models s2 and s3, HI would appear to be pervasive to an observer for a line of sight exceeding 200 pc, as suggested by the data (Liszt 1983, Lockman et al. 1986). The resemblance between the calculated and observed line profiles and structure of the ISM lend encouragement that this sort of model may ultimately reproduce a variety of features in the ISM and provide a suitable model for understanding the ISM.

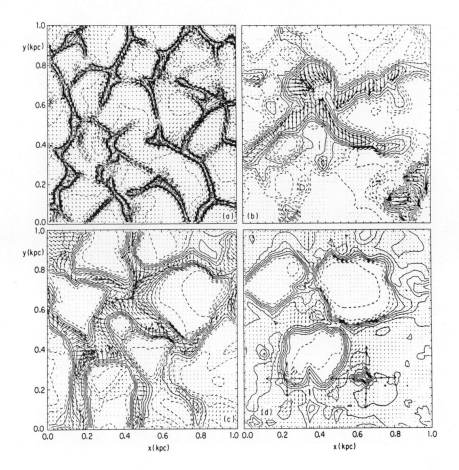

Fig. 1. The gas density contours and momentum vectors of the gas in a 1 kpc square grid are presented for the wind model and supernova models s1, s2, and s3 (a-d respectively). The wind model has twice the computational resolution of supernova models. The porosity of the gas decreases rapidly as the supernova rate is reduced by a factor of 9 from models s1 to s3.

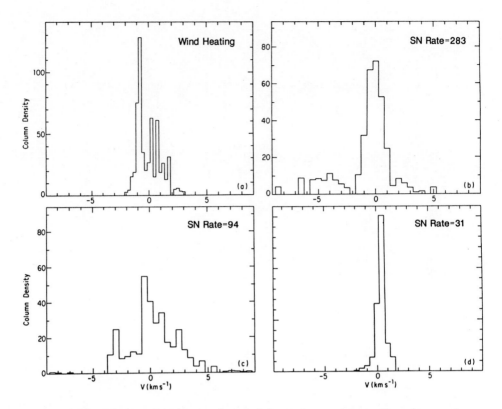

Fig. 2. The line profiles in arbitrary units are given for the wind model and supernova models s1, s2, and s3 (a-d respectively). The line width for the wind model and model s3 is about 2-4 km/sec while it is about 5 km/sec for model s2 and 5-10 km/s for model s1. Greater supernova rates lead to wider line profiles.

REFERENCES

Brinks, E., and Bajaja, E. 1986, Astron. and Astrophys, **169**, 14.

Chiang, W.-H., and Bregman, J.N. 1987, to be submitted to Ap.J.

Chiang, W.-H., and Prendergast, K.H. 1985, Ap.J., **297**, 507.

Cox, D.P. 1981, Ap.J., **245**, 534.

Cox, D.P., and Smith, B.W. 1974, Ap.J. Lett., **189**, L105.

Heiles, C. 1979, Ap.J., **229**, 533.

Heiles, C. 1984, Ap.J. Suppl., **55**, 585.

Liszt, H.S. 1983, Ap.J., **275**, 163.

Lockman, F.J., Hobbs, L.M., and Shull, J.M. 1986, Ap.J., **301**, 380.

McKee, C.F., and Ostriker, J.P. 1977, Ap.J., **218**, 148.

THE EFFECTS OF PROGENITOR MASS LOSS ON THE PROPERTIES OF A SUPERNOVA REMNANT

T.A. Lozinskaya
Sternberg Astronomical Institute, Moscow, USSR

Abstract: Observations of ring nebulae around WR and O_f stars, including those in OB associations, have been used to give information on the pre-supernova environment.

The nature and evolution of a supernova remnant (SNR) of any age is strongly dependent on the action of the progenitor on the ambient gas. The main factors modifying the pre-SN environment are:
1. ionizing radiation, which is probably accompanied by the reactive displacement of photo-evaporated cloudlets (McKee et al., 1984),
2. stellar winds (both 1. and 2. change in the course of the evolution of the star), and 3. ejection of a slow shell. In earlier work the influence of the progenitor on the ISM has been taken into account by assuming an idealized density distribution in the SN vicinity.

Observations of ring nebulae around WR and O_f stars make it possible to study directly the "preparation" of the surrounding interstellar gas by a massive progenitor before the SN explosion. Proceeding from the data of our investigations (Lozinskaya, 1982, 1983, 1986; Bochkarev et al., 1987) we can draw the following conclusions about the ISM parameters around WR and O_f stars, which are possible precursors of SNe of type II and of the Cassiopeia-A type.

1. Four types of nebulae excited by WR stars (Chu, 1981) and O_f stars (Lozinskaya, 1982) are observed. They are amorphous (R_a) and shell-like (R_s) HII regions, wind-blown bubbles (W) and stellar ejecta (E). Their parameters are listed in the following table.

Type	R_a	R_s	W	E
R (pc)	10 - 50	10 - 30	1 - 10	1 - 5
V_{exp} (km/s)	<10 (chaotic)	<10	20 - 100	20 - 50
Mass ($M_\odot$)	$10^2 - 10^3$	100 - 500	10 - 100	1 - 5
Structure	irregular	shell $\Delta R/R \sim 0.5$	filamentary shell $\Delta R/R \sim 0.1$	knotty shell

The kinematical age of W- and E- shells is found to be about (1 - 5)
10^4 yr. Soft X-ray emission found in NGC 6888 testifies to the fact
that, in accordance with the classical theory, wind-blown bubbles are
filled by hot plasma. Distinct shell-like structure (i.e. R_s-, W- or
E-type nebulae) is observed in about 40-50% of HII regions excited by
WR and O_f stars.

2. Multishell structure is often observed: small E- and W- type
shells are surrounded by extended R_s- or R_a- nebulae. The most
impressive example is provided by the system of four concentric shells
around the O_f star HD 148937: the stellar ejecta NGC 6164-5 (R~1 pc),
the wind-blown bubbles (R~6 pc), the limb-brightened HII region
(R~25 pc) and the outer dust shell (R~25-30 pc); see Fig. 1.

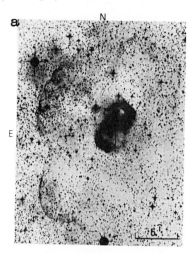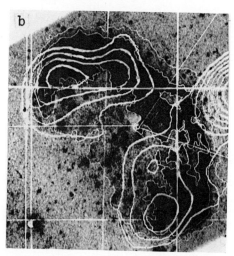

Figure 1: a - NGC 6164-5, two inner shells E and W types; image from
          Bruhweiler et al., 1981
          b - NGC 6888, OIII 5007 A image together with Einstein
          Observatory X-ray contours in the range 0.2-3.5 keV

Such a structure shows that mass loss in the form of stellar winds
and of slow ejecta occurs in the same star and that both processes are
not greatly separated in time. Thus the interstellar gas around a
massive SN-precursor is characterized by a complicated multilayer
structure with sharp density gradients formed by UV-radiation, stellar
winds and slow ejecta. The propagation of the SN-blast wave in such a
medium must be accompanied by formation of secondary (forward and
reflected) shock waves, see e.g. Itoh and Fabian, 1984. This might
manifest itself in the optical and X-ray emission of young and
middle-aged SNRs. Indeed, in Cassiopeia A we can distinguish the
two-component SN ejecta (fast-moving optical knots and the inner X-ray
shell) interacting with the two-component matter lost by the progenitor
(quasi-stationary flocculi which most likely represent the slow ejecta
of a WR star and the outer X-ray shell).

3. Wind-blown bubbles and ejecta are often characterized by an

elongated form; examples are NGC 6888, 2359, 6164-5 and the second
shell in the same system. In E-nebulae such a structure is most
probably the result of bipolar ejection. In W-nebulae it may be caused
by inhomogeneity of the ambient ISM as well as by a biconical stellar
wind. In particular, our latest investigations of the nebula NGC 6888
(in preparation) show that the structure and kinematics of the optical
filaments may be explained by inhomogeneity of the ISM (see also
Johnson and Songsathaporn, 1981). At the same time the X-ray picture
of two hot spots on opposite sides of the nebula (see Fig. 1), may be
caused by an asymmetrical stellar wind.

Irrespective of the formation mechanism such axially symmetric
distribution of the density in the environment of the progenitor may
provide an interpretation for the commonly observed "barrel-like" and
"biannular" morphology of radio supernova remnants (see Manchester,
1987). The other explanation which has been proposed by Manchester -
biconical relativistic particle flow from a central pulsar - is hardly
applicable to young shell type SNRs such as SN 1006 because of the
absence of a detectable synchrotron nebula. On the other hand, an
explanation invoking the asymmetries of swept-up or ejected gas around
the pre-SN star can be relevant to all SNRs.

It should be noted in this respect that bipolar flows from
pre-planetary stars and bipolar or annular forms of many planetary
nebulae indicate that a similar environment is typical also for the low
mass progenitors of SN I.

For old SNRs of large size the influence of radiation, mass loss
and previous SN explosions of nearby stars all become relevant. If a
SN has exploded in an OB association (where massive stars, the presumed
progenitors of supernovae, occur) these factors seem to determine the
pre-SN environment.

Shells and supershells around OB associations are a large subject
and I shall not dwell on it at length. I would only like to emphasize

Figure 2:
Schematic representation of
multishell formations around
OB associations.

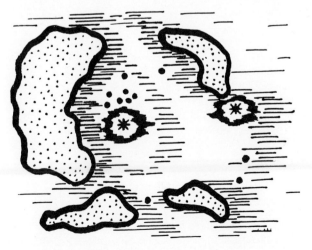

- represents dust
and molecular clouds,

- HII regions,

.. .- OB stars and
stellar clusters,

* * - WR, O_f stars.

that multishell hierarchical structure is observed here as well in the distribution of emission nebulae, dust and molecular clouds.

Fig. 2 shows the general scheme of such a shell hierarchy which we have detected around the associations Cas OB2 (Lozinskaya et al., 1986), Cep OB4 (Lozinskaya et al., 1987) and Cyg OB1 (Lozinskaya and Sitnik, in preparation). Inside the common shells around OB associations there are shell-like formations of smaller size related to young clusters and the smallest ring nebulae around the isolated sources of strong wind: WR and O_f stars. Such a multishell structure is observed more easily in the "face-on" LMC, the LMC 2 region being the most distinctive example (see Caulet et al., 1982).

References

Bochkarev, N.G., Lozinskaya, T.A., Piskunov, N.E., Pravdicova, V.V., Sitnik, T.G., 1987, Proc. of All-Union Meeting "WR stars and related objects", Tartu, 1986

Bruhweiler, F.C., Gull, T.R., Henize, K.G., Cannon, R.D., 1981, Ap. J., 251, 126

Caulet, A., Deharveng, L., Georgelin, Y.M., Georgelin, Y.P., 1982, Astr. Ap. 110, 185

Chu, Y.-H., 1981, Ap.J. 249, 197

Itoh, A., Fabian, A.C., 1984, MNRAS 208, 645

Johnson, P.G., Songsathaporn, R., 1981, MNRAS 195, 51

Lozinskaya, T.A., 1982, Astrophys. Space Sci. 87, 313

Lozinskaya, T.A., 1983, Sov. Astron. Lett. - Pisma Astron. J., 9, 469

Lozinskaya, T.A., 1986, "Supernovae and Stellar Winds; Interaction with the Interstellar Gas", Moscow, Nauka, 304 pages

Lozinskaya, T.A., Sitnik, T.G., Lomovskiy, A.I., 1986, Astrophys. Space Sci., 121, 357

Lozinskaya, T.A., Sitnik, T.G., Toropova, M.S., 1987, Sov. Astr. J. - Astron. Zh., 64, in press.

Manchester, R.N., 1987, Astr. Ap., 171, 205

McKee, C.F., Van Buren, D., Lazareff, B., 1984, Ap. J. Lett., 278, L115

RECENT X-RAY OBSERVATIONS OF SUPERNOVA REMNANTS AND THEIR INTERPRETATION

B. Aschenbach
Max-Planck-Institut für Physik und Astrophysik
Institut für extraterrestrische Physik
8046 Garching, W-Germany

Abstract: A review is given of recent observations of the X-ray emission from supernova remnants carried out on the Einstein, Tenma and EXOSAT satellites as well as from a few sounding rocket experiments. Our current interpretation of the high resolution images, high resolution energy spectra and the first few spectrally resolved images is discussed.

Introduction: Almost five years ago, in 1982 the IAU Symposium No. 101 on "Supernova Remnants and their X-ray Emission" was held in Venice. The proceedings edited by Danziger and Gorenstein include an extensive compilation of papers dealing with observations mainly made with the instruments on board of the Einstein satellite. It was in fact the Einstein observatory with its imaging and spectroscopic capability through which substantial progress has been made in supernova remnant research. Since then further contributions have come from the EXOSAT observatory of the European Space Agency, a preliminary account of which has been given by Aschenbach (1985), the Japanese Tenma satellite and a few sounding rocket experiments. Unlike radio and optical observations X-ray observations can be done exclusively from space and therefore substantial observational research is limited to the rare opportunities of X-ray astronomy satellites. Einstein, EXOSAT and Tenma terminated operation and the next block of observations will not come before 1990, when the German satellite ROSAT will be launched.

As of today there are about 150 galactic supernova remnants catalogued which have been identified as such by their radio morphology and their radio spectrum. From 40 of these objects X-ray emission has been detected. Thus the vast majority of the alleged supernova remnants have been seen only at radio frequencies and their true nature has still to be explored by future more sensitive observations. Outside the Galaxy 32 remnants have been found in the Large Magellanic Cloud and 6 in the Small Magellanic Cloud (respectively Mathewson et al., 1985 and Helfand, 1987). X-ray instrumentation has by now evolved to such a state that apart from this pure statistical information, which is however relevant to estimate supernova rates and supernova remnant lifetimes for instance, detailed studies of images, energy spectra and time variability of individual remnants can be made.

Images: High quality images became available first from the Einstein observatory with an angular resolution of typically 10 arcsec. The brighter remnants have subsequently been imaged by the smaller EXOSAT telescopes with poorer resolution. These images have been used to support the view of at least two main classes of remnants, which are the Crab-like remnants and the shell-type remnants. In his review Seward (1985) defines the Crab-like remnants as extended objects with a non-thermal X-ray spectrum with some evidence for a central energy source like an active pulsar. Crab-like remnants may appear as filled shells ("plerions") or as composite remnants which show both a radio shell and diffuse or point-like emission from the central region (Weiler, 1983). The recent EXOSAT observations have been presented by Davelaar and Smith (1986). Clearly, this class of remnants is intimately connected to neutron star research and a review on this topic has been given by Helfand and Becker (1984). Nomoto and Tsuruta (1986) have summarized the available data and upper limits on brightness and surface temperature of neutron stars in supernova remnants and have confronted it with current cooling theories. Clearly, more relevant in the context of this IAU Colloquium on the interaction of supernova remnants with the interstellar medium is the second class of remnants which is that of shells with a thermal spectrum. Unlike the Crab-like remnants which are most likely energized by a central compact object (Pacini and Salvati 1973, Rees and Gunn 1974, Aschenbach and Brinkmann 1975, Reynolds and Chevalier 1984, Brinkmann et al. 1986), the shell type remnants emit radiation from an optically thin plasma, which has been heated to X-ray temperatures by a shock wave. This shock wave may either be the blast wave associated with the stellar explosion (Heiles 1964), which interacts with the ambient interstellar or circumstellar material, or it may be a reverse shock wave (McKee 1974), which propagates inwards from the decelerated blast wave and raises the temperature of the stellar ejecta. In both cases the X-ray emitting shell (or shells) is expected to be perfectly circular and to have no spatial structure in surface brightness except a radial gradient for a spherically symmetrical explosion and homogeneous media. The X-ray images, however, have revealed quite the opposite with a great deal of structure present in all remnants. The young remnants Cas-A, Kepler, Tycho and SN1006 are still maintaining an approximate circular shape whereas older ones like Puppis-A are hard to reconcile with a shell at all (see figure 1).

Since X-ray emission scales with the square of density it is obvious to consider density variations within or across the extent of the remnant as the dominant source for the observed X-ray brightness structures. Large scale uniform density changes like the interstellar density gradient perpendicular to the galactic plane have been made responsible for the asymmetry in galactic latitude, observed in Puppis-A for instance (Petre et al. 1982). The increased density towards the galactic plane as well as,to a a smaller extent, the increased slow-down of the blast wave, gives rise to to higher X-ray surface brightness in the same direction. If a remnant has grown to such a size the blast wave has passed quite a number of individual cool, high density interstellar clouds of different sizes, as well as their warm, medium density envelopes, which are embedded in the general hot, low density intercloud medium (McKee 1981), and these structures will directly be imaged in X-rays if heated to the required temperatures. So, the interstellar medium structure will be visible in the X-ray images of relatively old remnants. Younger remnants show brightness variations as well, which are attributed to the clumpiness of the stellar ejecta heated by the reverse shock.

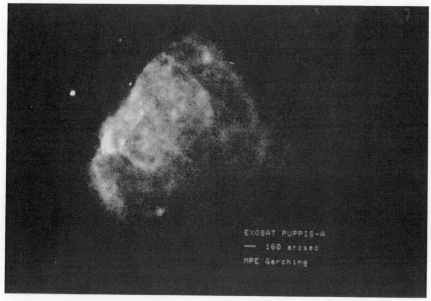

EXOSAT PUPPIS-A
— 160 arcsec
MPE Garching

Fig. 1. EXOSAT soft X-ray image of Puppis-A.

The detailed X-ray appearance of clouds in a low density medium depends on the physics of the interaction with the shock wave. Cowie and McKee (1977) and McKee (1981) have considered mass evaporation and hydrodynamic ablation to increase the density in the adjacent intercloud medium, thus raising the X-ray emission. An alternative approach to increase the X-ray emission near dense clouds has been proposed by Hester and Cox (1986). Led by an analysis of the X-ray and optical emission of the Cygnus Loop, they have proposed an additional compression of the down-stream X-ray emitting plasma by reflected or bow shocks around dense clouds. High resolution X-ray images of such clouds and their surroundings will help to discriminate the two effects, in particular if X-ray temperatures can be determined, which are supposed to be higher in the bow shock model. It is interesting to note (McKee 1981), that evaporation dominated remnants are affected on a large scale by the mass transfer from the clouds to the intercloud medium. In this case pressure and density will increase radially inwards contrary to the classical Sedov solution. Finally, non-thermal contributions to the X-ray emission may be not negligible in regions of electron acceleration and magnetic field amplification associated with the shock waves (Reynolds and Chevalier 1981).

X-ray images have been compared with optical and radio images, and the images of an individual remnant usually match each other quite well. In particular, the radio and X-ray images show very similar shapes, the outermost contours of which agree in remarkable detail down to the angular resolution limit. This indicates that heating of the plasma to X-ray energies and acceleration of electrons and magnetic field amplification sufficient for radio synchrotron radiation occur at least close to the shock front of the blast wave and on about the same time scale. Within the remnant the correlation is often less good. Radio bright regions sometimes coincide with

X-ray bright knots, but are more often slightly offset as in Puppis-A (Petre et al. 1982). In the majority X-ray enhancements have no radio counterpart and vice versa. The same picture holds for the comparison of X-
ray and optical images, which are dominated by filamentary and knotty
structure. The fact that some optically bright filaments are embedded in
X-ray bright regions, for instance in the Vela remnant (Kahn et al. 1985)
and in the Cygnus Loop (Hester and Cox 1986), indicates that the optical
emission is due to dense shocked clouds, which either partly evaporate or
produce bow shocks to enhance the ambient X-ray emission. Beyond this
the comparison of images taken in the different spectral domains has so
far not provided deeper insights in the physics of supernova remnants,
despite the high angular resolution now available in X-ray images.

Energy spectra: The thermal origin of the emission from shell type supernova remnants was for the first time established unambiguously by the
detection of the Fe-K emission line in the spectrum of Cas-A (Serlemitsos
et al. 1973). However, the clear detection of additional emission lines at
lower energies from other atomic species was severely hampered by the
low spectral resolution of the collimated proportional counter detectors.
This changed with the advent of the solid state spectrometer (SSS) which
was used in the focus of the Einstein observatory X-ray telescope. With an
improved resolution of about 160 eV over the nominal 0.5-4.5 keV detection band, numerous emission lines from highly ionized Mg, Si, S, Ar, and
Ca have been measured in the spectra of young supernova remnants including Cas-A, Kepler and Tycho (Holt 1983). Additional lines from highly
ionized N, O and Ne, as well as from Fe XVII have been discovered with
the Einstein Bragg Focal Plane Crystal Spectrometer (FPCS) in the older
remnant Puppis-A. With a resolving power of 100-1000, the forbidden,
intercombination and recombination lines of the He-like triplets from O VII
and Ne IX have been resolved for the first time (Winkler et al. 1981). The
spectral survey of emission lines from Puppis-A, which is the most detailed
result of non-solar X-ray spectroscopy so far, is displayed in figure 2.
Winkler et al. (1983) have pointed out in a very clear way how the various
line ratios can be used to perform detailed plasma diagnostics in determining electron temperature, ion population, ionization temperature and
ionization time.

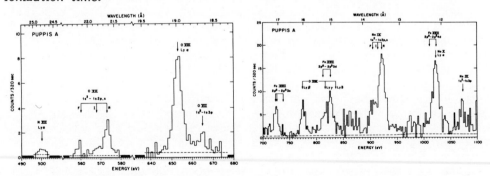

Fig. 2. Einstein FPCS spectra of Puppis-A (from Winkler et al. 1981).

The Einstein energy band extended up to about 4.5 keV and therefore the

Fe-K line emission as well as the high energy continuum, the existence of which was known from previous experiments, could not be covered simultaneously. Energy spectra over the 2-10 keV range with a spectral resolution better than the collimated proportional counter by typically a factor of 2 have become available from the non-imaging gas scintillation proportional counters (GSPC) flown on board of the EXOSAT and Tenma satellites. Figure 3 shows the EXOSAT GSPC spectrum of Tycho (Smith et al. 1987). The emission lines from transitions of He-like S, Ar, Ca and Fe ions are clearly resolved. The spectrum above about 5 keV is dominated by the Fe-K line and the continuum, the latter of which has been used to determine the electron temperature. Similiar spectra are available for Cas-A,

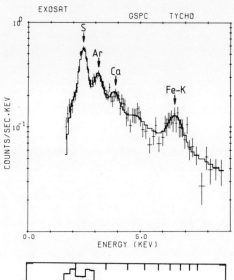

Fig. 3. EXOSAT GSPC spectrum of Tycho. The fit assumes a thermal bremsstrahlung continuum of kT=6.5 keV and emission lines of S, Ar, Ca and Fe superimposed (from Smith et al. 1987).

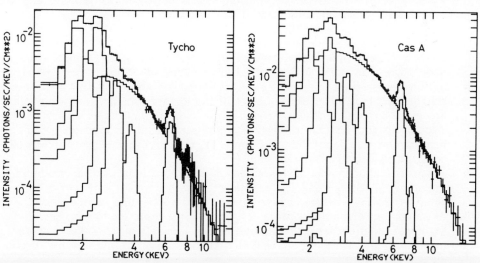

Fig. 4. Tenma GSPC spectrum of Cas-A and Tycho. The best fit assumes a single continuum spectrum and emission from 8 lines. The relative contribution of each component is shown in the lower histograms (from Tsunemi et al. 1986).

Kepler, RCW 103 and W49B and a detailed account of the EXOSAT spectra is given by Smith (1987). Tsunemi et al. (1986) have published the Tenma GSPC energy spectra of Cas-A and Tycho which extend to some lower energies to include the Si-K lines (see figure 4).

The emission from shell type remnants is considered to originate from an optically thin plasma, which has been heated to X-ray temperatures by shock waves. Model fits to the measured spectra attempt to determine X-ray temperature, total emission measure and elemental abundances, from which the density of the pre-shocked interstellar medium, the total swept-up mass and the explosion energy of the supernova can be derived using some 3-dimensional geometry and hyrodynamical solution for the expansion. If the contributions from the interstellar medium and the stellar ejecta can be disentangled, density and mass of the ejecta can be determined. In this way a link from the remnant to the progenitor star and the supernova type can be established in addition to the shock wave physics research. Early fits to the spectra have been based on the following assumption:

1.   the emitting plasma is in collisional equilibrium ionization;
2.   there is thermal equilibrium between ions and electrons implying the same temperature for both;
3.   the electrons have a Maxwellian velocity distribution;
4.   the plasma parameters including temperature, abundances, ionization stages, etc. are homogeneous over the entire remnant, at least over the field of view of the instrument;
5.   the hydrodynamical evolution of the remnants can be described by the self similarity Sedov solution, and
6.   the remnant expands into an ambient homogeneous medium.

The general results of these early analyses, which hold for almost any well studied remnant, can be summarized as follows:

1.   The spectrum is well described by the superposition of the emission from a two component plasma characterized by a low temperature of about 0.2 - 0.5 keV and a high temperature of about a few keV;
2.   a third, very high temperature component of about 30 keV has been suggested to be present in Cas-A (Pravdo and Smith 1979) which is, however, inconsistent with the EXOSAT data (Jansen et al. 1987). For Tycho, Pravdo and Smith derive a similarly high temperature component which could not be confirmed by EXOSAT because of inadequate sensitivity (Smith et al. 1987);
3.   elemental abundances from oxygen burning nucleosynthesis products including Si, S, Ar and Ca are largely overabundant compared with solar or cosmic values, whereas Fe tends to be less than solar;
4.   even young remnants contain a large X-ray emitting mass. For instance, Fabian et al. (1980) have found more than 15 solar masses in Cas-A, and Reid et al. (1982) estimate 15 solar masses for Tycho as well.

The validity of each of the six assumptions listed above was questioned, very early on, mainly from theoretical arguments. Now, there is growing observational evidence supporting more or less substantial modifi-

cations of the assumptions with subsequent revision of the parameters derived, the amount of which depends on age and environment of the individual remnant.

As pointed out by Gorenstein et al. (1974) the plasma in supernova remnants may not be in collisional ionization equilibrium because the time
scale to ionize the heavy elements to the equilibrium level by electron
collisions is of the order of $10^4/n_e$ years (Canizares 1984). For electron
densities n_e of about $1 - 10$ cm^{-2}, this is large compared with the age of
young remnants and has the effect that the heavy elements are underionized compared with their equilibrium population. Direct observational
evidence for non equilibrium ionization (NEI) has been obtained from the
analysis of the high resolution FPCS spectra of regions in Puppis-A (Canizares et al. 1983, Winkler et al. 1983, Fischbach et al. 1987) and in Cas-A
(Markert et al. 1987). The NEI conditions are derived from the weakness of
the forbidden lines relative to the resonance lines of OVII and NeIX (see
figure 2). Vedder et al. 1986 has recently published the FPCS spectrum
taken in the northern bright region of the Cygnus Loop (figure 5). From a
3σ upper limit of the forbidden to resonance line ratio of OVII they conclude that even in a remnant as old as the Cygnus Loop at least sections
exist within which collision equilibrium ionization has not been reached.
However, Gabriel et al. (1985) have pointed out that fast electrons in an
otherwise thermal plasma can mimic NEI conditions because they tend to
excite preferentially the resonance line, and therefore a weak forbidden-to-
resonance line ratio may not be conclusive. They compute that 20 keV
electrons having a proportion of 1 % of the total number of electrons are
sufficient to explain the OVII lines as observed in Puppis-A.

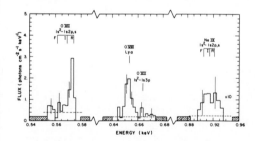

Fig. 5.Portions of the X-ray spectrum
of the northern bright region of the
Cygnus Loop measured with the FPCS
(from Vedder et al. 1986).

Further evidence for NEI conditions in Tycho and Cas-A has been obtained
from the EXOSAT (Jansen et al. 1987, Smith et al. 1987) and Tenma
(Tsunemi et al. 1986) observations of the Fe-K line. Both experiments have
measured consistently a line energy significantly lower than the value
expected from CEI which is based upon a temperature derived from the
high energy continuum. Applying NEI models both experiments agree that
Tycho is substantially more underionized than Cas-A, which is plausible
from the higher density in Cas-A.

Less direct but nonetheless evidence for NEI is deduced from the fact that

NEI models fit the observed X-ray spectra better than CEI models. NEI models have been constructed by numerous authors to describe time dependent ionization (Itoh 1977, 1979, Gronenschild and Mewe 1982, Shull 1982, Hamilton et al. 1983, Hamilton and Sarazin 1984, Nugent et al. 1984). The gross effect of NEI is an enhanced emission from lower ionization stages, which mimics a separate low temperature CEI plasma in addition to the temperature indicated by the high energy continuum. This is the reason that early X-ray spectra could be approximated by two temperature CEI models. However, with increasing better spectral resolution and broader energy coverage the fits became increasingly poorer. In order to explain the strong line emission from heavy elements observed in young remnants, a significant overabundance compared to solar values had to be adopted. Furthermore high emission measures implying high densities were needed to explain the high level of soft X-ray emission, which in turn led to high masses for the remnants. NEI models have recently been applied to some remnants with the result that the CEI based estimates about the total X-ray emitting mass have been refined. Using NEI emission to model the surface brightness distribution observed with the Einstein HRI, Gorenstein et al. (1983) derive a total of about 4 solar masses for Tycho shared one to one by the ejecta and swept up matter. This has to be contrasted with the result of about 15 solar masses which Reid et al. (1982) deduced from the Einstein IPC image and a CEI model. An even lower value of only 0.6 solar masses for the X-ray emitting mass of Tycho has been presented by Tsunemi et al. (1986), which they conclude from a NEI analysis of the Tenma energy spectrum. Figure 4 b shows the excellent fit. Although the mass has come down significantly the heavy elements are still a factor of 6-15 overabundant - but including iron - compared with solar values. The Tenma spectrum of Cas-A has also been fitted acceptably with a single component NEI model with a remarkable low value of 2.4 solar masses and element abundances very close to solar values. The Tenma analysis, however, is in conflict with the EXOSAT results analyzed by Smith et al. (1987) and Jansen et al. (1987) who claim that the spectra cannot be explained by a single component but require a two component NEI model. This disagreement is most evident in the values given for the continuum temperature, which unfortunately affects the elemental abundances and masses strongly. Thus, Smith et al. quote 3.3 solar masses for the swept up mass in Tycho. It remains to be seen how this conflict will be resolved in the future. This is interesting, since the Tenma results, for the first time, are clearly in line with respect to element abundances and mass with what is currently predicted from supernova explosion models advocating a type I for Tycho and type II for Cas-A.

Hamilton et al. (1986 b) have recently re-analysed a collection of spectra of Tycho which have been obtained with the instruments preceding Tenma and EXOSAT. They assume a two component NEI model, one each for the blast wave and the reverse shock. The composition is taken to be solar for the uniform interstellar medium and of pure heavy elements for the ejecta which are stratified according to atomic number. From the spectra which are shown in figure 6 they conclude that the X-ray emission is dominated by the shocked ejecta with an important contribution from the blast wave at high energies. The total ejected mass is 1.4 solar masses out of which

0.4 solar masses have not yet been shocked, and the mass shocked by the blast wave amounts to 1.3 solar masses. This is close to the results of Gorenstein et al. but significantly higher than the Tenma data analysis has shown. Hamilton et al. (1986 a) have successfully used this type of model to reproduce the featureless power-law like spectrum of SN1006 by putting SN1006 into a much lower ionization age than Tycho.

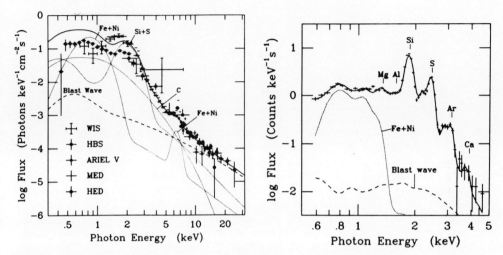

Fig. 6. Best fit of the two component NEI models of Hamilton et al. (1986) to low resolution (left) and high resolution (right) spectra of Tycho's supernova remnant.

The NEI models discussed above assume that the hydrodynamic evolution of the remnant can be described by the self similarity solutions of the Sedov type. Hughes and Helfand (1985) have instead used a numerical hydrodynamic shock code into which the time dependent ionization equations have been incorporated. Thus, the time dependent ionization structure can be computed simultaneously in both the blast wave heated ambient plasma as well as the reverse shock heated ejecta. Interestingly, they have found that the surface brightness distributions and the spectra of Kepler's supernova remnant as measured with the Einstein observatory instruments can be equally well reproduced if the emission originates predominantly either from the heated interstellar medium with the remnant in the Sedov phase or from the heated ejecta. Figure 7 shows the best fit to the Einstein SSS spectrum which is achieved with over-solar abundances of Si and S but under-solar abundance of Fe. The over-abundance is more pronounced for the reverse shock case but it is lower, by a factor of 2 to 3, than the results of two temperature CEI models of Becker et al.(1980). Whereas the Sedov case is compatible with a type I supernova event, the reverse shock heated ejecta model requires an ejecta mass of about 4.5 solar masses, which implies a massive progenitor star. It would certainly be useful to further constrain the model and include into the analysis the high energy spectrum to beyond the Fe-K line, which is now available from EXOSAT.

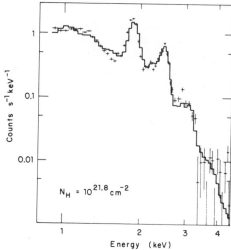

Fig. 7. Best fit of the hydrodynamic shock NEI model of Hughes and Helfand (1985) to the Einstein SSS spectrum of Kepler's supernova remnant.

In summary, spectral modelling has made progress in recent years by involving NEI in particular and better fits have been obtained with generally lower masses and lower elemental abundances have been obtained compared with CEI models. However, there seems to be no conclusion on the number of temperature components present in young remnants. It is striking that the analysis of relatively narrow band spectra like that by Tsumeni et al. for Tycho and Cas-A and that by Hughes and Helfand for Kepler favour a dominant single component. In contrast to this, the analysis of broad band spectra like that of Jansen et al. for Cas-A and Smith et al. for Tycho and that of Hamilton et al. for Tycho as well clearly require two NEI components which are associated with plasma heated by the blast wave and the reverse shock.

Spectrally resolved images: Spectrally unresolved images obtained with the Einstein HRI or IPC, or the EXOSAT CMA instrument, or even the earlier collimated scanning counters demonstrated a great deal of spatial structure to be present in many remnants. Structure like this may originate from different kinds of variations across the remnant, such as those of emission measure, i.e. density and depth of line of sight within the remnant, temperature, ionization structure, and even interstellar absorption column density towards remnants of significant extent. Clearly, these effects are important to be considered when determining elemental abundances and X-ray emitting mass, and therefore a spectrum spatially integrated over the whole remnant or over substantial portions of the remnant is of limited information (see for instance Brinkmann and Fink 1987).

Spectral variations across the remnant of Cas-A were first discovered in the Einstein IPC data, which show significantly different pulse height spectra in the south-west and the north- east section (Murray et al. 1979). Using the brightness distribution observed with the HRI, a two–circular–shell geometry and assuming pressure equilibrium, Fabian et al. (1980) has constructed a CEI temperature and density map revealing considerable variations within the remnant. In collaboration with the Naval Research Laboratory, the MPI group has obtained a spectrally resolved image on a sounding rocket flight (Aschenbach 1985). Using CEI emission and the element abundance of Becker et al. (1979), a temperature map has been derived, which confirms and expands the Einstein IPC results. The faint south-west region has a temperature in excess of 4 keV, whereas the bright northeast part is between 0.3 and 1 keV. The overall temperature distribution shows two distinct peaks at 0.5 keV and 5.4 keV. The hot component forms an almost complete shell along the outermost boundary at a radial distance of about 3 arcmin from the centre. The interior is rather uniform at the low temperature level. Assuming two spherical shells for the emission region, density and pressure maps have been constructed and the remnant is not found in pressure equilibrium, but with the highest pressure occurring along the outer boundary. Interestingly, the interior which seems to be in pressure equilibrium is separated from the outermost annulus by an about 30 arcsec wide pressure minimum, indicating a deceleration of the blast wave.

The EXOSAT imaging telescopes have been used to take very long exposures of Cas-A and to measure the spectral variations with high statistical accuracy, but with proportional counter type resolution (Jansen et al. 1987). They also find significant CEI temperature variations with the highest temperatures occurring in an outer annulus about 3 arcmin from the centre. Analysing the temperature distribution in radial sectors, the high temperature component is most pronounced in the smooth and faint regions whereas in the clumpy and bright regions the high temperature component appears reduced. Figure 8 shows two examples of temperature profiles reproduced from the paper of Jansen et al.

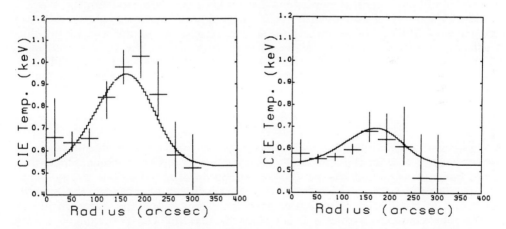

Fig. 8. CEI temperature profiles of the smooth and faint west south-west portion of Cas-A (left) and the clumpy and bright east south-east portion of Cas-A (right) as deduced from the EXOSAT PSD images by Jansen et al. (1987).

The Einstein IPC, the MPI rocket and the EXOSAT PSD results largely agree on the gross temperature distribution and demonstrate two temperature components in Cas-A one of which is apparently associated with the blast wave indicated by its high temperature, its outermost location and the underlying smooth and faint brightness distribution. As expected the low temperature component is associated with the reverse shock, which has heated the clumpy ejecta located more inside the remnant. So even in a remnant as young as Cas-A significant emission from heated ejecta has been found and this has independently been established by the remnant integrated spectra like those from the SSS, EXOSAT and Tenma as well as the spectrally resolved, but low energy images.

The first spectrally resolved images of the middle-aged remnant Puppis-A have been reported by the MPI group from a sounding rocket experiment (Pfeffermann et al. 1980). The CEI temperature map with a resolution of 3 arcmin shows a rather uniform temperature of $2\text{-}5.10^6$K for the interior, which agrees fairly well with the results from the Einstein high resolution FPCS CEI analysis (Winkler et al. 1981 a, b). In a subsequent paper, however, Canizares et al. (1983) showed that the spectrum taken with the 3 arcmin by 30 arcmin aperture of the FPCS cannot be explained by a CEI plasma but by a still ionizing plasma of an electron temperature in excess of 5.10^6K. The MPI rocket experiment revealed a second high temperature component of more than 10^7K which has been found earlier in spatially unresolved counter spectra. Unlike in Cas-A this component is not associated with the outer periphery but shows up in some pixels at the north eastern rim, along some filaments in the interior, but predominantly in the faint western parts, which is plausible since the region is presumably of a lower density of the interstellar medium and thus heated to a higher temperature by the blast wave. Furthermore it was found that the area of the bright eastern knot was by far the coolest part of the remnant, although the statistics were not sufficient to resolve the temperature of the knot itself. Significant spectral variations have also been detected among the 8 fields in Puppis-A observed with the 6 arcmin wide aperture of the SSS (Szymkowiak 1985). These variations are most obvious in the equivalent line widths from the He-like ions of heavy elements and in the soft X-ray part of the spectrum below 1 keV. The attempt to fit two temperature CEI models to the data failed and even fits with a single component NEI plasma are not convincing. Also, the pointings of the high resolution FPCS to various different fields in Puppis indicate spectral variations and possibly different temperatures (Fischbach et al. 1987).

A significant improvement in counting statistics and angular resolution has become possible with the EXOSAT absorption filter spectroscopy. Since the broad band transmission of the Lexan and boron filters used, is X-ray energy dependent, the ratio of the two X-ray fluxes depends on the source spectrum. In case of a CEI plasma of cosmic abundance the filter ratio scales approximately with log T and thus by dividing the two images a CEI temperature map of Puppis-A has been produced with a resolution of 1-arcmin and better (Aschenbach 1985). The map shows a great deal of temperature structure on scales even as small as the angular resolution limit and with temperatures between $5.9 \leq \log T \leq 7.1$. Beyond the upper limit the filter ratio is insensitive to temperature. There are 3 distinct regions forming the coolest parts within the remnants, which include the two bright eastern and northern knots, clearly resolved this time, and the outer section of the south-east elongated patch. It is interesting to note that the CEI X-ray temperature of the bright eastern knot is not uniform but varies between 6.0 and 6.4 for log T. Teske and Petre (1987) have reported CCD images of the region of the eastern knot in the forbidden red and green coronal iron lines, and under the assumption that the knot, which is supposed to be an interstellar cloud, is in CEI at a temperature of log T = 6.35 they derive a minimum mass of about 0.1 solar masses with a density in the range of 23 to 49 cm^{-3}. The EXOSAT results show that there is significantly cooler gas in the knot region, and thus they do not support Teske's and Petre's view that the visible FeXIV emission is from cooler inclusions in a still hotter medium. A comparison of the EXO-SAT CEI temperature map with the 6 arcmin diameter fields observed with the SSS and the 3 arcmin by 30 arcmin rectangular fields observed with the FPCS clearly reveals that the spectrum within each of the Einstein fields is not uniform but varies spatially (Jansen, Aschenbach and Bleeker, 1987), although these fields are already small compared with the extent of the remnants.

Spectrally resolved images have been taken also from the old Vela super-nova remnant and the Cygnus Loop with the Einstein IPC. For the Cygnus Loop, Ku et al. (1984) have pointed out that CEI best fits show lower temperatures along the limb compared with the centre, and that tempera-tures anticorrelate with intensity as expected for approximate pressure equilibrium. This view has been supported by Charles et al. (1985) by a detailed spectral study of two 1° wide fields located at the southern and western boundary of the remnant. Temperature and emission measure varia-tions on scales as small as 4 arcmin have been found, favouring the pre-sence of an inhomogeneous cloudy interstellar medium. Similar results have been obtained for the Vela remnant by Kahn et al. (1985), although the two remnants different significantly in their overall morphological appear-ance. However, the observations do not preclude pressure variations as large as a factor of 10.

Future prospects: Since the early years of proportional counters much progress has been made in supernova remnant research due to the availabi-lity of high resolution imaging, high resolution spectroscopy and broad band energy coverage. High resolution imaging made visible the cloudy interstel-lar medium and the clumpy ejecta, and it made possible the first compara-tive investigations between X-ray, optical and radio morphology; in particu-lar the first detailed observational studies of shocked clouds and possible thermal evaporation. The Einstein FPCS and SSS have shown that non-equilibrium ionization may be present in many remnants, even as old as the Cygnus Loop. However, it seems that the present generation of NEI models does not explain the broad band, high energy spectra as observed by EXOSAT for instance, unless multi NEI models with a sufficient number of free parameters are used. In order to better understand the images and spectra, hydrodynamical shock codes coupled with time dependent ionization are needed with an improved knowledge of the electron energy distribution which determines the ionization structure. These model calculations should take into account the results from the supernova explosion simulations including velocity, density and elemental abundances of the debris to pre-dict in detail the effects of the progenitor star. Similarly, the ambient medium into which the remnant expands has to be considered, including a stellar wind of the progenitor star, and a multi-component medium.

A key issue for further progress from an observational point of view is the future availability of spectrally resolved images which cover the energy region from 0.3 to about 10 keV. An energy band as broad as this is needed to disentangle the emission from the low temperature ejecta as well as from the high temperature interstellar medium heated by the blast wave and to determine independently elemental abundances for each component. A spectral resolving power of about 50 is adequate. The spatial resolution required is of course a small fraction of the remnant size. Hydrodynamical NEI model calculations of Brinkmann and Fink (1987) suggest that for a young remnant like Tycho a resolution of about 20 arcsec is adequate to resolve the ionization structure. Taking the degree of clumpiness, even lower values are indicated. For older remnants like Puppis-A, Vela or the Cygnus Loop, an angular resolution approaching the small structural scales present in the cloudy interstellar medium is appropriate. The bright eastern knot in Puppis-A for instance has a size of about 1-arcmin.

At present there are 5 missions being planned which carry imaging telescopes, i.e. the German ROSAT (1990) and Spectrosat (1993) missions, the Italian SAX (1992) mission, NASA's AXAF (1995) and XMM (1998) of the European Space Agency. The numbers in parenthesis give the currently envisaged launch dates. Except ROSAT, all other four missions await final approval. ROSAT will take images with high angular resolution (5 arcsec), and extremely high contrast due to the steep telescope point spread function and almost vanishing scattering. This will allow a search for compact objects within remnants at an significantly increased level of sensitivity compared with Einstein. With a throughput about 8 times greater than the Einstein HRI and a comparable background level, remnants can be studied to lower surface brightness within and outside our galaxy. Spectrally resolved images with an angular resolution of 20 arcsec and a spectral resolving power of about 2.5 will be taken with the position sensitive proportional counter in the energy band 0.1 - 2.2 keV. This spectral resolution is insufficient to resolve individual lines and to perform detailed plasma diagnostics but it will establish spectral variations across many remnants at an unprecedented level of angular resolution, and it will constrain the nonthermal component in Crab-like and composite remnants. Spectrosat will be a ROSAT follow-up modified by a transmission grating which will increase the resolving power to about 50-100.

At present, SAX will be the first mission which will carry imaging telescopes working up to 10 keV, with a spatial resolution of about 1 arcmin and a resolving power of about 10 at the Fe-K line. Very few years before the turn of the millenium, hopefully, the two great observatories AXAF and XMM will be launched into orbit. Both observatories cover the energy band from 0.1 to 10 keV. With AXAF sub-arcsecond imaging will become possible and very high spectral resolution as well, although at moderate throughput. XMM will perform low to high spectral resolution observations with high throughput but at the expense of angular resolution (30 arcsec).

Acknowledgements: I like to thank Dr. D. Helfand, F.A. Jansen, and Drs. A. Smith and H. Tsunemi, who provided me with reprints, preprints and data partially prior to publication.

References

Aschenbach, B., Sp.Sc.Rev. **40**, 447 (1985).

Aschenbach, B. and Brinkmann, W., Astron. Astrophys. **41**, 147 (1975).

Becker, R.H., Boldt, E.A., Holt, S.S., Serlemitsos, P.J., White, N.E., Ap.J. **237**, L77 (1980).

Becker, R.H., Holt, S.S., Smith, B.W., White, N.E., Boldt, E.A., Mushotzky, R.F., Serlemitsos, P.J. Ap.J. **234**, L73 (1979).

Brinkmann, W. and Aschenbach, B., Nature **313**, 662(1986).

Brinkmann, W. and Fink, H.H., this volume (1987).

Canizares, C.R., Proc. of the Intern. Symp. on "X-ray Astronomy '84", Oda, M. and Giacconi, R., eds., Bologna, 275 (1984).

Canizares, C.R., Winkler, P.F., Markert, T.H., Berg, C., "Supernova Remnants and their X-ray Emission", IAU Symp. No. 101, Danziger, J. and Gorenstein, P., eds., (D. Reidel), 205 (1983).

Charles, P.A., Kahn, S.M., McKee, C.F., Ap.J. **295**, 456 (1985).

Cowie, L.L. and McKee, C.F., Ap.J. **211**, 135 (1977).

Danziger, J. and Gorenstein, P., "Supernova Remnants and their X-ray Emission", Proc. of the IAU Symp. No. 101, Venice (D. Reidel) 1983.

Davelaar, J. and Smith, A., Proc. of the George Mason University workshop on "The Crab Nebula and Related Objects", Kafatos, M.C. and Henry, R.B.C., eds., Fairfax (Cambridge University Press), 219 (1986).

Fabian, A.C., Willingale, R., Pye, J.P., Murray, S.S., Fabbiano, G., M.N.R.A.S. **193**, 175 (1980).

Fischbach, K.F., Canizares, C.R., Markert, T.H., Coyne, J.H., this volume (1987).

Gabriel, A.H., Acton, L.W., Bely-Dubau, F., Faucher, P., Proc. ESA Workshop on "Cosmic X-Ray Spectroscopy Mission", ESA **SP-239**, 137 (1985).

Gorenstein, P., Harnden, Jr., F.R., Tucker, W.H., Ap.J. **192**, 661 (1974).

Gorenstein, P., Seward, F., Tucker, W., "Supernova Remnants and their X-ray Emission", IAU Symp. No. 101, Danziger, J. and Gorenstein, P., eds., (D. Reidel), 1, 1983.

Gronenschild, E.H.B.M. and Mewe, R., Astron. Astrophy. Suppl. **48**, 305ᵗ (1982).

Hamilton, A.J.S., Sarazin, C.L., Chevalier, R.A., Ap.J. Suppl. **51**, 115 (1983).

Hamilton, A.J.S. and Sarazin, C.L., Ap.J. **284**, 601 (1984).

Hamilton, A.J.S., Sarazin, C.L., Szymkowiak, A.E., Ap.J. **300**, 698 (1986a).

Hamilton, A.J.S., Sarazin, C.L., Szymkowiak, A.E., Ap.J. **300**, 713 (1986b).

Heiles, C., Ap.J. **140**, 470 (1964).

Helfand, D.J., (1987) this conference, not published.

Helfand, D.J. and Becker, R.H., Nature **307**, 215 (1984).

Hester, J.J. and Cox, D.P., Ap.J. **300**, 675 (1986).

Holt, S.S., "Supernova Remnants and their X-ray Emission", IAU Symp. No. 101, Danziger, J. and Gorenstein, P., eds., (D. Reidel), 17 (1983).

Hughes, J.P. and Helfand, D.J., Ap.J. **291**, 544 (1985).

Itoh, H., Publ. Astron. Soc. Japan **29**, 813 (1977).

Itoh, H., Publ. Astron. Soc. Japan **31**, 541 (1979).

Jansen, F.A., Smith, A., Bleeker, J.A.M., de Korte, P.A.J., Peacock, A., White, N.E., submitted to Ap.J. (1987).

Jansen, F.A., Aschenbach, B., Bleeker, J.A.M., this volume (1987).

Kahn, S.M., Gorenstein, P., Harnden, Jr., F.R., Seward, F.D., Ap.J. **299**, 821 (1985).

Ku, W.H.-M., Kahn, S.M., Pisarski, R., Long, K.S., Ap.J. **278**, 615 (1984).

Markert, T.H., Blizzard, P.L., Canizares, C.R., Hughes, J.P., this volume (1987).

Mathewson, D.S., Ford, V.L., Tuohy, I.R., Mills, B.Y., Turtle, A.J., Helfand, D.J., Ap.J. Suppl. **58**, 197 (1985).

McKee, C.F., Ap.J. **188**, 335 (1974).

McKee, C.F., "Supernova: A Survey of Current Research", Proc. NATO ASI Cambridge, Rees, M. and Stoneham, R.J., eds., (D. Reidel), 433 (1981).

Murray, S.S., Fabbiano, G., Fabian, A.C., Epstein, A., Giacconi, R., Ap.J. **234**, L69 (1979).

Nomoto, K. and Tsuruta, S., Ap.J. **305**, L19 (1986).

Nugent, J.J., Pravdo, S.H., Garmire, G.P., Becker, R.H., Tuohy, I.R., Winkler, P.F., Ap.J. **284**, 612 (1984).

Pacini, F. and Salvati, M., Ap.J. **186**, 249 (1973).

Petre, R., Canizares, C.R., Kriss, G.A., Winkler, P.F., Ap.J. **258**, 22 (1982).

Pfeffermann, E., Aschenbach, B., Bräuninger, H., Heinecke, N., Ondrusch, A., Trümper, J., Bull. Am. Astron. Soc. **11**, 789 (1980).

Pravdo, S.H. and Smith, B.W., Ap.J. **234**, L195 (1979).

Rees, M.J. and Gunn, J.E., M.N.R.A.S. **167**, 1 (1974).

Reid, P.B., Becker, R.H., Long, K.S., Ap.J. **261**, 485 (1982).

Reynolds, S.P. and Chevalier, R.A., Ap.J. **245**, 912 (1981).

Reynolds, S.P. and Chevalier, R.A., Ap.J. **278**, 630 (1984).

Serlemitsos, P.J., Boldt, E.A., Holt, S.S., Ramaty, R., Brisken, A.F., Ap.J. **184**, L1 (1973).

Seward, F.D., Comm. Astroph. XI, 1, 15 (1985).

Smith, A., Davelaar, J., Peacock, A., Taylor, B.G., Morini, M., Robba, N.R., submitted to Ap.J. (1987).

Smith, A., this volume (1987).

Shull, J.M., Ap.J. **262**, 308 (1982).

Szymkowiak, A.E., NASA Techn. Memor. 86169 (1985).

Teske, R.G. and Petre, R., Ap.J. **314**, 673 (1987).

Tsunemi, H., Yamashita, K., Masai, K., Hayakawa, S., Koyama, K., Ap.J. **306**, 248 (1986).

Vedder, P.W., Canizares, C.R., Markert, T.H., Pradhan, A.K., Ap.J. **307**, 269 (1986).

Weiler, K., The Observatory **103**, 85 (1983).

Winkler, P.F., Canizares, C.R., Clark, G.W., Markert, T.H., Petre, R., Ap.J. **245**, 574 (1981a).

Winkler, P.F., Canizares, C.R., Clark, G.W., Markert, T.H., Kalata, K., Schnopper, H.W., Ap.J. **246**, L27 (1981b).

Winkler, P.F., Canizares, C.R., Bromley, B.C., "Supernova Remnants and their X-ray Emission", IAU Symp. No. 101, Danziger, J. and Gorenstein, P., eds., (D. Reidel), 245 (1983).

X-RAY IMAGES OF SUPERNOVA REMNANTS

Frederick D. Seward

Harvard-Smithsonian Center for Astrophysics
60 Garden, Street, Cambridge, Massachusetts 02138

Einstein observations of supernova remnants have been reviewed and analyzed. Images of 44 galactic remnants have been reprocessed, merged when necessary, and collected into a catalog. Some bright remnants were viewed with both moderate and high resolution instruments (IPC with 1' resolution and HRI with 4" resolution). Some IPC images of nearby remnants have been separated into 2 energy bands, 0.2-0.6 keV and 0.6-4.5 keV; whereas most images cover the band 0.2-4.5 keV. The catalog consists of 72 images of the 44 remnants.

These images will be published in the form illustrated here. Contour levels are spaced geometrically as indicated below the figures and show the faintest observable features. The pictures are more linear and generally show only the brighter regions. Images are available now, however, in FITS format, on magnetic tape and may be obtained by writing to the author.

The x-ray morphology may be used to classify remnants. There are:
1) shell-like SNR with definite x-ray limb brightening.

2) filled-center SNR, brightest at the center with little or no emission from the limb, x-ray spectra are suspected to be thermal, radio images are shell-like.

3) SNR with internal neutron star or bright synchrotron nebula indicating the presence of a hidden pulsar. Other data sometimes aid in this classification; e.g. a hard continuum x-ray spectrum or a center-bright radio morphology.

4) SNR dominated by a bright central source, probably an accretion-powered binary.

5) SNR with irregular morphology. None of the above categories apply, or data are too crude to determine the morphology.

The 44 remnants are listed here with a _preliminary_ determination of the brightness as observed with the Einstein IPC.

Without the Einstein Observatory, these images of SNR would not exist. The observatory and imaging detectors were the result of the labors of R. Giacconi, H. Tananbaum, E. Schreier, L. Van Speybroeck, S. Murray, J.P. Henry, P. Gorenstein, F.R. Harnden, Jr.,and D. Fabricant. Thanks also to J. Brody, M. O'Shaughnessy, and L. Whitton who assisted in the preparation of material for this catalog.

SNR Included in Catalog

	Remnant	Common Name	IPC Rate (c/s)	Type	Comments
1	4.5+6.8	Kepler	7.3±0.4	shell	SN 1604
2	6.4-0.1	W 28	3.2±0.4	full	
3	11.2-0.3	-	1.00±0.1	shell	
4	21.5-0.9	-	0.49±.05	synchrotron	
5	27.4+0.0	Kes 73	1.07±0.1	central source	
5A		1E1838-049	-	-	not resolved
6	29.7-0.2	Kes 75	0.22±.03	synchrotron	
7	31.9+0.0	3C 391	0.24±.03	irregular	
8	33.7+0.0	Kes 79	0.44±.05	irregular	
9	34.6-0.5	W 44	3.3±0.3	full	
10	39.2-0.3	3C 396	.06±.01	irregular	
11	39.7-2.0	W 50	1.6 ave.	central source	
11A		SS433	1.2 ave.	-	variable
12	41.1-0.3	3C 397	0.75±0.1	irregular	
13	43.3-0.2	W 49B	0.67±.06	full	
14	49.2-0.7	W 51	0.9	irregular	
15	53.6-2.2	3C 400.2	0.80±0.1	full	
16	65.3+5.7	GKP	-	-	incomplete data
17	68.8+2.6	CTB 80	0.17±.03	central source	
17A		1E1951+327	0.14±.01		
18	74.3-8.5	Cyg Loop	620±40	shell	
19	74.9+1.2	CTB 87	.040±.01	synchrotron	
20	78.2+2.1	W 66	>0.6	-	incomplete data
21	82.2+5.3	W 63	0.3	irregular	
22	109.2-1.0	CTB 109	5.2±0.4	central source	
22A		1E2259+586	1.1±0.1	-	7s period
23	111.7-2.1	Cas A	61±2	shell	
24	119.5+9.8	CTA 1	0.9	full	
25	120.1+1.4	Tycho	22.3±1	shell	SN 1572
26	130.7+3.1	3C 58	0.35±.04	synchrotron	
27	132.7+1.3	HB 3	2.6	irregular	
28	160.4+2.8	HB 9	-	-	incomplete data
29	184.6-5.8	Crab	684	synchrotron	SN 1054
29A		PSR 0531+21	-	-	"age" 1240 yr
30	189.0+3.0	IC 443	12.8	irregular	
31	260.4-3.4	Pup A	230	shell	
32	263.8-1.7	Vela SNR	490	irregular	
32A	263.5-2.7	PSR 0833-45	2.1	-	"age" 13000 yr
33	290.1-0.8	MSH 11-61A	0.47±0.1	full	
34	291.0-0.1	MSH 11-62	0.22±.05	synchrotron	
35	292.0+1.8	MSH 11-54	9.1±1.0	irregular	
36	296.1-0.7	-	3.1	shell	
37	296.5+10.0	PKS 1209-52	2.6	shell	
38	315.4-2.3	RCW 86	8.5±1.0	shell	
39	320.3-1.2	MSH 15-52	2.40±0.2	synchrotron	
39A		PSR 1509-58	0.30±.04		"age" 1550 yr
40	326.3-1.8	MSH 15-56	1.0	irregular	
41	327.1-1.1	-	.085±.02	irregular	
42	327.4+0.4	Kes 27	0.40±0.1	full	
43	327.6+14.5	SNR 1006	11.1±1.0	shell	
44	332.4-0.4	RCW 103	9.3±1.0	shell	

4.5+6.8   Kepler's SN

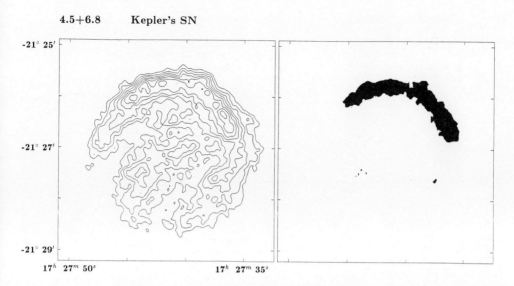

HRI, contour intervals are a factor of 1.3 in brightness.

34.6-0.5   W 44

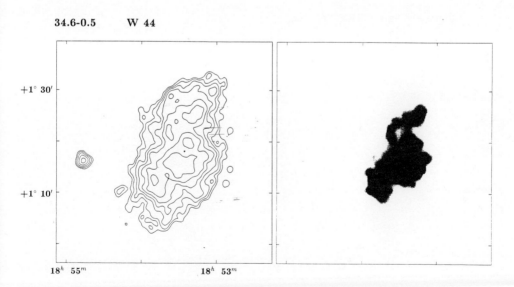

IPC, contour intervals are a factor of 1.4 in brightness.

49.2-0.7   W 51

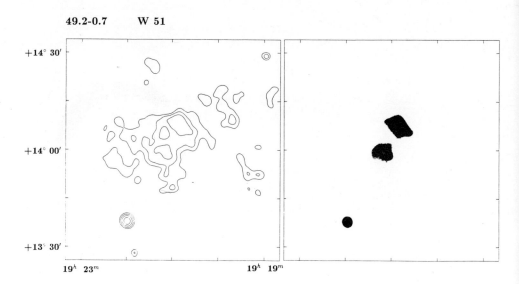

IPC, contour intervals are a factor of 1.5 in brightness.

296.5+10.0   PKS 1209-52

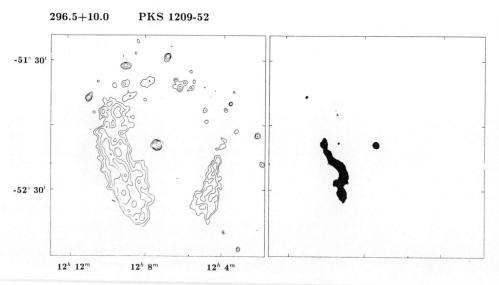

IPC, contour intervals are a factor of 1.5 in brightness.

EXOSAT OBSERVATIONS OF YOUNG SNRs

A. Smith

Space Science Department of ESA
ESTEC, NOORDWIJK, The Netherlands.

Introduction

In this work we study the 2 - 10 keV spectra of eight
relatively well known SNRs. These spectra were obtained
with the medium energy experiment on-board the X-ray
satellite EXOSAT. Details of the EXOSAT mission can be
found in Taylor et al 1981 and the ME experiment is
described in Turner, Smith and Zimmerman 1981. The ME
experiment is well suited for this study since it provides
a high sensitivity and a narrow field of view (45 ' FWHM).
These spectra are considered in the context of a model in
which the 2-10 keV continuum arises from shock heated
interstellar material.

Discussion

Fig. 1 shows the ME spectra of six of the eight objects
(Cas A, Tycho, Kepler, SN1006, G11.2-0.3, and RCW103. The
spectra obtained for RCW86 and W49B may be found in Klaas
et al 1987 and Smith et al 1985 respectively. A fuller
analysis of the total EXOSAT data for these objects may be
found in : Cas A - Jansen et al 1987; Tycho - Smith et al
1987a; Kepler - Smith et al 1987b; RCW103 - Peacock et al
1987; G11.2-0.3 - Peacock and Smith 1987. The data of
SN1006 has been independently examined in Jones et al 1984.
In fig.1 the spectrum above 3 keV has been fitted with a
simple thermal bremsstrahlung + emission line model. Below
3 keV the spectrum is far more complicated since it
includes emission lines due to S, Si, Fe-l (many lines) and
other elements. The situation is further confused by
uncertainties in the hydrogen column density and detailed
instrument response profile. For the ME this low energy
data appears as an unresolved excess.

The line often seen at about 6.5 keV is associated with Fe-
k emission. The exact energy and strength of this line
depends upon the temperature of the shock and the degree of
ionisation equilibrium between ions and electrons.
Throughout this work thermal equilibrium will be assumed
between the electron and ions, (i.e. the electrons are

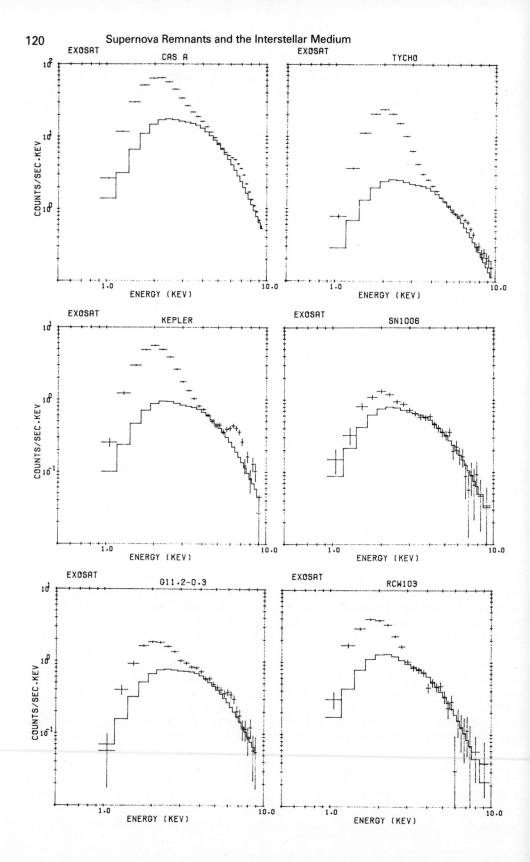

rapidly heated by the ions after the shock passage by some
non-coloumb process).   The results of this spectral fitting
are given in table 1.   In all cases the quality of the   fit
was   acceptable.   In   fig.1.   the   continuum fit   is   shown
without the inclusion of the line so that its strength   may
be   appreciated.   In   the brighter objects the fit   extends
beyond 10 keV but is not shown   in order to aid comparison.
It is interesting to note the relatively large variation in
the equivalent width of the Fe-k line,   (> 4000 eV in   W49B
to < 348 in SN1006).

Let   us examine first the x-ray continuum assuming that   it
arises   in shock heated ISM material and that the shock   is
just   the   primary   Sedov   shock.   Hamilton,   Sarazin   and
Chevalier   1983 is used as a basis for this   consideration.
We   will   adopt   the terminology used   in   this   paper,   in
particular η = E51*no^2 is the ionisation parameter. Taking
a   distance and   radius   of the shock   together   with   the
temperature   and strength of the continuum we can   estimate
the hydrodynamical parameters of the remnants. This is done
in table 2.   Note that the distance for Tycho is taken from
Braun 1985, other distance estimates etc. are referenced in
the table. The results seem quite reasonable. Note that the
line energies are generally in agreement   with observations
and   that the observed line strengths are never   less   than
that   prediced.   The   absence of an observable Fe   line   in
SN1006   is   just   related   to   the   low   ISM   density   and
consequently low η.

It   is also interesting to note that an   initial   explosion
energy   E51 = 0.184 .10^{51} (the weighted mean) is consistent
with all 8 objects.

Six   of   the   eight objects are   possibly   identified   with
historical   SN (the case for G11.2-0.3 is the weakest)   and
so   we know their ages.   These ages are compared with the
derived   hydrodynamical ages in   Table 3.   The agreement is
quite   good   except that for the youngest objects   (Cas   A,
Kepler and Tycho) the ages are over estimated and G11.2-0.3
seems   too   young to be associated with   SN386.   The   over
estimates   may   be   related to a period of   free   expansion
before   the onset of the Sedov phase.   Since the   estimated
values   of E51 are all consistent with a value of ~0.2   let
us use this   value to   refine the distance estimate   (and
other   hydrodynamical properties including age).   If   this
proves   to be a reliable way of determining distances   then
it   would be very useful since future x-ray   missions   will
include   imaging capabilities for this reason, with very high
sensitivity   and so greatly increase the number of   objects
in   this   sample.   The   distance   is   determined   from   the
following equation:-

$$D_{kpc} = (E51^2 * 0.0001098/(R^3 * F * Ts * exp(0.63/Ts)))^{1/5}$$

Where R is the radius in arc minutes,
Ts is the shock temperature in units of 10^7 degrees K and F
is the flux observed for E > 2keV (with n_H set to zero)

Note that the shock temperature is lower than the continuum
temperature by a factor of about 1.3.

Conclusions

The hypothesis that the 2-10 keV continuum is associated
with shock heated ISM material rather than reverse shock
heated enriched ejecta is consistent with observations for
these eight objects. It is interesting that the value of
E51 derived is usually around $0.2\ 10^{51}$ ergs and might
suggest a useful way of determining distances. However one
must be cautious since it is not certain that Cas A, Tycho
and Kepler are fully in a Sedov phase and so the estimate
of E51 for these objects may be in error. (For instance the
observed kinetic energy of the ejecta in Cas A is > 10^{51}
ergs! Jansen et al 1987)

Table 1.   SNR 2 - 10 keV spectral parameters.

SNR	Tc keV	Flux >2 keV ergs/cm^2/s * 10^{-11}	Fe-k EW eV	Fe-k E keV
Cas A	$3.74^{+.05}_{-.05}$	77.0	916^{+16}_{-16}	$6.62^{+.03}_{-.03}$
TYCHO	$6.5^{+.5}_{-.5}$	15.0	529^{+49}_{-10}	$6.48^{+.03}_{-.13}$
KEPLER	$5.3^{+3.0}_{-1.6}$	2.56	1910^{+640}_{-460}	$6.50^{+.05}_{-.14}$
RCW86	$3.36^{+.25}_{-.25}$	6.5	617^{+395}_{-395}	---
SN1006	$5.3^{+4.1}_{-1.8}$	3.9	<348	---
G11.2-0.3	$6.5^{+4.8}_{-2.4}$	$3.4^{+0.9}_{-0.4}$	513^{+574}_{-253}	$6.18^{+.30}_{-.33}$
RCW103	$3.1^{+2.5}_{-1.1}$	2.11	<1100	---
W49B	$1.81^{+.12}_{-.09}$	9.1	4700^{+200}_{-600}	$6.75^{+.01}_{-.04}$

Table 2. Hydrodynamical solutions

SNR	Cas A	Tycho
Diameter (′)	4.14 (+/- .2) [1]	7.54 (+/- .1) [1]
Distance (kpc)	2.90 (+/-.04) [1]	2.3 (+/- .2) [1]
Age (yrs) L_3	440 (402 - 480)	482 (438 - 531)
No (cm^{-3})	8.2 (6.5 - 9.1)	1.2 (1.1 - 1.3)
E51 (ergs)	.22 (.15 - .32)	.17 (.12 - .24)
neta51 η	15.2 (13.3-17.4)	.25 (.22 - .28)
Mswept (M$_o$)	6.6 (5.7 - 7.6)	2.9 (2.7 - 3.2)
EW (Fe-k) eV	~600	250
E (Fe-k) keV	6.59	6.48

	Kepler	RCW86
Diameter (′)	3.46 [1]	36 (+/- 2) [2]
Distance (kpc)	4.10 (+/- .9) [1]	2 (+/- 1) [3]
Age (yrs) L_3	437 (273 - 637)	2788 (1270-4586)
No (cm^{-3})	1.3 (1.0 - 1.9)	.120 (.087-.194)
E51 (ergs)	.086 (.020- .34)	.63 (.04 - 4.3)
neta51 η	.16 (.08 - .27)	.009 (.003-.019)
Mswept (M$_o$)	1.8 (1.2 - 2.3)	21 (3.5 - 59)
EW (Fe-k) eV	160 - 250	120 - 220
E (Fe-k) KeV	6.47 - 6.50	6.43 - 6.46

	SN1006	G11.2-0.3
Diameter (′)	31.4 (+/- 1) [4]	4.2 [5]
Distance (kpc)	1.5 (+/- 0.5) [4]	7.5 (+/- 2.5) [6]
Age (yrs) L_3	1450 (705 - 2461)	876 (443 -1460)
No (cm^{-3})	.10 (.06 - .17)	.77 (.50 - 1.21)
E51 (ergs)	.23 (.03 - 1.8)	.66 (.08 - 4.3)
neta51 η	.0024 (.0009-.0053)	.39 (.15 - .81)
Mswept (M$_o$)	4.9 (2.2 - 7.7)	11.1 (5.2 - 16.9)
EW (Fe-k) eV	<120	120 - 300
E (Fe-k) keV	<6.43	6.45 - 6.53

	RCW103	W49B
Diameter (′)	9.0 (+/- 1.0) [7]	3.6 (+/- 0.4)
Distance (kpc)	3.5 (+/- 0.5) [7]	4.8 (+/- 1.5) [8]
Age (yrs) L_3	1270 (722 - 2000)	911 (539 - 1363)
No (cm^{-3})	.43 (.24 - .74)	4.3 (3.1 - 6.4)
E51 (ergs)	.18 (.03 - 1.1)	.17 (.03 - .83)
neta51 η	.033 (.014 - .072)	3.2 (1.5 - 6.0)
Mswept (M$_o$)	6.3 (4.8 - 7.3)	10.3 (3.5 - 22.9)
EW (Fe-k) eV	150 - 400	1500
E (Fe-k) keV	6.44 - 6.50	6.55 - 6.56

[1] Braun 1985, [2] Caswell et al 1975, [3] Claas et al

1987, [4] Pye et al 1981, [5] Slee and Dulk 1974, [6]
Ilovaisky and Lequeux 1972, [7] Peacock et al 1987. [8]
Wilson 1970.

Table 3. SNR Ages and distances

SNR	True age yrs	Hydro. age yrs	Distance* kpc	Revised age* yrs
Cas A	307	440 (402 - 480)	2.58 - 2.75	(382 - 426)
Tycho	413	482 (438 - 531)	2.29 - 2.40	(467 - 517)
Kepler	381	437 (273 - 637)	5.09 - 5.86	(471 - 703)
RCW86	1800	2800 (1270 - 4586)	1.15 - 1.28	(1538-1805)
SN1006	979	1450 (706 - 2461)	1.19 - 1.48	(952 -1661)
G11.2	1599?	876 (443 - 1460)	3.86 - 4.97	(396 - 653)
RCW103	-	1270 (722 - 2000)	3.00 - 4.04	(849 -1769)
W49B	-	911 (539 - 1363)	4.59 - 5.32	(806 -1067)

* Assumes $E_o = 0.184 * 10^{51}$ ergs.

References

Braun, R, 1985, Thesis, Sterrewacht, Leiden, Netherlands.
Caswell, J.L., Clark, D.H., and Crawford, D.F., Austr. J.
Phys. Suppl., 37,39.
Claas, J. et al 1987, in preparation.
Hamilton, A.J.S., Sarazin, C.L. and Chevalier, R.A. 1983,
Ap.J.Suppl.,51, 115.
Ilovaisky, S.A., and Lequeux, J., 1972, Astr. Astro. 18,
169.
Jansen, F., Smith, A., Bleeker, J.A.M., deKorte, P.A.J.,
Peacock, A. and White, N.E., 1987 submitted to Ap.J.
Jones, L.R., Pye, J.P. and Culhane, J.L. Proc. Int. Symp.
on X-ray Astronomy, Bologna, 1984, p321.
Peacock, A., Smith, A., de Vries, J, Jansen, F., de Korte,
P.A.J. and Bleeker, J.A.M., 1987, in preparation.
Peacock, A. and Smith A. in preparation.
Pye, J.P., Pounds, K.A., Rolf, D.P. Seward, F.D., Smith, A.
and Willingale, R., M.N.R.A.S., 194, 569, 1981.
Slee, O.B., and Dulk, G.A., 1974, Galactic Radio Astronomy,
ed. F.K. Kerr and S.C. Simonson III, Reidel press p347
Smith, A., Jones, L.R., Peacock, A., and Pye, J.P. Ap.J.
296, 469, 1985.
Smith, A. Davelaar, J. Peacock, A. Taylor, B.G., Morini, M.
and Robba, N.R., 1987a, submitted to Ap.J.
Smith, A. Peacock, A. Arnaud, M. Ballet, J., Rothenflug, R.
and Rocchia, R. 1987b, in preparation.
Taylor, B.G., Andresen, R.D., Peacock, A. and Zobl, R.,
1981, Space Sci. Rev., 30, 479.
Turner, M.J.L., Smith, A. and Zimmerman, H.U., 1981, Space
Sci. Rev., 30, 513.
Wilson, T.L. 1970, ApJ. Letts. 7, 95.

X-RAY OBSERVATIONS OF SNR E0102.2-72.2 IN THE SMC

John P. Hughes
Harvard–Smithsonian Center for Astrophysics
Cambridge, Massachusetts 02138 USA

Abstract

The supernova remnant (SNR) E0102.2-72.2 is the brightest in the Small Magellanic Cloud (SMC) at X-ray wavelengths. This object, which is remarkable because of its high velocity ($\sim$4000 km s^{-1}) oxygen-rich optical emission, appears to be similarly remarkable at X-ray wavelengths. The high resolution imager (HRI) data can be quite well described by a thick ring with a radius of $\sim$19$''$ (6 pc at a distance of 63 kpc). The imaging proportional counter (IPC) X-ray spectral data can be best fit by a *single* emission line of energy $\sim$0.9 keV. It seems likely that this is the emission from a plasma of almost pure neon.

Introduction

SNR E0102.2-72.2 was discovered during the course of the *Einstein* IPC X-ray survey of the SMC (Seward and Mitchell 1981). The optical counterpart was discovered by Dopita, Tuohy, and Mathewson (1981) and shortly thereafter an optical velocity map was constructed by Tuohy and Dopita (1983). Follow up X-ray observations with the HRI were carried out by Inoue, Koyama, and Tanaka (1983). No higher resolution X-ray spectral observations have been made.

X-ray Morphology

The high resolution imager (HRI) X-ray data were summed in radial rings about the SNR center; figure 1 shows this average surface brightness profile and the best fit model for the emission. The data were compared to a simple geometric ring model characterized by three parameters: the outer radius (R_o), the radial thickness of the ring (ΔR), and its opening angle (θ). The plane of the ring was assumed to lie in the plane of the sky. With this parameterization a uniform spherical shell would have an opening angle of 90°. The models were convolved with the spatial spreading function of the HRI determined using a ground calibration image of a point source. In addition the statistical errors on the data were added in quadrature with a 1% systematic error.

A uniform spherical shell of emission yielded a reduced χ^2 of 6.66 for 17 degrees of freedom which can be rejected with high confidence. However, a thick ring with radius 19$''$, thickness 6.3$''$, and opening angle 67° gave an acceptable reduced χ^2 of 1.59 for 16 degrees of freedom (reject at between 90% and 95%). The observed emission at the center of the remnant, as well as beyond about 20$''$, can be attributed entirely to spreading due to the combined point response function of the HRI and *Einstein* X-ray mirrors. In particular there is no evidence

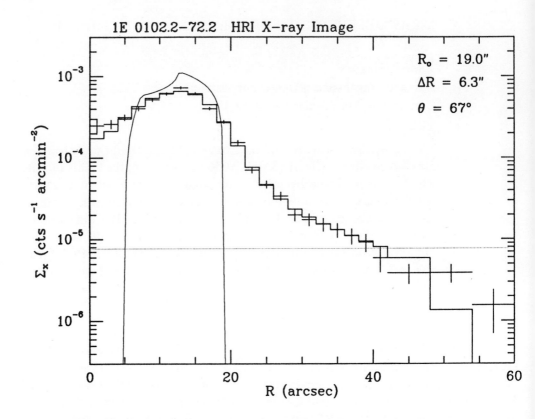

Fig. 1.–Plot of the average radial X-ray surface brightness profile of E0102.2-72.2. The thin solid line is the thick ring model seen in projection and the histogram is after convolving with the HRI point response function. The average background level is shown.

for any additional shocked component of interstellar or circumstellar material with a radius greater than 20″, although the precise limit on such a component depends on its assumed radius and X-ray spectrum.

Results of fits using the same model for the north, south, east and west quadrants confirm that the ring-like emission geometry persists throughout the remnant. There is some azimuthal variation in the fitted parameters, for example, a larger radius toward the north, but the only significant difference is in brightness (a factor of two variation from north to south).

X-ray Spectrum

The *Einstein* imaging proportional counter observed the SNR E0102.2-72.2 on three separate occasions. One observation occurred when the gain of the counter was rather high. Fits to the data for this pointing were inconsistent with the other observations and moreover yielded unacceptable χ^2 values for all models (the overall best fit had a reduced χ^2 of 3.97 for 9 degrees of freedom, which can

TABLE 1

Fits to IPC Data for E0102.2-72.2

Observations	Raymond and Smith Models			Single Line [a]	
	kT (keV)	N_H (10^{21} cm^{-2})	χ^2/ν	E (keV) [b]	χ^2/ν
I611, HRI	$0.53^{+0.18}_{-0.29}$	$1.7^{+4.6}_{-1.4}$	26.8 / 8	$0.887^{+0.026}_{-0.034}$	14.9 / 9
I7989	$0.41^{+0.39}_{-0.14}$	$1.5^{+3.2}_{-1.0}$	35.3 / 6	$0.863^{+0.017}_{-0.018}$	12.9 / 7

Notes:

[a] Width fixed at 0.1 keV.

[b] Statistical errors only. Error due to IPC gain uncertainty $\sim$0.1 keV.

be rejected at greater than 99.95%). Since the detector calibration is not well known for such high gain values, the results of this observation are considered suspect and will be neglected in the following discussion. The remaining two observations are consistent with each other.

The data were initially fit using solar abundance Raymond and Smith (1977) optically thin thermal emission models. Results of these fits are in Table 1. The best fit temperatures were about 0.5 keV and the best fit hydrogen column densities were $\sim$1.5$\times10^{21}$ atoms cm^{-2}. However these fits have unacceptably high χ^2 values and can be rejected at greater than the 99.9% confidence level. Acceptable fits can be obtained with a model of a single narrow emission line (with a width of less than $\sim$0.15 keV) at an energy of about 0.9 keV.

In order to investigate the significance of this result, the IPC data for a sample of several other SNRs were analyzed in the same manner as for E0102.2-72.2. The remnants chosen were the galactic remnants Kepler and Tycho, and the remnants N132D and N62A in the Large Magellanic Cloud. These represent a broad cross-section of remnant types both as regards morphology and spectrum. Although the spectra of these objects are known to be complicated by nonstandard elemental abundances and nonequilibrium ionization, nevertheless it is possible to obtain acceptable fits to the IPC data (for three of the four remnants) using solar abundance Raymond and Smith plasma models. Tycho's remnant, the sole exception, has very small statistical errors, covers a large spatial region of the detector ($\sim$10'), and has the most severe nonequilibrium ionization effects of the remnants considered. None of these remnants' spectra, however, can be fit by a single emission line model; the χ^2 values for this model are all unacceptable by rather large amounts. This analysis supports the proposed single emission line model for E0102.2-72.2.

What might be the origin of the 0.9 keV line emission for E0102.2-72.2? In the energy range around 0.9 keV the prominent emission lines are those of neon and iron. Specifically, for neon there are the forbidden, resonance, and intercombination lines of the helium-like ion (Ne IX) at about 0.91 keV and the Lyman α line of the hydrogen-like ion (Ne X) at 1.02 keV. In addition there are many iron L-shell transitions in this region. The brightest iron lines, at temperatures of about 0.5 keV, are at 1.01, 0.82, and 0.73 keV and occur in Fe XVII.

However, it seems unlikely for several reasons that the observed X-ray emission is coming from a pure iron plasma. First, the complex of iron L-shell lines covers a larger range in energy than do the neon lines and hence would be less likely to resemble a single narrow emission line. Second, the optical spectrum shows no evidence for any elemental constituents with a higher atomic number than that of neon, such as magnesium, silicon, or calcium. Hence a pure neon plasma is an attractive choice. Such a plasma under equilibrium ionization with a temperature of ~1 keV or less would have sufficiently strong line emission from the helium- and hydrogen-like species to satisfy the IPC data. In addition, higher temperature plasmas in a nonequilibrium state would have similarly strong line emission from these ions.

Conclusions

1. The X-ray emission of E0102.2-72.2 comes from a thick ring with an outer radius of 19" (~6 pc), radial thickness of 6.3" (~2 pc), and opening angle of 67°, which lies nearly in the plane of the sky. A uniform spherical shell is definitely precluded by the data.

2. The IPC X-ray spectrum is fitted best by a single narrow emission line at an energy of ~ 0.9 keV. This can best be described as the emission from a plasma of almost pure neon.

References

Dopita, M. A., Tuohy, I. R., and Mathewson, D. S. 1981, *Ap. J. (Letters)*, **248**, L105.

Inoue, H., Koyama, K., and Tanaka, Y. 1983, in *IAU Symp. 101, Supernova Remnants and Their X-Ray Emission*, ed. I .J. Danziger and P. Gorenstein (Dordrecht: Reidel), p. 535.

Raymond, J. C., and Smith, B. W. 1977, *Ap. J. Suppl.*, **35**, 419.

Seward, F. D., and Mitchell, M. 1981, *Ap. J.*, **243**, 736.

Tuohy, I. R., and Dopita, M. A. 1983, *Ap. J. (Letters)*, **268**, L11.

EINSTEIN BRAGG CRYSTAL SPECTROMETER OBSERVATIONS OF CAS A - A NONEQUILIBRIUM IONIZATION INTERPRETATION

T.H. Markert, P.L. Blizzard, C.R. Canizares
M. I. T., Cambridge, Mass., U.S.A.
J.P. Hughes
Center for Astrophysics, Cambridge, Mass., U.S.A.

Abstract: We use *Einstein* FPCS observations of lines of highly ionized neon, silicon and sulfur to constrain the parameters of the supernova remnant Cas A.

We observed the supernova remnant Cas A on several occasions in 1979 - 1981 with the Focal Plane Crystal Spectrometer (FPCS) on the *Einstein* Observatory (Canizares *et al.* 1979). The FPCS uses Bragg crystals to obtain high spectral resolutions (E/ΔE > 100) over the energy range 500 - 3500 eV. We used the FPCS to study features in the Cas A spectrum in the range 800 - 1250 eV (where the dominant lines are from highly ionized iron and neon) and around the energies of the lines of helium- and hydrogen-like silicon (1800 - 2100 eV) and sulfur (2400 - 2700 eV).

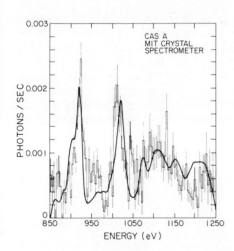

Figure 1 shows the spectrum we observed in the range 850 - 1250eV. Overlaid on the data is a model spectrum based on the observations of the sun compiled by Doschek and Cowan (1984). We have been unable to "fit" any simple model to the data, but show this result to illustrate that many of the lines and temperatures are known approximately although a detailed understanding of Cas A is not yet available.

Figure 1

It is not surprising that not all of the high resolution spectral data can be explained in terms of any simple model. The data contain a great deal of information and Cas A is known to be a complex object. For example, there are at least two regions in Cas A which are undergoing shock heating (the primary and reverse shocks). There are undoubtedly non-cosmic abundances in the X-ray emitting matter (because Cas A is a supernova remnant). It is a young object (about 300 years) so it is unlikely that ionization equilibrium has been established. The material is expanding at several thousand km s^{-1} so the lines are Doppler broadened. Finally, Cas A is clumpy - the irregular spatial appearance is probably indicative of spectral complexities as well.

In order to learn something about this object without attempting to explain every feature of the spectrum, we developed a straightforward and relatively simple non-equilibrium model which we applied to a few of the brighter and less ambiguous spectral features. In this model we assume that the X-ray emission from Cas A arises from a thin plasma which is instantaneously shock-heated to an X-ray emitting temperature T_e, (the electron temperature).

The first step in the model is to solve the ionization balance equations for each of the elements of interest, i.e.

$$\frac{dF_i}{dt} = n_e \times \{\alpha_{i-1}F_{i-1} - [\alpha_i + R_{i-1}]F_i + R_iF_{i+1}\} \qquad [1]$$

where F_i is the fraction of a given atomic species in the ith ionization state, α_i is the ionization rate by electron collision from state i (into state i + 1) and R_i is the recombination rate to state i (from state i + 1). The equations were solved using the method of Hughes and Helfand (1985). The rate coefficients are those used by Hughes and Helfand (which were taken from Raymond and Smith 1977 and subsequent revisions).

The ion structure in this model is a function of T_e and the ionization time parameter τ ($\equiv n_e$ [electron density] $\times$ t [time]). For a given ionization structure (corresponding to a particular value of T_e and τ) we then computed the expected emissivity for the X-ray lines of interest. The emissivity formulae were taken from Mewe and Gronenschild (1981) and included rate coefficients for processes of inner-shell ionization, electron-impact excitation, radiative recombination and dielectronic recombination. Once the emissivities are computed, the flux in a particular line is given by

$$f_i \;=\; V/4\pi d^2 \times \exp(-\sigma n_H) \times n_Z n_e \varepsilon_i(T_e,\tau) \qquad \text{photons cm}^{-2}\ \text{s}^{-1} \qquad [2]$$

where d   = distance to the object, V   = volume of emitting plasma, σ = cross-section at energy E_i for absorption of the X-rays by intervening matter (Morrison and McCammon 1983), N_H = hydrogen column density, n_Z = density of element Z, and ε_i = the emissivity of line i (photons $cm^3\ s^{-1}$).

Ideally, we would like to measure precise fluxes for a large number of lines and use the information to compute the various parameters on the right side of equation [2]. Since we only have a few flux measurements, however, we have chosen to take ratios of the measured values. If chosen wisely, many of the unknown parameters will cancel and the resulting ratio will be a function of only two uncertain parameters.

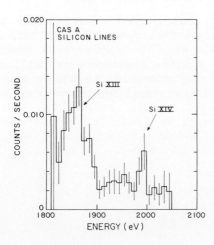

COUNTS / SECOND

CAS A
SILICON LINES

Si XIII

Si XIV

0.020

0.010

0

1800   1900   2000   2100

ENERGY (eV)

Figure 2

This technique is illustrated by our results for observations of the silicon lines. The portion of the Cas A spectrum between 1800 and 2100 eV is shown in Figure 2. The helium-like silicon (actually, a blend of the three n=2 to n=1 transitions of Si XIII) and the hydrogen-like silicon (Si XIV Lyman α) are indicated. We measured the flux of the two lines and computed the ratio.

We found $\dfrac{f_{Si\ XIV}}{f_{Si\ XIII}}$ = 0.42 ± 0.13

$$= \exp((\sigma(2006\ eV) - \sigma(1860\ eV))\ N_H)\ \frac{\varepsilon_{Si\ XIV}(T_e,\ \tau)}{\varepsilon_{Si\ XIII}(T_e, \tau)} \qquad [3]$$

For most plausible values of N_H, the first term in the ratio is nearly unity, so that the ratio is essentially a function only of T_e and τ. We generated a table of values of the Si ratios as predicted by our model for ranges of the parameters T_e and τ. Figure 3(a) shows a contour plot of the table generated for a large region of T_e and τ space. The contours are the 1σ and 2σ uncertainty intervals of the silicon line ratios. It is clear that the observations constrain the values of the plasma parameters. Note that at equilibrium (τ large), the electron temperature is in the range $6.8 < \log T_e < 7.4$.

We performed a similar analysis using observations of the sulfur lines (SXVI at 2521 eV and SXV at 2460 eV). The ratio of observed fluxes constrained the (T_e, τ) space to a region nearly identical to that of the silicon line ratio.

In figure 3(a) we show a small rectangle that overlaps our silicon-sulfur contours. This is the result obtained by Tsunemi *et al.* (1986) using the gas scintillation proportional counter (sensitive above about 1.5 keV) on *Tenma*. The *Tenma* group analyzed their data using a technique quite similar to the FPCS method presented here and were thus able to determine an allowed region of the (T_e,τ) parameter space. By using their measurement of the X-ray continuum above about 2 keV to find T_e directly, they were able to obtain a still smaller parameter region, as shown in the figure.

In Figure 3(b) we show a (T_e, τ) contour plot obtained from the ratio of fluxes of hydrogen-like to helium-like neon (Ne X at 1022 eV and the Ne IX complex in the range 905-922 eV). It is clear from figures 3(a) and (b) that the low-energy (neon) and the higher energy (silicon, sulfur, *Tenma*) results are inconsistent. We interpret the higher-energy (and, for most values of τ, the higher temperature) region as resulting from the primary shock moving into the interstellar medium. The lower-energy region we then interpret as resulting from the reverse shock propagating backward into the ejecta.

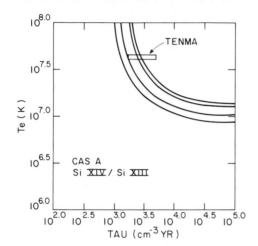 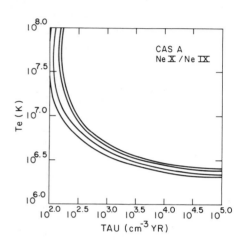

Figure 3   *Contours are 1 and 2σ regions of the non-equilibrium parameter space constrained by the observations of silicon lines (figure 3(a) on the left) and neon lines (figure 3(b) on the right). Observations of sulfur lines (not shown) are consistent with the silicon results. The results obtained with the Tenma satellite are shown in figure 3(a).*

This work was supported in part by NASA grant NAG 8-494.

REFERENCES

Canizares, C.R., Clark, G.W., Markert, T.H., Berg, C., Smedira, M., Bardas. D. Schnopper, H., and Kalata, K. 1979, *Ap. J. (Letters)*, 234, L33.

Doschek, G.A. and Cowan, R.D. 1984, *Ap. J. Suppl.*, 56, 67.

Hughes, J.P. and Helfand, D.J. 1985, *Ap. J.*, 291, 544.

Jansen, F.A., Smith, A., and Bleeker, J.A.M. 1987, this conference.

Raymond, J.C. and Smith, B.W. 1977, *Ap. J. Suppl*, 35, 419.

Mewe, R. and Gronenschild, E.H.B.M. 1981, *Ast. and Ap. Suppl.*, 45, 11.

Morrison, R. and McCammon, D. 1983, *Ap. J.*, 270, 119.

Tsunemi, H., Yamashita, K., Masai,K., Hayakawa, S., and Koyama, K. 1986, *Ap. J.*, 306, 248.

MODELS FOR X-RAY EMISSION FROM TYCHO'S REMNANT

Barham W. Smith and Eric M. Jones
Los Alamos National Laboratory
Los Alamos, New Mexico, USA

Abstract: We reexamine the X-ray emission from Tycho's
remnant using results from hydrodynamic models computed
with a detailed spherically symmetric code. The observed
synchrotron radio contours (Green and Gull 1983) appear to
require a cloudy circumstellar medium (Dickel and Jones, 1985;
Dickel, Eilek, and Jones 1987), thus we explore the X-ray
emission properties of similar models. We find that they tend
to produce broad shells of X-ray emission that resemble the
observed X-ray map of Tycho (Seward, Gorenstein, and Tucker
1983). A simple hydrodynamic model can satisfy both radio and
X-ray observations, but it has little similarity to the evolution
of remnants in cloudy media dominated by thermal conduction
(McKee and Ostriker 1977). More work needs to be done to
ensure that the spectrum as well as the X-ray map can be
modeled with the same cloudy circumstellar medium, although
we believe it will not be difficult to obtain as good a statistical
agreement with the spectral data as other models have
achieved (e.g. Hamilton, Sarazin, and Szymkowiak 1986).

Many authors have considered the effect of a cloudy medium on the
evolution of supernova remnants, and we will not repeat their arguments.
Instead we offer a progress report on a project to reproduce all the general
radio and X-ray features of remnants such as Tycho with a simple
hydrodynamic model that demonstrates the need for a cloudy circumstellar
medium.

Dickel, Eilek, and Jones (1987) have shown that many small cloudlets are
required in circumstellar space to reproduce the general morphology of the
radio maps of Tycho's remnant. We ask what the X-ray emission will be
from the models that satisfy the radio maps. By computing the
nonequilibrium ionization balance in every cell of a hydrocode on every
timestep, we can compute X-ray snapshots of the models both in spectrum
and surface brightness. We use improved versions of programs from the
atomic physics and X-ray spectrum code system of Raymond and Smith
(1977). A joint fitting process is required, since elemental abundances in
different parts of the remnant will lead to different X-ray emissivities per
gram in different parts of the remnant, so that a change in abundances to
fit the spectrum changes the map. This joint procedure has not yet been
carried out, but it is straightforward.

The hydrodynamic calculation discussed here is the same as discussed by
Dickel, Eilek, and Jones (this Colloquium), a presupernova stellar model
exploded into a cloudy interstellar medium. For details of the models, please

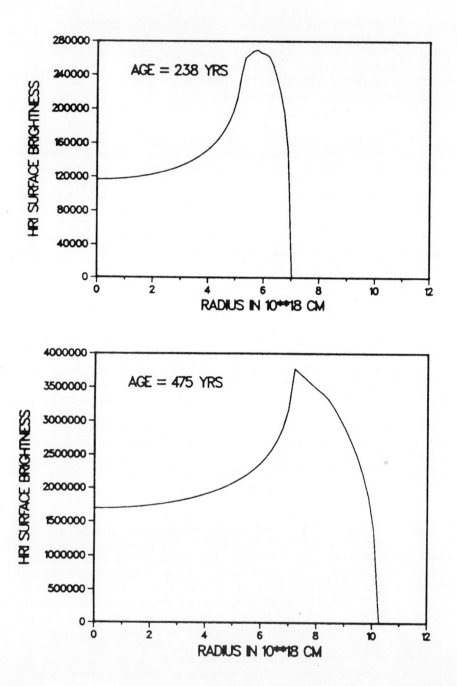

Figure 1:   X-ray surface brightness profiles computed from a hydrodynamic computation of the evolution of a supernova remnant in a cloudy medium. The units of surface brightness are arbitrary.   The time since explosion is shown in each panel.

see their paper, but note here that there is no thermal conduction in these models (cf. McKee and Ostriker 1977), since it does not necessarily affect the dynamics of remnants.   In the Figure we present the X-ray surface brightness of this model at two times, computed by folding the X-ray emissivity in each cell of the hydrocode (i. e. in each spherical shell) through the response of the High Resolution Imager on the Einstein Observatory.   The shell structure in these profiles is the region between the forward and reverse shocks, while the falloff toward the center has a shape dictated by the limb brightened shell.   These profiles can be compared to the HRI map of Tycho (Seward, Gorenstein, and Tucker 1983) for radial profiles.   The X-ray emission in the map is distinctly broadened in radius near the edge, a characteristic of cloudy models.   Our spherically symmetric model cannot reproduce the tangential clumping without further tricks to represent the average behavior of clumps, but we have modeled the overall features by including fine radial zones.

Several simple statements can be made about these models.   They should have no trouble satisfying the best X-ray spectral data on Tycho, which is probably the data from the Solid State Spectrometer on the Einstein Observatory (Becker et al. 1980) and from HEAO-1 (Pravdo and Smith 1979; Pravdo et al. 1980).   Recently Hamilton, Sarazin, and Szymkowiak (1986) gave a detailed discussion of fits to the spectral data, and we see no difficulty in principle to matching their goodness of fit.   However, in spite of the great effort made by those authors to fit the spectrum of Tycho, they comment that their model does not predict the observed expansion parameter of the remnant.   One way to ameliorate this serious problem is to replace models with homogeneous circumstellar media with cloudy models.

In summary, we have used a simple hydrodynamic model to show that it can satisfy all the following observed features of Tycho: The observed expansion rate at the current epoch, the general radio and X-ray morphology, and the detailed X-ray spectrum.

References

Becker, R. H., Holt, S. S., Smith, B. W., White, N. E., Boldt, E. A., Mushotsky, R. F., and Serlemitsos, P. J. 1980, Ap. J. (Letters), 235, L5.

Dickel,J. R., Eilek, J. A., and Jones, E. M. 1987, this Colloquium.

Dickel, J. R., and Jones, E. M. 1985, Ap. J., 288, 707.

Green, D., and Gull, S. 1983, in "Supernova Remnants and their X-ray Emission", ed. J. Danziger and P. Gorenstein (Dordrecht: Reidel).

Hamilton, A. J. S., Sarazin, C. L., and Szymkowiak, A. E. 1986, Ap. J., 300,713.

McKee, C. F., and Ostriker, J. P. 1977, Ap. J., 218, 148.

Pravdo, S. H., and Smith, B. W. 1979, Ap. J. (Letters), 234, L195.

Pravdo, S. H., Smith, B. W., Charles, P. A., and Tuohy, I. R. 1980, Ap. J. (Letters), 235,L9.

Raymond, J. C., and Smith, B. W. 1977, Ap. J. Suppl., 35, 419.

Seward, F., Gorenstein, P., and Tucker, W. 1983, Ap. J., 266, 287.

NON- EQUILIBRIUM IONISATION IN SUPERNOVA REMNANTS
- A case like Tycho -

W. Brinkmann and H.H. Fink
Max-Planck-Institut für Physik und Astrophysik, Institut für
Extraterrestrische Physik, 8046 Garching, F.R.G.

Abstract: X-ray observations of young supernova remnants
(SNR) provide the most direct tool to study their evolution,
their chemical composition, and their interaction with the
interstellar medium. We will show for a SNR with the characteristics of Tycho that great care has to be taken in interpreting spectral data obtained with X-ray detectors with low
spatial resolution.

I. Introduction

Observations of historical SNR of known age, especially those of type I
explosions with well defined circular appearance should allow direct conclusions about the explosion itself and the structure of both the star and the
surrounding interstellar medium. Although much progress has been made in
recent years, particularly by observations with the Einstein, EXOSAT and
Tenma satellites, there are still rather large discrepancies in the interpretation of the data. This is due to insufficient theoretical modelling of the
SNR as well as to deficiencies in current observational techniques. The
purpose of this paper is to demonstrate that high spectrally and spatially
resolved observations are necessary to get reliable information about the
remnant's properties.
As an example we use parameters for a SNR with the characteristic properties of Tycho, well studied by Gorenstein et al. (1983), Hamilton et al.
(1986), and Tsumeni et al. (1986). We follow the hydrodynamic evolution of
the remnant using a 1-D Lagrangean hydrocode with a mixed artificial
viscosity given by Wilkins (1980). The initial profile is that of an exploding
white dwarf, analytically approximated by Chevalier (1982), expanding into
a uniform ambient medium. The physical parameters were taken similar to
those of the above papers on Tycho, i.e., $E_0 = 7 \cdot 10^{50}$ ergs, $n_0 = 0.5$ cm^{-3}
and a mass of the ejecta of $M = 1.4 M_{\odot}$.
For both the ejecta and the interstellar medium we use solar abundances
for the 13 elements H, He, C, N, O, Ne, Mg, Si, S, Ar, Ca, Fe, and Ni
and assume temperature equilibrium between electrons and ions throughout
the calculation. For SNR of type I explosions, i.e. Carbon deflagration of
white dwarfs, ejecta material composed of mainly heavier elements is
expected (Nomoto et al., 1984). A proper modelling of the Tycho SNR
would require the use of ejecta material with greatly non solar abundances,
which is currently under way. However, in this case a different hydrodynamical structure of the remnant is expected as the electron pressure
becomes strongly dependent on the ionisation structure.
The ionisation structure was calculated with an implicit method at every
time step for every zone in the hydrocode (for details see Brinkmann and
Fink, 1987) using the rate coefficients given by Arnaud and Rothenflug
(1985). The hydrocode was run until the age of Tycho (~410 years) was
reached.

II. Equilibrium Ionisation

In a first step we assumed that the matter was everywhere in ionisation equilibrium at the corresponding temperature. The photon spectrum integrated over the whole remnant is shown in Figure 1a for the energy range from 1 to 10 keV.

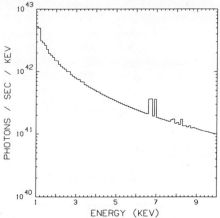

A best fit to this spectrum (excluding the line regions) with a thermal bremsstrahlungs law would result in a temperature of $kT \simeq 14$ keV. Fitting a thermal ionisation equilibrium spectrum to the data would result in a temperature of $kT \simeq 9$ keV. In both cases there is some excess at higher energies, which would be naturally interpreted as an additional hard component.

Fig.1a: Integrated equilibrium spectrum

Assume now that we have an X-ray detector with high spatial resolution and sufficient spectral resolution. Then line-of-sight spectra as given in Fig. 1b, c taken at the remnant's centre and near the rim, on the shock, could be obtained.

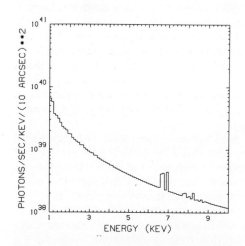

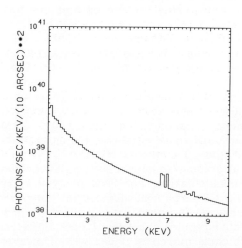

Fig.1b: Line-of-sight equilibrium spectrum near remnant's centre

Fig.1c: Line-of-sight equilibrium spectrum on shock

The differences in the spectra are evident and bremsstrahlung fits would result in temperatures of $T \simeq 5$ keV at centre and of $T \geq 20$ keV on the shock. This shows that fits to a spectrum integrated over the whole remnant (see above) give only limited information on the remnant's nature. However, none of these temperatures is representative for the remnant. The hydrodynamical simulations show that the shock- and reverse shock temperatures are different, the fitted values are just "weighted" temperature averages. In particular, the various line ratios in the spectrum could be interpreted as being due to non-equilibrium effects. Therefore, it seems hard to draw clear physical conclusions from X-ray observations with low spatial resolution, even for the idealized case where the emitted spectrum is locally an equilibrium spectrum.

III. Non-equilibrium Ionisation

In a second run we evolved the hydrodynamic structure simultaneously with the ionisation structure. As we used solar abundances, where H and He were always completely ionized, the local electron density is close to the equilibrium case and the dynamical evolution of the two models is nearly identical.
At the age of 410 years the ionisation structure in the outer regions of the remnant is in a very strong non-equilibrium state as can be seen from the total spectrum in Fig. 2a.

Bremsstrahlungs or thermal equilibrium fits with temperatures of $kT \simeq$ 10 keV would represent the continuum of the spectrum above $\geq$ 4 keV reasonably well, but the line strengths do not match such temperatures at all: the iron line at 6.7 keV is much stronger in equilibrium models, whereas the large equivalent widths of the S, Si, and Ar lines seem to indicate the presence of a much cooler gas with a temperature of $kT \simeq$ 1 keV.

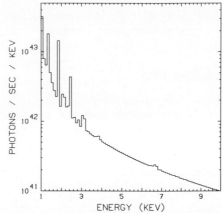

Fig.2a: Integrated non-equilibrium spectrum

The best way of getting a better insight into the physical parameters of the remnant would be to take spatially well resolved spectra at different radii from the remnant's centre. Figure 2b and 2c show two line-of-sight spectra, one taken at the centre (b), the other on the shock (c).

The continuum slopes are very complex and would require more component fits, which do not allow reliable conclusions on the remnant's temperature structure. A good diagnostic tool represents the iron line at about 6.7 keV as it is the dominant line for plasma temperatures above ~ 1 keV. Fig. 2b shows that, although the gas temperature at the shock is $\geq 10^8$ K, the iron line has not formed yet. In appears only further in, i.e. at times much later than the crossing time of the shock. Its maximum equivalent width is reached near the position of the reverse shock. Near the centre (fig. 2c) its equivalent width is smaller again due to averaging effects. As the radial variations of the equivalent widths of the lines correspond to their temporal evolution, i.e. their ionisation time $n_e t$, their measurement together with a well defined continuum flux give strong model constraints on the temperature-density-structure of the remnant.

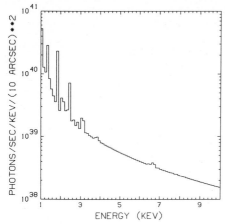 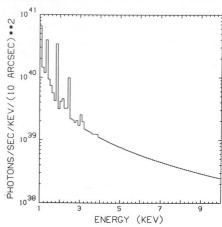

Fig.2b: Line-of-sight non-equilibrium   Fig.2c: Line-of-sight non-equili-
        spectrum near remnant's centre   brium spectrum on shock

IV. Conclusions

We have demonstrated that X-ray spectra integrated over the whole remnant do not reveal with the required accuracy the physical parameters of a SNR. Radially resolved spectra, together with a detailed hydrodynamic simulation and a simultaneous calculation of the ionisation structure seems to be the only way to obtain "physically reliable" parameters for the supernova remnants.

References
Arnaud, M., and Rothenflug, R.,1985, Astron.Astroph.Suppl. **60**, p.425.
Brinkmann, W., and Fink, H.H.,1987, in preparation
Chevalier, R.A.,1982, Ap.J. **259**, L59.
Gorenstein,P., Seward,F., and Tucker,W.,1983, Proc.IAU Symposium **101**
Hamilton,A.J.S., Sarazin,C.L., and Szymkowiak,A.E.,1986 Ap.J. **300**,p.713
Nomoto, K., Thielemann, F.K., and Yokoi, K.,1984, Ap.J. **286**, p.644
Tsunemi,H., Yamashita,K., Masai,K., and Hayakawa,S.,1986 Ap.J. **306**,p.248
Wilkins, M.L.,1980, J.Comp.Phys. **36**, p.281.

A HYDRODYNAMICAL MODEL OF KEPLER's SUPERNOVA REMNANT CONSTRAINED BY X-RAY SPECTRA

J.Ballet[1], M.Arnaud[1], J.P.Chieze[2], B.Magne[3], R.Rothenflug[1]

1-Service d'astrophysique, CEN Saclay
   91191 GIF SUR YVETTE CEDEX, FRANCE
2-Centre d'etudes de Bruyeres le Chatel
   BP 12, 91680 BRUYERES LE CHATEL, FRANCE
3-Centre d'etudes de Limeil Valenton
   BP 27, 94190 VILLENEUVE ST GEORGES, FRANCE

Abstract: We present NEI hydrodynamical models of Kepler's SNR compared with the EXOSAT and EINSTEIN X-ray spectra.

I-INTRODUCTION

The remnant of the historical supernova observed by Kepler in 1604 was recently observed in X-rays by the EXOSAT satellite up to 10 keV. A strong Fe K emission line around 6.5 keV is readily apparent in the spectrum (Smith et al, 1987). From an analysis of the light curve of the SN, reconstructed from historical descriptions, Baade (1943) proposed to classify it as type I. Standard models of SN I based on carbon deflagration of a white dwarf predict the synthesis of about 0.5 M_Θ of iron in the ejecta. Observing the iron line is a crucial check for such models.

Alternatively Dogget and Branch (1985) argue that the light curve of SN II-L is very similar to that of SN I and that the original observations are compatible with either type. In view of this uncertainty we have run a hydrodynamics-ionization code for both SN II and SN I remnants.

II-HYDRODYNAMICAL MODEL

The detailed comparison between theory and observation is far from straightforward: the hydrodynamical evolution of the shocked interstellar medium and of the 'reverse shocked' ejecta, together with the ionization structure throughout the SNR, must be computed in order to be able to construct synthetic spectra to be compared to the observations (e.g. Itoh, 1977; Nugent et al, 1984).

Here we use a 1-D lagrangian hydrodynamical code. A detailed description is given in a forthcoming paper (Ballet et al, 1987). Our model assumes an initially constant density in the ejecta, spherical symmetry, a uniform ambient medium and temperature equipartition between electrons and ions. The ionization structure is followed using the ionization and recombination rates of Arnaud and Rothenflug (1985). The emission model is from Mewe et al (1985). Synthetic spectra are computed separately for each element, allowing their abundances to vary.

Indeed we always assume that heavy elements are trace elements:
- For an SN II remnant we assume homogeneous ejecta with 90% hydrogen + 10% helium in number.
- On the other hand current SN I models predict pure heavy elements in the ejecta with a stratified structure (e.g. Nomoto et al, 1984). Hamilton et al

(1985a,b), using an analytical approximation for the hydrodynamics, already worked on such a picture and compared it favorably with the two other presumably type I historical supernovae remnants (SN 1006 and Tycho). In this preliminary work we consider 1.4 $M_\odot$ of pure Helium ejecta. This assumption simplifies the computations (no feedback between the ionization structure and the density and temperature) and ensures a nearly correct A/Z ratio. A more sophisticated model is in progress.

III—OBSERVATIONAL CONSTRAINTS

The age of the Kepler SNR is evidently known (380 years). Its angular radius is measured from X-ray and radio maps: about 100". We fixed the interstellar column density to $2.8\ 10^{21}$ Hcm^{-2} (Danziger and Goss, 1980).

The observed X-ray spectra are depicted in figures 1a and 1b: the spectrum obtained by the SSS on board the EINSTEIN observatory (Becker et al, 1980) and the spectrum obtained with the Medium Energy proportional counter of EXOSAT (Smith et al, 1987). The two low energy points on figure 1b correspond to the fluxes obtained with the LE+CMA experiment, with the Lexan 3000 and Boron filters. The EXOSAT instrument response we adopted is depicted in Smith et al (1987) and the SSS one in Ballet et al (1987): in particular we let as a free parameter the amount of ice on the SSS.

The main features of these spectra are:
1—the Fe L lines below 1.3 keV, main component in the CMA energy range.
2—the Si and S lines around 1.85 and 2.4 keV, whose centroids and strengths are well determined due to the good resolution of the SSS.
3—a cool continuum
4—a hard continuum (essentially free-free emission for a standard medium)
5—the Fe K lines (around 6.5 keV)
Only these last two components are important above 4 keV and the shape and strength of this continuum together with the centroid and intensity of the line are determined by the EXOSAT observation.

Therefore the quality of the fit will be very sensitive to the following model outputs: temperature and density of the hottest component (the main shock), the ionization structure of Si,S,Fe and their abundances.

This combined analysis can of course constrain the model better than previous ones based on the SSS data alone (Hughes and Helfand, 1985).

III—2 MAIN RESULTS

SNII model: We adopted the ejecta mass: $M_* = 5\ M_\odot$.
The continuum at 6.5 keV is essentially due to bremsstrahlung on fully stripped hydrogen and helium and can be computed by means of a pure hydro code. It depends on 2 free parameters: n_0 (outside density) and E_0 (explosion energy), the age and angular radius of the SNR being independently 'known'. Thus a relation between n_0 and E_0 is rapidly obtained by comparing such computations with the observed value. Then we made full hydro – ionization computations for 3 cases consistent with that first constraint: (0.7cm^{-3},7.10^{50}ergs); (0.7cm^{-3},10^{51}ergs); (0.55cm^{-3},1.4 10^{51}ergs).
Our main conclusions are:
—The explosion energy must be high enough ($E_0 > 10^{51}$ergs) to supply a reverse shock hot enough to produce a sufficient continuum at low energy (below 4 keV).

- Si and S are rapidly ionized in the ejecta for such high values of E_0, the temperatures increasing with the explosion energy. The centroids of the Si

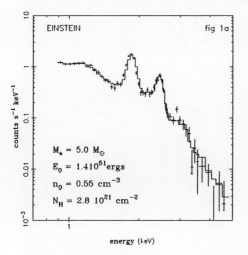

EINSTEIN   fig 1a

$M_* = 5.0 M_\odot$

$E_0 = 1.4 10^{51}$ ergs

$n_0 = 0.55$ cm^{-3}

$N_H = 2.8 \ 10^{21}$ cm^{-2}

counts s^{-1} keV^{-1}

energy (keV)

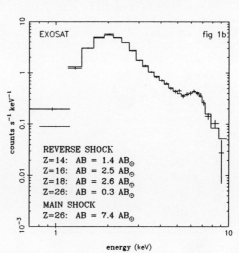

EXOSAT   fig 1b

REVERSE SHOCK
Z=14:   AB = 1.4 AB$_\odot$
Z=16:   AB = 2.5 AB$_\odot$
Z=18:   AB = 2.6 AB$_\odot$
Z=26:   AB = 0.3 AB$_\odot$
MAIN SHOCK
Z=26:   AB = 7.4 AB$_\odot$

counts s^{-1} keV^{-1}

energy (keV)

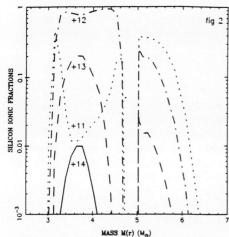

fig 2

+12

/+13

+11

+14

SILICON IONIC FRACTIONS

MASS M(r) (M$_\odot$)

and S K lines originating in this medium are higher than observed. This is illustrated in figure 1a, which depicts the best fit obtained for the $E_0=1.4 \ 10^{51}$ ergs model, assuming normal abundances in the ambient medium and letting free the ejecta ones. Figure 2 shows the silicon ionic profile obtained. Notice that Si (and S) is less ionized in the main shock due to a higher ionization delay (lower density). Therefore a better fit is obtained if we allow an overabundance of Si and S in the outside medium , but values as high as 8 and 10 times solar are required. An alternative is that the lines are redshifted. A redshift of .75% (v= 2250 km s^{-1}, consistent with the shock velocity) is required.

- The ratio FeK/FeL is very low, Fe being underionized in most layers. A correct ratio is only obtained if iron is overabundant in the main shock (see figure 1). Indeed, in spite of a stronger ionization delay, the main shock temperature is so high that the FeK/FeL ratio is larger there than in the reverse shock.

SNI model: The conclusions are essentially the same. We checked the following cases consistent with the continuum flux at 6.5 keV: $n_0=0.5$cm^{-3}, $E_0=0.5,0.8,1.4 \ 10^{51}$ ergs and $n_0=0.65$cm^{-3}, $E_0=0.3 \ 10^{51}$ ergs.
-The reverse shock continuum is stronger for this helium plasma and always accounts for the continuum observed at low energy.
-Si and S are more ionized than in the SNII model, excluding models with $E_0>5.10^{50}$ergs. Figure 3 shows the best fit spectrum for $E_0= 3.10^{50}$ergs

assuming normal abundances in the ambient medium and letting free the ejecta ones. Notice that the energy centroids of the Si and S lines are still too high. No reasonable redshift can lower their centroids enough.

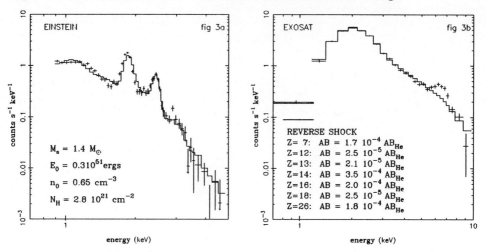

- A very faint FeK line is produced in that case (see figure 3b). As above, an overabundance of Fe (~10 times the solar value) is required in the outside medium to match the observation.

IV- CONCLUDING REMARKS

The mean ionization of a given element depends on its spatial distribution, as each shell has a different ionization history. In the present models, iron was uniformly distributed inside the ejecta. In this case, the observed FeK/FeL ratio implies unrealistic ambient overabundances. Relaxing this assumption, we presently study models in which iron is confined in the inner shells. Preliminary results are encouraging.

REFERENCES

Arnaud M., Rothenflug R., 1985, Astr.Ap.Suppl.,60, 425
Baade W., 1943, Ap.J., 97, 119
Ballet J., Arnaud M., Chieze J.P., Magne B., Rothenflug R., 1987, Astr.Ap., to be submitted
Becker R.H., Boldt E.A., Holt S.S., Serlemitsos P.J., White N.E., 1980, Ap.J.(Letters), 237, L77
Dogget J.B., Branch D., 1985, Astron.J., 90, 2303
Hamilton A.J.S., Sarazin C.L., Szymkowiak A.E., 1986a, Ap.J., 300, 698
                                                  1986b, Ap.J., 300, 713
Hugues J.P., Helfand D.J., 1985, Ap.J., 291, 544
Itoh H., 1977, Publ.Astron.Soc.Japan, 29, 813
Mewe R.,Gronenschild E.H.B.M.,Van den Oord G.H.J.,1985,Astr.Ap.Suppl.,62,197
Nomoto K., Thielemann F.K., Yokoi K., 1984, Ap.J., 286, 644
Nugent J.J, Pravdo S.H., Garmire G.P., Becker R.H., Tuohy I.R., Winkler P.F., 1984, Ap.J., 284, 612
Smith A., Peacock A., Arnaud M., Ballet J., Rothenflug R., Rocchia R., 1987, Ap.J., to be submitted

THE INFLUENCE OF EQUILIBRATION ON THE X-RAY INTENSITY OF SEDOV MODELS

D. Jerius and R. G. Teske
Department of Astronomy
University of Michigan
Ann Arbor, Michigan

Abstract: We have synthesized a set of SNR X-ray and coronal iron surface brightness profiles to determine obser- vational constraints to help distinguish how the remnants' ion-electron energy equilibration time-scales compare to their ionization time-scales. The profiles are based on Sedov-Taylor blast wave models under the two scenarios (1) equal T_{ion} and T_e or (2) Coulomb equilibration of T_e and T_{ion}. Resultant spectra were convolved with *Einstein* HRI and IPC sensitivity curves to simulate satellite observa- tions of actual remnants.

Introduction: Previous studies of X-ray emission from SNR's in the adiabatic phase have indicated the importance of the ion-electron energy equilibration process behind the shock front in determining the electron- ion temperature structure. Models of the Cygnus Loop (Raymond *et al.* 1983; Fesen and Itoh 1985) favor equipartition by Coulomb interactions while models for Tycho's remnant (Hamilton *et al.* 1986b) and SN1006 (Hamilton *et al.* 1986a) indicate the presence of a possible rapidly equilibrated component. Complicating the picture is the possibility of reverse shocks setting up electron populations of different temperatures. A cooler electron component is also present due to secondary ionizations.

In order to study these processes, we have synthesized a set of X-ray and forbidden coronal iron surface brightness profiles. Teske (1984) pointed out that the forbidden coronal iron lines are a good indicator of which equilibration process is at work. However, it is predicted that in most remnants the [Fe X] red line will be weak or undetectable even though the [Fe XIV] green line is detectable. In these cases, the X-ray emission might prove a suitable substitute for the red line. In addition, the difference in spectral sensitivity between the various imaging instruments might shed some light on the problem (Hamilton and Sarazin 1984) since the equilibration processes tend to produce distinguishable spectra.

Models: We have computed a set of blast wave models for the shock speeds v_s = 275, 400, 525, 650 km/s and n_0 = 0.01, 0.1, 1 cm^{-3}. The models include time-dependent, non-equilibrium ionization of C, N, O, Ne, Mg, Si, S, Ca, Fe, Ni, Ar, Al, and Na. The models were used to generate X-ray spectra (both line and continuous emission) at 99 posi- tions along a projected radius of a remnant. The spectra were then convolved with instrumental responses and interstellar extinction to produce synthetic surface brightness profiles. Similar profiles were produced for the two coronal lines.

Results: We find that at low shock speeds, profiles for the two equilibration processes are essentially identical. At higher speeds, the
electron and ion temperatures grow more disparate in the Coulomb equilibration models, producing a lower degree of ionization and thus a softer
spectrum behind the shock front in the Coulomb case. However, because
the spectrum observed at a point in the remnant is the integral of the
spectra at all points (at different radii) along that line of sight, the
resultant brightness profiles depend upon conditions across the whole
remnant. Geometrical factors can mask any differences.

Unfortunately the X-ray profiles for extinction free models
(Fig. 1) are only slightly sensitive to the equilibration processes
(their peak values varying by less than 50%), so that fitting observed
profiles to models may not give dependable results. Since the *Einstein*
IPC and HRI have different spectral sensitivities, a comparison of peak
intensities as seen by the two instruments might lead to the differentiation of the equilibration processes. We find, however, that only in
cases with interstellar extinction is there an appreciable difference,
and then only for shocks with $v_S > 400$ km/s and low or moderate densities
(Fig. 2).

The applicability of the iron line emissions to this problem
depends on whether or not they are observable. Lucke *et al.* (1979)
find a lower limit of detectibility of 2×10^{-8} erg cm^{-2} s^{-1} sr^{-1} by
aperture photometry while Teske and Petre (1987) quote a limit of
1×10^{-7} ergs cm^{-2} s^{-1} sr^{-1} as done by CCD observations. In terms of the
synthesized profiles, the [Fe X] line is observable only at lower shock
speeds with $n_0 > 0.2$ cm^{-3} (see Fig. 3). In this region the [Fe XIV]
green line is observable for $400 \leq v_S \leq 650$ km/s (both of these predictions assume no interstellar extinction). Comparisons of peak iron to
X-ray intensities indicate that they might differentiate between the
equilibration process in this region. Ambiguities due to the influence
of interstellar extinction on the X-ray emission may compromise this
procedure, since shocks with different speeds and equilibration processes
may mimic each other. We believe, however, that with high signal to
noise observations or independent measurements of the shock speeds that
the iron line intensities will prove invaluable. Examination of the
spectra produced at the location of maximum X-ray emission reveals that
several lines of N-like ions, Li-like Ne VIII and Ne-like Fe XVII are
fairly sensitive to the equilibration process and may serve as yet
another means of determining which process is at work.

Application to Observed Remnants: We have applied our X-ray models to
two remnants, the Cygnus Loop and Puppis A. For the Cygnus Loop, we
used the IPC data of Ku *et al.* (1984) and found that our best fit to the
data used blast parameters determined by the optical study of Fesen and
Itoh (1985); $v_S = 220$ km/s, $n_0 = 0.54$ cm^{-3} and an interstellar column
density of 4×10^{20} cm^{-2} (see Fig. 4). With these parameters, there is no
distinction between the two types of equilibration. Our attempts to fit
the data with Ku's parameters were not at all successful, leading us to
believe that the IPC pulse-height spectrum is sensitive to different

shock speeds than the integrated IPC radial profiles (mainly because the pulse-height spectrum emphasizes $E_{h\nu} \geq 0.2$ keV, which will favor higher speed shocks).

Data for Puppis A were kindly provided by R. Petre. We fitted a model to a radial cut taken through regions studied by Szymkowiak (1985) with the *Einstein* SSS, choosing a profile running across the shock front at position angle 42° through Szymkowiak's region N. We find that models using Szymkowiak's spectra-based parameters do not fit the radial profile very well. This is attributable to both departures in spherical symmetry in Puppis A and to the high probability that the ISM is non-uniform (Petre *et al.*, 1982; Teske and Petre 1987).

Conclusion: Preliminary comparisons of our X-ray models to published data have not produced a concrete differentiation between the two ion-electron energy equilibration processes. Comparisons with other data are planned. The additional information provided by further observations of SNR's in the forbidden coronal iron lines should provide enough information to distinguish these processes.

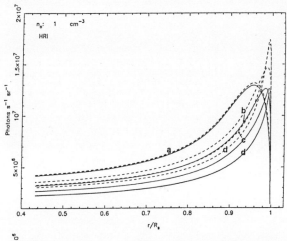

Fig 1. X-ray surface brightness profiles, no interstellar extinction. Solid curves; rapid equil. Dashed curves: Coulomb equil. Shock speeds: a 275 km/s; b 400 km/s; c 525 km/s; d 650 km/s.

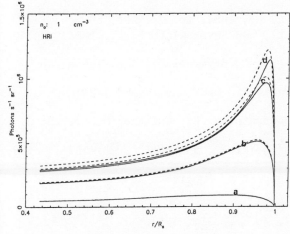

Fig 2. X-ray surface brightness profiles, with an interstellar column density of 3×10^{21} cm^{-2}. Other details are the same as in Figure 1.

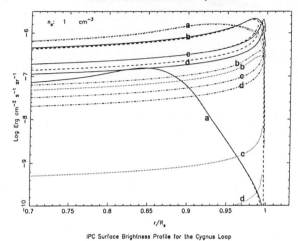

Fig 3. Forbidden coronal iron surface brightness profiles. Solid: [Fe XIV], rapid equil. Dashed: [Fe XIV], Coulomb equil. Dotted: [Fe X], rapid equil. Dash-dotted: [Fe X], Coulomb equil. Shock speeds are labelled as in Figure 1.

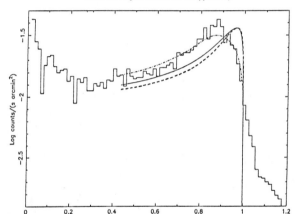

Fig 4. Theoretical IPC profiles for the Cygnus Loop. Histogram: Data of Ku *et al.* (1984). Models with Ku's parameters -- solid curve: Coulomb equil.; dashed: rapid equil. Model with V_s = 220 km/s, n_o = 0.54 cm^{-3} -- dash-dot curve.

References

Fesen, R. A., and Itoh, H. 1985, *Ap. J.*, 295, 43.

Hamilton, A.J. S., and Sarazin, C. L., 1984, *Ap. J.*, 284, 601.

Hamilton, A. J. S., Sarazin, C. L., and Szymkowiak, A. E., 1986a, *Ap. J.*, 300, 698.

Hamilton, A. J. S., Sarazin, C. L., and Szymkowiak, A. E. 1986b, *Ap. J.*, 300, 713.

Ku, W. H.-M., Kahn, S. M., Pisarski, R., and Long, K. S. 1984, *Ap. J.*, 278, 615.

Lucke, R. L., Zarnecki, J. C., Woodgate, B. E., Culhane, J. L., and Socker, D. G., 1979, *Ap. J.*, 228, 615.

Raymond, J. C., Blair, W. P., Fesen, R. A., and Gull, T. R. 1983, *Ap. J.*, 275, 636.

Szymkowiak, A. E. 1985, "X-Ray Spectra of Supernova Remnants", NASA Technical Memorandum 86169.

Teske, R. G. 1984, *Ap. J.*, 277, 832.

Teske, R. G., and Petre, R. 1987, *Ap. J.*, 314, 673.

THE X-RAY EMISSION FROM SN 1987A AND THE REMNANT OF SN 1572 (TYCHO)

Hiroshi Itoh[1], Kuniaki Masai[2], and Ken'ichi Nomoto[3]

[1]Department of Astronomy, University of Kyoto, Sakyo-ku, Kyoto 606, Japan
[2]Institute of Plasma Physics, Nagoya University, Chikusa-ku, Nagoya 464, Japan
[3]Department of Earth Science and Astronomy, University of Tokyo, Meguro-ku, Tokyo 153, Japan

Abstract: The thermal X-ray emission from SN 1987A may be enhanced to a detectable level when the blast shock hits the circumstellar medium which has formed in the red-supergiant stage of the progenitor.

The X-ray spectrum of Tycho observed with the satellite Tenma can be explained approximately within the context of a carbon deflagration model for Type Ia supernovae, if the ejecta are assumed to be mixed partially.

SN 1987A: A theoretical analysis of the light curve of SN 1987A indicates that the supernova progenitor had a fairly small radius of the order of 10^{12} cm (Shigeyama et al. 1987). It is likely that the progenitor once evolved to a red supergiant, lost a substantial fraction of its envelope material, and contracted. We have calculated the dynamical evolution and nonequilibrium X-ray emission of a supernova remnant (SNR) in such an environment, using a spherically symmetric hydrodynamic code (Itoh 1977). We assume that SN 1987A has exploded in a cavity, which is bounded at a radius R_c by the wind material coasting radially with a speed, v_w, of 10 km s^{-1} and a mass loss rate, $\dot{M}_w$, of 5 x 10^{-5} $M_\odot$ yr^{-1}. By assuming that the explosion occurred at a time t_c since the end of the intense mass loss, R_c is written as $v_w t_c$. The value of t_c is very uncertain, and tentatively set equal to 10^3 yr in the present calculation. The ejecta are assumed to have a mass of 5 $M_\odot$ and an initial kinetic energy of 2 x 10^{51} erg. The density distribution in the outer regions of the initial ejecta is taken to be proportional to r^{-7}, where r is the distance from the centre of the star, and is truncated at a point where the expansion speed reaches $(v_{ex} =)$ 2 x 10^4 km s^{-1}. The initial expansion of the ejecta is assumed to be homologous. Nonequilibrium ionization and X-ray emission are calculated for H, He, C, N, O, Ne, Na, Mg, Al, Si, S, Ar, Ca, Fe, and Ni. The rate coefficients are taken mainly from Arnaud and Rothenflug (1985) and Mewe et al. (1985). Only ions are assumed to be heated substantially across the shock front and electrons are assumed to be heated through Coulomb collisions with the ions.

The calculated time evolution of the 2-20 keV luminosity is shown in figure 1. The blast shock reaches R_c at a time, t = $t_h \sim$ $(v_w/v_{ex})t_c$, since the explosion. It can be shown that the volume emission measure of the shocked wind attains a maximum of about $(\dot{M}_w/mv_w)^2/4\pi R_c$ with m being the mean atomic mass, when the blast-shock

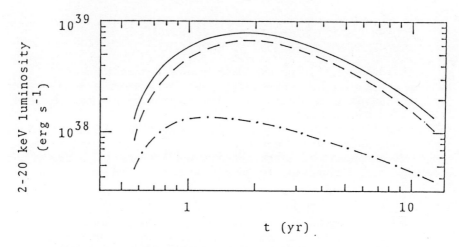

Figure 1:   Time evolution of the 2-20 keV luminosities of the whole remnant (——————), the ejecta (— — .—), and the wind material (— . — . —).

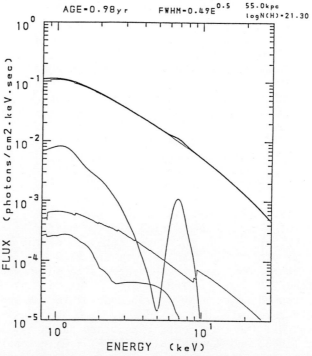

Figure 2:   The spectrum at t = 0.98 yr. The total, bremsstrahlung, recombination, two-photon, and line emissions are shown. The total and line spectra are convolved with a Gaussian to a resolution of 0.49 $E^{0.5}$ keV (FWHM), where E is the photon energy in keV. The distance and interstellar hydrogen column density to the supernova are assumed to be 55 kpc and 2×10^{21} cm^{-2}, respectively.

radius is about $2R_c$. If the expansion is nearly free, this occurs at t
$\sim 2t_h$.   The reverse-shocked ejecta are several times as luminous in
this energy range as the blast-shocked wind, owing to the relatively
high densities. Figure 2 shows the X-ray spectrum at t = 0.98 yr. The
bremsstrahlung dominates the spectrum.   The line emission is generally
weak, because abundant metal elements are ionized nearly completely.
The Fe K-line blend is an exception, and can be used for plasma
diagnostics.   The thermal X-ray emission can be distinguished from
emissions of interior origins (e.g., ^{56}Co decay or a pulsar), because
the spectrum of the former is less affected by photoelectric absorption
than those of the latter.   The enhanced X-ray emission may be
detectable by the satellite Ginga if the maximum volume emission
measure mentioned above is large enough. More detailed discussion will
be presented elsewhere (Itoh et al. 1987).

The Remnant of SN 1572 (Tycho):   Type Ia supernovae are likely to be
due to the deflagration of a C/O white dwarf.   However, this has not
been fully confirmed by the observations, in particular at   X-ray
wavelengths, of SNRs.   Using the numerical code described above, we
have calculated the dynamical evolution and nonequilibrium X-ray
emission of an SNR on the basis of model W7 constructed by Nomoto et
al. (1984) and Thielemann et al. (1986).   The results have been
compared with the observations of the remnant of SN 1572 (Tycho).

The observed outer radius and thickness of the reverse shock wave
(Seward et al. 1983) and the observed continuum spectrum (Tsunemi et
al. 1986) are reproduced approximately for an age of 411 yr, by
assuming an ambient density of 1.3 amu cm^{-3}.   Then the distance to
Tycho is obtained as 2.43 kpc.   It is found that the calculated
equivalent width of the Fe K-line blend is sensitive to the mixing of
the ejecta.   Since the reverse shock has not propagated deep into the
iron-dominated inner regions of the ejecta, transport of iron atoms
into the outer regions by mixing results in a significant increase in
the Fe line emission.   We assume that the composition of the ejecta
with the initial velocities less than 1.33×10^4 km s^{-1} is homogenized
by mixing.   The upper boundary of the velocity range corresponds
approximately to the location at which the convective deflagration
front freezes. As shown in figure 3, the resultant spectrum agrees well
with the spectrum observed with the satellite Tenma (Tsunemi et al.
1986), if the interstellar hydrogen column density, N_H, to Tycho is
taken to be about 1×10^{22} cm^{-2}. For a conventional value of $N_H = 3 \times 10^{21}$
cm^{-2}, the calculated photon fluxes below 2.5 keV are larger than the
observed fluxes by up to a factor of 1.8.   In either case, the carbon
deflagration model is indicated to apply to Tycho's supernova.   The
numerical results will be discussed in more detail elsewhere.

Acknowledgments: Numerical computations were made at the Institute of
Plasma Physics of Nagoya University and at the Data Processing Centre
of Kyoto University.

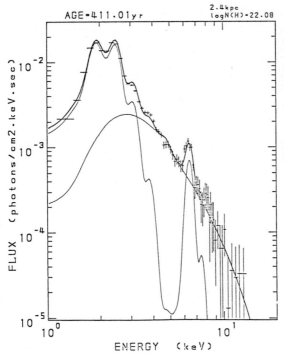

Figure 3: A model spectrum for Tycho. The total, continuum, and line emission are convolved with the efficiency and response function of the GSPC aboard Tenma. The distance and interstellar hydrogen column density to Tycho are assumed to be 2.43 kpc and 1.2×10^{22} cm^{-2}, respectively. The crosses denote the spectrum observed with Tenma (Tsunemi et al. 1986).

References

Arnaud, M., and Rothenflug, R. 1985, Astron. Astrophys. Suppl., 60, 425.

Itoh, H. 1977, Publ. Astron. Soc. Japan, 29, 813.

Itoh, H., Hayakawa, S., Masai, K., and Nomoto, K. 1987, submitted to Publ. Astron. Soc. Japan.

Mewe, R., Gronenschild, E.H.B.M., and van den Oord, G.H.J. 1985, Astron. Astrophys. Suppl., 62, 197.

Nomoto, K., Thielemann, F.-K., and Yokoi, K. 1984, Astrophys. J., 286, 644.

Seward, F., Gorenstein, P., and Tucker, W. 1983, Astrophys. J., 266, 287.

Shigeyama, T., Nomoto, K., Hashimoto, M., and Sugimoto, D. 1987, submitted to Nature.

Thielemann, F.-K., Nomoto, K., and Yokoi, K. 1986, Astron. Astrophys., 158, 17.

Tsunemi, H., Yamashita, K., Masai, K., Hayakawa, S., and Koyama, K. 1986, Astrophys. J., 306, 248.

PLASMA DIAGNOSTICS WITH X-RAY LINES
OF OXYGEN IN PUPPIS A

Kathryn F. Fischbach, Claude R. Canizares,
Thomas H. Markert, Joan M. Coyne
M.I.T., Cambridge, Mass., U.S.A.

Abstract: High resolution X-ray spectral observations of
Puppis A were performed with the FPCS on *Einstein*. We use
plasma diagnostics of lines from OVII and OVIII to constrain
the values of temperature, ionization timescale, and
hydrogen column density.

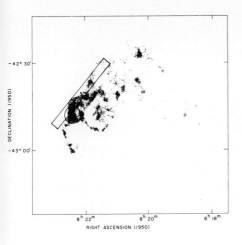

Figure 1

During November and December
of 1979, the shock front of the
Puppis A supernova remnant was
observed using the Focal Plane
Crystal Spectrometer (FPCS) on the
Einstein Observatory (Canizares *et
al.* 1979). A 3 × 30 arc minute
aperture was aligned with the shock
front, and RAP and TAP crystals were
used to scan the energy ranges near
the lines of hydrogen-like oxygen (O
VIII Lyman α, 630-670 eV and O VIII
Lyman β, 750-800 eV) as well as the
resonance, intercombination and
forbidden lines (n=2 to n=1
transitions) of helium-like oxygen
(O VII 550-580 eV).

Figure 1 shows the FPCS aperture position for the observation of
the shock front (rectangle) as well as the aperture position used for an
observation of a bright knot (Winkler *et al.* 1983, circle) overlaid on
the Puppis A image obtained by Petre *et al.* (1981) with the *Einstein*
HRI. The raw FPCS spectra for the shock front appear in Figure 2, below.
Note the differing background levels for the different observations.

The X-ray emission lines detected provide useful diagnostics of
conditions in the shock front. For transitions i → g and j → k of
ionization states +x and +y of an element Z, the ratio of X-ray fluxes
is given by:

$$\frac{f_{ig}}{f_{jk}} = \frac{\Omega_{ig} \times n_{+x} \times \exp(-\sigma(E_{ig})N_H) \times \exp(-E_{ig}/(kT_e))}{\Omega_{jk} \times n_{+y} \times \exp(-\sigma(E_{jk})N_H) \times \exp(-E_{jk}/(kT_e))}$$

where Ω_{ig} is the effective collision strength for transition $i \rightarrow g$ (see Vedder *et al.* 1986), E_{ig} is the transition energy, T_e the electron temperature, n_{+x} the density of ion x, σ the cross-section per hydrogen atom for photoelectric absorption at the energy E_{ig}, and N_H the hydrogen column density. It is difficult to make very many precise measurements so as to be able to solve for all of the unknowns in the problem. The approach we have taken is to measure a few carefully chosen line intensities well so that as many parameters as possible will cancel.

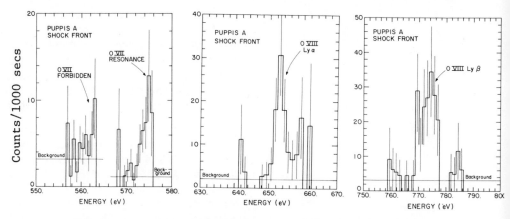

Figure 2 - FPCS Observations of Puppis A Shock Front

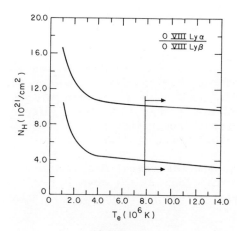

Figure 3

For example, a comparison of two shock front lines from the same ion, O VIII Ly α and O VIII Ly β, yields an allowed region in the parameter space of column density N_H and electron temperature T_e (see Figure 3). We have used collision strengths from Shull (1981). We do not have a second ratio that serves to further confine the possible values of N_H, but, as shown below, we are able to exclude temperatures below $\approx 7.9 \times 10^6$ Kelvin, so that N_H is restricted to a range of 3 to 10×10^{21} cm^{-2}.

The region allowed by the data is between the two curved lines and to the right of the vertical line.

In order to use the other oxygen lines in Figure 2 to learn more about the shock front, it is necessary to perform a non-equilibrium analysis. This is required because the relative abundances of the various oxygen ions, which are functions of the time since the plasma was shocked, are parameters in the formulae for line fluxes. (For the analysis summarized in Figure 3 a non-equilibrium approach is unnecessary since both lines are from the same ion and collisional excitation will dominate in all plausible conditions.)

The ionization structure is determined by solving a set of $Z + 1$ simultaneous differential equations. The flux ratios are then determined by using the resulting ion abundances explicitly in the equations for the line emissivities. We have used the technique of Hughes and Helfand (1985) to solve the ion balance equations. The complete analysis technique is discussed in more detail in Markert *et al.* (1987).

The line ratios of the oxygen lines constrain the various parameters of the non-equilibrium model. For the lines we have chosen these are T_e, N_H, and τ. Here τ ($\equiv$ electron density $\times$ time since the shock) is the ionization timescale and measures the extent to which ionization equilibrium has been attained.

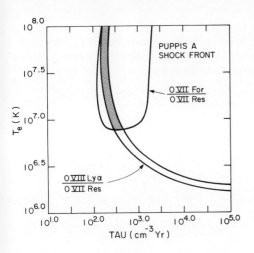

Figure 4

In Figure 4 we show $\pm 1\sigma$ contours in (T_e, τ) parameter space which show the regions allowed by FPCS measurements of the line ratios O VIII Lyman α to O VII Resonance and O VII Forbidden to O VII Resonance. The overlapping (shaded) region is approximately the 90% confidence region for the values of T_e and τ. For this analysis we have assumed $N_H = 6 \times 10^{21}$ cm^{-2}, consistent with Figure 3, although in fact the results for (T_e, τ) are not particularly sensitive to the choice of N_H.

In Table 1 (below) we list ranges of the parameters we have determined in the analyses shown in Figures 3 and 4. We also include values for the parameters determined previously for two other regions of Puppis A: a bright knot of emission along the shock front (Winkler *et al.* 1983) and a region in the interior of the remnant (Winkler *et al.* 1981).

Table 1

	N_H $(10^{21} cm^{-2})$	$T_e (10^6 K)$	$n_e \times t$ (yr cm^{-3})
Shock Front[1]	3 to 10	$\geqslant 7.9$	150 to 400
Eastern Bright Knot[2]	≈ 2	≈ 7	≈ 1000
Interior[3]	2 to 6	$\geqslant 1.5$	–

1 This paper
2 Winkler *et al.* 1983
3 Winkler *et al.* 1981

Petre *et al.* 1981, have concluded that the eastern knot is a region where the shock front has encountered a cloud of density ≈ 10 cm^{-3}, implying (from the table) a time since shocking of 100 years. If the eastern knot and the shock front have the same age (a reasonable assumption because of the proximity of the two features), then we conclude that the density of the shock front is in the range 1.5 to 4.0 cm^{-3}.

We thank Leah Bateman, Peter Vedder and Meg Urry for assisting in manuscript preparation, data analysis and for helpful discussions. We are indebted to Jack Hughes for supplying his computer code. This work was supported in part by NASA grant NAG 8-494.

REFERENCES

Canizares, C.R., Clark, G.W., Markert, T.H., Berg, C., Smedira, M., Bardas. D., Schnopper, H., and Kalata, K. 1979, *Ap. J. (Letters)*, 234, L33.

Hughes, J.P. and Helfand, D.J. 1985, *Ap. J.*, 291, 544.

Markert, T.H., Blizzard, P.L., Canizares, C.R. and Hughes, J.P. 1987, this conference.

Petre, R., Canizares, C.R., Kriss, G.A. and Winkler, P.F. 1981, *Ap. J.*, 258, 22.

Shull, J.M. 1981, *Ap. J. Suppl.*, 46, 27.

Vedder, P.W. Canizares, C.R., Markert, T.H. and Pradhan, A. 1986, *Ap. J.*, 307, 269.

Winkler, P.F., Canizares, C.R., Clark, G.W., Markert, T.H., Petre, R. 1981, *Ap. J.*, 245, 574.

Winkler, P.F., Canizares, C.R., and Bromley, B.C. 1983, in *Supernova Remnants and Their X-ray Emission*, eds. J. Danziger and P. Gorenstein, (Dordrecht: Reidel), 245.

X-RAY IMAGES OF PKS1209-52 AND ITS CENTRAL COMPACT X-RAY SOURCE

Yutaka Matsui
Dept. of Physics & Astronomy, Northwestern Univ.
Evanston, IL 60201 USA

Knox S. Long
Center for Astrophysical Sciences, Johns Hopkins Univ.
Baltimore, MD 21218 USA

Ian R. Tuohy
Mount Stromlo and Siding Spring Observatories,
Australian National Univ., ACT 2606, Canberra, Australia

Abstract: A complete X-ray image of the SNR PKS1209-52 (= G296.5+10.0) was obtained with the IPC and HRI on the *Einstein Observatory*. The remnant has a shell-like X-ray morphology much like its appearance at radio wavelengths, while a compact X-ray source is clearly detected near the center of the remnant. The flux observed from the X-ray nebula $F_{(0.1-4.5\ keV)}$ is 8×10^{-11} ergs cm^{-2} s^{-1}, which corresponds to a luminosity $L_{(0.1-4.5\ keV)} = 8 \times 10^{35}$ ergs s^{-1} for a distance of 2 kpc. Applying a simple shell model to the X-ray emission distribution, we derived an ambient interstellar medium $n_0 = 0.08$ H atoms cm^{-3}, total X-ray emitting plasma mass 150 M$\odot$, and thermal energy 1.2×10^{50} ergs. The flux from the compact X-ray source $F_{(0.15-4.5\ keV)}$ is $\sim 2 \times 10^{-12}$ ergs cm^{-2} s^{-1}. There are no obvious optical counterparts brighter than $m_V \sim 22$ within the 3.3" radius HRI error circle. If the object is a hot neutron star, the HRI/IPC count rate ratio implies a surface temperature of 1.6×10^6 K for $N_H = 3.2 \times 10^{21}$ cm^{-2}.

Introduction: The SNR PKS1209-52 was first detected as a soft X-ray source with *HEAO-1* by Tuohy et al. (1979). They fitted the X-ray pulse height distribution with a model spectrum of cosmic abundance plasma at collisional equilibrium (Raymond and Smith 1977) with a temperature of 1.9×10^6 K and a column density of 3.2×10^{21} H atoms cm^{-2}. Applying a simplified Sedov solution, they derived an ambient ISM density ~ 0.4 cm^{-3}, an initial supernova energy 7×10^{50} ergs, and an age of 20000 years for a distance of 2 kpc. Recently Kellett et al. (1987) observed the bright eastern portion of the remnant using the *EXOSAT Observatory* and obtained a soft X-ray image and a spectrum. The X-ray spectrum was fitted with the Raymond and Smith thin hot plasma model with solar abundances. The best fit was obtained for a temperature of 1.7×10^6 K and a column density of 1.4×10^{21} atoms cm^{-2}. Thus PKS1209-52 is the SNR whose X-ray emitting plasma indicates the lowest characteristic temperature ever observed.

The distance to PKS1209-52 is poorly known since it has been estimated only from the radio Σ - D relation. The distance estimates are in the range from 1.1 kpc (Milne 1979) to 1.9 kpc (Caswell and

Lerche 1979). On the other hand, Mills (1983) has suggested that the distance to galactic SNRs is more accurately given by $d_{kpc} = (1280/S_{408 \text{ MHz}})^{1/2}$, where $S_{408 \text{ MHz}}$ is the total flux at 408 MHz in Jy. This gives a distance of 3.9 kpc for PKS1209-52. We adopt a distance of 2 kpc for the analysis following.

Einstein Observations and Results: Observations of PKS1209-52 were carried out with the Imaging Proportional Counter (IPC) and the High Resolution Imager (HRI) on the _Einstein Observatory_ (Giacconi et al. 1979). Eight IPC exposures were obtained to map the total extent of the SNR; exposure times ranged from 1570 to 2880 sec. In addition, a follow-up 5930 sec. observation of the SE limb was obtained. Using only the central 36' x 36' of each field to avoid severe vignetting, the individual images were merged into a single map shown in Figure 1. The remnant appears as a heavily fragmented but roughly circular ring at X-ray wavelengths. In addition, there is a compact X-ray source near the center of the ring. The point-like nature of this source was subsequently confirmed with the HRI. The HRI image (Figure 2) was obtained on January 22, 1981 with 3930 sec. exposure. The compact source is located at $\alpha = 12^h 07^m 23.^s 50$, $\delta = -52°09'49"$ (1950) with an error radius of 3.3" (90% confidence). The IPC count rate was 0.085 ± 0.005 cts s^{-1} on July 11, 1979 and the HRI rate was 0.015 ± 0.002 cts s^{-1}. The IPC count rate corresponds to an energy flux ~2 x 10^{-12} ergs cm^{-2} s^{-1} between 0.15 and 4.5 keV. The X-ray image is similar to the radio map at 5 GHz obtained by Milne and Dickel (1975). The only noticeable difference between the X-ray and radio images is in the NW, where there is very weak X-ray emission despite the moderately strong radio emission. As there is a correlation between X-ray and radio surface brightnesses in SNRs (Matsui et al. (1984); Berkhuijsen (1986)), the X-ray appearance of PKS1209-52 is mostly intrinsic. The outer edge of the X-ray and radio emission and the optical filaments in the north roughly follow a circle of radius 36' centered at $\alpha = 12^h 06^m 47.^s 3$, $\delta = -52°12'29"$ (1950), which we identify as the geometrical center of the remnant.

Spectral analysis of our IPC data without reprocessing is difficult due to the complex spatial gain variation of the detector. So we adopt the plasma model and the interstellar absorption derived from the _HEAO-1_ data. For this model the IPC count rate of 3.61 cts s^{-1} for the remnant (excluding the point source) corresponds to an energy flux of 7.8 x 10^{-11} ergs cm^{-2} s^{-1} between 0.1 and 4.5 keV, and an intrinsic X-ray luminosity $L_{(0.1-4.5 \text{ keV})} = 7.8 \times 10^{35}$ ergs s^{-1}. For the adopted temperature and N_H, apparent X-ray emissivity $\varepsilon_x = 5.0 \times 10^{-14}$ IPC cts cm^5 s^{-1} (see Figure 11 of Leahy et al. 1985). Taking the geometrical center as the center of the SNR, we obtained the radial X-ray surface brightness distribution of the entire remnant. For a uniform density shell, the observed maximum limb-brightening of 3.7 corresponds to a ratio of shell thickness to outer radius $\Delta R/R = 0.16$. Since the maximum surface brightness, 1.23 x 10^{-3} cts arcmin^{-2} s^{-1}, is at R - ΔR = 29', we have R = 35' and ΔR = 6'. Then we have a density in the shell $n_e = 0.23$ cm^{-3} and an ambient medium density $n_o = 0.08$ H cm^{-3}. The mass of the plasma producing the total X-ray emission is 150 M⊙ and the thermal energy content $E_{th} = 1.2 \times 10^{50}$ ergs.

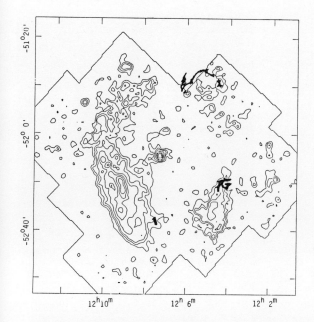

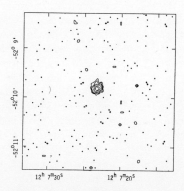

Fig. 2. HRI image of the central compact X-ray source.

Fig. 1. Smoothed IPC image of PKS1209-52. Contours are 0.6, 1.1, 1.6, 2.6, 3.6, 4.6, 5.6, and 6.6 x 10^{-3} cts arcmin^{-2} s^{-1}. FWHM of the pont spread function is ~2.8'. The positions of optical nebulosity are indicated by dark areas.

<u>The Compact X-ray Source near the Center of PKS1209-52</u>: The HRI error circle of the point X-ray source is shown in Figure 3, overlaid on the ESO blue plate. Only one star (designated A) is visible within the error circle, 2.2" to the north of the X-ray centroid. We obtained a low resolution (~10 Å) spectrum of this star using the Anglo-Australian Telescope on June 18, 1981. The spectrum of star A corresponds to that of a normal K subdwarf with V = 17.5, U-B = 1.4, B-V = 1.2, and V-R = 0.5. An identification of star A with the point X-ray source is not plausible in view of the ratio log(F_X/F_v) = 1.0, which is at least 3 orders of magnitude greater than that expected from a K star (Vaiana et al. 1981). A recent radio map of PKS1209-52 obtained at a frequency of 843 MHz with 40" beam sets an upper limit of 4 mJy for a source at the position (Roger 1986). A pulsar search made by Manchester, D'Amico, and Tuohy (1985) yielded an upper limit of 1 mJy at 1.4 GHz. For typical pulsar parameters (e.g. Tuohy et al. 1983), the limit implies a radio luminosity between 10^8-10^9 Hz of < 1.2 x 10^{28} ergs s^{-1}. This value is a factor of 3 below that of the weakest radio pulsar (PSR1509-58) associated with a SNR (see Table 1 in Tuohy et al. 1983).

Helfand and Becker (1984) estimated a probability of 0.04 for finding an X-ray source of the observed intensity in an area of the SNR. However, this probability reduces to only 0.0006 for finding an X-ray source within 6.1' of the geometrical center. For an assumed age of 20000 y, the transverse velocity required to move 6.1' is ~170 km s^{-1}, which is not at all unreasonable compared with measured pulsar

velocities of 20-500 km s^{-1} (e.g.
Helfand, Chanan, and Novick 1980).

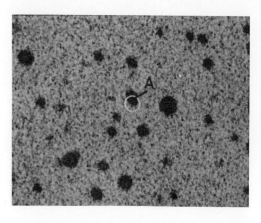

A hot neutron star emitting
blackbody radiation is a viable
possibility for the X-ray source.
Here we adopt the model of a
neutron star with a stellar radius
R_o = 16.1 km and a mass M = 1.31
M⊙ proposed by Pandharipande,
Pines, and Smith (1976). The
gravitational redshift factor at
the surface of the neutron star
$1 + z = (1 - 2GM/c^2R_o)^{-1/2}$ = 1.15.
We calculated the expected IPC and
HRI count rates from a hot neutron
star and found that the observed
HRI/IPC count rate ratio is consistent with a surface temperature
T_o = 1.6 x 10^6 K if N_H = 3.2 x
10^{21} cm^{-2}. The surface temperature
derived above is close to the best-

Fig. 3. HRI error circle of the
compact X-ray source superposed
on the ESO blue plate.

fit blackbody temperature of 1.8 x 10^6 K derived by using the *EXOSAT
Observatory* (Kellett et al. 1987). Such a temperature is commensurate
with the surface temperature of a neutron star predicted by Nomoto and
Tsuruta (1986) even for an age of ~2 x 10^4 years.

References
Berkhuijsen, E.M. 1986, Astr. Astrophys., **166**, 257.
Caswell, J.L. and Lerche, I. 1979, M.N.R.A.S., **187**, 201.
Giacconi, R. et al. 1979, Ap. J., **230**, 540.
Helfand, D.J. and Becker, R.H. 1984, Nature, **307**, 215.
Helfand, D.J., Chanan, G.A., and Novick, R. 1980, Nature, **283**, 337.
Kellett, B.J. et al. 1987, M.N.R.A.S., **225**, 199.
Leahy, D.A., Venkatesan, D., Long, K.S., and Naranan, S. 1985, Ap. J.,
  294, 183.
Manchester, R.N., D'Amico, N., and Tuohy, I.R. 1985, M.N.R.A.S.,
  212, 975.
Matsui, Y., Long, K.S., Dickel, J.R., and Greisen, E.R. 1984, Ap. J.,
  287, 295.
Mills, B.Y. 1983, Supernova Remnants and their X-ray Emission,
  ed. J. Danziger and P. Gorenstein (Dordrecht: Reidel), p. 551.
Milne, D.K. 1979, Aust. J. Phys., **32**, 83.
Milne, D.K. and Dickel, J.R. 1975, Aust. J. Phys., **28**, 209.
Nomoto, K. and Tsuruta, S. 1986, Ap. J. (Letters), **305**, L19.
Pandharipande, V.R., Pines, D., and Smith, R.A. 1976, Ap. J., **208**, 550.
Raymond, J.C., and Smith, B.W. 1977, Ap. J. Suppl., **35**, 419.
Roger, R.S. 1986, Private Communication.
Tuohy, I.R., Garmire, G.P., Manchester, R.N., and Dopita, M.A. 1983,
  Ap. J., **268**, 778.
Tuohy, I.R. et al. 1979, Ap. J. (Letters), **230**, L27.
Vaiana, G.S. et al. 1981, Ap. J., **245**, 163.

LINE EMISSION PROCESSES IN ATOMIC AND MOLECULAR SHOCKS

J. Michael Shull
Center for Astrophysics and Space Astronomy, and
Joint Institute for Laboratory Astrophysics
University of Colorado and National Bureau of Standards

ABSTRACT. This review discusses the observations and theoretical models of interstellar shock waves in diffuse and molecular clouds. After summarizing the relevant gas dynamics, atomic, molecular and grain processes, and physics of radiative and magnetic precursors, I describe observational diagnostics of shocks. I conclude with a discussion of two new topics: unstable or non-steady shocks and thermal conduction in metal-rich shocks.

1   INTRODUCTION

Because the physics of interstellar shock waves has recently been been the subject of a comprehensive review (Shull and Draine 1987), this review will concentrate on selected aspects of shock line emission, models, and diagnostics. In particular, I will cover the basic physics of atomic shocks in diffuse clouds and multi-fluid MHD shocks with magnetic precursors, and then turn to some recent problems of non-steady shocks and the role of thermal conduction in metal-rich shocks associated with young supernova remnants.

The environments of these shocks range in density from diffuse clouds, which contain mostly atomic gas ($n_H = 0.1 - 10^3$ cm^{-3}) and cool by optical and ultraviolet emission lines of H, He, and atomic ions, up to molecular clouds ($n(H_2) = 10^3 - 10^8$ cm^{-3}) which cool by optical, infrared, sub-millimeter, and millimeter wavelength lines of atoms and molecules. Emission lines are widely used by astronomers as diagnostics of density, temperature, abundance, and excitation mechanism (*e.g.*, shock excitation vs. photoionization). The techniques of observation have resulted in a natural division between shocks in diffuse atomic gas and dense molecular clouds. In this review, I follow a similar division, based on the theoretical distinctions which govern diffuse and molecular shock models. In §2, I discuss single-fluid shocks in diffuse gas, the gas dynamic, atomic, molecular, and grain processes involved in their study, and the relevant observations. In §3, I discuss multi-fluid MHD shocks with magnetic precursors. In §4, I discuss two recent topics: non-steady shocks and the role of thermal conduction.

2   SINGLE-FLUID SHOCKS IN DIFFUSE CLOUDS

2.1   Gas Dynamics

A shock is sometimes described as a "hydrodynamic surprise". A fluid element is suddenly accelerated from an initial pre-shock velocity to a post-shock velocity. Thus, shocks are supersonic disturbances driven by thermal pressure (a "piston"), radiative acceleration, or other mechanical sources. The most common sources of high velocity gas in the interstellar medium (ISM) are supernova remnants (SNRs), stellar winds, molecular outflows from pre-main-sequence stars, and infalling H I (21-cm) clouds.

At this point, we should make an important distinction between "adiabatic" and "radiative" shock waves. The sudden, discontinuous jump in density, flow velocity, and temperature is characteristic of an adiabatic shock. No energy is lost in the front (hence the term adiabatic), and the gas is heated to a large temperature subject to the constraints of mass and momentum conservation. All single-fluid shocks contain an adiabatic shock transition. If the post-shock gas can radiate away its energy in a time short compared to the flow time, the temperature drops and the post-shock gas is compressed to maintain approximately constant total pressure – this forms the radiative shock wave. A typical compression in a strong, radiative shock, limited by magnetic pressure, is about a factor of 100. Examples of adiabatic shocks occur at the peripheries of young SNRs in the Sedov-Taylor evolution phase. The strong X-ray line emission observed toward Tycho, Kepler, and Cas A (Becker *et al.* 1979, 1980a,b) has been interpreted as shocked metal-rich ejecta (Shull 1982; Gronenschild and Mewe 1982; Hamilton, Sarazin, and Chevalier 1983). Fast non-radiative shocks, which encounter H^0 more rapidly than it can be pre-ionized, are believed responsible for weak Balmer line emission (Chevalier and Raymond 1978; Chevalier, Kirshner, and Raymond 1980) and ultraviolet lines and two-photon continuum from filaments just outside the main optical filaments in the Cygnus Loop (Raymond *et al.* 1983; Fesen and Itoh 1985).

The structure of a "radiative shock" can approximately be divided into three regions: (1) a radiative precursor in which the ambient gas is moderately heated and partially ionized by ultraviolet photons produced in the shocked layer; (2) the "adiabatic shock front", a thin layer in which the pre-shock gas is accelerated and heated by dissipative processes; and (3) a much broader layer, in which inelastic collisions produce radiative cooling, emission, recombination, and further compression downstream from the front. The state of the gas beyond the last layer depends on boundary conditions at the driving source and on the total column density of shocked gas. Magnetic fields, thermal conduction, and ambient UV radiation often play a role in determining the density and temperature of this interface. Recent theoretical studies of steady-state radiative shocks include Raymond (1979), Shull and McKee (1979), Seab and Shull (1983, 1985), Dopita *et al.* (1984), and Cox and Raymond (1985), while Innes, Giddings, and Falle (1987a,b,c) recently examined non-steady shocks.

The radiative precursors of fast shocks, with $V_s > 110$ km s^{-1}, produce singly ionized H and He ahead of the front (Shull and McKee 1979). This results in a shock front in which the dissipation is governed by plasma instabilities rather than collisions, and the "collisionless shock front" has a negligible thickness. For slower shocks, the gas is only partially ionized (in the absence of external sources of ionizing radiation). The ionized component still undergoes a collisionless shock, and the large H^0-H^+ charge exchange cross section $\sigma_{in} = 3 \times 10^{-15}$ cm^2 (Dalgarno and Yadav 1953, Dalgarno 1958) ensures that the ions and neutrals remain coupled. The front structure in slower ($V_s < 20$ km s^{-1}) shocks is determined by elastic H^0-H^0 collisions, with a scale length $\lambda_{nn} = (n\sigma_{nn})^{-1} \approx (10^{15}$ cm$)n_0^{-1}$ set by the density of neutral particles and their elastic cross section. Since the gas "jumps" discontinuously from its pre-shock to post-shock conditions, such shocks are called "J-shocks" (Draine 1980). I will discuss multi-fluid (continuous) "C-shocks" in §3. The existence of an adiabatic shock transition requires that the relative kinetic energies of pre-shock and post-shock gas be dissipated into heat over a scale length small compared to the cooling length. A radiative shock occurs when radiative cooling occurs faster than dynamical times. Thus, the shock velocity must be sufficiently slow or the gas sufficiently dense so that the cooling time of the post-shock gas is less than the dynamical expansion time.

The physical variables of density, velocity, pressure, and temperature behind a shock are determined by hydrodynamical equations of mass, momentum, and energy conservation for "steady flow". (Steady flow assumes a homogeneous and steady source of pre-shock gas, as well as a shock front which does not slow appreciably in the time for gas to flow through the full cooling zone.) For quantitative analyses of radiative shocks, it is convenient to consider a frame of reference in which the adiabatic front is stationary. The pre-shock gas of density ρ_1 and pressure P_1 then flows into the front at velocity $v_1 = V_s$, is compressed discontinuously to density $\rho_2 = 4\rho_1$ (for strong shocks with $\gamma = 5/3$), decelerated to velocity $v_2 = v_1/4$, and heated to post-shock temperature $T_2 = T_s \gg T_1$. (Physical variables ahead of the front are subscripted 1, those immediately behind the front are subscripted 2, while those at general positions downstream are unsubscripted.) If a magnetic field is present (we consider the simple case of a field $\vec{B}$ perpendicular to the flow), a sufficient level of ionization assures that B is coupled to the matter, leading to "flux freezing" in which vB and B/ρ are constant. Conservation of mass and momentum, plus the flux freezing condition, gives the downstream pressure as a function of density ρ,

$$P(\rho) = P_1 + \rho_1 v_1^2 \left[1 - \rho_1/\rho\right] + (B_1^2/8\pi) \left[1 - \rho^2/\rho_1^2\right]. \tag{1}$$

The temperature follows from the ideal gas law, $T = [\mu P(\rho)/\rho k]$, where μ is the mean mass per particle (ions, electrons, and neutrals). Conditions immediately behind the front are derived by combining eq. (10) with the adiabatic gas law, $P/\rho^\gamma = $ constant. The density jump, ρ_2/ρ_1, at the front is then given by the solution to the quadratic equation:

$$2(2 - \gamma) \left(\frac{\rho_2}{\rho_1}\right)^2 + [(\gamma - 1)M^2 + 2\gamma(1 + \beta)] \left(\frac{\rho_2}{\rho_1}\right) - (\gamma + 1)M^2 = 0, \tag{2}$$

where we define the quantities, $c_{s1}^2 = P_1/\rho_1$; $M = v_1/c_{s1}$; and $\beta = (B_1^2/8\pi P_1)$. Here, c_{s1} is the isothermal sound speed in the pre-shock gas, $v_{A1} = B_1/(4\pi\rho_1)^{1/2}$ is the Alfvén speed, $M = (\rho_1 v_1^2/P_1)^{1/2}$ is the isothermal Mach number, and $\beta = (v_{A1}/2c_{s1})^2$ is the ratio of magnetic pressure to gas pressure. We characterize the strength of the interstellar magnetic field by the dimensionless parameter b, so that $B_1 = (10^{-6} \text{ G})n_1^{1/2}b$ and $v_A = (1.84 \text{ km s}^{-1})b$, with n_1 measured in hydrogen nuclei cm^{-3}. In interstellar clouds the parameter b is typically of order unity (Mouschovias 1976; Troland and Heiles 1986), and diffuse-cloud interstellar shocks are "super-Alfvénic" $(V_s > v_{A1})$ as well as supersonic $(V_s > c_{s1})$. In the strong-shock limit $(M \gg 1)$ the Rankine-Hugoniot jump conditions reduce to:

$$\left(\frac{\rho_2}{\rho_1}\right) = \left(\frac{v_1}{v_2}\right) = \left(\frac{\gamma + 1}{\gamma - 1}\right) \to 4 \tag{3}$$

$$P_2 = \left(\frac{2\rho_1 v_1^2}{\gamma + 1}\right) \to \frac{3\rho_1 v_1^2}{4} \tag{4}$$

$$kT_2 = \left[\frac{2(\gamma-1)}{(\gamma+1)^2}\right]\mu_s v_1^2 \rightarrow \frac{3\mu_s v_1^2}{16}. \tag{5}$$

The last numbers are evaluated for $\gamma = 5/3$. For strong shocks ($M \gg 1, b \approx 1$) magnetic fields do not appreciably alter these jump conditions.

2.2   Shock Structure

Models of radiative shocks are usually parameterized by several quantities: the shock velocity V_s; the pre-shock density n_1, temperature T_1, and magnetic field B_1; and the set of elemental abundances (e.g., H, He, C, N, O, Ne, Mg, Si, S, Fe). The pre-shock ionization states of these elements are also required, but in the absence of external ionizing radiation, these may be specified self-consistently by computing the structure of the radiative precursor (Shull and McKee 1979). A new ingredient to shock models (Seab and Shull 1983, 1985) is a pre-shock grain model, specifying constituents and size distributions of grains and the initial depletions of the heavy elements which compose them, primarily C, O, Si, Mg, and Fe.

In steady, plane-parallel flow, one assumes $\partial/\partial t = \partial/\partial x = 0$, so that $(d/dt) = v(d/dx) = (\rho_1 v_1/\rho)(d/dx)$ is the Lagrangian derivative following a parcel of fluid. The post-shock density in the cooling zone is derived from an energy equation,

$$\frac{d}{dx}\left[\rho v\left(\frac{v^2}{2} + U + \frac{P}{\rho}\right) + \left(\frac{B^2 v}{4\pi}\right)\right] + n^2 \mathcal{L}(T) = 0. \tag{6}$$

Here, the total (specific) internal energy U includes internal quantum states of excitation. The total loss function (cooling minus heating) is given by,

$$n^2 \mathcal{L}(T) = n^2(\mathcal{L}_{rad} + \mathcal{L}_{dis}) - H_{ext}$$
$$+ \sum_j n_j \left[n_e(\alpha_j E_{r,j} + C_j I_j) - 4\pi \int_{\nu_j}^{\infty} \sigma_j(\nu)(1 - \nu_j/\nu)J_\nu d\nu\right], \tag{7}$$

where $\mathcal{L}_{rad}$ and $\mathcal{L}_{dis}$ are the cooling rate coefficients for collisionally excited radiative transitions and molecular dissociations, and H_{ext} is any external heating source. For species (j) of density n_j, including all ion states of all elements and molecules, α_j is the recombination rate coefficient for ion state $(j+1) \rightarrow j$. The mean energy of recombining electrons is $E_{r,j}$, the collisional ionization rate coefficient is C_j, the radiation intensity is J_ν, the ionization threshold is $I_j = h\nu_j$, and $\sigma_j(\nu)$ is the photoionization cross section. The ionization state and cooling rate behind radiative shocks are far from equilibrium, and $\mathcal{L}(T)$ differs from the radiative cooling coefficient $\Lambda(T)$.

Downstream from the shock front, radiative cooling results in a large compression ($\rho \gg \rho_1$), while the total pressure ($P + \rho v^2 + B^2/8\pi$) remains constant. For no magnetic field, the thermal pressure P varies by only 33%, from its post-shock value of $3\rho_1 v_1^2/4$ to the full value of the "ram pressure" $\rho_1 v_1^2$ when $\rho \gg \rho_1$ (eq. [1]). When $B = 0$, the final compression of the shock can be quite large,

$$\left(\frac{\rho_f}{\rho_1}\right) = M^2\left(\frac{T_1}{T_f}\right), \tag{8}$$

where ρ_f and T_f are the final density and temperature and M is the isothermal Mach number. However, the compression is limited by a realistic initial magnetic field (eq. [12]), since the magnetic pressure eventually dominates the momentum flux ($B^2 \propto \rho^2$, whereas $P \propto \rho T$ and $\rho v^2 \propto \rho^{-1}$). Thus, the maximum compression in a strong magnetized shock is set by the relation $\rho_1 v_1^2 \approx B_f^2/8\pi = (B_1^2/8\pi)(\rho_f/\rho_1)^2$, or

$$\left(\frac{\rho_f}{\rho_1}\right) = \left(\frac{8\pi\rho_1 v_1^2}{B_1^2}\right)^{1/2} = 2^{1/2}\left(\frac{v_1}{v_{A1}}\right) \approx (77)\left(\frac{v_{s7}}{b}\right), \tag{9}$$

where $v_{s7} = (V_s/100 \text{ km s}^{-1})$, where $b \approx 1$ is the magnetic field parameter, and ρ_f and B_f are final (maximum) values of post-shock density and magnetic field. A typical compression is about a factor of 100.

2.3   Atomic and Grain Processes

The post-shock structure of radiative shocks depends on a variety of atomic processes, the most important of which are collisional ionization, photoionization, radiative and dielectronic recombination, ion charge exchange with H^0 and He^0, and radiative cooling. The emissivity in lines and continuum is dominated by electron-impact excitation of resonance, semi-forbidden, and forbidden lines of H^0, He^0, He^+ and ions of abundant elements (mostly C and O ions). The rates of these processes are temperature dependent and involve heavy element abundances, gas-grain interactions, and radiative transfer.

Immediately behind an adiabatic shock of $V_s = (100 \text{ km s}^{-1})v_{s7}$, the temperature is $T_s = (3\mu_s V_s^2/16k) = (1.44 \times 10^5 \text{ K})v_{s7}^2$, where we have assumed that He/H = 0.1 and that H and He are singly ionized by the radiative precursor ($\mu_s = 0.636m_H$). At these temperatures, the radiative cooling is dominated by collisional ionization of He^+ and excitation of permitted and semi-forbidden lines of $He^+(\lambda 304)$ and carbon and oxygen ions. In general, the degree of ionization of these species is lower than it would be in coronal ionization equilibrium, and the initial radiative cooling rate exceeds equilibrium values by factors of 10 to 100.

The non-equilibrium ionization fractions, $f_i = n_i/n_{tot}$, of the elements are determined by integrating time-dependent differential equations of the form,

$$\left(\frac{df_i}{dt}\right) = f_{i-1}[n_e C_{i-1} + G_{i-1}] - f_i[n_e(C_i + \alpha_{i-1}) + n(H^0)Z_i + G_i]$$
$$+ f_{i+1}[n_e\alpha_i + n(H^0)Z_{i+1}], \tag{10}$$

where $C_i(T), \alpha_i(T)$, and $Z_i(T)$ are rate coefficients (cm^3 s^{-1}) for collisional ionization *from*, recombination *to*, and charge exchange *from* ionization state (i), and G_i is the photoionization rate (s^{-1}) from state (i). The coefficient $C_i(T)$ is dominated by electron impact and includes both direct (valence shell) ionization as well as autoionization following inner-shell excitation. The latter is particularly important at high temperatures for ions with 1 or 2 electrons outside a closed shell. The recombination coefficients $\alpha_i(T)$ include both radiative and dielectronic recombination; dielectronic recombination dominates over radiative by a substantial factor at high temperatures ($T > 20,000$ K for most ions). Tables of $\alpha_i(T)$ and $C_i(T)$ may be found in Shull and Van Steenberg (1982).

Photoionization cross sections may be found in Reilman and Manson (1979) and Clark *et al.* (1985). Charge exchange collisions with H^0 (and sometimes He^0) are often the most effective means of reducing the ion state in shocks containing a substantial population of neutrals (Shull and McKee 1979; Butler and Raymond 1980). Charge exchange of H° with multiply ionized species dominates dielectronic recombination when the neutral fraction $n(H^0)/n(H_{tot})$ exceeds 1 to 5%. Charge exchange rate coefficients are discussed by Dalgarno and Butler (1978), McCarroll and Valiron (1976), Butler and Dalgarno 1980; Heil, Butler, and Dalgarno (1980), Butler, Heil and Dalgarno (1980), Baliunas and Butler (1980), and Dalgarno, Heil, and Butler (1981). Generally, the rates with ions of charge $z \geq +3$ are fast ($> 10^{-9}$ cm^3 s^{-1}). Rates for doubly charged ions are mixed: C III, S III, and Ne III are slow ($\sim 10^{-12}$ cm^3 s^{-1}), while N III, O III, and Si III are fast. Charge exchange of N II is slow ($\sim 10^{-12}$ cm^3 s^{-1}), but resonant charge exchange between O II and H I effectively couples the O and H ionization fractions, (O II/O I) $\approx$ (8/9)(H II/H I).

Electron collisions dominate the excitation of the permitted, semi-forbidden, and optical forbidden lines of atoms and ions. Infrared fine structure lines are excited by collisions with electrons, H^+ and H^0 (Dalgarno and McCray 1972). The electron impact excitation rate coefficient C_{ij}(cm^3 s^{-1}), for a transition (i-j) of energy E_{ij}, is parameterized by the dimensionless "collision strength" Ω_{ij}:

$$C_{ij} = (8.616 \times 10^{-6} \text{ cm}^3 \text{ s}^{-1}) \left(\frac{\Omega_{ij}}{g_i}\right) T^{-1/2} \exp\left(\frac{-E_{ij}}{kT}\right), \tag{11}$$

where g_i is the statistical weight of the lower state and T is the temperature (K). References for excitation of H^0, He^0, and He^+ and ions of heavy elements are found in Shull and McKee (1979). Other recent tabulations of collision strengths include: Osterbrock (1974, with revisions), Raymond and Smith (1977), Shull (1981), and Cox and Raymond (1985). Compilations of electron impact excitation data for atomic ions are available as scientific reports from Los Alamos (Merts *et al.* 1980) and JILA (Gallagher and Pradhan 1985).

Figure 1 shows the temperature profiles in three 100 km s^{-1} shock models (Seab and Shull 1985). The post-shock column density, N_H, is a convenient measure of post-shock distance or Lagrangian flow time, independent of pre-shock density n_1. By the constancy of mass flux (or nv) in one-dimensional flow, $N_H = n_1 V_s t$, where t is the flow time for a parcel of fluid to reach column N_H. The three cooling profiles in Fig. 1 represent models in which heavy elements are: (i) depleted from gas phase; (ii) initially depleted, but allowed to re-enter gas phase via grain processing; and (iii) fully in gas phase (undepleted). Evidently, the post-shock abundance of atomic coolants such as C, O, Si and Fe, can have an important effect on the total column and thus the strengths of emission lines.

Grain processing in shocks comes from grain-grain collisions and thermal and non-thermal gas-grain sputtering (more details are found in Seab 1987). Modeling is complicated by the need to specify the grain constituents and size distribution, uncertainties in the sputtering yields in He-grain collisions (Barlow 1978; Draine and Salpeter 1979), and the rates of vaporization, partial vaporization, and shattering in grain-grain collisions (Seab and Shull 1983, 1985; McKee *et al.* 1987). In fast shocks ($V_s > 150$ km s^{-1}) grain collisions with hot post-shock ions (primarily He) dominate the sputtering of small grains, whereas grain-grain collisions and non-thermal sputtering are relatively more important

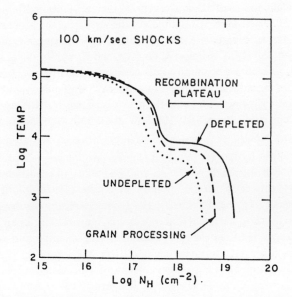

Figure 1: Post-shock temperature profiles versus column density $N_H = n_0 V_s t$ (cm^{-2}).

for larger grains in lower velocity shocks. Crucial to these "non-thermal" processes are the large gyrovelocities of charged grains generated by "betatron acceleration" as the magnetic field is compressed with the gas in the cooling zone. Most of the non-thermal grain destruction is produced in the strongly cooling layers around 10^4 K.

As the gas recombines and cools, the radiative cooling rate falls and the temperature reaches a plateau near 6000 K. Here, ionizing photons produced in the hot post-shock layer deposit their energy by photoionizing the newly recombined H^0 and He^0. The cooling immediately behind the front is due primarily to electron-impact collisional ionization of H^0, He^0, or He$^+$, depending on shock velocity, plus electron-impact excitation of resonance and semi-forbidden lines of H^0($Ly\alpha$), He^0(λ584,626), He$^+$(λ304), and ions of abundant heavy elements, primarily C and O. Below 20,000 K, the cooling is dominated by forbidden and semi-forbidden lines of heavy elements, such as [O III] λ5007, [O II] λ3727, C III] λ1909, [S II] λ6716,6731 and C II] λ2326. Because a forbidden line may be collisionally de-excited when the electron density exceeds the line's "critical density", $n_{cr} = A_{21}/C_{21}$ ranging from 10^2 to 10^6 cm^{-3}, the cooling scale may be lengthened in shocks of higher density. In the gas below 10^3 K, infrared fine structure lines dominate the cooling – for example [Si II] 34.8μm, [O I] 63 and 145μm, and various lines of [Fe II] (1.27, 1.6, 5.0, and 26μm). If the gas contains a fraction of H$_2$, the rotational lines are also important coolants.

2.4   Molecular Processes

A full discussion of molecular chemistry in diffuse clouds is beyond the scope of this paper. Useful reviews on related subjects are: molecular abundances in hydrostatic cloud models (van Dishoeck and Black 1986), chemistry in molecular shocks (Hollenbach and McKee 1979; McKee and Hollenbach 1980), and a general review of interstellar molecular hydrogen (Shull and Beckwith 1982). Here we confine our discussion to the processes of H_2 formation and destruction. Because radiative association of two H atoms is forbidden by dipole selection rules, interstellar H_2 is believed to form most rapidly on grain surfaces. When two H^0 atoms collide with a grain and stick, they migrate and eject an H_2 molecule with substantial kinetic, vibrational, and rotational energy (Hollenbach and Salpeter 1971). In grain-free or pre-galactic environments, H_2 may also form by slower gas-phase reactions with H^- or H_2^+ (Lepp and Shull 1984; MacLow and Shull 1986; Shapiro and Kang 1987).

Dissociation of H_2 occurs either by a two-step process initiated by absorption of a UV photon in one of the Lyman ($\lambda < 1120$Å) or Werner bands ($\lambda < 1021$Å) or by collisions with H^0, H^+, or e^- (Hollenbach and McKee 1980). The photodissociation rate may be diminished by "self-shielding" in the Lyman lines (Jura 1974; Shull 1978) or by dust opacity. At low density ($n_H < 10^5$ cm^{-3}) the rate of collisional dissociation can also be reduced by "radiative stabilization" (Roberge and Dalgarno 1982; Lepp and Shull 1983), in which radiative decays decrease the populations of vibrationally and rotationally excited H_2 levels which are subject to large collisional dissociation rates in thermal (Boltzmann) populations. Rate coefficients for radiative decay of vibrational and rotational states of H_2 are given by Turner, Kirby-Docken, and Dalgarno (1977), and for H^0-H_2 collisional excitation and dissociation by Lepp and Shull (1983), revised at high temperature by MacLow and Shull (1986).

Molecular cooling in shocks arises from the excitation of rotational and vibrational states of H_2, CO, H_2O, and other abundant molecules. Since molecular shocks are often slower and the gas denser and more neutral than in diffuse clouds, the excitation comes from collisions with H^0 and H_2 as well as from electrons. In addition, magnetic fields and multi-fluid effects play an important role (see §3). Here, we restrict our discussion to J-shocks with velocities of order 10 km s^{-1}, in which the post-shock temperature is $T_s \approx (2900K)(V_s/10$ km s$^{-1})^2$. These temperatures are sufficient to excite many rotational states (J) and several ($v = 1$ and 2) vibrational states of H_2. For small J and v, the H_2 excitation temperatures are $T_e(J) = E(J)/k = (85K)J(J+1)$ and $T_e(v) = (6300K)v$.

2.5   Observations and Line Diagnostics

Many authors have remarked on the spectral signatures of shock waves, as distinguished from H II regions and other photoionized regions (Baldwin, Phillips, and Terlevich 1981; Fesen, Blair, and Kirshner 1985). Generally, SNRs are characterized by strong optical forbidden line emission over a wide range of ionization states. For example, [S II]/Hα is stronger in SNRs than in H II regions. The main observational features attributed to radiative shock waves in the optical are:

1. Strong emission lines, relative to Hβ, of underionized species, e.g., [O I] 6300, [N I] 5200, [O II] 3726,3729, [S II] 6716,6731.

2. A high excitation temperature ($T > 20,000$; K) measured from the intensity ratio of [O III] lines, [4363/(5007 + 4959)].

3. The presence of a range of ionization states, e.g., [O I], [O II], [O III], [Ne III], [Ne V].

4. Large ratios of [O I]/Hβ and [O II]/Hβ , relative to H II regions.

The fourth effect has been demonstrated empirically for SNRs in the Galaxy and M31/M33 (Fesen *et al.* 1985). New wavelength bands have opened up other shock discriminants. In the ultraviolet, shocks waves produce strong resonance and semi-forbidden lines of C II] 2326, C III] 1909, O III] 1663, N III] 1750, O IV] 1402, C IV 1549, and N V 1240. In the infrared, the fine structure lines of [O I] 63 μm, [Si II] 34.8 μm, [Ne II] 12.8 μm, and [Fe II] (26, 1.6, 1.27, 5 μm) are strong. While these features are not unique (photoionized gas at high densities can mimic some of the optical line ratios), the combination of lines from several ion stages and varying excitation temperatures can be used to attribute the power source to shocks.

Certain line ratios can be used to constrain the shock velocity V_s, the pre-shock density n_0, and the abundances. The emission lines of [O III], [Ne III], C III], C IV, and N V are the best "speedometers" since their intensities rise steeply with V_s. The temperature in the post-shock "recombination zone" may be gauged by the intensity ratios of [O III] [4363/(5007+4959)] or [Ne III] 3342/3869. Density sensitive line ratios include [O II] 3729/3726 and [S II] 6716/6731, as well as certain infrared fine structure lines. Shocks into diffuse clouds generally have recombination-zone densities less than 10^3 cm^{-3}. The reason for the absence of higher densities is clear: shocks which propagate from a lower density intercloud medium into a dense cloud are slowed by a factor equal to the square root of the density contrast (momentum flux ρv^2 is conserved). The emission from these much slower shocks ($V_s = 10$ km s^{-1}) would prove difficult to detect optically, although infrared lines of [O I], [Ne II], [Si II], and [Fe II] might be detected with a new generation of detectors.

Abundance determinations from emission lines are fraught with uncertainties. For example: (1) Changes in heavy element abundance are difficult to distinguish from differences in density, magnetic field, and velocity. (2) Intensities of strong forbidden lines "saturate" with increasing velocity or abundance, since there is only a fixed amount of energy available for these cooling lines in the recombination zone. (3) Strong UV lines such as C III] 1909 and C IV 1549 become insensitive to V_s as the post-shock temperature rises above their ionization and excitation temperature. (4) Many resonance lines (C II 1035, 1335; C IV 1549) become optically thick, and their emergent intensities are reduced by scattering in the shock layer. (5) Grain disruption by sputtering and grain-grain collisions introduces another degree of freedom: variable gas-phase abundances of C, O, Si, and Fe; (6) the shocks may be unstable or non-steady. Despite these uncertainties, one may still make progress with carefully chosen forbidden and semi-forbidden lines of ionized species (Raymond *et al.* 1981; Raymond 1984; Dopita *et al.*, 1984).

The Cygnus Loop is the best studied of the older SNRs. It is generally pictured (Raymond 1984; Hester and Cox 1986) as a 400 km s^{-1} blast wave, propagating in an intercloud medium of density $n_H \approx 0.2$ cm^{-3} and driving 100 km s^{-1} shocks into clouds of density $2 - 10$ cm^{-3}. The faster, non-radiative shocks produce the observed X-rays while the slower (radiative) shocks produce the bright optical filaments. Spectra of several bright filaments show line ratios which disagree radically with radiative shock models having normal abundances and full recombination zones. Some filaments show [O III]/Hβ as large as 10-25, whereas current shock models do not allow ratios greater than 3, owing to rapid O^{+2} - H^0 charge exchange in the recombination zone (see Table I). More generally,

Fesen *et al.* (1982) showed that the distribution of line ratios, *e.g.*, [O III]/Hβ versus [O II]/Hβ or [O II]/Hβ versus [O I]/Hβ, bears little resemblance to those predicted by standard radiative shock models.

The remedy to this disagreement may be to "truncate" the shocks' recombination zones. A full radiative shock, complete with recombination zone, requires a column density $N_{rec} \approx 10^{19}$ cm^{-2}, corresponding to a flow time of $(3 \times 10^4$ yr$)(1$ cm$^{-3}/n_0)(100$ km s$^{-1}/V_s)$. If the zone with $T < 10^4$ K is missing, as a result of an inhomogeneous pre-shock medium or thermal instability, then the region which produces Hβ recombination lines will be missing and the O III charge exchange will be insignificant in the more ionized gas. A physical realization of this scenario for the optical filaments requires either many small ($< 10^{16}$ cm) cloudlets engulfed by the blast wave (Fesen, Blair, and Kirshner 1982), or variations in line-of-sight surface brightness produced by a few wavy thin sheets (Hester and Cox 1986; Hester 1987).

A final observational topic concerns grain destruction by shocks. Emission lines from Si and Fe are particularly affected by the grain sputtering and grain-grain collisions in the cooling zone. Seab and Shull (1983) discussed effects on the UV selective extinction curves of shock processed gas, and McKee *et al.* (1987) have recently addressed the question of "grain history", coupling theoretical shock models of grain processing with models of the multi-phase structure of the ISM.

The *Copernicus* satellite found that significant fractions of many heavy elements are depleted from the interstellar gas, presumably locked up in dust grains. Quantitatively, we define the "depletion factor" d_i of an element (i) by,

$$\log d_i = \log \left(\frac{N_i}{N_H} \right) - \log \left(\frac{N_i}{N_H} \right)_\odot \tag{12}$$

where N_i and N_H are the column densities of (i) and of hydrogen, and where $(N_i/N_H)_\odot$ is the solar or cosmic abundance (Withbroe 1971; Grevesse 1983). The influence of shocks is believed to explain the correlation of refractory element depletions with cloud velocity. The optical observation (Routly and Spitzer 1952; Siluk and Silk 1974) that interstellar clouds with high velocities show systematically higher ratios of Ca II/Na I has been interpreted as evidence of selective grain destruction, which returns the highly depleted calcium back to gas phase. The same correlation of depletion with cloud velocity has been seen in UV absorption studies of Si and Fe (Shull, York, and Hobbs 1977). Both data and theoretical models are consistent with the general conclusion that clouds with velocities greater than about 20 km s^{-1} have larger gas-phase abundances of refractory elements than low-velocity gas. Grain processing may also be responsible for the correlation (Fig. 2) of heavy element depletions with mean line-of-sight column density $\bar{n} = N(\text{H})/r$. A possible physical interpretation is that lines of sight with low $\bar{n}$ are more likely to have had extensive shock processing of grains.

3   MHD SHOCKS WITH MAGNETIC PRECURSORS

The fundamental concept underlying the following discussion of MHD shocks in gas of low fractional ionization is that the matter in the shocked regions may be thought of as consisting of several distinct, interpenetrating fluids. Normally one thinks of three fluids:

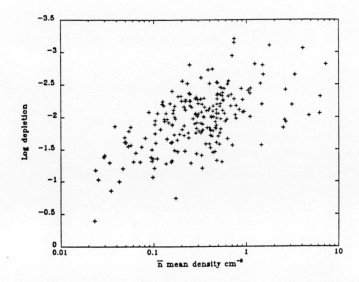

Figure 2: Depletion factors d_i for interstellar Fe toward 225 OB stars observed by IUE (Van Steenberg and Shull 1987) are correlated with mean hydrogen density $\bar{n}$. This effect may result from shock destruction of grains along low-$\bar{n}$ lines of sight.

(i) the neutral particles; (ii) the ions; and (iii) the electrons. Under some circumstances it may be useful to consider the charged dust grains to constitute a fourth fluid. The motivation for this conceptual decomposition is that under some circumstances (e.g., in a shock transition) these fluids may develop appreciably different flow velocities and temperatures.

A necessary condition for a shock wave to occur in an initially quiescent medium is that a compressive disturbance be advancing into the medium at a velocity greater than the signal speed. Otherwise signals will travel ahead of the shock and inform the quiescent medium that a compression is approaching. In a fluid consisting of neutrals, ions, and electrons, a wave of sufficiently long wavelength (low frequency) must have the neutrals, ions, and electrons moving together. If no magnetic field is present, the compressional signal speed is just the sound speed $c_{s,nie} = (5P/3\rho)^{1/2}$ where $P = k(n_n T_n + n_i T_i + n_e T_e)$ is the total gas pressure, and ρ is the total density. In a magnetized fluid there are two distinct compressional modes, referred to as the "fast" and "slow" magnetosonic modes; the speed v_f of the fast mode may be derived (Spitzer 1962) from the Alfvén speed $v_{A,ie} = B/(4\pi\rho_i)$ and the thermal sound speed $c_{s,ie} = [5k(n_i T_i + n_e T_e)/3\rho_i]^{1/2}$. A shock will occur if the compressive disturbance is advancing with a velocity $V_s > v_{f,nie}$.

How large are the Alfvén velocities? Existing observations of interstellar magnetic field strengths (Troland and Heiles 1986) are consistent with an empirical "scaling law" $B = (1\mu G)(n_H/\text{cm}^{-3})^{1/2}$ for densities $10\text{ cm}^{-3} < n_H < 10^6\text{ cm}^{-3}$. This relation implies relatively large values of $v_{A,ie}$ in predominantly neutral clouds containing heavy ions of mass $\sim 20 m_H$:

$$v_{A,ie} \approx (50 \text{ km s}^{-1}) \left[\frac{n_i/n_H}{10^{-4}}\right]^{-1/2} \left[\frac{20 m_H}{\rho_i/n_i}\right]^{1/2}. \tag{13}$$

For example, a dense molecular cloud with $n_i/n_H = 10^{-7}$ would have $c_s \approx 1$ km s^{-1} and $v_{f,nie} \approx v_{A,ie} \approx 1500$ km s^{-1}. Compressive disturbances with $V_s > v_{f,nie}$ will be shock waves, since long-wavelength signals cannot travel faster than the disturbance. More complete discussions of MHD shocks may be found in a recent review (Shull and Draine 1987) and in the literature (Draine, Roberge, and Dalgarno 1983; Draine 1986; Chernoff 1987).

Consider MHD shocks for which $V_s < v_{A,ie}$, so that the magnetized plasma is "sub-magnetosonic". There are two basic classes of solutions: (1) C-type ("continuous") shocks, and (2) J-type ("jump") shocks with magnetic precursors. In C-type shocks, no discontinuity is present: all the flow variables vary continuously through the shock transition, and ordinary molecular viscosity plays no role. The J-type shocks resemble single fluid shocks in that there is a "jump" transition in which molecular viscosity effects an irreversible change in the neutral flow variables on a length of order one molecular mean free path; for our purposes such a change is treated as a discontinuity, with the flow variables across the discontinuity related through the Rankine-Hugoniot jump conditions. However, ahead of this "J-front" the neutral gas is accelerated and heated by collisions with streaming ions – i.e., the shock has a "magnetic precursor". In the frame of reference of the shock, the neutral gas is flowing supersonically upstream from the shock, and subsonically immediately downstream from the shock.

The dominant processes for cooling the neutral gas in C-type MHD shocks include emission from rotationally- and vibrationally-excited H_2, rotationally excited CO, rotationally excited H_2O, and the excited fine structure levels of C I, C II, and O I. When the neutral gas temperature and density are high enough (Lepp and Shull 1983), collisional dissociation of H_2 can become an important sink for thermal energy. The electron gas can become significantly hotter than the neutrals. The electrons are cooled by elastic collisions with the neutrals, and by collisional excitation of the same atomic, molecular, and ionic excited states which are important for cooling the neutral gas. In addition, the electron gas may sometimes be hot enough to collisionally populate excited electronic states of abundant atoms and molecules. The power per area converted into heat in a strong shock is of order $\rho_{n0} V_s^3/2$, where ρ_{n0} is the preshock mass density. At the present time it appears possible to detect H_2 line emission only from shocks with $n_H > 10^4$ cm^{-3}.

4   NEW RESEARCH PROBLEMS

In the remainder of this review, I would like to discuss two areas of research which have opened new avenues for the interpretation of interstellar shock waves. The first of these is the topic of unstable or "non-steady shocks", and the second is the role of thermal conduction in metal-rich shocks associated with "fast-moving knots" in Cas A and other oxygen-rich SNRs.

Since the pioneering work of Cox (1972), most numerical models of radiative shocks have assumed steady flow. However, observations of radiative filaments near some SNRs

(Fesen, Blair, and Kirshner 1982; Hester and Cox 1986) exhibit line ratios which differ from theoretical predictions. These discrepancies may arise because the shock's recombination zone is truncated, because the shocks are unstable, or because the shock flow is not steady. Theoretical considerations which support these interpretations include: (1) the ISM is inhomogeneous on small scales, and some clouds may have columns less than that required ($N_{rec} \approx 10^{19} cm^{-2}$) for a complete recombination zone; (2) for velocities V_s exceeding 150 to 200 km s^{-1}, the shock front decelerates on a timescale shorter than the time for a parcel of fluid to traverse the radiatively cooling layer; (3) high-velocity shocks may be dynamically unstable.

Several authors have shown that radiative cooling can produce instabilities (Falle 1981; Chevalier and Imamura 1982; Bertschinger 1986). If the logarithmic slope, $\alpha = d(log\Lambda)/d(logT)$ is less than a critical value between 0.5 and 1.5, a radiative shock never becomes steady, even if it is driven by a constant-velocity piston. Innes *et al.* (1987b,c) have constructed realistic models of the non-equilibrium ionization and radiative cooling, and conclude that shocks with $V_s > 150$ km s^{-1} are unstable to small perturbations which affect the emission line intensities. There has also been considerable debate about dynamical and gravitational instabilities of shocked gaseous layers (Elmegreen and Lada 1977; Elmegreen and Elmegreen 1978; Vishniac 1983). In recent work, Voit (1987) has shown that in an incompressible fluid, the symmetric and anti-symmetric modes in an unaccelerated slab transform continuously into Rayleigh-Taylor and gravity-wave modes as deceleration becomes more important. Thus, a slab of decelerating gas compressed by a SNR can undergo gravitational collapse and result in accelerated star formation. Even before the slab becomes gravitationally unstable, though, a two-dimensional dynamic instability (Vishniac 1983; Bertschinger 1986) may disrupt the state of the shell. For observers, the key question is whether the non-linear state of the instability is so chaotic that one cannot use idealized models to derive meaningful abundances from the data. I believe that we do not yet have the answer to this question.

A second new area of shock research concerns the role of thermal conduction in metal-rich shocks associated with young SNRs such as Cas A. The fast-moving knots of Cas A (Kirshner and Chevalier 1979) exhibit strong emission lines of oxygen, sulfur, argon, and calcium, but evidently no hydrogen or helium. These emissions have been interpreted as shock-heated gases composed mainly of oxygen (Chevalier and Kirshner 1978, 1979; Itoh 1981). However, these shocks have a number of physical features that make it unwise to extrapolate from cosmic-abundance models:

- For a given velocity, the post-shock temperatures are ~ 16 times greater than cosmic-abundance shocks because of the larger atomic weight of oxygen.

- Heavy-element abundances are enhanced by factors $> 10^3$, resulting in strong radiative cooling, steep temperature gradients, and strong thermal conduction.

- The strong radiative cooling causes temperature decoupling between the ions and electrons, even if $T_e = T_i$ at the front.

- Collisionally excited oxygen lines produce a much stronger photoionizing flux.

Itoh (1981) has computed models of pure-oxygen shocks, neglecting thermal conduction. At $V_s \approx 100$ km s^{-1}, the incoming oxygen atoms are mostly preionized by the radiative precursor. In the post-shock region, the intense radiative cooling results in ion-electron temperature decoupling ($T_e \ll T_i$), and the electron temperature drops to $\sim 10^2 K$ before the ions have a chance to recombine. In these models, one has the amazing result that O IV, O V, and O VI exist at $T \sim 100K$. However, these models are almost certainly incorrect if one includes the effects of thermal conduction.

In new models of metal-rich shocks, Borkowski and Shull (1987) find that over 90% of the energy flux is carried by thermal conduction. The energy equation (eq. [6]) is modified with a new conductive term but omitting the magnetic field,

$$\frac{d}{dx}\left[\rho v \left(\frac{v^2}{2} + U + \frac{P}{\rho}\right) +\right] + n^2 \mathcal{L}(T) - \frac{d}{dx}\left(\kappa(T)\frac{dT}{dx}\right) = 0. \tag{14}$$

where $\kappa(T) = \kappa_0 T^{5/2}$ is the coefficient of thermal conductivity (Spitzer 1962) and $n^2 \mathcal{L}(T)$ is the total loss function (radiative cooling minus photoelectric heating – see eq. [7]). One can define two scale lengths for cooling and conduction,

$$L_{cool} = \left(\frac{\frac{5}{2}P_s V_s}{n^2 \mathcal{L}(T)}\right); \tag{15}$$

$$L_{cond} = \left(\frac{\kappa(T_s)T_s}{\frac{1}{2}\rho_1 v_1^3}\right), \tag{16}$$

where $v_1 = V_s$ is the shock velocity and P_s and T_s are the post-shock pressure and temperature. The dimensionless ratio of these two lengths, $\alpha = L_{cond}/L_{cool}$ may be used to gauge the importance of conduction. Since $\mathcal{L}(T) \propto \Lambda(T)$ in the absence of heating, $T_s \propto V_s^2$, and $\kappa \propto T^{5/2}$, the parameter $\alpha \propto T_s^{1/2}\Lambda(T_s)$. For temperatures $5.2 < logT < 7.3$, the radiative cooling coefficient $\Lambda(T) \propto T^{-1/2}$ and α is approximately constant. For normal (cosmic) metal abundances and equilibrium cooling, $\alpha \approx 0.005$ for $V_s \approx 100$ km s^{-1}. However, non-equilibrium cooling behind the front can be $\sim 10^2$ times greater than equilibrium values, and oxygen-rich shocks have cooling rates enhanced by over three orders of magnitude. The conclusion is that thermal conduction *must* be included for the Cas A shocks, and it can lead to interesting effects in normal-abundance shocks as well.

Unfortunately, thermal conduction complicates the numerical solution of radiative shocks. Because one must specify not only the post-shock temperature T_s but also the temperature gradient $(dT/dx)_s$, the problem becomes a "two-point boundary value problem". With special assumptions ($T_i = T_e$, constant ionization fraction, and a specified cooling function), Borkowski and Shull (1987) have separated the second-order energy equation into two integrable equations for $q(T)$ and $T(x)$. A solution yields the conductive flux $q(T) = \kappa(T)(dT/dx)$ in terms of the temperature T, which yields $T(x)$. The conductive flux also affects the Rankine-Hugoniot jump conditions. Figure 3 shows several cosmic-abundance shock solutions for the dimensionless flux $\hat{q} = q(T)/(\frac{1}{2}\rho_1 v_1^3)$ in terms of normalized temperature, $\tau = T/T_{s0}$, where T_{s0} is the post-shock temperature in the absence of conduction. Evidently the effects of conduction are to decrease the post-shock temperature ($\tau < 1$) and flatten the temperature gradients by conducting heat to the cooler post-shock layers. The conductive effects in metal-rich shocks are even more dramatic. Further work is underway, including more realistic non-equilibrium cooling,

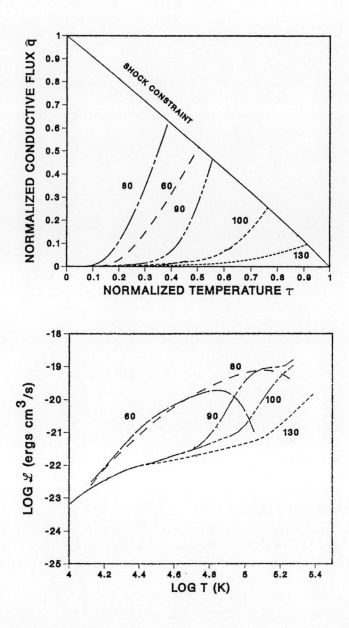

Figure 3: Cosmic-abundance shock trajectories of normalized conductive flux $\hat{q}$ versus normalized temperature $\tau = T/T_{s0}$, based on non-equilibrium cooling rates from Shull and McKee (1979). Shock velocities are given in km s^{-1}. Conductive flux dominates radiative cooling when the trajectory's slope exceeds 45°. The shock constraint represents the front, with conductive effects included in the jump conditions. Clearly, conduction cannot be neglected in the presence of strong non-equilibrium cooling or metallicity enhancements.

ionization, and radiative transfer. The hope is that we will be able use these models to better define the abundances of O, S, Ar, and Ca in these fast-moving knots.

This work was supported by Astrophysical Theory grants from NASA (NSG-7128 and NAG5-193) at the University of Colorado. I thank Dr. Rob Fesen for his comments on the manuscript and Kazimierz Borkowski for discussions on thermal conduction.

REFERENCES

Baldwin, J.A., Phillips, M.M., and Terlevich, R. 1981, *P.A.S.P.*, **93**, 5.

Baliunas, S.L., and Butler, S.E. 1980, *Ap. J. Letters*, **235**, L45.

Barlow, M. 1978, *M.N.R.A.S.*, **183**, 367.

Becker, R., *et al.* 1979, *Ap. J. Letters*, **234**, L73.

Becker, R., *et al.* 1980a, *Ap. J. Letters*, **235**, L5.

———— 1980b, *Ap. J. Letters*, **237**, L77.

Bertschinger, E. 1986, *Ap. J.*, **304**, 154.

Borkowski, K., and Shull, J.M. 1987, in preparation.

Butler, S.E., and Dalgarno, A. 1980, *Ap. J.*, **241**, 838.

Butler, S.E., Heil, T.G., and Dalgarno, A. 1980, *Ap. J.*, **241**, 442.

Butler, S.E., and Raymond, J.C. 1980, *Ap. J.*, **240**, 680.

Chernoff, D. F. 1987, *Ap. J.*, **312**, 143.

Chevalier, R.A., and Kirshner, R.P. 1978, *Ap. J.*, **219**, 931.

Chevalier, R.A., and Kirshner, R.P. 1979, *Ap. J.*, **233**, 154.

Chevalier, R.A., Kirshner, R.P., and Raymond, J.C. 1980, *Ap. J.*, **235**, 186.

Chevalier, R.A., and Raymond, J.C. 1978, *Ap. J. Letters*, **225**, L27.

Clark, R.E.H., Cowan, R.D., and Bobrowicz, F.W. 1986, *Atomic and Nuclear Data Tables*, **34**, 415.

Cox, D.P. 1972, *Ap. J.*, **178**, 143.

Cox, D.P., and Raymond, J.C. 1985, *Ap. J.*, **298**, 651.

Dalgarno, A. 1958, *Phil. Trans. Roy. Soc., Ser. A*, **250**, 426.

Dalgarno, A., and Butler, S. 1978, *Comments Atom. Molec. Phys.*, **7**, 129.

Dalgarno, A., Heil, T.G., and Butler, S.E. 1981, *Ap. J.*, **245**, 793.

Dalgarno, A., and McCray, R. 1972, *Ann. Rev. Astr. Ap.*, **10**, 375.

Dalgarno, A., and Yadav, H.N. 1953, *Proc. Phys. Soc.*, **A66**, 173.

Dopita, M., Binette, L., d'Odorico, S., and Benvenuti, P. 1984, *Ap. J.*, **276**, 653.

Draine, B. T. 1980, *Ap. J.*, **241**, 1021 [Erratum 1981, **246**, 1045].

Draine, B. T. 1986, *M.N.R.A.S.*, **220**, 133.

Draine, B. T., Roberge, W. G., and Dalgarno, A. 1983, *Ap. J.*, **264**, 485.

Draine, B.T., and Salpeter, E.E. 1979, *Ap. J.*, **231**, 77.

Elmegreen, B.G., and Elmegreen, D.M. 1978, *Ap. J.*, **220**, 1051.

Elmegreen, B.G., and Lada, C.J. 1977, *Ap. J.*, **214**, 725.

Fesen, R.A., and Itoh, H. 1985, *Ap. J.*, **295**, 43.

Fesen, R.A., Blair, W.P., and Kirshner, R.P. 1982, *Ap. J.*, **262**, 171.

———— 1985, *Ap. J.*, **292**, 29.

Gallagher, J., and Pradhan, A.K. 1985, *JILA Information Center Report*, No. 30, University of Colorado.

Grevesse, N. 1983, *Physica Scripta*, **T8**, 49.

Gronenschild, E.H.B.M., and Mewe, R. 1982, *Astr. Ap. Suppl.*, **48**, 305.

Hamilton, A.J.S., Sarazin, C.L., and Chevalier, R.A, 1983, *Ap. J. Suppl.*, **51**, 115.

Heil, T.G., Butler, S.E., and Dalgarno, A. 1980, *Phys. Rev. A*, **23**, 1100.

Hester, J.J. 1987, *Ap. J.*, **314**, 181.

Hester, J.J., and Cox, D.P. 1986, *Ap. J.*, **300**, 675. *Ap. J. Letters*, **252**, L21.

Hollenbach, D. J., and McKee, C. F. 1979, *Ap. J. Suppl.*, **41**, 555.

———— 1980, *Ap. J. Letters.*, **241**, L47.

Hollenbach, D.J., and Salpeter, E.E. 1971, *Ap. J.*, **163**, 155.

Innes, D.E., Giddings, J.R., and Falle, S.A.E.G. 1987a, *M.N.R.A.S.*, **224**, 179.

———- 1987b, M.N.R.A.S., submitted, *Dynamical Models of Radiative Shocks. II. Unsteady Shocks.*

———- 1987c, M.N.R.A.S., submitted *Dynamical Models of Radiative Shocks. II. Spectra.*

Itoh, H. 1981, *Publ. Astr. Soc. Japan*, **33**, 1.

Jura, M. 1974, *Ap. J.*, **191**, 375.

Lepp, S., and Shull, J. M. 1983, *Ap. J.*, **270**, 578.

———— 1984, *Ap. J.*, **280**, 465.

McCarroll, R., and Valiron, P. 1976, *Astr. Ap.*, **53**, 83.

MacLow, M.M., and Shull, J.M. 1986, *Ap. J.*, **302**, 585.

McKee, C.F., and Hollenbach, D.J. 1980, *Ann. Rev. Astr. Ap.*, **18**, 219.

Merts, A.L., Mann, J.B., Robb, W.D., and Magee, N.W. 1980, *Los Alamos Informal Report* No. LA-8267-MS.

Mouschovias, T. Ch. 1976, *Ap. J.*, **207**, 141.

Osterbrock, D.E. 1974, *Astrophysics of Gaseous Nebulae*, (San Francisco: Freeman).

Phillips, A.P., Gondhalekar, P.M., and Pettini, M. 1982, *M.N.R.A.S.*, **200**, 687.

Raymond, J.C. 1979, *Ap. J. Suppl.*, **39**, 1.

———— 1984, *Ann. Rev. Astr. Ap.*, **22**, 75.

Raymond, J.N., Black, J.H., Dupree, A.K., Hartmann, L.H., and Wolff, R.S. 1981, *Ap. J.*, **246**, 100.

Raymond, J.N., Blair, W.P., Fesen, R.A., and Gull, T.R. 1983, *Ap. J.*, **275**, 636.

Raymond, J.C., and Smith, B.W. 1977, *Ap. J. Suppl.*, **35**, 419.

Reilman, R.F., and Manson, S.T. 1979, *Ap. J. Suppl.*, **40**, 815.

Roberge, W.G., and Dalgarno, A. 1982, *Ap. J.*, **255**, 176.

Routly, P.M., and Spitzer, L. 1952, *Ap. J.*, **115**, 227.

Seab, C.G. 1987, in *Interstellar Processes*, eds. D.J. Hollenbach and H. Thronson, proceedings of Tetons Summer School, (Dordrecht: Reidel), in press.

Seab, C.G., and Shull, J.M. 1983, *Ap. J.*, **275**, 652.

———— 1985, "Shock Processing of Interstellar Grains", in *Interrelationships among Circumstellar, Interstellar, and Interplanetary Grains*, NASA CP-2403, p 37.

Shapiro, P., and Kang, H. 1987, *Ap. J.*, in press.

Shull, J.M. 1978, *Ap. J.*, **219**, 877.

———— 1981, *Ap. J. Suppl.*, **46**, 27.

———— 1982, *Ap. J.*, **262**, 308.

Shull, J.M., and Beckwith, S. 1982, *Ann. Rev. Astr. Ap.*, **20**, 163.

Shull, J.M., and Draine, B.T. 1987, in *Interstellar Processes*, eds. D.J. Hollenbach and H. Thronson, proceedings of Tetons Summer School, (Dordrecht: Reidel), in press.

Shull, J.M., and McKee, C.F. 1979, *Ap. J.*, **227**, 131.

Shull, J.M., and Van Steenberg, M.E. 1982, *Ap. J. Suppl.*, **48**, 95 [Erratum **49**, 351].

Shull, J.M., York, D.G., and Hobbs, L.M. 1977, *Ap. J. Letters*, **211**, L139.

Siluk, R., and Silk, J. 1974, *Ap. J.*, **192**, 51.

Spitzer, L. 1962, *Physics of Fully Ionized Gases* (2d ed.; New York: Interscience) *Ann. Rev. Astr. Ap.*, **13**, 133.

Troland, T. H., and Heiles, C. 1986, *Ap. J.*, **301**, 339.

Turner, J., Kirby-Docken, K., and Dalgarno, A. 1977, *Ap. J. Suppl.*, **35**, 281.

van Dishoeck, E.F., and Black, J.H. 1986, *Ap. J. Suppl.*, **62**, 109.

Van Steenberg, M.E., and Shull, J.M. 1987, *Ap. J.*, in press.

Vishniac, E.T. 1983, *Ap. J.*, **274**, 152.

Voit, G.M. 1987, this conference (and *Ap. J.*, in preparation).

Withbroe, G. 1971, in "The Menzel Symposium", NBS SP-53, ed. K. Gebbie, (Washington: Government Printing Office).

SHOCKED MOLECULAR GAS IN THE SUPERNOVA REMNANT IC 443: MODELS WITH AN ENHANCED IONIZATION RATE

George F. Mitchell
Saint Mary's University
Halifax, N. S., Canada, B3H3C3

Abstract: The high molecular abundances seen in IC443 may be due to hot chemistry in shocked gas or to an increase in the cosmic ray ionization rate. New calculations reported here show that shock chemistry dominates over the relevant short time scale.

1. INTRODUCTION: IC 443 is the only supernova remnant which we know to be interacting with interstellar molecular gas. Evidence for an encounter of the remnant IC 443 with molecular material is various: (1) Emission from vibrationally excited molecular hydrogen indicates hot postshock gas (Treffers 1979; Burton et al. 1987). (2) The species HI, CO, OH, CS, HCN, and HCO^+ have extremely broad linewidths, typically 30-40 km s^{-1} (DeNoyer 1979a,1979b; Dickenson et al. 1980; Fesen and Kirschner 1980; DeNoyer and Frerking 1981; White et al. 1987). Such broad lines may indicate shock acceleration. (3) The abundances of OH, CS, HCN, and HCO^+ derived from the observations are anomalously high (DeNoyer and Frerking 1981; White et al. 1987).

Mitchell and Deveau (1983) showed that the high abundances of OH and HCO^+ could be explained by chemical processes in shocked gas. Elitzur (1983), on the other hand, suggested that the high observed HCO^+ abundance could be due to an increase in the ionization rate from cosmic rays trapped in the supernova remnant. The suggestion of Elitzur is plausible, but has not been tested by a detailed calculation. The purpose of the present work is to see whether shock models are consistent with recent molecular observations and to assess the effects of an increased cosmic ray ionization rate on molecular abundances. New calculations of molecular abundances behind shocks have been carried out using improved chemistry and using a range of ionization rates.

2. THE OBSERVATIONS: The extraction of a column density from a line intensity requires the excitation temperature, a quantity whose value is often not known. The dependence of abundances on excitation temperature can be partially removed by taking an abundance ratio. Abundances relative to CO, N(X)/N(CO), are often used. Table 1 lists abundances relative to CO for twelve species at four different positions in IC 443. The positions are given in a footnote to Table 1. The abundances at position 1 are from DeNoyer and Frerking (1981) for an excitation temperature of 100 K. All other abundances in Table 1 are from White et al. (1987). A Large Velocity Gradient (LVG) model has been applied by White et al. to their observations at positions 2 and 4 of Table 1. The LVG entries in Table 1 were obtained using a kinetic temperature of 100 K. A comparison of the optically thin and LVG abundances of HCO^+, HCN, and CS for position 2 is instructive: The LVG abundance ratios for these species are 10 to 50 times larger than the abundance ratios for optically thin transitions. White et al. (1987) conclude that the assumption of optical thinness is incorrect. There are, of course, large uncertainties in the LVG abundances themselves. For example, the shocked gas is found to be clumpy on very small scales (Burton et al. 1987), so that the use of a single density is an oversimplification. Also, the deduced LVG abundances are quite sensitive to the kinetic temperature.

3. MODEL ABUNDANCES COMPARED TO OBSERVATIONS: Table 2 gives shocked column densities relative to CO for a model in which a shock of 10 km s^{-1} propagates into gas with an initial density of 100 cm^{-3}. The cosmic ray ionization rate, ζ_0, is 10^{-17} s^{-1}. Abundances are given for species which have been detected in IC 443 and for species with observational upper limits to abundances. Column density ratios are tabulated for three postshock times, 100 years, 400 years, and 1200 years. These numbers represent possible times since the observed gas was shocked. For IC 443, we do not know the time since the remnant overtook the molecular cloud. A recent estimate of the age of IC 443, 5000 years (Mufson et al. 1986), puts an upper limit on the postshock time.

A comparison of Tables 1 and 2 shows that the model for a 400 year old shock gives the best overall agreement with observations. In particular, the model abundances for HCO^+ and CS are remarkably close to the LVG abundances of White et al. (1987). The model ratio, OH/CO = 0.24, is very close to the observed ratio, 0.32, at position 1. For HCN, the model ratio at 400 years is somewhat high, being four times higher than the LVG ratio at position 4 and ten times higher than the LVG ratio at position 2. The model is consistent with the observational upper limits for CH_3OH, N_2H^+, SO, SO_2, OCS, and HCS^+. The model is in disagreement with observed upper limits to H_2CO and H_2CS: The model produces too much H_2CO (by a factor of 50) and too much H_2CS (by a factor of 60).

TABLE 1
IC 443 OBSERVATIONS: COLUMN DENSITIES RELATIVE TO CO

SPECIES	Position 1[*] opt. thin	Position 2 opt. thin	Position 2 LVG	Position 3 opt. thin	Position 4 LVG
HCO^+	7.0(-4)	8.2(-4)	7.7(-3)	8.2(-4)	6.3(-3)
HCN	2.3 (-4)	<8.8(-5)	5.0(-3)	4.4(-4)	1.3(-2)
OH	0.32	-----	-----	-----	-----
H_2CO	<6.1(-5)	-----	-----	-----	-----
CH_3OH	-----	<7.0(-3)	-----	<9.0(-3)	-----
N_2H+	<1.1(-5)	-----	-----	-----	-----
CS	3.2(-4)	6.8(-5)	3.2(-3)	3.8(-5)	-----
SO	<1.2(-4)	<2.2(-4)	-----	<2.6(-4)	-----
SO_2	-----	<3.0(-3)	-----	<3.0(-3)	-----
H_2CS	-----	<8.0(-5)	-----	<1.4(-4)	-----
OCS	<7.2(-4)	-----	-----	-----	-----
HCS^+	-----	<9.4(-5)	-----	<2.0(-4)	-----

[*] Position 1: RA(1950) = 6^h 14^m 15^s, Dec(1950) = 22° 27' 50"
Position 2: RA(1950) = 6^h 14^m 41.9^s, Dec(1950) = 22° 22' 40"
Position 3: RA(1950) = 6^h 14^m 44.1^s, Dec(1950) = 22° 23' 30"
Position 4: RA(1950) = 6^h 14^m 42.0^s, Dec(1950) = 22°33' 40"

Formaldehyde is produced by the reactions $CH_3 + O \rightarrow H_2CO + H$ and $CH_2 + OH \rightarrow H_2CO + H$. A lower H_2CO would follow if less CH_3 and CH_2 were formed. This is not expected, however, since CH_3 and CH_2 are formed in hot gas which has either appreciable C or appreciable C^+. A low H_2CO abundance is difficult to understand since C is observed to be abundant in dense clouds and C^+ is abundant in lower density gas. Enhancement of formaldehyde is a rather robust prediction of shock models because it can be formed via C or C^+. Further observations of formaldehyde in IC 443 would be useful. Thioformaldehyde is formed in the postshock gas from methane via $S^+ + CH_4 \rightarrow H_3CS^+ + H$, followed by dissociative recombination. Less H_2CS would be produced if the preshock gas contained less S^+ (i. e. if the preshock gas were denser) or if methane did not become abundant in the postshock gas. It remains, of course, possible that one or more rapid destruction processes for H_2CO and/or H_2CS have not yet been recognized.

4. AN INCREASED COSMIC RAY IONIZATION RATE: Calculations were carried out for the shock models described above but with larger cosmic ray ionization rates. To illustrate the effect of the ionization rate on abundances, Figure 1 shows fractional abundances as a function of postshock time for the ions HCO^+ and N_2H^+ and the neutrals OH and HCN. Abundances are given for two ionization rates, the accepted mean value of $\zeta_o = 10^{-17}$ s^{-1}, and an enhanced ionization rate of 10^{-15} s^{-1}. It can be seen from Figure 1 that abundances begin to be affected by the change in ζ_o after several hundred years. N_2H^+ is more abundant in the high ionization case at early times, but does

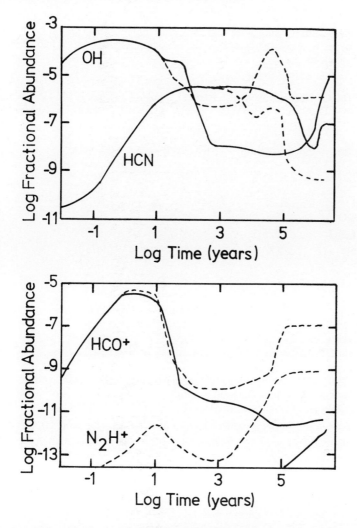

Fig. 1. Fractional abundances, $n(X)/n$, as a function of postshock time for several species. Solid curves represent a model with $\zeta_o = 10^{-17}$ s^{-1} and dashed curves represent a model with $\zeta_o = 10^{-15}$ s^{-1}. Shock speed is 15 km s^{-1} and initial gas density is 100 cm^{-3}.

not become abundant until some 10^5 years after the shock. The fractional abundance of HCO$^+$ is larger in the high ionization model, being about 10 times larger by 5,000 years. The shocked *column density* after 5,000 years is, however, dominated by HCO$^+$ near the shock front: The shocked column density of HCO$^+$ after 5,000 years is essentially identical for the two ionization rates. The abundance of OH becomes much larger, in the high ζ_o model, after several hundred years. Again, however, the column density of OH is dominated by the abundant OH near the shock front (i. e. at early times in Figure 1): The shocked column density of OH is not influenced by the ionization rate until $\sim 10^4$ years after the shock. At 10^4 years, the column density of OH in the $\zeta_o = 10^{-15}$ s^{-1} model is ~ 3 times larger than in the standard ($\zeta_o = 10^{-17}$ s^{-1}) model. For the higher ionization model, the fractional abundance of HCN becomes *lower* after several hundred years. Again, the column density of HCN is little affected by the change in ζ_o until several thousand years after the shock: At 5,000 years, the column density of HCN is ~ 3 times lower with the higher ionization rate. Because the encounter of IC 443 with the molecular gas probably began less than 5,000 years ago, any increase in the cosmic ray ionization rate has not yet had an appreciable effect on column abundances.

TABLE 2

SHOCKED COLUMN DENSITIES RELATIVE TO CO FOR A MODEL WITH
AN INITIAL DENSITY OF 100 CM^{-3} AND A SHOCK SPEED OF 10 KM S^{-1}

| Species | N(X)/N(CO) | | |
	100 years	400 years	1200 years
HCO$^+$	2.9(-2)	5.4(-3)	1.6(-3)
HCN	5.0(-2)	5.3(-2)	5.0(-2)
OH	1.2	0.24	7.1(-2)
H$_2$CO	2.8(-3)	3.0(-3)	2.8(-3)
CH$_3$OH	9.2(-6)	1.6(-5)	3.2(-5)
N$_2$H$^+$	1.8(-9)	3.6(-10)	1.2(-10)
CS	2.3(-3)	7.8(-3)	1.1(-2)
SO	7.8(-5)	1.1(-4)	6.1(-5)
SO$_2$	1.7(-6)	3.0(-6)	4.1(-6)
H$_2$CS	9.3(-4)	6.5(-3)	9.0(-3)
OCS	2.2(-6)	4.5(-6)	2.7(-6)
HCS$^+$	3.4(-5)	1.7(-5)	4.3(-6)

5. CONCLUSIONS: (1) The observed high abundances of HCO$^+$, OH, and CS in IC 443 can be explained by shock models; (2) The models produce more HCN than is observed, by factors of 4 to 10; (3) The models are consistent with observed upper limits to CH$_3$OH, N$_2$H$^+$, SO, SO$_2$, OCS, and HCS$^+$; (4) The models overproduce H$_2$CO and H$_2$CS by a factor of 50; (5) If the supernova event occurred about 5,000 years ago, as models suggest, any enhancement in the cosmic ray ionization rate has had a negligible effect on molecular column abundances.

REFERENCES

Burton, M. G., Geballe, T. R., Brand, P. W. J. L., and Webster, A. S. 1987, preprint.
DeNoyer, L. K. 1979a, Ap. J. (Letters), 228, L41.
DeNoyer, L. K. 1979b, Ap. J. (Letters), 232, L165.
DeNoyer, L. K. and Frerking, M. A. 1981, Ap. J. (Letters), 246, L37.
Dickenson, F. D., Kuiper, E. N. R., Dinger, A. S., and Kuiper, T. B. H. 1980, Ap. J. (Letters), 237, L43.
Elitzur, M. 1983, Ap. J., 267, 174.
Fesen, R. A. and Kirschner, R. P. 1980, Ap. J., 242, 1023.
Mitchell, G. F. and Deveau, T. J. 1983, Ap. J., 266, 646.
Mufson, S. L., McCollough, M. L., Dickel, J. R., Petre, R., White, R., and Chevalier, R. 1986, A. J., 92, 1349.
Treffers, R. B. 1979, Ap. J. (Letters), 233, L17.
White, G. J., Rainey, R., Hayashi, S. S., and Kaifu, N. 1987, Astron. Ap., 173, 337.

[Ni II] EMISSION IN SUPERNOVA REMNANTS

Richard B.C. Henry, University of Oklahoma

and

Robert A. Fesen, University of Colorado

ABSTRACT: Combining new spectrophotometric data on the Orion and Crab Nebulae with detailed calculations, we show that strong [Ni II] λ7378 emission observed in numerous SNR's may be associated with gas having $N_e > 50,000 \; cm^{-3}$.

INTRODUCTION: Strong [Ni II] emission at λ7378 in the Crab Nebula was first observed by Miller (1978). Subsequent spectrophotometric measurements by Péquignot and Dennefeld (1983) and Henry, MacAlpine, and Kirshner (1984) indicated that the strength of the λ7378 line rivals that of Hβ throughout the entire filament system. Initial interpretations of these findings suggested that the ratio of nickel to iron exceeds its solar value by more than an order of magnitude (see Henry 1984a,b; Péquignot and Dennefeld 1983). Recently, observations by Dennefeld (1986) of two Galactic and three extragalactic supernova remnants indicated the presence of strong [Ni II] emission in these objects as well, leading again to inferred $\frac{Ni}{Fe}$ which is roughly an order of magnitude above solar. Such an abundance scenario would be compatible with supernova ejecta which contain large amounts of material synthesized under neutron-rich conditions, ie. deep inside the core of the precursor star. However, the overabundance interpretation is seriously challenged by the ubiquitous presence of relatively strong [Ni II] emission in objects where elemental enhancements are not expected, such as the SNR IC 443 (Fesen and Kirshner 1980), Seyfert galaxies (Halpern and Oke 1986), and Herbig-Haro objects (Brugel et al. 1981).

THE APPARENT NICKEL ABUNDANCE IN ORION: Henry (1984a,b) found an apparent abundance ratio of $\frac{Ni}{Fe}$ in Orion which was roughly an order of magnitude higher than the solar value, based upon observations by Grandi (1975) of [Ni II] λ7378 and [Fe II] λ8617 at one location in that Nebula. (The *apparent* abundance ratio refers to the value derived from a straightforward interpretation of line strengths, ignoring any unusual mechanisms which could alter the observed ratio.) In order to substantiate this conclusion, we have determined the apparent abundance of Ni at three additional locations in the Orion, this time with respect to O, since the abundance of the latter is well established. We employed the Coudé feed telescope as well as the IIDS attached to the 2.1m telescope, both at KPNO, to obtain spectrophotometric measurements of [Ni II] λ7378 and [O II] λ7325. The three locations correspond roughly to Peimbert and Torres-Peimbert's (1977) positions 1a, 3a, and 4a.

In order to determine the apparent $\frac{Ni}{O}$ abundance ratio at each of the positions, a general relation between the observed line ratio, $\frac{I(\lambda 7378)}{I(\lambda 7325)}$, and the number abundance ratio, $\frac{Ni}{O}$, was established for H II regions by calculating numerous photoionization models over a range of stellar effective temperatures and total nebular densities. We employed the code recently used by Henry and Shipman (1986) but with the addition of nickel to the elements included in the calculations. Photoionization cross-sections for nickel were taken from Reilman and Manson (1979).

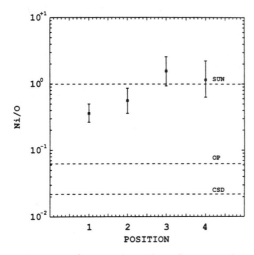

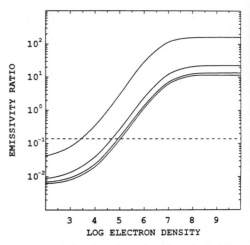

Fig. 1 - Ni/O number abundance ratio, normalized to the solar value, for four positions observed in Orion. Dashed lines toward the bottom are observed Fe/O ratios (see text).

Fig. 2 - Calculated ratio of emissivities of [Ni II] $\lambda7378$/[S II] $\lambda6724$, as a function of electron density for 1000K (top curve), 5000K, 10000K, and 15000K respectively. The dashed line shows the mean observed ratio.

Since radiative and dielectronic recombination coefficients for nickel are not available in the literature, we employed the values for iron published by Woods, Shull, and Sarazin (1981). This is a reasonable substitute, due to the close similarities of the nickel and iron photoionization cross-sections, from which recombination cross-sections are calculated. In addition, we included a charge exchange rate for the $Ni^{+2} + H^o \rightarrow Ni^+ + H^+$ reaction calculated by Neufeld (1985). Finally, the collision strengths and transition probabilities for Ni^+ from Nussbaumer and Storey (1982) were used.

Using a relation derived in this way, we determined the apparent $\frac{Ni}{O}$ ratio at each of our three positions plus that of Grandi (1975) using observed line strengths from our and Grandi's observations, as well as relevant data from Peimbert and Torres- Peimbert (1977) which were necessary to determine the temperature and density at each of the locations. Our resulting abundance ratios are presented in Fig. 1, where we have plotted the $\frac{Ni}{O}$ ratio, normalized to its solar value, for each position. The errors in the abundances include contributions from estimated observational uncertainties in line strengths, temperatures, densities, and the abundance-determining technique itself. The two dashed lines toward the bottom of the plot indicate the $\frac{Ni}{O}$ levels which would be expected in Orion if Ni and Fe are depleted by the same fractions. The levels shown correspond to determinations of the Fe abundance by Cosmovici et al. (1980) and Olthoff and Pottasch (1975). We adopted the O abundance of Peimbert and Torres-Peimbert (1977), who found O to be present at a level that was very nearly solar.

Our results indicate that the apparent nickel abundance in Orion is roughly the same as it is in the sun, ie. more than an order of magnitude greater than the expected level. This finding agrees closely with the much simpler analysis in Henry (1984a,b). Unfortunately, no other strong lines of [Ni II] are observable in the optical, UV, or near IR which would allow independent checks on this result.

PRODUCTION OF [Ni II] λ7378 EMISSION: The chemical properties of nickel and iron are very similar, due to the similarity of the electron structures of the two elements. In the absence of evidence to the contrary, we must conclude that nickel and iron form grains at roughly equal rates. Thus, the possibility that iron fractionates out of gas and into the solid phase more readily than nickel does seems an unlikely explanation for the apparent overabundance of gas-phase nickel. We therefore proceed under the assumption that the apparent nickel excess is the result of either: 1) an enlarged Ni^+ zone in the nebula; 2) an excitation process other than pure inelastic collisions; 3) erroneous atomic data; or 4) unusual temperature or electron density conditions which favor λ7378 production.

An enlargement of the Ni^+ zone would be produced if: 1) the photoionization cross-section of Ni^o were larger; 2) the total recombination rate onto Ni^+ were larger; or 3) the total recombination rate onto Ni^o were smaller. (The total recombination rate refers to the sum of radiative, dielectronic, and charge exchange recombination rates.) In any of the three cases, the effect is to increase the relative amount of Ni^+ at the expense of other ions of nickel. Since we are searching for order-of-magnitude effects, in order to enlarge the Ni^+ zone sufficiently, we must presume that the photoionization or recombination cross-sections are in error by an order of magnitude. For charge exchange, the coefficients calculated by Neufeld (1985) would have to be underestimated by a similar amount. On the other hand, the emissivity of λ7378 could be increased above inelastic collision levels by excitation of the line through recombination. At the same time, order of magnitude errors in the published values for collision strengths and transition probabilities would result in an underestimation of λ7378 emissivity. Finally, photon pumping mechanisms appear unimportant, since none of the excitation energies from the ground state of [Ni II] coincide with strong nebular lines.

To test the effects of many of the above processes, we began by calculating a standard photoionization model for the Crab Nebula filaments which roughly fit the average line strengths observed by Davidson (1987) and Fesen and Kirshner (1982). We employed the solar ratio by number of $\frac{Ni}{H}$. This model produced a strength of λ7378 which was more than a factor of 15 below observations. Next, we experimented with large changes in the relevant cross-sections, rate coefficients, or other atomic data discussed above and noted the effects such changes had on the relative strength of λ7378. All experiments failed to uncover a process which is capable of explaining the strong observed λ7378 emission in the Crab.

We then abandoned the use of models and considered physical conditions under which [Ni II] λ7378 emissivity, $j_{\lambda 7378}$, would be expected to be high with respect to lines formed in the same region of the nebula by ions of more abundant elements. In particular, considerations of ionization potentials as well as ionization structure information gleaned from our photoionization models indicated that λ7378 should be produced coincidentally with [S II] λ6724. Therefore, we calculated the expected ratio of emissivities for a solar value of $\frac{Ni}{S}$ over a large range of temperatures and densities. Our results are shown in Fig. 2, where the dashed horizontal line indicates the mean observed value of the ratio of these two lines. From observational and model results, we expect the electron temperature of the gas to be between 5000K and 10,000K. Therefore, observations of $\frac{I(\lambda 7378)}{I(\lambda 6724)}$ in the Crab imply densities between $10^{4.7}$ and 10^5. These densities exceed the critical density of λ6724 ($\sim 10^4 cm^{-3}$) but are below that of λ7378, ($\sim 10^7 cm^{-3}$), and thus under such conditions [Ni II] λ7378 is produced more efficiently. However, since [S II] densities observed in the Crab are normally $\sim 10^3 cm^{-3}$, we speculate that the low-ionization regions of the filaments may actually be composed of two components: a low density component $(N_e \sim 10^3 cm^{-3})$ which produces most of the [S II] emission, and a second component with a density exceeding $10^{4.7} cm^{-3}$ which produces the [Ni II] emission.

Certain observations will help to clarify this problem somewhat. We have attempted measurements of other optical lines of [Ni II] such as $\lambda 4326$, $\lambda 3999$, and $\lambda 6666$ in an effort to check results against predicted ratios of these lines to $\lambda 7378$ for verification that the observed emission in the latter is due primarily to collisional excitation. However, because of the weakness of these lines, definite conclusions cannot be drawn at the present time. Of considerable value for helping to confirm the excitation mechanism of $\lambda 7378$ would be the measurement of the strengths of [Ni II] lines at 1.19μ, 6.63μ, and 10.7μ, lines which are predicted by Henry (1987) to rival the strength of $\lambda 7378$. In addition, Nussbaumer and Storey (1982) have shown that $\frac{I(\lambda 7412)}{I(\lambda 7378)}$ is an effective temperature diagnostic, and Henry's (1987) results imply that $\frac{I(1.19\mu)}{I(\lambda 7378)}$ is a particularly good probe of electron density. Thus observations of these ratios could help to establish the conditions in the gas that is responsible for the strong [Ni II] emission observed in the Crab Nebula and other SNRs.

REFERENCES

Brugel, E.W., Böhm, K.H., and Mannery, E. 1981, *Ap.J. Suppl.*, **47**, 117.
Cosmovici, C.B., Strafella, F., and Dirscherl, R. 1980, *Ap.J.*, **236**, 498.
Davidson, K. 1987, preprint.
Dennefeld, M. 1986, *Astr. Ap.*, **157**, 267.
Fesen, R.A., and Kirshner, R.P. 1980, *Ap.J.*, **242**, 1023.
———. 1982, *Ap.J.*, **258**, 1.
Halpern, J.P., and Oke, J.B. 1986, *Ap.J.*, **301**, 753.
Grandi, S.A. 1975, *Ap.J. Letters*, **199**, L43.
Henry, R.B.C. 1984a, in *Stellar Nucleosynthesis*, ed. C. Chiosi and A. Renzini (Dordrecht: Reidel), p. 43.
———. 1984b, *Ap.J.*, **281**, 644.

———. 1987, *Ap.J.*, **322**, in press.
Henry, R.B.C., MacAlpine, G.M., and Kirshner, R.P. 1984, *Ap.J.*, **278**, 619.
Henry, R.B.C., and Shipman, H.L. 1986, *Ap.J.*, **311**, 774.
Miller, J.S. 1978, *Ap.J.*, **220**. 490.
Neufeld, D.A. 1985, private communication.
Nussbaumer, H., and Storey, P.J. 1982, *Astr. Ap.*, **110**, 295.
Olthof, H., and Pottasch, S.R. 1975, *Astr. Ap.*, **43**, 291.
Peimbert, M., and Torres-Peimbert, S. 1977, *M.N.R.A.S.*, **179**, 217.
Péquignot, D., and Dennefeld, M. 1983, *Astr. Ap.*, **120**, 249.
Reilman, R.F., and Manson, S.T. 1979, *Ap.J. Suppl.*, **40**, 815.
Woods, D.T., Shull, J.M., and Sarazin, C.L. 1981, *Ap.J.*, **249**, 399.

THE OXYGEN-RICH SUPERNOVA REMNANT IN
THE SMALL MAGELLANIC CLOUD

William P. Blair
The Johns Hopkins University, Baltimore, Md., U.S.A.

John C. Raymond
Harvard-Smithsonian Center for Astrophysics, Cambridge, Ma., U.S.A.

John Danziger and Francesca Matteucci
European Southern Observatory, Garching bei Munchen, West Germany

Abstract: We report ultraviolet and optical spectra of 1E 0102-7219, the oxygen-rich supernova remnant in the Small Magellanic Cloud. The UV data contain strong lines of oxygen, carbon, neon, and magnesium. OI recombination lines in the optical and UV permit the relative line intensities to be determined from 1200Å to 1 micron. Models assuming shock excitation and X-ray photoionization have been calculated and compared with the observations.

I. Introduction

The supernova remnant 1E 0102-7219 was first identified in the *Einstein* X-ray survey of the Small Magellanic Cloud (SMC) by Seward and Mitchell (1981). Measurements with the Imaging Proportional Counter on *Einstein* showed the object to be the second brightest X-ray source in the SMC and to have a relatively soft X-ray spectrum, similar to SNRs in our Galaxy and the Large Magellanic Cloud (LMC). However, the IPC count rate was ≥ 10 times that of the next brightest SMC remnant.

Optically, Dopita, Tuohy and Mathewson (1981) identified the remnant with a faint ring of filaments ~ 24 arcsec in diameter visible only in the light of [O III] $\lambda 5007$, adjacent to the H II region N76A. At a distance of 59 kpc for the SMC, the diameter of the [O III] ring is 6.9 pc. The spectroscopic peculiarities of the SNR were demonstrated by Tuohy and Dopita (1983): spectra of individual knots showed strong lines of oxygen and neon, but nothing else. Velocity mapping indicated an overall expansion velocity of ~ 6500 km s^{-1}, perhaps distributed in a twisted ring morphology. These characteristics imply an expansion age for the remnant (assuming no deceleration) of 2150 years.

Objects with inferred large abundances of such elements as oxygen, neon, sulfur, etc., are thought to represent the explosions of massive (Population I) stars. Nuclear burning in the progenitor stars had progressed to or perhaps even beyond the oxygen-burning stage when the SN explosion took place. Other members of the class include N132D and 0540-69.3 in the LMC , Cas A and G292.0 + 1.8 in our Galaxy, and the powerful young SNR in the irregular galaxy NGC 4449. More recently, some knots dominated by oxygen emission have been identified in Puppis A (Winkler and Kirshner 1985), although this remnant is thought to be considerably older.

Some heavy elements such as carbon, silicon, and magnesium have no strong lines in the optical and their abundances are unconstrained by optical spectra alone. Also, so few lines are available in the optical that it is difficult to constrain models for the emission. Since the reddening to 1E 0102-7219 is fairly small, we have used the IUE Observatory to obtain UV spectra of this remnant. New optical data have also been obtained, extending spectral coverage to 1 micron. These new data provide considerable insight into the physical conditions and possible emission mechanisms in this remnant.

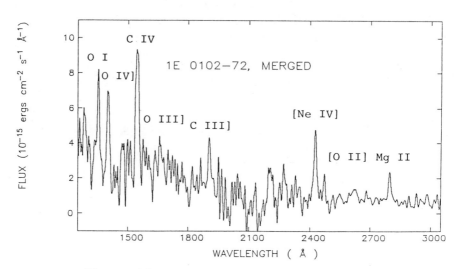

Figure 1: Merged SWP+LWP spectrum of 1E 0102-7219.

Table 1

Optical/UV Line Intensities for 1E 0102-7219

Ion	λ	$F(\lambda 5007{=}100)$	$I(\lambda 5007{=}100)$
O I]	1356	12.1	20.2
Si IV/O IV]	1400	19.6	31.5
C IV	1549	35.7	52.0
O III]	1664	6.0	8.4
C III]	1909	8.0	10.6
[Ne IV]	2425	15.2	18.3
[O II]	2470	4.9	5.8
Mg II	2798	5.9	6.4
[Ne V]	3426	9.5	10.8
[O II]	3727	150	164
[Ne III]	3869	19.0	20.7
[O III]	4363	4.6	4.8
[O III]	5007	100	100
[O I]	6300	5.0	4.7
[O II]	7320,30	10.0	9.0
O I	7774	3.6	3.2

Table Notes: 1) UV to optical scaling assumes a theoretical ratio of $I(1356)/I(7774) = 5.9$.
2) Reddening correction assumes $E(B-V) = 0.04$ (galactic) plus $E(B-V) = 0.04$ (SMC).

II. Observations

For the IUE observations, we combined pairs of back-to-back ESA and US1 shifts to obtain long (14+ hour) SWP and LWP exposures of the SNR. The observations were scheduled to permit the long dimension of the large aperture to be roughly aligned along the bright southeastern rim of the [O III] ring visible in Tuohy and Dopita's (1983) photograph. Because of the long exposures, the background levels are fairly high, but emission lines from the SNR are clearly visible on the photowrites as diffuse patches of emission.

We have extracted the spectra from the line-by-line data in order to optimize the signal-to-noise ratio of the resulting spectra. The merged SWP+LWP spectrum is shown in Figure 1 after removal of hits and reseau marks and after smoothing. The strongest line is C IV λ1550, and C III] λ1909 and the Si IV/O IV] blend at λ1400 are also seen protruding from a noisy continuum. In the LWP range, only three lines are clearly seen: [Ne IV] λ2425, [O II] λ2470 and Mg II λ2800. A surprising feature in this spectrum is the strong line at $\sim$ 1354 Å which we identify as a recombination line of O I. Conspicuous by its near absence is the O III] λ1664 line, which was the only line detected in the UV spectrum of the SNR in NGC 4449 (Blair *et al.* 1984); interestingly, no carbon lines were seen in the NGC 4449 remnant.

Our optical data were obtained with long slit CCD spectrographs on the ESO 3.6 m telescope and the 4 m telescope at CTIO. In the 7-10000 Å region, several permitted O I transitions are seen, the strongest of which is at 7774 Å. The relative intensities of these lines indicate partially optically thin conditions in the recombining gas. Since the 7774 Å line feeds the 1356 Å line seen in the UV, the line intensities throughout the optical/UV region can be scaled. An abbreviated list of the most important lines is given in Table 1.

III. Models

Shocks are a natural explanation for the observed emission in that the ejecta encounter a 'reverse' shock when they overtake the decelerating high pressure shell of swept-up interstellar material. While most of the material passing through the reverse shock becomes so hot it emits at X-ray, rather than optical, wavelengths, the slower shocks in dense clumps of ejected gas can produce optical emission. Chevalier and Kirshner (1978) interpreted the Fast Moving Knots in Cas A as such shocks, and this picture is supported by the sudden appearance and disappearance of many of these knots (*cf.* van den Bergh and Kamper 1983).

However, only a few shock models have been calculated for abundances appropriate to oxygen-rich remnants. Itoh (1981) constructed models of radiative shocks in pure oxygen and extended them to include O I recombination (Itoh 1987). Dopita, Binette and Tuohy (1984; hereafter DBT) computed models for a mixture of C, O and O-burning products expected from a 25 $M_\odot$ precursor. We have used a modified version of the radiative shock code described in Cox and Raymond (1985) to compute a few models with higher or lower shock velocity than those of Itoh or DBT, models with different abundance sets, and models which include the observed X-ray emission of 1E 0102-7219. All these models assumed instant electron-ion equilibration. Aside from a general confirmation of the published models, we find that the X-ray emission alone can completely ionize any gas less dense than about 1 cm^{-3}, so that even precursors of very slow shocks can be ionized.

Problems arise in attempting to match the observed high ionization species Ne IV and Ne V, and in matching the ratios of UV lines to the optical. Models which produce [O II]/[O III] and [O I]/[O III] comparable to the observed values predict no emission at all in [Ne IV] or [Ne V], because the gas never reaches high ionization states. Models which collisionally ionize oxygen to O^{++} have high enough temperatures to excite strong UV emission, but the observed UV lines are rather weak. The [Ne IV] and [Ne V] problem could be explained by assuming a wide range of shock velocities within the observed volume, but any shocks which produce these lines would give a C IV λ1550/C III] λ1909 ratio far larger than observed. Also, this would not alleviate the problem with the O III UV to optical line ratio.

The X-ray luminosity of 1E 0102-7219 is 1.5×10^{37} ergs s^{-1}, so X-ray photoionization may be considerable. Since detailed measurements of the X-ray spectrum are not available, we have used the X-ray emission code of Raymond and Smith (1977) with updated atomic rates to generate model X-ray spectra covering the 0-2 KeV range at 4 eV resolution. Either the DBT abundance set or a pure oxygen gas was assumed. Temperatures of 2.5×10^6 and 6

$\times 10^6$ K were chosen. The spectra were then scaled to match the observed flux from the remnant. Each X-ray spectrum was used to compute the ionization state and heating rate of gas at various densities, and a temperature at which heating balances cooling was found.

From inspection of the models, photoionized gas with a range of densities emits strongly in [O III]. It can produce enough [O I] $\lambda 6300$ and O I $\lambda 7774$ emission in some of the denser regions, and it naturally produces the observed modest UV to optical line ratios. There are two difficulties, however. First, the predicted [O II] $\lambda 3727$/[O III] $\lambda 5007$ intensity ratio is less than 0.3 in all cases, while the observed ratio varies between about 1.3 and 2. Second, the [O III] emission is dominated by 10-15000 K gas, while the temperature-sensitive [O III] 4363/5007 ratio indicates a temperature of 25000 K. The O III line ratio might be explainable by a recombination contribution to the $\lambda 4363$ line. The weakness of the predicted [O II] seems to be a more fundamental problem, reflecting the fact that at densities high enough that O^+ is more abundant than O^{++}, the temperatures are too low to overcome the 3.3 eV excitation potential of the $\lambda 3727$ lines.

The fault probably lies in the assumption of ionization equilibrium. While the heating and cooling time scales are a few years, the ionization time scale is ~ 2000 years, which is roughly the expansion age of the remnant. Thus the gas may be less highly ionized than predicted by the equilibrium models for a given density, and it will probably be warmer, since the power provided by X-ray heating is shared among fewer electrons. A time-dependent model would probably give an [O II]/[O III] ratio in agreement with that observed. It could probably also account for the high ionization stages--C IV, Ne IV and Ne V--because only two Auger ionizations are required.

Although we have been unable to determine a unique model, the combined optical/UV data shed light on several aspects of this remnant. Many of the general characteristics of the observed emission can be accounted for with the X-ray photoionization models, but shocks may also play an important role. Oxygen is the dominant element, but substantial amounts of neon, carbon and magnesium are also present. The relative abundances are crudely like those expected from a star of about 20 $M_\odot$. Further refinements in the relative abundances should permit the mass of the precursor star to be determined more accurately.

References

Blair, W.P., Raymond, J.C., Fesen, R.A., and Gull, T.R. 1984, *Ap. J.*, **279**, 708.

Chevalier, R.A. and Kirshner, R.P. 1978, *Ap. J.*, **219**, 931.

Cox, D.P. and Raymond, J.C. 1985, *Ap. J.*, **298**, 651.

Dopita, M.A., Binette, L., and Tuohy, I.R. 1984, *Ap. J.*, **282**, 142.

Dopita, M.A., Tuohy, I.R., and Mathewson, D.S. 1981, *Ap. J. (Letters)*, **248**, L105.

Itoh, H. 1981, *Pub. Astr. Soc. Japan*, **33**, 1.

----------. 1987, *Pub. Astr. Soc. Japan*, , in press.

Raymond, J.C. and Smith, B.W. 1977, *Ap. J. Suppl.*, **35**, 419.

Seward, F.D. and Mitchell, M. 1981, *Ap. J.*, **243**, 736.

Tuohy, I.R. and Dopita, M.A. 1983, *Ap. J. (Letters)*, **268**, L11.

van den Bergh, S. and Kamper, K.W. 1983, *Ap. J.*, **268**, 129.

Winkler, P.F. and Kirshner, R.P. 1985, *Ap. J.*, **299**, 981.

LIMITS ON THE PRESENCE OF SULFUR IN THE YOUNG SNRs LMC N132D AND SMC 1E0102.2-7219

Barry M. Lasker
Space Telescope Science Institute
Baltimore, MD, USA.

Abstract: A new area-spectroscopy search for sulfur as a tracer of the "Si Group" of elements in the ejecta of the young SNRs, LMC N132D and SMC 1E0102.2-7219, yields negative results for [S II]$\lambda\lambda$6719, 6731 and [S III]$\lambda\lambda$9064, 9532.

Observations of very young supernova remnants (SNRs) offer the opportunity to observe interstellar material consisting of relatively uncontaminated stellar ejecta whose composition and kinematics are directly pertinent to the progenitor star. Members of this relatively rare class are the galactic objects Cas A, Puppis A, and G292.0+1.8; N132D and 0540−69.3 in the LMC; 1E0102.2-7219 in the SMC; and an unresolved object in NGC 4449 (*cf.*, summary table in Winkler and Kirshner 1985). Of these, we have selected the Magellanic Cloud objects, N132D and 1E0102.2-7219, which have known distances, lie in relatively unobscured (albeit cluttered) fields, and have angular sizes of the order of 20″, for further study with the CTIO 1.5 m telescope*.

Previous works establish that these SNRs consist of ring-like structures a few pc in diameter expanding with velocities in the range 2000-3000 km/sec and having kinematical ages of the order of 1000 yr, that the expanding material contains no hydrogen or helium but consists primarily of oxygen, and that stationary material of relatively normal composition is also present (Dopita and Tuohy 1984; Lasker 1980).

The absence of spectroscopic evidence for products of nucleosynthesis from the "Si Group," particularly S, which we may reasonably expect at the level of $\approx 0.1 M_\odot$ (compared to $3 M_\odot$ for O) from models of stellar evolution (*e.g.*, Johnston and Yahil 1984) motivates a further search. (Abundances from the Johnston-Yahil model are much higher for the dominant species, Si, which does not have any convenient lines in the optical, and about a factor of 10 lower than S for Ca and Ar.)

In order to take observations as independent as possible from specific models of the ionization and excitation conditions in the ejecta, we included observations of both accessible ionization states of sulfur (S$^+$ and S^{++}), the ionization potentials of which bracket that of O$^+$, which is known to be present. We used the CTIO 1.5 m telescope with the cassegrain spectrograph and the GEC CCD to obtain an area map covering the spectral regions $\lambda\lambda$6200–7700, sensitive to [S II]$\lambda\lambda$6719, 6731 and $\lambda\lambda$8300–9900, sensitive to [S III]$\lambda\lambda$9064, 9532. The pixel size was 2″ along the slit,

---

* The younger LMC object, 0540−69.3 was excluded from this program because of its smaller angular size; however, one does note that Dopita and Tuohy (1984) have detected Doppler-shifted [S II] and Ca I] in it.

which was oriented E-W, and 6″ perpendicular to it. In each SNR an area the size of the expanding ring was searched with a grid of 600 sec exposures; additionally, one location in each, selected for the presence of rapidly moving [O III]λ5007, was examined for [S III] with exposures totaling 6000 sec.

No new rapidly moving spectral features were detected in a preliminary inspection of the spectra. The corresponding upper limits, which vary with the exposure time, the assumed area of the rapidly moving material, and the presence of night-sky emission lines at specific search wavelengths, are in the range 10^{-4} to 10^{-6} erg sec^{-1} sr^{-1} and further reductions to improve the limits are in progress. Depending on physical conditions in the ejecta, the lower of these values may be relevant to stellar evolution models.

Space Telescope Science Institute is operated by the Association of Universities for Research in Astronomy, Inc. (AURA), under contract to the National Aeronautics and Space Administration. The Cerro Tololo Interamerican Observatory is operated by AURA under contract to the National Science Foundation.

REFERENCES

Dopita, M. A., and Tuohy, I.R. 1984 *Ap.J.*, **282**, 135.
Johnston, M. D., and Yahil, A. 1984 *Ap.J.*, **285**, 587.
Lasker, B. M. 1980, *Ap.J.*, **237**, 765.
Winkler, P. F., and Kirshner, R. P. 1985 *Ap.J.*, **299**, 981.

LONG SLIT ECHELLE SPECTROSCOPY OF SUPERNOVA REMNANTS IN M33

William P. Blair
The Johns Hopkins University, Baltimore, Md., U.S.A.

You-Hua Chu
The University of Illinois, Urbana, Il., U.S.A.

Robert C. Kennicutt
The University of Minnesota, Minneapolis, Mn., U.S.A.

Abstract: We have obtained long slit echelle spectroscopy for 10 of the brightest supernova remnants in M33 using the KPNO 4 m telescope. The profiles at Hα indicate bulk motions in the range 100-350 km s^{-1} in these remnants. Nearly all of the objects show signs of contamination by low velocity H II emission at some level. This affects the line intensities measured from low resolution data and may affect diameter measurements of these remnants.

I. Introduction and Background

Samples of supernova remnants (SNRs) in nearby galaxies are important for studies of SNR evolution and also provide probes of the galaxy in which they reside. The number of SNRs and their relative sizes can be related to the supernova rate and models of the interstellar medium. Relative emission line intensities can be used to estimate physical conditions and chemical abundances; a sample of remnants can then be used to investigate variations in abundances and abundance gradients in spiral galaxies.

The most effective method for finding SNRs in nearby galaxies has proven to be comparison of [S II] and Hα imagery: SNRs have much higher [S II]/Hα ratios than H II regions and can often be identified even in somewhat confused regions of nearby galaxies. However, many SNRs are expected in giant H II regions, but confusion problems make application of the [S II]/Hα criterion impossible. We believe SNRs embedded in giant H II regions can be identified kinematically (*cf.* Chu and Kennicutt 1986), but there are no *integrated* velocity profiles available for comparison.

The advent of more realistic shock model calculations (Cox and Raymond 1985; Dopita *et al.* 1984) makes it possible to interpret the relative line intensities of SNRs in terms of chemical abundances (*cf.* Blair and Kirshner 1985). However, since the relative line intensities in H II regions and SNRs are different, any *contamination* of low resolution SNR spectra by H II emission can affect the abundance determinations. H II contamination might also cause problems with SNR diameter estimates, which are needed for SNR evolution studies.

These problems have motivated us to obtain echelle observations of SNRs in M33. At the scale of M33 (1 arcsec $\approx$ 3.5 pc), many of the brighter SNRs fall almost entirely in the slit, permitting measurements of integrated line profiles. Also, at high resolution and with a long slit, effects of H II contamination can be assessed. Below we describe these observations and some preliminary results.

II. Observations

We have used the echelle spectrograph in long slit mode on the 4 m telescope at Kitt Peak to obtain spectrograms of ten of the brightest SNRs in M33. A 79 line mm^{-1} grating and an interference filter were used to isolate the order containing both Hα and [N II] $\lambda\lambda$6548,6584, while the long slit oriented E-W across the SNRs provided spatial information. The 800 $\times$ 800 TI-4 chip was binned by 2 in the spatial direction (to decrease read-out noise)

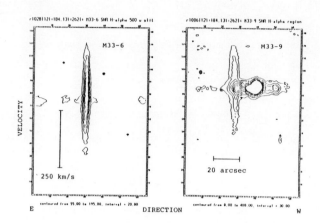

Figure 1: Representative velocity profiles for two M33 remnants. The object on the left shows almost no H II contamination while the object on the right is embedded in the edge of a bright H II region.

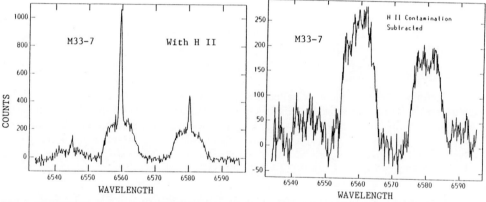

Figure 2: Extracted spectra for SNR M33-7, before and after removal of H II contamination. Note the square shape of the broad SNR profile.

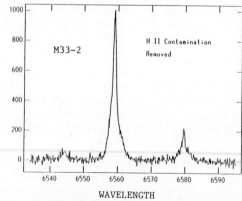

Figure 3: Line profiles for a centrally-peaked SNR, M33-2. Note asymmetry of profile.

yielding 0.755 arcsec pixels: Vignetting of the long slit limited spatial coverage to about 3 arcmin, but this was much larger than any of the SNRs. The 500μ slit width corresponded to ~ 3 arcsec, and measurement of comparison spectra indicates a velocity resolution of ~ 25 km s^{-1} was achieved. The slit was wide enough that most if not all of the light from each remnant entered the slit: Thus, integrated velocity profiles were obtained. Seven of the objects were also observed at [O III] $\lambda 5007$ with the same set-up, but at lower signal-to-noise.

The data have been reduced using IRAF at KPNO and at Johns Hopkins. Instrumental curvature was quite apparent in the comparison lines, so the data were rectified so that spatial information was exactly along the lines and the dispersion direction exactly along the columns. Bad columns and cosmic rays were removed from the regions of interest, and a general "background" (including night sky and sometimes diffuse emission from M33) was subtracted. One dimensional spectra corresponding to the SNRs or adjacent H II emission could then be extracted and measured. Representative examples of the two dimensional data and extracted profiles are shown in Figures 1 - 3.

Table 1
Derived Properties of SNRs[a]

Object	V_{bulk} (km s^{-1})	FWHM (km s^{-1})	% H II Contam.	D_{ech} (arcsec)	Size[b] (arcsec)
M33-2	263	59	<1	2.0	2.0
M33-4	183	52	12	7.7	5.1
M33-6	345	237	2	<1.0	2.6
M33-7	331	443	21	1.7	<1.7
M33-8	313	151	34	3.0	3.1
M33-9	272	256	25	2.7	2.0
M33-11	185	100	4	3.2	2.0
M33-16	224	91	54	1.5	1.7
M33-18	208	142	37	7.3	10.9
M33-20	103	82	>70	3.5-5	7.1

[a]Measurements shown pertain to the Hα profile.
[b]From Blair and Kirshner (1985).

III. Results

The presence of H II emission which potentially contaminates the SNR spectra was judged by inspection of CCD imagery of the remnants (Long *et al.*, this conference) as well as the presence of low velocity emission adjacent to or overlapping the high velocity SNR profiles. In addition, the [N II] $\lambda 6584$/Hα ratio is considerably lower in M33 H II regions than in the SNRs, so the ratio of these lines in the extracted data also provides clues as to whether low velocity emission is photoionized or simply SNR emission with low velocity along the line of sight. Nearly all of the objects we observed showed effects of contamination at some level, ranging from essentially negligible to as much as 70% of the total flux measured at the SNR position. To the extent possible, the H II contamination has been subtracted, yielding broad profiles which represent the integrated bulk velocity profiles for each remnant. (This is not strictly true for some of the larger diameter remnants, where a somewhat smaller fraction of the total light was sampled.) The effects of this contamination on the abundance estimates using low resolution spectral data (Blair and Kirshner 1985) have not been assessed, but it will be considerable for several of the objects. A summary of some measured parameters for each remnant is given in Table 1.

The integrated profiles show a variety of shapes, ranging from almost square (see Figure 2) to centrally peaked with faint, broad wings (see Figure 3). The bulk velocities listed in Table 1 were determined by measuring the full width of the line profile at zero intensity and dividing by two. Some information on the profile shape can be obtained by comparing the FWHM with the bulk velocity of each object: the larger the FWHM, the more "boxy" the profile. Square profiles probably indicate fairly uniform expansion of a thick shell, where the front and back of the shell produce the sides of the box and the material moving in the plane of the sky fills in the center. Centrally peaked profiles with faint, broad wings can be modeled assuming a shock wave passing through a cloudy medium (*cf.* Norman *et al.*, this conference): shocks are driven into the clouds and accelerate them to a significant fraction of the shock velocity (*cf.* McKee, Cowie and Ostriker 1978).

This may help explain the apparent discrepancy between the large bulk velocities indicated for the remnants in Table 1 and the need for relatively slow shocks to explain the relative line intensities in (corrected) low resolution spectra. However, bulk motions of 200-300 km s^{-1} imply even faster shocks in the intercloud region. These shocks might be unstable (*cf.* Innes, Giddings and Falle 1987), but would produce intense UV emission. In the two M33 SNRs detected with IUE, neither appeared to be due to a high velocity shock (Blair and Raymond 1984). These objects should also be strong soft X-ray sources, but only one M33 SNR (#9) corresponds roughly in position with an *Einstein* source (Long *et al.* 1981; Markert and Rallis 1983). However, only sources with L$_x \geq 10^{37}$ ergs s^{-1} would have been seen.

We have estimated the diameters of the SNRs from the echelle spectra in the following way: the lines containing the red or blue wing of the high velocity SNR emission were summed and a FWHM measured for the profile across the dispersion. This was then deconvolved from the seeing profile which was measured to be ~ 2 arcsec. These diameters, shown as D$_{ech}$ in Table 1, actually refer only to the E-W dimension. Our measurements range from considerably smaller to considerably larger than diameters measured directly from images, and cause some concern about the reliability of diameter estimates from images alone.

IV. Future Work

In addition to comparing the profiles at [N II] λ6584 and [O III] λ5007 to Hα, we need to obtain echelle profiles of these SNRs at [S II] λλ6717,6731. Since these lines are density sensitive, we will not only be able to obtain the average SNR density but will also get the density of the H II component into which many of the remnants appear to be expanding. This will permit a much better understanding of the interaction of each remnant with the surrounding interstellar medium. We also hope to use our integrated line profiles to identify SNRs embedded in giant H II regions in M33.

References

Blair, W. P., and Kirshner, R. P. 1985, *Ap. J.*, **289**, 582.

Blair, W.P., and Raymond, J.C. 1984, in *Six Years of IUE Research*, NASA CP-2349, p. 103.

Chu, Y.-H., and Kennicutt, R. C., Jr. 1986, *Ap. J.*, **311**, 85.

Cox, D. P., and Raymond, J. C. 1985, *Ap. J.*, **298**, 651.

Dopita, M. A., Binette, L., D'Odorico, S., and Benvenuti, P. 1984, *Ap. J.*, **276**, 653.

Innes, D. E., Giddings, J. R., and Falle, S. A. E. G. 1987, *M.N.R.A.S.*, **224**, 179.

Long, K.S., D'Odorico, S., Charles, P.A., and Dopita, M.A. 1981, *Ap. J. (Letters)*, **246**, L61.

Markert, T. H., and Rallis, A. D. 1983, *Ap. J.*, **275**, 571.

McKee, C. F., Cowie, L. L., and Ostriker, J. P. 1978, *Ap. J. (Letters)*, **219**, L23.

NEW SUPERNOVA REMNANTS IN M33

Knox S. Long and William P. Blair
The Johns Hopkins University, Baltimore, Md., U.S.A.

Robert P. Kirshner
Harvard-Smithsonian Center for Astrophysics, Cambridge, Ma., U.S.A.

P. Frank Winkler, Jr.
Middlebury College, Middlebury, Vt., U.S.A.

Abstract: Existing catalogues of supernova remnants (SNRs) in external galaxies are very incomplete. Potentially however, such samples are of great importance in understanding SNRs, since the distances to objects in a given sample are essentially the same and since absorption is small (compared to galactic SNRs). We have recently obtained Hα+[NII], Hβ, [SII], [OIII], and 6100 Å continuum CCD images of nine selected areas in M33 using the KPNO 4m. In addition to the six SNRs already known to exist in the fields we have surveyed, we have identified 21 other nebulae with [SII]:Hα+[NII] ratios which may be SNRs. Spectra of seven of these nebulae were obtained subsequently and show that the majority are indeed SNRs. A more detailed analysis of regions containing significant HII region contamination and a search for very small diameter remnants is currently underway.

I. Introduction

SNRs are difficult to identify in nearby galaxies because they are faint and hard to distinguish from H II regions. Although X-ray (Long *et al.* 1981) and radio (D'Odorico *et al.* 1982, Cowan and Branch 1985) searches may eventually be more effective at identifying extragalactic SNRs, most known SNRs have been found optically. The most successful searches have been based upon interference filter imagery with follow-up spectroscopy. Emission nebulae with [S II]:Hα ratios exceeding 0.4 are generally confirmed to be SNRs when spectroscopic observations are carried out (Dopita *et al.* 1980, Blair *et al.* 1981). Such ratios are expected for radiative shocks ($\sim$100 km s^{-1}) propagating through cloudlets in the ISM.

In M33, 19 SNRs have been identified by this technique previously (see D'Odorico, Dopita and Benvenuti 1980). They range in diameter from 3.5 to 70 pc and have Hα luminosities of order 10^{36} to 10^{37}ergs s^{-1} (Blair and Kirshner 1985). The previous surveys, conducted primarily with photographic plates, are very incomplete. The SNRs that have been found obey a number-diameter (N($<$D)) relationship that is consistent with free expansion to large diameter. The thermal energy content of the remnants appears to grow with diameter. Both of these results conflict with our conventional understanding of supernova remnant theory. However, both may be associated with the lack of a complete sample to any radius (as well as inaccurate diameter estimates). A larger sample of SNRs with well-defined sensitivity limits will help resolve these problems.

M33 is an ideal galaxy for pursuing such questions because it is nearby (1 arcsec = 3.5 pc), relatively face on (i$\sim$50°), and not very affected by Galactic absorption. As a result we have begun an interference filter CCD survey of M33. CCDs are ideal for extending traditional identification techniques for SNRs because they are linear and have large dynamic range and high quantum efficiency. In addition, CCD images can be used to provide the first set of integrated optical line fluxes for a significant sample of SNRs. This is information which has not been available previously since spectroscopic observations are usually made only of an isolated portion of a SNR. We are following up our imaging with spectroscopy which will permit a study of the structure and abundances of the ISM in M33.

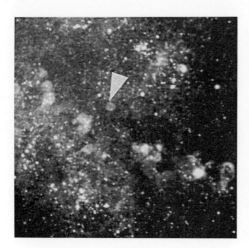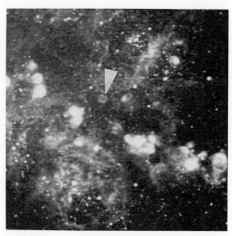

Figure 1: The raw [S II] and Hα+[N II] images of field 'h'. Notice how much brighter most of the emission nebulae are in Hα+[N II] compared with [S II]. An exception is the SNR candidate near the center of the field (indicated with an arrow).

II. Observations

The survey consists of observations in the light of Hα+[N II], [S II], [O III], Hβ and the 6100 Å continuum. The [S II], Hα+[N II] and 6100 Å continuum filter images are used to distinguish ISM-dominated SNRs, H II regions and stars. The [O III] and Hβ images are useful in searching for ejecta-dominated SNRs and in measuring the reddening of individual nebulae. In an initial observing run in 1986 Sept 12-14 using the prime focus CCD on the KPNO 4m, we observed nine fields, each 3.7 arc minutes square, covering the nucleus and a portion of the inner spiral arms of M33 to the south and west of the nucleus. The field centers are listed in Table 1.

Table 1: Field Centers				
Field	RA (1950)	DEC (1950)	No. of New Candidates	No. of Known SNRs
a	01 30 13.0	+30 24 13	3	3
b†	01 31 01.7	30 24 15	3	0
c	01 31 01.7	30 17 18	2	1
d	01 31 01.6	30 20 45	4	0
e	01 30 45.4	30 24 11	4	0
f	01 30 29.2	30 24 17	3	0
g	01 30 45.4	30 20 46	0	1
h	01 30 46.0	30 17 13	2	1
i	01 30 29.0	30 13 45	0	0
†galactic nucleus				

Until our analysis is complete we cannot quantify the sensitivity of the survey in detail. However, the faintest nebulae in our images have surface brightnesses less than 10^{-16} ergs cm^{-2} s^{-1} arcsec^{-2}. Our survey region contained six SNRs in the list of D'Odorico *et al.* (1980). All were easily detectable in all of the emission line images. As illustrated in

Figure 2b, [S II]:Hα+[N II] ratios through our filters for previously identified SNRs range from 0.36 to 0.52, averaging 0.45. Bright H II regions have [S II]:Hα+[N II] ratios which cluster near 0.14 (Figure 2c).

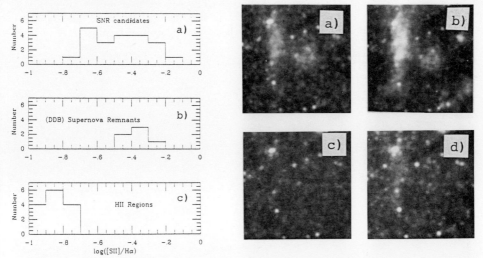

Figure 2: [S II]:Hα+[N II] ratios through our filters for (a) new candidates, (b) previously identified SNRs and (c) H II regions in M33. Figure 3: (a) [S II], (b) Hα+[N II], (c) 6100 Å, and (d) [O III] images of a one arcminute region in field 'c' containing a diffuse patch of emission selected as a candidate SNR as a result of its strong [S II] emission.

III. Results

A detailed analysis of these images is in progress. However, based upon an initial visual inspection of the images, we have selected 21 relatively isolated nebulae exhibiting relatively strong [S II] emission which we suspect are SNRs. A typical example of one of the candidates appears in Figure 3. These objects exhibit a wide range of morphologies from diffuse spots (such as the object in Figure 3) to limb-brightened emission regions (see Figure 1). The majority but certainly not all of the objects are located in regions where there is substantial confusion with nearby stars or H II regions. As one might expect, the candidates are on average fainter than the previously identified SNRs. They range in size from 10 to 80 pc.

Accurately determining line fluxes from faint nebulae is difficult because there are so many sources of background, many of which vary on about the same scale as the nebulae whose line fluxes one is trying to measure. Sources of "unwanted" background include individual stars and star clusters, a diffuse stellar continuum, diffuse emission line gas, and in some cases moonlight. Ultimately we intend to subtract stars individually from the images and model the diffuse contribution in the emission line images using the 6100 Å continuum images as a template and then extract accurate line fluxes from the reduced line images. It is important to analyze the images in this way because we need a quantitative detection criterion in order to estimate the completeness of the survey. Here however we have adopted a rough approach to estimate the line fluxes and ratios. We have simply found the excess flux associated with a nebula in the raw image by subtracting the average flux in an adjacent visually selected background region. We have then corrected the emission line fluxes for stellar contamination by scaling from the observed excess (or deficit) in the continuum image using scaling ratios obtained from moderately bright stars in the image. The [S II]:Hα+[N II] ratios we obtain for all these nebulae are shown in Figure 2a. All of these nebulae have ratios which are higher than typical H II regions, although it is not clear, on the basis of this plot alone, that all are SNRs.

In an attempt to verify that we are in fact detecting new SNRs we have obtained spectra of seven of the 21 objects. Two examples, obtained with the IDS on the 2.5m Isaac Newton telescope at La Palma are shown in Figure 4. These particular objects are certainly SNRs. The total [S II] flux exceeds that in Hα. In fact, five of the seven candidates for which we have spectra appear to be SNRs, on the basis of their [S II]:Hα ratios.

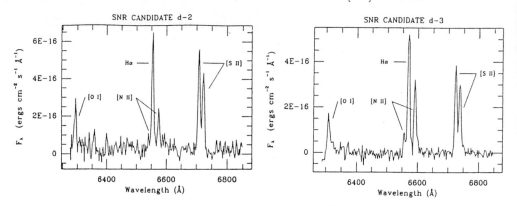

Figure 4: Spectra of two candidates in field 'd'. Both show strong [S II] emission relative to Hα which is characteristic of radiative shocks found in SNRs.

In conclusion, although the analysis of these data is not complete, it is apparent that SNRs can be discovered efficiently in this manner in M33. If the survey is extended to include all of M33, the number of known SNRs will be doubled or even tripled. This will lead to a more accurate estimate of the SN rate in M33 and enable us to determine the relative numbers of SNRs in the spiral arms and interarm regions. It will be important to investigate the fainter SNRs to insure that we understand "typical" M33 SNRs. Most of the previously known SNRs are very high surface brightness objects (reminiscent of N49 in the Large Magellanic Cloud); echelle observations indicate surprisingly high shock velocities in many of these SNRs (Blair et al., this conference). This new sample should provide SNRs with a wider range of shock velocities for comparison with models and lead to more reliable interpretation of chemical abundances than is possible currently.

References

Blair, W. P. and Kirshner, R. P. 1984, *Ap.J*, **289**, 582.

Blair, W. P., Kirshner, R. P., and Chevalier, R. 1981, *Ap. J.*, **247**, 879.

Cowan, J. J. and Branch, D. 1985, *Ap.J.*, **293**, 400.

D'Odorico, S., Dopita, M. A., and Benvenuti, P. 1980, *Astr. Ap. Suppl.*, **40**, 67.

D'Odorico, S., Goss, W. M., and Dopita, M. A. 1982, *M.N.R.A.S.*, **198**, 1059.

Dopita, M. A., D'Odorico, S. and Benvenuti, P. 1980, *Ap. J.*, **236**, 638.

Long, K. S., Helfand, D. J., and Grabelsky, D. A. 1981, *Ap. J.*, **248**, 925.

SUPERNOVA REMNANTS IN GIANT HII REGIONS

You-Hua Chu
Astronomy Department, University of Illinois
1011 W. Springfield Avenue, Urbana, IL 61801

Robert C. Kennicutt, Jr.
Astronomy Department, University of Minnesota
116 Church Street, S.E., Minneapolis, MN 55455

I. Introduction

Giant HII regions contain large numbers of massive stars, and hence are expected to contain large numbers of SNRs. Until recently, however, only a few SNRs have been identified in extragalactic giant HII regions. Moreover, most of these SNRs are located at the outskirts of HII regions, instead of the core where most of the stars are located. The low detection rate and the outlying locations of the SNRs may be due to: 1) observational difficulties - the background HII regions are much more luminous than the SNRs in both optical line emission and radio continuum; 2) intrinsic invisibility of SNRs - stellar wind and SNRs may have created a supershell (Mac Low and McCray 1987), and the core of a giant HII region is filled with hot tenuous coronal gas; or 3) a genuine deficiency of supernovae and SNRs in the HII regions (Sramek and Weedman 1986).

We have begun to use high dispersion emission-line spectroscopy to supplement the traditional methods in the search for SNRs in large HII regions. Such observations offer a very sensitive means of detecting SNRs in these regions, and the potential of studying the interaction of remnants with the interstellar medium in complex environments. Here we present a brief progress report on this program, and summarize what is known about SNRs in the largest HII regions in several nearby galaxies.

II. SNRs in Giant HII Regions

SNRs are actually present in the largest HII regions of many nearby spiral and irregular galaxies. Table 1 lists nearby galaxies with confirmed SNRs in giant HII regions. These include previously identified SNRs, and objects identified in our spectroscopic observations (SMC, M101).

The SNR W49B in W49, probably the largest HII region in the Galaxy, is so heavily obscured that no optical observations are possible. The radio properties of W49B are typical for SNRs: a nonthermal shell source with a spectral index of -0.5 (Green, Baker, and Landecker 1975).

Table 1. SNRs in Giant HII Regions

Galaxy	Distance (kpc)	HII Region	Ionizing Flux (ion. photon/s)	SNR
Galaxy	--	W49	7 x 10^{50}	W49B
LMC	55	30 Dor	5 x 10^{51}	30 Dor B
SMC	70	N66	2 x 10^{50}	anon
M33	720	NGC 604	1.5 x 10^{51}	M33-16
M101	6000	NGC 5471	1.7 x 10^{52}	(3 SNRs)

30 Dor B is the only positively identified SNR in 30 Doradus (the largest HII region in the Large Magellanic Cloud), although there is kinematic evidence for several other possible SNRs, as discussed later. 30 Dor B is located at ~100 pc from the core of the HII region. Its velocity structure is well described by a complete but irregular shell, expanding at ~180 km/s. To the east of this shell, there is high-velocity material moving at +120 km/s relative to the average background nebular velocity; the relation between this high-velocity material and the SNR 30 Dor B is not clear. 30 Dor B is Crab-like in the radio, but its slower expansion velocity (~180 km/s) and larger linear size (diameter ~10 pc) make it much older than the Crab. We derive a dynamic age of about 10^4 yr, which is considerably longer than the previous estimates (Long and Helfand 1979; Gilmozzi et al. 1983). The radio emission is more extended to the west of the optical components (the [S II]/Hα enhanced region), and there is high velocity material to the east of the remnant. The statistics of Crab-like SNRs are minimal enough to prohibit meaningful comparisons between 30 Dor B and isolated SNRs of similar type.

N66 is the largest HII region in the Small Magellanic Cloud. It contains an X-ray source and nonthermal radio sources (Ye and Turtle, 1986, private communication), but no optical SNRs were previously identified. The comparison of a radio map at 843 MHz (Mills et al. 1982) and a short-exposure Hα picture of the core region in N66 shows that the brightest peaks in the radio map do not have corresponding Hα features. A long-slit Hα echellogram taken at this region shows high-velocity clumps, up to ~ +170 km/s with respect to the background HII region velocity. The coincidence of nonthermal radio emission and an X-ray source at this fast expanding shell makes it a good SNR candidate.

NGC 604, the brightest HII region in M33, contains a known SNR (D'Odorico, Dopita, and Benvenuti 1980). SNRs in M33, at a distance of 720 kpc, are hardly resolved by ground based telescopes. The echellogram of NGC 604 shows expanding shells with a range of sizes and velocities in the central HII region, and an unresolved broad component in the SNR. The SNR has similar surface brightness and kinematic structures as the other SNRs in M33 (Blair, Chu, and Kennicutt 1987). Its radio flux density is about 0.6 mJy at 1420 MHz (Sramek and Weedman 1986), within the typical range for the other SNRs in M33 (D'Odorico, Goss, and Dopita 1982).

NGC 5471, the brightest HII region complex in M101, contains at least 3 SNR candidates (Chu and Kennicutt 1986). All have large velocity widths (300-460 km/s at the base) which are similar to those of isolated SNRs in M33 and the LMC. In addition, spectrophotometry and CCD images show enhanced [SII] emission at the SNRs. The most remarkable properties of these SNRs are their high optical and radio luminosities. The brightest SNR in NGC 5471 B, first identified by Skillman (1985), has an Hα luminosity of $6.3 \pm 3 \times 10^{38}$ erg/s, and radio flux density of 2.3 ± 0.2 mJy at 1415 MHz. The size (< 30 pc), kinematic structures, and Hα luminosity imply a middle-aged SNR (~a few x 10^4 yr), but the radio luminosity is only matched by very young SNRs (a few x 10^2 yr). The SNR in NGC 5471 A probably has similar properties. It is possible that these properties resulted from the peculiar interstellar environment in the dense core of NGC 5471; it is also possible that multiple SNR events are seen there. However, it must be noted that any search for SNRs at this large a distance will detect only unusually bright objects. Similar SNRs in other giant HII regions have to be found and studied before this issue can be settled.

III. How to Search for SNRs in Giant HII Regions

Conventionally, SNRs are identified using [SII]/Hα imagery, X-ray, and radio surveys. However, in distant luminous HII regions the environments are so unusual that unconventional methods have to be used to search for SNRs. [S II]/Hα imagery may be effective if the background HII region is not too luminous, or if the SNRs are very luminous, especially if CCD detectors are used. The radio surveys are good for picking out nonthermal sources if the spatial resolution is sufficient. As we have demonstrated in N66 and NGC 5471, high dispersion 2-D spectroscopy is an alternative and a very effective way to search for SNRs in complex emission-line regions. The combination of these methods, especially a high resolution radio survey with high dispersion 2-D spectroscopy, is probably the most effective way to search for SNRs in giant HII regions.

Why bother searching for SNRs in such complex, high background regions when there are plenty of relatively isolated, easily observed SNRs available? For the study of the interaction of SNRs with the interstellar medium, the remnants in giant HII regions provide an opportunity to investigate how the evolution of SNRs are altered in the relatively dense ionized nebular environment. Conversely it should be possible to establish whether the SNRs significantly influence the global kinematics and dynamics of the giant HII regions and their surroundings. Very luminous and energetic SNRs, such as the object in NGC 5471 B, may require unusually massive progenitor stars, and it is of great interest to identify and measure a larger sample of these unusual objects. Finally the detection and study of SNRs in regions such as giant HII regions, starburst galaxies, and galactic nuclei is important in its own right, since supernovae may be dominant contributions to the energetics and radio emission of these objects.

In order to address some of these questions we are following up our initial survey on several fronts. Comparable high-dispersion spectra of isolated SNRs in the Magellanic Clouds and M33 are being obtained (Chu and Kennicutt 1987; Blair, Chu, and Kennicutt 1987), in order to compare the kinematic properties of the SNRs in HII regions with isolated ones. These data also provide the best available data on the integrated kinematic properties and nebular environments of SNRs in general. We also hope to enlarge our sample of giant HII regions surveyed, in order to set firmer limits on the number of very luminous SNRs. Finally we have obtained high-resolution data at multiple positions in the nearest supergiant HII region, 30 Doradus in the LMC, in order to conduct a sensitive search for SNRs (and other kinematic structures) in the dense nebular core of this HII region.

IV. Fast Expanding Shells in the Core of 30 Doradus

Our long-slit echelle spectra at several positions in 30 Doradus reveal a number of faint shells, expanding at 150-200 km/s, in the bright nebular core. Features with similar expansion velocities have been reported and interpreted as wind-blown bubbles by Meaburn (1984). We find these shells very intriguing, since expansion velocities of isolated wind-blown bubbles have never been observed to be greater than 100 km/s. These expansion velocities are more in accord with the other SNRs in the Magellanic Clouds (Chu and Kennicutt 1987), and the presence of dense high-velocity clumps in the shells is also much more characteristic of SNRs. From kinematic evidence alone, these fast expanding shells could conceivably be either stellar wind blown bubbles or SNRs. It is important to examine closely the stellar content and the radio properties of these shells in order to determine their real nature.

References:

Blair, W. P., Chu, Y.-H., and Kennicutt, R. C. 1987, in this volume.
Chu, Y.-H., and Kennicutt, R. C., 1986, Ap. J., 311, 85.
Chu, Y.-H., and Kennicutt, R. C., 1987, in preparation.
D'Odorico, S., Dopita, M. A., and Benvenuti, P. 1980, Astr. Ap., 40, 67.
D'Odorico, S., Goss, W. M., and Dopita, M. A., 1982, M. N. R. A. S., 198, 1059.
Gilmozzi, R., Murdin, P., Clark, D. H., and Malin, D. 1987, M. N. R. A. S., 202, 927.
Green, A. J., Baker, J. R., Landecker, T.L., 1975, Astr. Ap., 44, 187.
Long, K. S., and Helfand, D. J. 1979, Ap. J. (Letters), 234, L77.
Mac Low, M.-M., and McCray, R. 1987, Ap. J., in press.
Meaburn, J. 1984, M. N. R. A. S., 211, 521.
Mills, B. Y., Little, A. G., Durdin, J. M., and Kesteven, M. J., 1982, M. N. R. A. S., 200, 1007.
Skillman, E. D. 1985, Ap. J., 290, 449.
Sramek, R. A., and Weedman, D. W., 1986, Ap. J., 302, 640.

SUPERNOVA REMNANT SHOCKS IN AN INHOMOGENEOUS INTERSTELLAR MEDIUM

Christopher F. McKee
Departments of Physics and of Astronomy
University of California, Berkeley CA 94720

Abstract. The inhomogeneity of the interstellar medium (ISM) has a profound effect on the propagation of the interstellar shock generated by a supernova and on the appearance of the resulting supernova remnant (SNR). Low mass supernovae produce remnants that interact with the "pristine" ISM, which has density inhomogeneities (clouds) on a wide range of scales. The shock compresses and accelerates the clouds it encounters; inside the blast wave, the clouds are hydrodynamically unstable, and mass is injected from the clouds into the intercloud medium. Embedded clouds interact thermally with the shock also, adding mass to the hot intercloud medium via thermal evaporation or subtracting it via condensation and thermal instability. Mass injection into the hot intercloud medium, whether dynamical or thermal, leads to infrared emission as dust mixes with the hot gas and is thermally sputtered. The remnants of massive supernovae interact primarily with circumstellar matter and with interstellar material which has been processed by the ionizing radiation and wind of the progenitor star. After passing through any circumstellar material which may be present, the shock encounters a cavity which tends to "muffle" the SNR. The remnants of massive supernovae therefore tell us more about the late stages of the evolution of massive stars than about the ISM.

1. INTRODUCTION

The ISM is observed to be highly inhomogeneous: most of the mass is concentrated in diffuse atomic clouds ($T \sim 100\mathrm{K}$) and dense molecular clouds ($T \sim 10\mathrm{K}$) which occupy a small fraction of the volume, whereas much of the volume is occupied by low density, warm HI and HII ($T \sim 10^4\mathrm{K}$) and by very low density, hot coronal gas ($T \sim 10^6\mathrm{K}$). This inhomogeneity is a direct consequence of the the energy flow from stars through the thermally unstable ISM and into the intergalactic medium (McKee 1986). If the rate of stellar energy injection were very low, then the ISM would settle into a thin, quiescent HI disk; in order to be stable against gravitational collapse, the mean density of the medium would have to be low. At the opposite extreme of a high energy injection rate, the ISM would heat up to $T \gg 10^6$ K and blow away in a galactic wind. For a broad range of intermediate energy injection rates, the ISM settles into a multiphase medium such as we observe (Ikeuchi, Habe, and Tanaka 1984).

Stars inject energy into the ISM in the form of radiation, supernova explosions, and winds. By far the dominant constituent of this energy flow is starlight. Most of this radiation escapes into the intergalactic medium without interacting with the interstellar gas, although it may heat interstellar dust *en route*. A fraction of the far–ultraviolet radiation between about 6 eV and the ionization limit of hydrogen at 13.6 eV heats the gas by ejecting photoelectrons from dust grains; this may well be the dominant heating mechanism for the HI, and it leads naturally to a two–phase medium with 100 K clouds embedded in a 10^4 K intercloud medium (deJong 1980). Ionizing stellar radiation ($h\nu > 13.6$ eV) is quite effective at heating the gas. Its effects tend to be localized in HII regions, but it has a profound effect on the evolution and appearance of the SNRs which occur there.

After starlight, the most important contributor to the stellar energy flux is supernova explosions. Supernovae are a particularly important form of energy injection because they probably determine the velocity distribution of interstellar clouds (e.g., Spitzer 1968a), they are responsible for the acceleration of cosmic rays, and they can produce large volumes of hot coronal gas (Cox and Smith 1974). If the filling factor f_h of the hot gas is large ($f_h \gtrsim \frac{1}{2}$), the ISM is a three phase medium with cold and warm clouds embedded in the hot gas; SNRs expand primarily in the hot gas and determine the pressure of the ISM (McKee and Ostriker 1977). Although f_h is high in the local ISM, its value at a typical point in the galactic disk remains controversial (Cox 1987). It is quite likely that f_h depends on both the nature of the galaxy and the location within it. In principle, observations of SNRs could provide the means to measure f_h, but it would first be necessary to determine which SNRs are interacting with the "pristine" ISM, unaffected by the progenitor star. As we shall see below, this is not a straightforward task.

Energy injection via stellar winds is negligible for low mass stars. For massive stars, the total wind energy injection during core hydrogen burning is less than that in the final supernova explosion unless the the star is initially very massive, $M \gtrsim 60 M_\odot$ (Abbott 1982). Stars initially above about 40 $M_\odot$ are believed to go through a Wolf–Rayet stage after exhausting their central hydrogen (Chiosi and Maeder 1986), during which the wind energy injection is comparable to that of a supernova (Abbott 1982). Comparing the total supernova energy injection with the total wind energy injection in the Galaxy, Abbott (1982) concluded that supernovae dominate by a factor of about five. Winds do have two important effects, however: First, the interstellar bubbles of hot, low density gas produced by these winds (Castor, McCray, and Weaver 1975) "muffle" the final supernova explosion, making the resulting SNR more difficult to detect. Second, a significant amount of mass can be ejected by low velocity winds, which are energetically unimportant but observationally critical, since the interaction of the SNR blast wave with this circumstellar matter is readily observable.

The flow of stellar energy through the ISM thus leads to two qualitatively different types of inhomogeneity: In the general ISM, the medium is characterized by clouds embedded in a lower density intercloud medium, whereas near massive stars the ionizing radiation and winds reconfigure the ambient ISM into a radially stratified medium. Supernovae from low mass stars (Type Ia) explode in the pristine ISM and should provide a powerful probe of its structure. Supernovae from massive stars (Type II and, presumably, Type Ib—e.g., Wheeler and Levreault 1985) cannot "see" the pristine ISM until they have expanded to very large sizes and become difficult to observe. Here we shall explore the consequences of both types of inhomogeneity on the structure and appearance of an SNR, and how SNRs in turn affect the ambient medium.

2. SNR SHOCKS IN A CLOUDY ISM

2.1 Cloud Crushing

An SNR in a cloudy medium expands through the intercloud medium, driving shocks into the clouds it overtakes. There is a simple relation between the velocity of the cloud shock, v_s, and that of the SNR blast wave, v_b. Let ρ_{i0} be the initial density of the intercloud medium, and let ρ_{c0} be the density of the cloud. We assume that the shocks are strong: the pressure behind the shocks greatly exceeds that ahead, so that they are highly supersonic. Then the pressure behind the blast wave shock is $\sim \rho_{i0} v_b^2$, and that behind the cloud shock is $\sim \rho_{c0} v_s^2$. Since the pressure of the shocked intercloud gas drives the shock into the cloud, these pressures must be comparable, and we conclude that (Bychkov and Pikel'ner 1975, McKee and Cowie 1975)

$$v_s \simeq (\rho_{i0}/\rho_{c0})^{1/2} v_b. \tag{1}$$

A more precise evaluation, which allows for the decrease in v_s and v_b with time as the SNR expands, shows that for a Sedov–Taylor blast wave this simple result should be accurate to within a factor of about 1.5 (*cf.* McKee *et al.* 1987). However, an important *caveat* should be kept in mind: the derivation is based on the assumption that the blast wave is non–radiative, so that the ram pressure and the thermal pressure behind the blast wave shock are comparable. If, on the other hand, the blast wave is radiative, then the cloud shock velocity is $v_s \simeq (\rho_{i1}/\rho_{c0})^{1/2} v_b$, where ρ_{i1}, the density of of the shocked intercloud medium, may be up to 100 times greater than ρ_{i0}. We focus on non–radiative blast waves here.

The fate of a cloud engulfed by a blast wave depends on its size. For simplicity, assume that the cloud is approximately spherical, with radius a. Let R be the distance between the cloud center and the site of the SN explosion. The blast wave expands as $R_b \propto t^\eta$, where $\eta = 2/5$ for a Sedov–Taylor blast wave. Let

$$\chi \equiv \frac{\rho_{c0}}{\rho_{i0}} \tag{2}$$

be the cloud/intercloud density ratio; we assume $\chi \gg 1$. The three relevant time scales are the cloud crushing time,

$$t_{cc} \equiv \frac{a}{v_s} = \frac{\chi^{1/2} a}{v_b}; \tag{3}$$

the intercloud crossing time,

$$t_{ic} \equiv \frac{2a}{v_b}; \tag{4}$$

and the age of the SNR,

$$t \equiv \frac{\eta R}{v_b} = \frac{2}{5} \frac{R}{v_b}, \tag{5}$$

where the second equality in equation (5) is for Sedov–Taylor blast waves. The use of a rather than $2a$ in the definition of the cloud crushing time t_{cc} is rather arbitrary, but it does allow for the fact that the cloud is crushed from both sides.

Our assumption that the density ratio χ is much greater than 1 ensures that $t_{cc} > t_{ic}$. Clouds then come in one of three sizes, small, medium, or large:

* *Small clouds:* $t > t_{cc} \implies a < \eta R/\chi^{1/2} = 0.4R/\chi^{1/2}$. The SNR does not evolve significantly during the cloud crushing. For $t \gg t_{cc}$, the cloud crushing is completed within a short distance behind the blast wave shock. Because the cloud is crushed promptly, the pressure driving the cloud shock, and hence v_s, are approximately constant.

* *Medium clouds:* $t_{cc} > t > t_{ic} \implies \eta R/\chi^{1/2} < a < \eta R/2$. The SNR does not evolve significantly while the blast wave in the intercloud medium crosses the cloud, but it does evolve while the cloud is being crushed. Since the pressure driving the cloud shock drops as the SNR expands, the cloud shock decelerates.

* *Large clouds:* $t_{ic} > t \implies a > \eta R/2 = 0.2R$. The blast wave weakens significantly between the time it first encounters the cloud and the time it has crossed the cloud; since R_b increases by at least a factor $1.2/0.8 = 1.5$ in crossing a large cloud, the blast wave pressure $P \propto R_b^{-3}$ drops by at least a factor 3. The force exerted on the cloud is impulsive, so the cloud shock decelerates significantly. If the cloud is somewhat larger than the minimum large cloud, it will affect the overall expansion of the blast wave; a cloud with $a > 0.5R$ occupies more than about 1/16 of the volume of the SNR.

In sum, the crushing of a small cloud is like being squeezed by a vise, whereas that of a large cloud is like being struck by a hammer.

Cloud crushing occurs in several well–defined stages (*cf.* Nittman, Falle, and Gaskell 1982; Heathcote and Brand 1983). We shall focus on small clouds, since most of the theoretical work has concentrated on this case.

i) *Initial transient.* When the blast wave first strikes the cloud, a transmitted shock propagates into the cloud and a reflected shock propagates back upstream into the shocked intercloud gas. So long as these shocks are approximately plane parallel, the pressure between them is about 6 times that behind the initial blast wave shock (for a $\gamma = 5/3$ gas and $\chi \gg 1$: Silk and Solinger 1973). In a time of order $a/v_b = t_{ic}/2$, the reflected shock in the intercloud medium settles into a standing bow shock. The peak pressure between the shocks is then only about 3 times that behind the initial blast wave shock (McKee and Cowie 1975).

ii) *Shock compression.* After a time of order t_{ic}, the flow around the cloud converges on the axis behind the cloud, producing a reflected shock in the intercloud gas and driving a shock into the rear of the cloud (Woodward 1976). The shocks compressing the cloud from the sides are weaker than those compressing it fore and aft because the pressure in the flow around an obstacle has a minimum at the sides (Nittman *et al.* 1982). The result is that the cloud is crushed into a thin pancake, with its lateral dimensions reduced by about a factor 2; the main cloud shock propagating in from the nose of the cloud is stronger than the secondary shock propagating in from the rear

(Woodward 1976). If the cloud shock is fast enough ($v_s \gtrsim 200$ km s^{-1}) it will be non–radiative; otherwise it will be radiative and the cloud will be compressed to high density (Sgro 1975). The collision of the main and secondary shocks results in yet greater compression.

iii) *Re-expansion.* When the main cloud shock reaches the rear of the cloud, a strong rarefaction is reflected back into the cloud, causing the cloud to expand downstream (Woodward 1976); the low pressure at the sides of the cloud induces it to expand sideways also (Nittman *et al.* 1982).

iv) *Final fate.* Nittman *et al.* (1982) and Heathcote and Brand (1983) have suggested that hydrodynamic instabilities will subsequently destroy the cloud; Nittman *et al.* also invoked the lateral expansion of the cloud as a destruction mechanism. The possible stabilizing effects of magnetic fields have yet to be considered. Even if the cloud did break up into a spray of very small cloudlets as a result of instabilities, the spray might behave as a cloud until either thermal evaporation caused it to merge with the intercloud medium, or expansion reduced its mean density to that of the intercloud medium. The final fate of a small cloud in a SNR thus remains an open question; medium and large clouds are unlikely to be destroyed.

The net acceleration of the cloud by the blast wave is due to two effects, shock acceleration and ram pressure acceleration (McKee, Cowie, and Ostriker 1978). The main cloud shock accelerates the gas up to $v_1 = (3/4, 1)v_s$ if the shock is (non–radiative, radiative), respectively; the average velocity of the entire cloud is slightly less than this because of the secondary shock propagating in from the the rear of the cloud. The ram pressure acceleration is due to the flow of intercloud gas past the cloud, which exerts a drag on the cloud. In the absence of magnetic fields, the drag for subsonic flow depends only on the Reynolds number Re of the flow. Using Draine and Giuliani's (1984) value for the viscosity ν of an ionized plasma of cosmic abundance, we have

$$Re = \frac{av_i}{\nu} = 15.8 \frac{n_i \mathcal{M} a_{\mathrm{pc}}}{T_{i7}^2}, \tag{6}$$

where $\mathcal{M} \equiv v_i/C_i = v_i(\rho_i/P_i)^{1/2}$ is the isothermal Mach number of the intercloud gas, n_i is the density of hydrogen nuclei in the intercloud gas, $T_{i7} \equiv T_i/10^7$ K, and a_{pc} is the cloud radius in parsecs. In terms of the proton–proton mean free path $\lambda_p = 6.00 \times 10^{17} T_{i7}^2/n_i$ cm (see Spitzer 1962), this is

$$Re = 3.07(a/\lambda_p)\mathcal{M}. \tag{7}$$

Hence, in the absence of magnetic fields, the ion mean free path inside an SNR is quite long and the flow can be quite viscous. Magnetic fields reduce the mean free path by many orders of magnitude, however, with a corresponding reduction in the viscosity and increase in the Reynolds number: for typical conditions in the intercloud medium ($n_i = 0.2$ cm^{-3}, $B = 3$ μG), the Reynolds number perpendicular to the magnetic field increases by a factor $2 \times 10^{19} T_{i7}^3$! (For a general discussion of viscosity in a magnetized plasma, focusing on viscous heating, see Hollweg [1986].) Although this reduction in the viscosity occurs only perpendicular to the field, the full effect of the reduction is likely to occur in the problem at hand because the flow of the gas past the cloud will tend to drape the field lines along the boundary layer between the cloud and intercloud medium. If the magnetic field in the boundary layer becomes dynamically important, it could suppress the turbulence which sets in there for $Re \gtrsim 10^5$ in the absence of a field. For flow past a sphere, the drag coefficient C_D (the ratio of the drag to $0.5\rho_i v_i^2 \pi a^2$) is about 0.4 for $Re \sim 10^5$; for flow past a circular disk, similar to the configuration the cloud reaches near the end of the shock compression phase, $C_D \simeq 1.1$ over a wide range of Re (Batchelor 1967). For supersonic flow, such as we are considering here, the drag is greater because of the emission of sound waves; for flow past a sphere, $C_D \sim 1.0$ (Chernyi 1961), and for flow past a plate C_D is somewhat larger still. McKee et $al.$ (1978) adopted $C_D = 2$. In terms of the nomenclature introduced here, they found that small clouds are strongly affected by ram pressure acceleration, whereas medium and large clouds are not. Nittman et $al.$ (1982) have argued that ram pressure acceleration is unimportant because the lateral expansion of the cloud will destroy it before substantial acceleration can occur. Their argument implicitly assumes that the cloud shock is non–radiative, however, and does not apply to the more common case of radiative cloud shocks. Furthermore, even in the case of non–radiative cloud shocks, substantial ram pressure acceleration can occur as the lateral expansion reduces the mass per unit area of the cloud, making it more like a sail than a cannonball. Medium and large clouds are accelerated primarily by the cloud shock, which decelerates as it propagates through the cloud. The dynamics of the decelerating cloud shock, and the pressure and velocity distribution behind it, have been discussed by McKee et $al.$ (1987); at late times, the cloud shock is approximately momentum–conserving. Because the blast wave also decelerates, the medium and large clouds can coast beyond the blast wave shock at late times, an effect dramatically portrayed in the numerical simulations of Tenorio–Tagle and Rozcyzka (1986) and Rozcyzka and Tenorio–Tagle (1987).

The interaction of the blast wave with a cloud is subject to both Rayleigh–Taylor and Kelvin–Helmholtz instabilities. When the blast wave first strikes the cloud, it subjects the dense cloud gas to an instantaneous acceleration by the light intercloud gas. This results in a transient Rayleigh–Taylor instability which produces a variation in the velocity of the interface of $\delta v \simeq (ka_0)v_1$ for $\chi \gg 1$, where a_0 is the initial amplitude of the displacement of the interface, k is the wavenumber of the displacement, and v_1 is the velocity of the shocked cloud gas (Richtmyer 1960). The derivation of this result assumes that $ka_0 \ll 1$; hence, the velocity variation δv is small compared to the velocity v_1 imparted to the cloud gas, and the instability significantly alters the shape of the cloud only after the cloud has undergone substantial compression. For a planar cloud, no further instability would result. However, the finite size of the cloud leads to a convergence and acceleration of the cloud shock, and hence a further weak Rayleigh–Taylor instability (Woodward 1976). The Kelvin–Helmholtz instability due to the flow of the intercloud gas past the cloud strongly distorts the surface of the cloud and can ultimately lead to substantial ablation of the cloud (Nulsen 1982; see below).

Because of the complex nonlinear nature of the blast wave–cloud interaction, it appears that only numerical simulation can determine the ultimate fate of the cloud: What is the total momentum transferred to the cloud? How much mass is lost from the cloud? Is the cloud disrupted, distorted, or driven into gravitational collapse? What is the effect of the interstellar magnetic field? To date, no conclusive answer is available to any of these questions. Even if we assume that the magnetic field is dynamically unimportant, there are two major problems which have tended to confound efforts at numerical simulation. The first and most obvious problem is numerical diffusion at the cloud–intercloud interface. Remarkably, the only calculation which overcomes this problem was published over a decade ago, by Woodward (1976; see also Woodward 1979). He used a code which had an Eulerian treatment of the intercloud medium and a Lagrangian treatment of the cloud, thereby permitting accurate treatment of vorticity in the intercloud gas and of the radiative shock in the cloud. The problem he addressed was the collision of a density wave shock with a cloud; since the pressure behind the shock was time–independent, the cloud was effectively "small". The pressure jump was only a factor 8, considerably weaker than in a typical SNR shock; as a result, the flow behind the shock was subsonic relative to the cloud and a bow shock did not form. He found that most of the cloud was shocked by the main cloud shock, resulting in a pancake–shaped cloud near the end of the compression phase. The cloud was strongly distorted by Kelvin–Helmholtz and Rayleigh–Taylor instabilities, and the effort involved in manually rezoning the interface forced him to terminate the calculation before the end of the compression phase. Subsequent calculations (e.g., Nittman *et al.* 1982, Tenorio–Tagle and Rozcyzka 1986, Rozcyzka and Tenorio–Tagle 1987) appear to have strong numerical diffusion at the boundary, making it difficult to disentangle cloud/intercloud mixing due to instabilities from mixing due to numerical effects.

The second problem confronting the numerical simulator is the disparity between the actual Reynolds number of the flow and the effective Reynolds number in the simulation. For example, Woodward's (1976) simulation had a grid size of 0.25 pc in the intercloud gas; if this is taken to be the effective mean free path, then for $\mathcal{M} \sim 1$ we have $Re \sim 180$ from equation (7). By contrast, in the absence of a magnetic field, the proton mean free path is about 3×10^{13} cm, giving $Re \sim 5 \times 10^6$. If this were a simulation of non–magnetic, subsonic flow past a rigid sphere, the discrepancy in Re would be significant because the boundary layer and wake would be turbulent in the physical case and laminar in the simulation, so that, for example, the drag in the physical case would be about half that in the model (Batchelor 1967). In fact, however, it was a simulation of nearly sonic flow past a sphere roughened by instabilities; the interstellar magnetic field would probably not have been dynamically important, but it would have increased the Reynolds number perpendicular to the field by about 10 orders of magnitude. My own feeling is that the Reynolds number disparity will not significantly limit our ability to determine the final fate of a blasted cloud, although it will limit our ability to follow the details of the flow in the intercloud medium.

One of the major motivations for studying the blast wave–cloud interaction has been to determine the conditions under which shock compression can induce gravitational collapse. Woodward (1976) found that dense globules formed by the combined action of the Kelvin–Helmholtz and Rayleigh–Taylor instabilities became gravitationally bound and presumably would have formed stars, although an unrealistically low magnetic field ($B < 1\ \mu G$) would have been required. Nittman (1981) was the first to include a magnetic field in the simulation. He demonstrated that a dynamically important field would not only limit the density reached by the shocked cloud, but also reduce the shear which led to the Kelvin–Helmholtz in the first place. A more direct method of driving gravitational collapse, and one less subject to numerical difficulties, has been studied by Hillebrandt and his collaborators: namely, to compress the entire cloud to the point of instability. Krebs and Hillebrandt (1983) considered large clouds ($a = 10$ pc, $R = 30$ pc) initially close to the maximum stable mass M_J; indeed, their Model A started with $M > M_J$ (the ratio of central to mean density was about 10, larger than the maximum value of 5.78 allowed for stable configurations [Spitzer 1968b]). The remaining two models they considered had $M \sim 0.4 M_J$, and they found that a 10^{51} erg supernova induced gravitational collapse, whereas a 10^{50} erg one did not. Subsequently, Oettl, Hillebrandt, and Müller (1985) studied the stabilizing effects of a magnetic field on Model A.

2.2 Mass Exchange

If most of the mass of the interstellar medium is tied up in clouds, processes which transfer this mass into the intercloud medium can have a dramatic effect on the dynamics and appearance of an SNR blast wave. Correspondingly, processes which transfer mass from the intercloud medium back into the clouds, which tend to act when the blast wave becomes weak, are important in maintaining the multi–phase medium.

In this review we shall focus on mass exchange driven by thermal conduction. Inside the blast wave the temperature ratio of the shocked intercloud medium and the shocked clouds is of order $\chi \gg 1$ if the cloud shocks are non-radiative, and larger yet if they are radiative. Thermal conduction will act to smooth out the resulting sharp temperature gradients. How it does so depends on the ratio of the electron mean free path to the cloud size and on the topology of the magnetic field. For the moment we neglect the magnetic field, returning to its effects below. The effects of a finite mean free path can be measured in terms of the saturation parameter

$$\sigma'_0 \equiv \frac{2}{25} \frac{\kappa_i T_i}{\rho_i C_i^3 a} = 1.67 \frac{\lambda_{ee}}{a} = 0.393 \frac{T_{i7}^2}{n_i a_{pc}} \tag{8}$$

(Cowie and McKee 1977, Balbus and McKee 1982), where $\kappa_i = 5.6 \times 10^{-7} T_i^{5/2}$ erg s^{-1}K^{-1}cm^{-1}is the classical thermal conductivity for a dilute plasma of cosmic abundance (Draine and Giuliani 1984) and λ_{ee} is the mean free path for electron–electron energy exchange. The saturation parameter is inversely proportional to the Reynolds number: Recall that the Prandtl number measures the relative importance of viscosity and conduction, and for an ideal gas it is

$$Pr = \frac{5}{2} \frac{k\nu\rho}{\kappa\mu} \tag{9}$$

(Landau and Lifschitz 1959). Neutral gases have $Pr \sim 1$, but for an ionized cosmic plasma $Pr = 0.0321$ is small because the heat is carried by the electrons whereas the viscosity is due to the slower ions. Combining equations (6), (8), and (9) then shows that σ'_0 is inversely proportional to Re:

$$\sigma'_0 Re = \frac{\mathcal{M}}{5Pr} = 6.23 \mathcal{M}. \tag{10}$$

Provided radiative losses are negligible, the heat flux from the inter-cloud medium into the cloud heats the cloud surface and drives an evaporative flow away from the cloud. For $\sigma'_0 \lesssim 1$, thermal conduction is approximately described by classical theory with the heat flux $= \kappa \nabla T$. Balancing the heat flux $\kappa_i T_i / a$ against the energy flux $5Pv/2$ gives an evaporation rate $\dot{m} = 4\pi a^2 \rho v \sim 8\pi a \kappa \mu / 5a$ for spherical clouds. More precise evaluation gives

$$\dot{m} = \frac{16\pi\mu\kappa a}{25k} = 2.50 \times 10^4 T_i^{5/2} a_{\rm pc} \qquad g \ s^{-1} \tag{11}$$

for $0.03 \lesssim \sigma'_0 \lesssim 1$ (Cowie and McKee 1977; the numerical coefficient is slightly lower than given there because we have used Draine and Giuliani's [1984] value for κ). The generalization to ellipsoidal clouds, including disks and needles, was obtained by Cowie and Songaila (1977). Balbus (1985) has further extended the theory to cover clouds of arbitrary shape and to cover systems of clouds. It must be emphasized that steady evaporative flow is possible only in 3 dimensions; in the planar case, the temperature gradient relaxes until radiative losses become important and the evaporative flow switches over to a cooling flow of the type considered by Doroshkevich and Zel'dovich (1981).

In general, we can write

$$\dot{m} = \rho_i C_i A_c F(\sigma'_0), \tag{12}$$

where A_c is the surface area of the cloud and F is a number of order unity; for classical evaporation of spherical clouds, $F = 2\sigma'_0$. If we define an ablation time $t_a \equiv m/\dot{m}$, then just behind the blast wave shock ($\rho_i = 4\rho_{i0}, C_i^2 = 3v_b^2/16$) we have

$$\frac{t_a}{t_{cc}} = \frac{\chi^{1/2}}{3^{3/2}F}; \tag{13}$$

for a density ratio $\chi \sim 100$, clouds are crushed before they are ablated if $F \lesssim 2$.

When the electron mean free path becomes comparable to the cloud radius, the classical theory of heat conduction breaks down and heat transport becomes non-local: the heat flux at a given point depends not only on the temperature in the immediate neighborhood but also on the global temperature structure, since electrons have sufficiently long mean free paths to reach the point from substantial distances (e.g., Luciani, Mora, and Pellat 1985). Nonetheless, it is possible to develop a simple phenomenological model that captures the essential physics: In the limit of a very steep temperature gradient, collisions are negligible and the heat flux q should saturate at a value proportional to that which would result from all the electrons streaming freely down the temperature gradient,

$$q_{\rm sat} = f n_e k T_e (k T_e / m_e)^{1/2}, \tag{14}$$

where the "flux limit" f is a numerical factor to be determined. The actual heat flux is then taken as the minimum of the classical and saturated values, or, if a smoother estimate is desired, as the harmonic mean,

$$q = \frac{q_{\rm cl}}{1 + (q_{\rm cl}/q_{\rm sat})}, \tag{15}$$

where q_{cl} is the classical heat flux (Balbus and McKee 1982, Giuliani 1984). In evaporative flows, the maximum value of q_{cl}/q_{sat} is comparable to σ_0'. Cowie and McKee (1977) argued that $f \simeq 0.32$. Laser fusion experiments are consistent with a somewhat smaller value, $f \sim 0.1$ (see Max, McKee, and Mead 1980 and references therein). Recent analytic theory which allows for the streaming of suprathermal electrons (Campbell 1984) agrees with equation (15) to within a factor 1.4 with $f = 0.32$; if attention is restricted to weakly saturated flows, $0.8 \gtrsim q/q_{cl} \gtrsim 0.4$, the agreement improves with $f = 0.15$. Campbell suggests that plasma instabilities could reduce q by up to a factor 2 in the highly saturated case. The saturated heat flux can be expressed in terms of hydrodynamic variables as $q_{sat} = 5\phi_s \rho C^3$, where $\phi_s = 3.5f$ for an ionized cosmic plasma. The range $0.3 \gtrsim f \gtrsim 0.1$ then corresponds to $1 \gtrsim \phi_s \gtrsim 0.3$. Cowie and McKee (1977) solved for the evaporation rate allowing for saturation of the heat flux in the range $100 \gtrsim \sigma_0' \gtrsim 1$; for higher σ_0' the theory breaks down because the hot electrons can penetrate all the way to the cloud surface. In this range the factor $F(\sigma_0')$ in equation (12) is a slowly increasing function of σ_0', reaching a maximum of $(4, 15)$ for $\phi_s = (0.3, 1)$. Numerical integrations by Giuliani (1984) confirmed the analytic calculations of Cowie and McKee, the primary difference being the use of equation (15) by Giuliani instead of the simpler $\min(q_{cl}, q_{sat})$ used by Cowie and McKee. He and Draine and Giuliani (1984) also explored the effects of viscosity on evaporative flows. Saturation of the heat flux is important in the study of solar flares as well (e.g., Smith 1986).

When the electron mean free path becomes so large that the electrons can easily penetrate into the cloud, the uncertainty in the theory of saturated thermal conduction becomes less important. Such suprathermal evaporation is as though the cloud were bathed in a gas of cosmic rays; the cloud pressure is determined by the evaporative outflow and by the requirement that the cloud itself be in thermal balance with cosmic ray heating (Balbus and McKee 1982). This regime corresponds to $\sigma_0' \gtrsim 10^3$, and values of F up to 10^3 are possible.

In the opposite limit of small mean free path, radiative losses become important. The condition that radiative losses balance conductive heating is $n^2 \Lambda(T) \sim \nabla \cdot \kappa \nabla T$, where $\Lambda(T)$ is the cooling function; this defines a length scale

$$\lambda_F \equiv \left(\frac{\kappa T}{n^2 \Lambda} \right)^{1/2}. \tag{16}$$

For clouds small compared to λ_F, evaporation is unaffected by radiative losses, whereas for clouds large compared to λ_F, radiative losses dominate and gas condenses onto the clouds from the intercloud medium. The dividing line between evaporation and condensation occurs at a cloud radius $a \simeq 0.3\lambda_F$ (McKee and Cowie 1977). In the temperature range $10^{7.5} \gtrsim T \gtrsim 10^5$ K the cooling function varies approximately as $T^{-1/2}$, so that this condition defines a critical value of the saturation parameter, $\sigma_0' \simeq 0.03$. Classical evaporation is thus restricted to the range $1 \gtrsim \sigma_0' \gtrsim 0.03$, and condensation occurs for smaller values of σ_0'. Calculations of the non–equilibrium ionization in evaporative flows have been made by Ballet, Arnaud, and Rothenflug (1986), and self–consistent calculations of the dynamics of evaporative flows including radiative losses have been carried out recently by Böhringer and Hartquist (1987). Relatively little work has been done on condensation (Graham and Langer 1973, Doroshkevich and Zel'dovich 1981, Begelman and McKee 1987).

Magnetic fields can, in principle, reduce thermal conduction by an enormous factor, just as in the case of viscosity. There is an important distinction between the two, however: heat can be conducted along field lines even when they are dynamically unimportant, but viscous stress cannot. For evaporative flows in a *stationary* medium, Cowie and McKee (1977) stated that magnetic fields would have relatively little local effect on the evaporation rate; this was based on the idea that the outflow would comb out the field so that the heat from the external medium could more easily reach the cloud surface, and Balbus (1986) has recently confirmed this. However, blasted clouds are embedded in a moving ambient medium which draws the field out behind the cloud, significantly reducing the heat flux to the cloud (although this is partly compensated by the resulting increased temperature gradient). Magnetic reconnection should eventually allow the cloud to come back into thermal contact with its surroundings, and once the flow velocity past the cloud falls below the evaporative velocity (which is of order $\sigma_0' C_i$ for classical evaporation), normal evaporation could resume. The global effects of magnetic fields on evaporation remain to be determined. Balbus (1986) argues that each cloud is in a one–dimensional flux tube, so that evaporation will cease once the temperature relaxes to the point that radiative losses become important. A more general way of looking at the problem is to consider each cloud to occupy a "basin" defined by the condition that heat flows toward the cloud in its basin (Begelman and McKee 1987). In the absence of a magnetic field, the size of the basin is of order the mean intercloud separation. With a field, the shape of the basin will be altered; the rate of evaporation will be reduced insofar as the mean distance to a point in the basin is increased, vanishing altogether when this distance becomes comparable to λ_F.

Hydrodynamic instabilities, particularly the Kelvin–Helmholtz instability, also lead to mass exchange between clouds and the intercloud medium. Nulsen (1982) has argued that the instability will saturate when the rate at which mass is stripped from the cloud is large enough to smooth out the velocity gradient driving the instability, which will occur when the momentum flux of the stripped gas is comparable to that impinging on the cloud. This gives a mass loss rate of order $\dot{m} = \pi a^2 \rho_i v_i$, corresponding to $F = \mathcal{M}/4$ in equation (12). He terms this process "turbulent viscous stripping." This process is effective when evaporation is not: it is independent of the orientation of the magnetic field, so long as the field is weak; it works best when the relative velocity is high (but subsonic relative to the intercloud gas); and, since $\dot{m} \propto T^{1/2}$ rather than $T^{5/2}$, it is relatively more important at moderate temperatures. Pieces of the cloud torn off by the instability will eventually merge with the intercloud medium by thermal evaporation once they are comoving and become magnetically connected with the intercloud medium.

This discussion has concentrated on the processes by which mass is transferred from clouds into the intercloud medium. The reverse process occurs by radiative cooling. For hot gas with $\Lambda = 1.6 \times 10^{-19} T^{-1/2}$ erg cm^3 s^{-1}, the cooling time is proportional to the entropy variable $s \equiv T^{3/2}/n$:

$$t_{\text{cool}} = 6.3 \times 10^{-5} s \qquad \text{yr} \qquad\qquad (17)$$

(Kahn 1976, McKee 1982), which is independent of any expansion or contraction the gas experiences. Mass injection into the intercloud medium reduces s and hence the cooling time. The cooling gas can condense onto pre–existing clouds, or, because it is thermally unstable (Field 1965; McCray, Stein, and Kafatos 1975) it can form new clouds. We have only begun to unravel the processes which determine the mass balance in the multiphase ISM, and it will be a challenge, both theoretical and observational, to improve our understanding.

2.3 SNR Evolution

Because this review is focused on the physical processes associated with SNRs in an inhomogeneous medium, we shall provide only a brief list of some of the references which consider the effects of mass exchange and cloud crushing on SNR evolution. Early studies (Chevalier 1975, McKee and Cowie 1975) concentrated on the effects of thermal conduction in young SNRs without, however, allowing for saturation effects. More recently, Hamilton (1985) has analyzed the expansion of a clumpy young SNR under the extreme assumption that the clumps experience no ram pressure deceleration, only ablation. The effects of evaporation on SNRs in their subsequent evolution was first studied by McKee and Ostriker (1977); Chieze and Lazareff (1981) obtained the similarity solution for this problem, and Cowie, McKee, and Ostriker (1981) carried out a numerical simulation. SNR evolution with arbitrary mass injection is discussed by Ostriker and McKee (1987). Clouds also affect SNR evolution by draining momentum from the blast wave due to cloud drag. Cox (1979) was the first to consider such impeded blast waves, under the assumption that the interior is isothermal; Ostriker and McKee (1987) have considered the opposite case of an adiabatic interior. Energy losses due to cloud crushing were considered by Cowie *et al.* (1981), and analytic solutions have been obtained by Ostriker and McKee (1987). The emission of sound waves due to the interaction of an SNR blast wave with embedded clouds and the consequent heating of the ISM has been discussed by Spitzer (1982) and Spitzer and Ikeuchi (1984).

Despite the fact that SNRs with mass injection are fundamentally different from those without, it has proven difficult to determine whether a given remnant is dominated by mass injection. The primary reason for this sad state of affairs is that most known remnants are about the age of the Cygnus Loop or younger, and for such remnants the observational distinctions between evaporative and Sedov–Taylor remnants are relatively minor (Cowie *et al.*, 1981). Furthermore, the evaporative solution applies to SNRs in a three–phase ISM unaffected by the progenitor star, whereas massive SNRs occur in a radially stratified medium quite different from the normal ISM (see below). Indeed, the Cygnus Loop appears to be such an SNR, and it is no longer tenable to consider it as a possible evaporative remnant, as did Cowie *et al.* A promising technique for studying mass injection in SNRs is through observations of the infrared emission from collisionally heated dust grains injected into the hot intercloud medium; Dwek (1981) has studied the dust emission expected from evaporating clouds, and more generally one would expect the distribution of dust emission from a remnant dominated by mass injection to be quite different from one that is not. High resolution radio observations provide a good means for studying cloud crushing in SNRs, since most of the radio emission from older remnants is believed to originate from crushed clouds (Blandford and Cowie 1982).

3. SNR SHOCKS IN A RADIALLY STRATIFIED ISM

3.1 HII Regions and Stellar Wind Bubbles in a Cloudy ISM

Just as massive stars are the progenitors of supernovae, so, in a sense, the HII regions around these stars are the progenitors of the resulting SNRs. As discussed in the Introduction, the outpouring of energy from a massive star profoundly alters the surrounding medium, converting it from a cloudy medium into one which is radially stratified around the progenitor star. The evolution and appearance of an SNR in such a medium is quite different from that in either a homogeneous medium or a uniform, cloudy medium.

We begin by reviewing the structure of the medium around a massive star embedded in an initially homogeneous medium. The ionizing radiation creates a Strömgren sphere extending out to a radius $R_i = 66.9(S_{49}/n_i^2)^{1/3}$ pc, where n_i is the density of ionized gas inside R_i and S_{49} is the rate of production of ionizing photons in units of 10^{49} photons s^{-1}; $S_{49} \simeq (0.04, 1, 8)$ for a (B0V, O6.5V, O4V) star, respectively (Panagia 1973). The pressure of the ionized gas causes the HII region to expand as $R \propto t^{4/7}$ after a time $R_i/C_{\rm II}$, where $C_{\rm II}$ is the isothermal sound speed in the ionized gas (Spitzer 1968a). Massive stars have strong stellar winds, with kinetic luminosities $L_w = \dot{M}_w v_w^2/2$ of about 10^{35} erg s^{-1} for a B0V star and 10^{36} erg s^{-1} for an O6.5V star (Abbott 1982). Such a wind creates a bubble of hot ($T \sim 10^6$ K) gas inside the HII region, and the bubble expands as $R_b \propto (E/\rho_0)^{1/5} t^{2/5} \propto (L_w/\rho_0)^{1/5} t^{3/5}$ (Castor $et\ al.$ 1975). Because the medium is homogeneous, the radiative losses are small and the bubble can grow quite large by the end of the star's life; e.g., for an O6.5 star, $R_b \lesssim 75 n_m^{-1/5}$ pc, where n_m is the ambient density.

The evolution of the ISM around a massive star is quite different if the medium is initially cloudy. The ionizing radiation causes the clouds to lose mass ("photoevaporation"), and since the mass loss is asymmetric, the cloud rockets away from the ionizing star (Oort and Spitzer 1955, Kahn 1969). When ionizing radiation first strikes a cloud, it drives an ionization–shock front into the cloud which crushes the cloud, just as if it were overtaken by an SNR blast wave. Provided the cloud is not too small, the ionized gas flowing away from the cloud has sufficient opacity to shield the cloud from the full brunt of the incident ionizing photons. Balancing ionizations and recombinations in this layer gives $\alpha n_e^2 a \delta = S/4\pi r^2$, where α is the hydrogen recombination coefficient for excited states and δ is the fractional thickness of the layer. Calculations show that if n_e is measured at the isothermal sonic point and most of the ionizing photons are absorbed in the recombination layer, then $\delta \sim 1/6$ (Bertoldi 1987). The ratio of n_e to the ambient HII region density n_i is then

$$\frac{n_e}{n_i} = \left(\frac{R_i^3}{3r^2 a \delta} \right)^{1/2} = \frac{774}{r_{\rm pc} a_{\rm pc}^{1/2}} \left(\frac{S_{49}}{n_i^2} \right)^{1/2}. \tag{18}$$

Since the cloud is assumed to lie within the Strömgren radius, this ratio is necessarily greater than unity, provided only that the recombination layer exists. The pressure inside the cloud is $P_c = 2(2n_e k T_e)$, so that for $T_e = 10^4$ K,

$$\frac{P_c}{k} = 3.1 \times 10^7 \frac{S_{49}^{1/2}}{r_{\rm pc} a_{\rm pc}^{1/2}} \quad {\rm cm}^{-3}{\rm K}, \tag{19}$$

which, for small clouds close to the star. can be comparable to the pressure inside the Cygnus Loop ($P/k \sim 10^{6-7}$ cm^{-3}K [Raymond *et al.* 1987]). The crushing of the cloud by the ionization–shock front is a radiation–driven implosion, which in principle can drive the cloud into gravitational collapse, thereby inducing star formation (Sandford, Whitaker, and Klein 1982).

Once the cloud has been crushed by the ionization–shock front, it settles into a comet–shaped equilibrium (Bertoldi and McKee 1987), such as that observed for ionized globules in HII regions (Reipurth 1983). The time for the cloud to be destroyed by photoevaporation is $t_{\text{ion}} \simeq 2a/3v_D$, where $v_D = C_I^2/2C_{II}$ is the velocity of the ionization front and C_I^2 is P/ρ in the neutral cloud, including the magnetic pressure. The numerical value of t_{ion} depends on the structure of the magnetic field in the crushed cloud, since that determines both C_I^2 and the reduction in the cloud radius a. Just as in the case of SNRs, photoevaporating clouds come in several different sizes: tiny clouds are fully ionized before they are crushed; small clouds reach the cometary equilibrium, but are destroyed before the rocket effect can move them very far; medium clouds are displaced a substantial distance before destruction; and large clouds are not crushed before the star dies.

Elmegreen (1976) showed that when a massive star turns on in a cloudy medium, the radius of the Strömgren sphere is initially quite large because the density which determines the radius is that of the intercloud medium. The Strömgren sphere rapidly contracts, however, as the photoevaporated debris from the clouds raises the density in the HII region. Cloud destruction and the rocket effect combine to clear the clouds from a region around the star. McKee *et al.* (1984) generalized this analysis to the case of a distribution of cloud masses and allowed for cloud crushing. They showed that if the number of clouds more massive than m varies as $1/m$, then the radius R_h of the cleared zone grows as $t^{4/7}$, just as does that of a Strömgren sphere in a homogeneous medium. The gas inside R_h is homogeneous and has a low density, so it is difficult to see. Much of the photoevaporated gas accretes into an HI shell outside the HII region. If the star is approximately stationary and of spectral type B0 or earlier, then at the end of the star's life the mass of the shell is $M_h \sim 2.4 \times 10^4 n_m^{0.1}$ $M_\odot$, and its radius is $R_h \simeq 56 n_m^{-0.3}$ pc, where n_m is the mean density of the ambient medium, including clouds. In contrast to an HII region in a homogeneous medium, which is bounded by an ionization front, an HII region in a cloudy medium is bounded by a recombination front because the central star is unable to keep all the photoevaporated gas ionized (Shull *et al.* 1985). The presence of clouds in the HII region dramatically alters the evolution of the stellar wind bubble, since photoevaporated gas causes catastrophic cooling if the bubble attempts to expand beyond the homogenization radius R_h (McKee *et al.* 1984); as a result, a stellar wind bubble is regulated so that it is intermediate between an adiabatic bubble (Castor *et al.* 1975) and a momentum–conserving one (Avedisova 1972; Steigman, Strittmatter and Williams 1975). The resulting picture of the evolution of an HII region is consistent with Chu's (1981) and Lozinskaya's (1982) classification of nebulae around Wolf–Rayet stars and Of stars.

3.2 Massive SNRs

When an isolated, massive star explodes, we expect the ambient medium to consist of one or more shells of circumstellar material close to the star; next, a very low density ($n \sim 10^{-2} - 10^{-3}$ cm^{-3}) stellar wind bubble ($R < R_b$); then a region of low density, homogenized HII ($R_b < R < R_h$); a region of HII in which large clouds survive ($R_h < R < R_i$); and finally a dense HI shell at $R \sim R_i$. Even in the absence of a stellar wind, the fact that the ionizing radiation from the star homogenizes the ambient medium out to a radius $R_h \simeq 56 n_m^{-0.3}$ pc (for $M \gtrsim 20 M_\odot$) implies that any dense gas within $R \sim 20$ pc of the star at the end of its life must be circumstellar (i.e., ejected from the star) rather than interstellar; in particular, much of the nebulosity that Chu (1981) observed around Wolf–Rayet stars must be circumstellar (McCray 1983; McKee et al. 1984). The details of the evolution of interstellar bubbles in this environment remain to be worked out, but it is likely that the bubbles around massive stars ($M \gtrsim 20 M_\odot$) will extend out to $\gtrsim 20$ pc as well. We conclude that *young, massive SNRs interact primarily with circumstellar material.* This implies that the dust to gas ratio in these objects may not be standard. Lozinskaya (private communication, 1987) has pointed out that the difference between "naked" and "clothed" Crabs may simply reflect differences in the timing and amount of circumstellar mass ejection.

While the SNR blast wave is interacting with the circumstellar matter, it will deviate from free expansion ($v_b =$ const.), but unless the circumstellar mass M_{cir} greatly exceeds the mass M_{ej} ejected by the supernova, it will not enter the Sedov–Taylor stage. Once the blast wave enters the stellar wind cavity, it will return to free expansion. Hence, *young, massive SNRs remain in free expansion out to*

$$R_A \simeq 2.1 \left(\frac{M_{ej} + M_{cir}}{M_\odot n} \right)^{1/3} \quad \text{pc}, \tag{20}$$

which for $M_{ej} + M_{cir} = 10 M_\odot$ and $n = 10^{-2}$ cm^{-3}is 20 pc. The low density in the bubble and in the surrounding homogenized HII region reduce the luminosity of the SNR below the value it would otherwise have had: *massive SNRs are muffled.* For example, if the wind bubble extended out to 20 pc, beyond which the density jumped to $n = 1$ cm^{-3}, then the SNR would never be bright in X-rays after the emission from the shocked ejecta and circumstellar matter faded out: the density inside the bubble is too low to give detectable X–ray emission, and after the blast wave strikes the shell the post–shock temperature would be less than 10^6 K. Note that *all three effects discussed here—the circumstellar interaction, the free expansion, and the muffling—also apply to the large number of massive SNRs which occur in stellar associations.* These effects can be reduced if the star is moving through a relatively high density medium so that the Strömgren radius is small: in the direction of motion of the star, the Strömgren radius is reduced to $R_i = 5.8/(n_{m2}^2 v_{*6}^4)^{1/3}$ pc, where v_{*6} is the stellar velocity in units of 10^6 cm s^{-1}and $n_{m2} = n_m/(10^2$ cm$^{-3})$.

The observed morphology of optical filaments in SNRs such as the Cygnus Loop is a problem of long standing: if the blast wave is interacting with interstellar clouds, then one would expect SNRs to show a scalloped appearance as shocks wrap around the clouds (McKee and Cowie 1975; Tenorio–Tagle and Rozyczka 1986; Rozyczka and Tenorio–Tagle 1987), whereas in fact the filaments are curved in the opposite sense and tend to follow the curvature of the blast wave. This problem led McKee and Cowie (1975) to suggest that the clouds near the Cygnus Loop are smaller than the thickness of the shell ($a < 1$ pc), so that the scalloped appearance would occur only on small scales, but careful observations (Hester, Parker, and Dufour 1983) show that this interpretation is no longer tenable. Instead, the blast wave appears to be interacting with large, roughly spherical sheets, just as expected for an SNR in a medium which has been radially stratified by its progenitor star. McCray and Snow (1979) suggested that the wind from the progenitor would lead to this result, and we now see that the effects of photoionization are at least as important. Careful studies of the optical morphology of other middle–aged SNRs would be quite valuable.

Our discussion has concentrated on SNRs from massive progenitors ($M \gtrsim 20\ M_\odot$) because such SNRs spend their entire observable life interacting with either circumstellar matter or interstellar matter that has been processed by the progenitor. However, the number of SNRs from lower mass B stars is expected to be comparable to that from more massive stars. Their behavior will resemble that of their more massive cousins so long as the blast wave radius is less than R_i, and then will gradually approach that of Type Ia SNRs. The Cygnus Loop is a likely example: Charles *et al.* (1985) have argued that the absence of clouds embedded in the Loop implies that it is the remnant of a high mass star, and that the presence of the filaments within 20 pc of the center limits the spectral type of the progenitor to later than B0. Two examples have been suggested in the LMC: N49 (Shull *et al.* 1985) and N132D (Hughes 1987). Shull *et al.* have developed a detailed model for N49 based on the hypothesis that the progenitor was a B star which processed the ambient cloudy medium by photoionization. They point out the pressure in the HII region dropped when the B star evolved into a red supergiant, allowing the inner edge of the HI shell to encroach into the HII cavity, and they suggest that this effect is apparent in the X-ray data.

4. CONCLUSIONS

The inhomogeneous structure observed in the ISM is due in no small part to SNRs, and observations of SNRs provide a potentially powerful probe of this structure. Supernovae from massive progenitor stars ($M \gtrsim 20 M_\odot$), including both Type II and probably Type Ib, occur in a medium which has been completely transformed by mass ejection and energy injection from the progenitor, and the resulting remnants therefore tell us more about stellar evolution than about the ISM. In their youth, massive SNRs are most readily observed while interacting with circumstellar material. Stellar wind bubbles allow such SNRs to remain in approximate free expansion out to relatively large radii. Once they expand beyond the circumstellar material, they are "muffled" by the low density in the wind bubble and homogenized HII region; this can be accentuated in stellar associations. Even in old age, such SNRs interact primarily with matter processed by their progenitor stars, unless a high progenitor velocity or a nearby dense molecular cloud permits ambient interstellar matter to lie near the supernova. SNRs from lower mass progenitors—corresponding to both Type Ia SN and Type II SN from B stars of intermediate mass— can probe the structure of the "pristine" ISM, but we are only beginning to unravel the complex physical processes associated with the interaction of a blast wave with an interstellar cloud, and to distinguish such remnants from their more massive counterparts.

Acknowledgments. I wish to thank Sam Falle for several provocative remarks, and Frank Bertoldi and Denis Cioffi for comments on the manuscript. This research is supported in part through NSF grant 86-15177.

REFERENCES
Abbott, D.C. 1982, *Ap.J.*, **263**, 723.
Avedisova, V.S. 1972, *Sov. Astron. A.J.*, **15**, 708.
Balbus, S.A. 1985, *Ap.J.*, **291**, 518.
——. 1986, *Ap.J.*, **304**, 787.
Balbus, S.A., and McKee, C.F. 1982, *Ap.J.*, **252**, 529.
Ballet, J., Arnaud, M., and Rothenflug, R. 1986, *Astron. Astrophys.*, **161**, 12.
Batchelor, G. 1967, *An Introduction to Fluid Dynamics*, (Cambridge: Cambridge
        University Press).
Begelman, M.C., and McKee, C.F. 1987, in preparation.
Bertoldi, F. 1987, in preparation.
Bertoldi, F., Klein, R., McKee, C.F., and Sandford, M.T. 1987, in preparation.
Bertoldi, F. and McKee, C.F. 1987,in preparation.
Blandford, R.D., and Cowie, L.L. 1982, *Ap.J.*, **260**, 625.
Böhringer, H., and Hartquist, T. 1987, *Astron. Astrophys.*, (submitted).
Bychkov, K.V., and Pikel'ner 1975, *Sov. Astr. Letters*, **1**, 14.
Campbell, P.M. 1984, *Phys. Rev. A.*, **30**, 365.
Castor, J., McCray, R., and Weaver, R. 1975, *Ap.J. (Letters)*, **200**, L107.
Charles, P., Kahn, S., and McKee, C.F. 1985, *Ap.J.*, **295**, 456.
Chernyi, G.G. 1961, *Introduction to Hypersonic Flow*, (NY: Academic Press).
Chevalier, R.A. 1975, *Ap.J.*, **200**, 698.
Chieze, J.P., and Lazareff, B. 1981, *Astron. Astrophys.*, **95**, 194.
Chiosi, C., and Maeder, A. 1986, *Ann Rev. Astron. Astrophys.*, **24**, 329.
Chu, Y.H. 1981, *Ap.J.*, **249**, 195.
Cowie, L.L., and McKee, C.F. 1977, *Ap.J.*, **211**, 135.
Cowie, L.L., McKee, C.F., and Ostriker, J.P. 1981, *Ap.J.*, **247**, 908.
Cowie, L.L., and Songaila, A. 1977, *Nature*, **266**, 501.
Cox, D.P. 1979, *Ap.J.*, **234**, 863.
——. 1987, this volume.
Cox, D.P., and Smith, B.W. 1974, *Ap.J. (Letters)*, **189**, L105.
de Jong, T. 1980, in *Highlights of Astronomy* Vol. 5, ed. P.A. Wayman (Reidel:
        Dordrecht), p. 301.
Doroshkevich, A.G., and Zel'dovich, Ya. B. 1981, *Sov. Phys. J.E.T.P.*, **53**, 405.
Draine, B.T., and Giuliani, J.L. 1984, *Ap.J.*, **281**, 690.
Dwek, E. 1981, *Ap.J.*, **246**, 430.
Elmegreen, B.G. 1976, *Ap.J.*, **205**, 405.
Field, G.B. 1965, *Ap.J.*, **142**, 531.
Giuliani, J.L. 1984, *Ap.J.*, **277**, 605.
Graham, R., and Langer, W.D. 1973, *Ap.J.*, **179**, 469.
Hamilton, A.J.S. 1985, *Ap.J.*, **291**, 513.
Heathcote, S.R., and Brand, P.W.J.L. 1983, *M.N.R.A.S.*, **203**, 67.
Hester, J.J., Parker, R.A.R., and Dufour, R.J. 1983, *Ap.J.*, **273**, 219.
Hollweg, J.V. 1986, *Ap.J.*, **306**, 730.
Hughes, J.P. 1987, *Ap.J.*, **314**, 103.
Ikeuchi, S., Habe, A., and Tanaka, Y.D. 1984, *M.N.R.A.S.*, **207**, 909.
Kahn, F.D. 1969, *Physica*, **41**, 172.
——. 1976, *Astron. Astrophys.*, **50**, 145.
Krebs, J., and Hillebrandt, W. 1983, *Astron. Astrophys.*, **128**, 41.

Landau, L.D., and Lifschitz, E.M. 1959, *Fluid Mechanics*, (Reading: Addison Wesley).
Lozinskaya, T.A. 1982, *Ap. and Sp. Sci.*, **87**, 313.
Luciani, J.F., Mora, P., and Pellat, R. 1985, *Phys. Fluids*, **28**, 835.
Max, C.E., McKee, C.F., and Mead, W.C. 1980, *Phys. Fluids*, **23**, 1620.
McCray, R. 1983, *Highlights of Astronomy*, **6**, 565.
McCray, R., and Snow, T.P. 1979, *Ann. Rev. Astron. Astrophys.*, **17**, 213-40.
McCray, R., Stein, R.F., and Kafatos, M. 1975, *Ap.J.*, **196**, 565-70.
McKee, C.F. 1982, in *Supernovae: A Survey of Current Research*, M.J. Rees and R.J. Stoneham (eds.), (Reidel: Dordrecht, Holland), p. 433.
——. 1986, *Astrophys. and Sp. Sci.*, **118**, 383.
McKee, C.F., and Cowie, L.L. 1975, *Ap.J.*, **195**, 715.
——. 1977, *Ap.J.*, **215**, 213-25.
McKee, C.F., Cowie, L.L., and Ostriker, J.P. 1978, *Ap.J. (Letters)*, **219**, L23.
McKee, C.F., Hollenbach, D., Seab, C., Tielens, A. 1987, *Ap.J.*, (in press).
McKee, C.F., and Ostriker, J.P. 1977, *Ap.J.*, **218**, 148.
McKee, C.F., Van Buren, D., and Lazareff, B. 1984, *Ap.J. (Letters)*, **278**, L115.
Nittman, J. 1981, *M.N.R.A.S.*, **197**, 699.
Nittman, J., Falle, S., and Gaskell, P. 1982, *M.N.R.A.S.*, **201**, 833.
Nulsen, P.E.J. 1982, *M.N.R.A.S.*, **198**, 1007.
Oettl, R., Hillebrandt, W., and Müller, E. 1985, *Astron. Astrophys.*, **151**, 33.
Oort, J.H., and Spitzer, L. 1955, *Ap.J.*, **121**, 6.
Ostriker, J.P., and McKee, C.F. 1987, *Rev. Mod. Phys.* (submitted).
Panagia, N. 1973, *Astron. J.*, **78**, 929.
Raymond, J.C., Hester, J.J., Cox, D., Blair, W.P., Fesen, R.A., and Gull, T.R. 1987, *Ap.J.*, submitted.
Reipurth, B. 1983, *Astron. Astrophys.*, **117**, 183.
Richtmyer, R.D. 1960, *Com. Pure Appl. Math*, **13**, 297.
Rozyczka, M., and Tenorio-Tagle, G. 1987, *Astron. Astrophys.*, **176**, 329.
Sandford, M.T., Whitaker, R.W., and Klein, R.J. 1982, *Ap.J.*, **260**, 183.
Sgro, A.G. 1975, *Ap.J.*, **197**, 621.
Shull, P., Dyson, J.E., Kahn, F.D., and West, K. 1985, *M.N.R.A.S.*, **212**, 799.
Silk, J., and Solinger, A. 1973, *Nature*, **244**, 101.
Smith, D.F. 1986, *Ap.J.*, **302**, 836.
Spitzer, L. 1962, *Physics of Fully Ionized Gases*, (New York: Wiley).
——. 1968a, *Diffuse Matter in Space*, (New York: Wiley).
——. 1968b, in *Stars and Stellar Systems VII: Nebulae and Interstellar Matter* ed. B. Middlehurst and L. Aller (Chicago: U. Chicago Press), p. 1.
——. 1982, *Ap.J.*, **262**, 315.
Spitzer, L., and Ikeuchi, S. 1984, *Ap.J.*, **283**, 825.
Steigman, G., Strittmatter, P.A., and Williams, R.E. 1975, *Ap.J.*, **198**, 575.
Tenorio-Tagle, G., and Rozyczka, M. 1986, *Astron. Astrophys.*, **155**, 120.
Wheeler, J.C., and Levreault, R. 1985, *Ap.J. (Letters)*, **294**, L17.
Woodward, P.R. 1976, *Ap.J.*, **207**, 484.
——. 1979, in IAU Symp. 84, *The Large Scale Character of the Galaxy*, ed. W.B. Burton (Dordrecht: Reidel), p. 159.

COLLECTIVE EFFECTS IN SHOCK PROPAGATION THROUGH A CLUMPY MEDIUM

M.L. Norman[*], J.R. Dickel, M. Livio and Y.-H. Chu
University of Illinois at Urbana-Champaign,
Urbana, IL 61801

Abstract : A numerical simulation of shock propagation in a clumpy medium with a weak magnetic field is presented which illustrates a number of dynamical processes of potential importance for explaining spectral line width and radio polarization measurements in supernova remnants.

Introduction: There is considerable evidence to support the suggestion by McKee and Cowie (1975) that the blast wave in an expanding supernova remnant (SNR) encounters density irregularities in the surrounding medium. Dense clouds will get swept up and eventually evaporate in their hotter shocked surroundings. The data include both optical and radio images showing numerous small-scale features with sizes ranging down to less than the instrumental resolution. Recently Shull (1983a,b) has found spectral line widths of over 100-200 km/s in sections of SNRs N49 and N63A in the Large Magellanic Cloud indicating a large velocity dispersion within a single cloud; this might represent random motions of small cloudlets within the larger feature. Another prominent feature of SNR is an apparent net radial stretching of their magnetic field lines which could be caused by the distortion and wrapping around of the edges of clouds encountered by the outward moving shock. Theoretical estimates of the energy and mass involved in the interactions and evaporation of clouds give parameters which match the observed expansion (Cowie, McKee and Ostriker 1981). Finally, a 1-dimensional hydrodynamic model has been used to show that the observed shell thickness and radial brightness variations in young SNR can be explained by interactions with a number of little clumps (Dickel, Eilek and Jones 1987).

The true situation will involve the interactions of the initial shock with a collection of randomly placed clumps. The result will be a complex pattern of colliding shocks which have been both reflected and transmitted at different speeds through the different densities encountered. The process will distort the density, pressure, magnetic field structure, and other parameters of the clump and interclump regions. To assess the resultant structures and confirm the viability of such a phenomenon to actually explain the observations we have utilized a 2-dimensional magnetohydrodynamics code to follow all the physical variables as the shock propagates through the clumpy medium.

Numerical Methodology: The initial conditions for the simulation consist of a uniform background gas to which a number of gaussian-shaped dense clumps are added, randomly distributed within the two-dimensional problem domain ($0 \leq X \leq 600$; $0 \leq Y \leq 400$). The clumps have a maximum density of twenty times the interclump gas density, but are in pressure balance with the interclump gas by virtue of their lower temperature. In addition, a uniform weak magnetic field is initialized oriented parallel to the X axis. A Mach number=10 planar shock is injected along the Y=0 boundary at t=0 and driven in time through an appropriate choice of inflow boundary conditions. An outflow boundary condition is used at Y=400 so that the shock may exit the problem domain. Periodic boundary conditions are applied at X=0 and X=600. A γ=5/3 ideal gas equation of state is assumed for both the clump and interclump gas. As we assume radiative cooling to be unimportant, the results of the simulation are scale invariant and are wholly determined by the dimensionless parameters of the problem.

* also, National Center for Supercomputing Applications

The equations of ideal magnetohydrodynamics are advanced in time subject to these initial and boundary conditions using the ZEUS code under development at the National Center for Supercomputing Applications, University of Illinois, Urbana-Champaign. ZEUS is an outgrowth of an earlier, purely hydrodynamical code (Norman and Winkler 1986). The magnetic vector potential which describes the magnetic field is evolved along with the conservation laws for mass, momentum and energy in an inviscid gas, taking into account the JxB Lorentz force on the fluid trajectories (Clarke, Norman and Burns 1986). ZEUS is an Eulerian finite-difference code which employs the second order-accurate monotonic upwind advection scheme of Van Leer (1977) for all field variables except for the vector potential, which is advected using the third order-accurate PPM advection scheme of Colella and Woodward (1984). This is necessary since the magnetic field strength, which is given by the curl of the vector potential, is required to be second order-accurate in keeping with the other field variables. Shock waves are treated using the Von Neumann-Richtmyer artificial viscosity. In general, the dynamical equations are time-differenced in an explicit fashion, the exception being the use of a time-implicit energy equation for flows with strong cooling. ZEUS has been extensively tested against a variety of flow problems where the solution is either known analytically or has been determined experimentally. A grid of 300x200 uniform zones is used for the present simulation, which required approximately 2 hours of cpu time on a CRAY XMP.

Results: The shock front corrugates as it propagates through regions of different densities and sound speeds introduced by the clumps (Fig. 1a). When the incident shock wave encounters a clump, it is transmitted more slowly than in the interclump gas, producing the dimples seen in Fig. 1b. A reflected shock wave is also produced upstream of each clump. These merge to form the intersecting bow shocks seen in Fig. 1b. The effect of the transmitted shock wave on a clump is to flatten and compress it while imparting to it a velocity normal to the local shock surface. By virtue of the shock front corrugations, a range of transverse velocities are imparted since it can happen that an individual clump is struck at an angle to the overall direction of shock propagation.

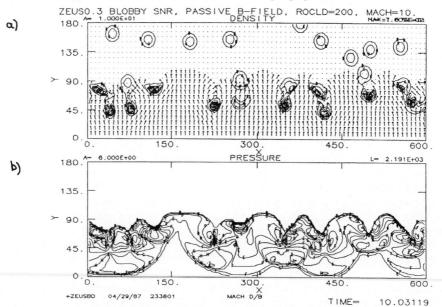

Fig.1 Density contours and velocity vectors (a) and pressure contours (b) showing the corrugation of the shock front as it encounters the clumps (small circles in Fig. 1a).

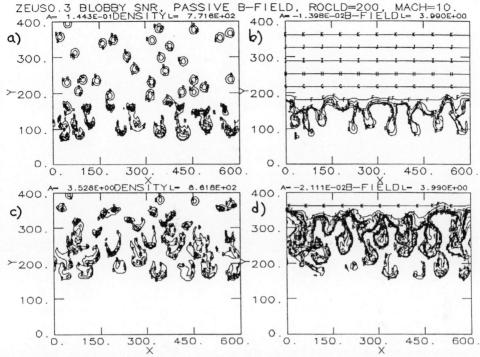

Fig. 2 Distribution of gas density and magnetic field lines at two instants in time [(a,b) t=20; (c,d) t=40], showing the development of horseshoe clump geometry and vertical stretching of magnetic field lines.

With time, the shocked clumps assume a characteristic horseshoe shape, as shown in Fig. 2a,c. This shape is produced by the cross-stream variation of the pressure difference between the upwind and lea sides of the clump. Magnetic field lines, which are anchored to the clumps via flux freezing, are stretched downstream in the interclump gas, as shown in Figs. 2b,d. Downstream of each clump, the magnetic field geometry is that of oppositely directed field lines -- ideal conditions for magnetic reconnection. Although our code assumes zero resistivity, numerical truncation errors introduce a grid resistivity which snaps magnetic field lines on the resolution scale of the calculation. As each clump is resolved by thirty or more zones in each direction, the large scale magnetic field geometry is faithfully simulated, however.

By the time the shock wave has crossed the problem domain, it has interacted with each clump, setting them all in motion. An individual clump will be accelerated in a direction determined by the local shock normal, which in turn is determined by the location of neighboring clumps. As a result of this shock-mediated collective interaction of clumps, a range of longitudinal and transverse velocities are produced. To quantify this effect, theoretical line profiles were calculated by binning in velocity space the square of the gas density moving at a particular longitudinal (parallel to the overall shock propagation direction) or transverse speed. These two distributions were then normalized and plotted in Fig. 3. As can be seen, each distribution consists of a narrow central core and broad wings. Notice that the velocity widths in both the core and wings are comparable between the two orientations, indicating an effective thermalization of shock directed kinetic energy into clump random kinetic energy.

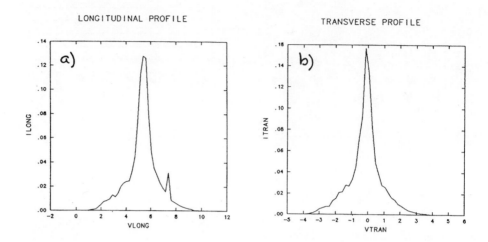

Fig. 3 Theoretical normalized line profiles at two viewing angles to the initial shock propagation direction, (a) parallel, (b) perpendicular. V_{long} and V_{tran} are in units of the interclump soundspeed (=1); the shock velocity = 10. Note the narrow core and broad wings have comparable velocity widths in both projections, indicating an effective thermalization of shock directed kinetic energy into clump random kinetic energy via collective effects.

Conclusions: Line widths comparable to the speed of the SNR blastwave can be produced via a shock-mediated collective interaction of dense clumps provided that their filling factor is sufficiently high. The assumed 11% filling factor in this 2-D simulation would translate into a 3% filling factor in 3-D, which is not unrealistic. Theoretical line profiles consist of a narrow central core surrounded by broad wings which, in this simulation, are quite insensitive to viewing angle. We would not expect this to be the case at lower clump filling factors. Weak magnetic field lines permeating the unshocked clumpy medium will be stretched in the direction of shock propagation, which could account for the net radial polarization seen at centimeter wavelengths in many SNRs (e.g., Cas A).

This work was supported by an internal research grant at the National Center for Supercomputing Applications, University of Illinois at Urbana-Champaign.

References:
Clarke, D.A., Norman, M.L. and Burns, J.O. 1986, Ap.J. Lett., **311**, L63.
Colella, P. and Woodward, P.R. 1984, J. Comp. Phys., **54**, 115.
Cowie, L.L., McKee, C.F. and Ostriker, J.P. 1981, Ap.J., **247**, 908.
Dickel, J.R., Eilek, J.A., and Jones, E.M. 1987, *this volume*.
McKee, C.F. and Cowie, L.L. 1975, Ap.J., **195**, 715.
Norman, M.L. and Winkler, K.-H.A. 1986, in *Astrophysical Radiation Hydrodynamics*,
    NATO ASI C188, eds. K.-H.A. Winkler and M.L. Norman, (Reidel:Dordrecht), 187.
Shull, P. 1983a, Ap.J., **275**, 611.
_____. 1983b, Ap.J., **275**, 592.
VanLeer, B. 1977, J. Comp. Phys., **23**, 276.

THE DOMINANT INTERACTIONS BETWEEN SNR AND THE ISM

Robert Braun
National Radio Astronomy Observatory*,
P.O. Box O, Socorro, New Mexico 87801, USA

Abstract: The interactions of supernovae ejecta with the surrounding medium which have a dominant influence on the resulting SNR dynamics and brightness are considered for both young and evolved objects. The important roles of clumpy ejecta in determining the brightness of young SNR and of shock impingement on a pre-existing shell in influencing both the brightness and subsequent dynamics of evolved SNR are stressed.

Introduction: The basic types of interactions which might be expected to occur between the ejecta of a supernova and the surrounding environment have been postulated for some time (e.g. McKee 1983). These are: the interaction of diffuse ejecta with 1) diffuse ISM/CSM either with or without a power law density profile, 2) high density clumps ($\leq$ .1 pc) of ISM/CSM, 3) high density clouds ($\geq$ 1 pc) of ISM and 4) high density walls of a pre-existing cavity, as well as the interaction of clumpy ejecta with each of the preceding components. The relatively short timescales for the dissipation and/or thermalization of clumpy ejecta make it more probable that only its interaction with 5) diffuse shocked ejecta/ISM/CSM may be commonly observed. While all of these processes are likely to occur to a greater or lesser extent, it is interesting to consider which might play a dominant role in the development of observable (shell-type) remnants.

Dominant Interactions in Young SNR: The dynamics of the youngest known galactic SNR appear to be governed primarily by process 1) above. The diffuse components remain dynamically coupled and generate a classical spherical shock front albeit with reverse shock propagation back through the ejecta. The slow moving optical knots seen in Cas A and Kepler represent interactions of type 2) above, while evidence for interstellar clouds (process 3) is seen in the Balmer line filaments of Tycho and SN1006. However, the mass of diffuse ejecta that is involved in powering the shock front appears to be surprisingly low. IRAS observations have allowed a good estimate of the shocked ISM mass to be made in the case of Tycho and (perhaps with somewhat higher uncertainty due to a possibly abnormal gas/dust ratio) Cas A (Braun 1987, Dwek 1987), while current expansion measurements have allowed determination of the dynamical age. Together these quantities imply a diffuse ejecta mass of about 0.4 $M_{\odot}$ with an associated SN energy, $E_o \leq 2 \times 10^{50}$ erg for both SNR. These values imply that most of the ejecta mass in these systems, $\sim$1 $M_{\odot}$ in Tycho and $\geq$ 5 $M_{\odot}$ in Cas A, resides in clumps. Furthermore, since the shock velocity has now decreased from an initial value of $\sim$10,000 km s^{-1} to $\sim$2,000 km s^{-1}, the remaining ejecta mass will contribute only 1/25 the energy per unit mass of the fast diffuse component. Total energies are thus likely to be $\leq$ 3 x 10^{50} erg. Observational evidence for clumpy ejecta is seen in the bright clumpy radio/X-ray ring interior to the outer shock in Tycho (e.g. Seward, Gorenstein and Tucker 1983) which dominates the emission from this source. Even more direct evidence for clumpy ejecta is seen in Cas A where both high-density clumps, the optically observed FMK (e.g. Van den Bergh and Kamper 1983), and more tenuous clumps, the bow-shock producing blobs which dominate the radio structure (Braun, Gull and Perley 1987), are seen continuously puncturing the shell of decelerated diffuse ejecta and shocked ISM.

It seems that the dynamics of young SNR are dominated by diffuse interactions of type 1) and to a lesser extent type 3), and it is these processes which are responsible for Balmer line optical filaments and collisionally heated dust emission. High brightness X-ray and radio emission (and high velocity oxygen line emission) on the other hand is closely tied to the secondary shocks

---

* The National Radio Astronomy Observatory is operated by Associated Universities Inc., under contract with the National Science Foundation.

and turbulence caused by the interception of clumpy ejecta (wherein the majority of the pre-cursor mass resides) by
the decelerated reverse/outer shock zone.
**Dominant Interactions in Evolved
SNR:** By the time a supernova remnant has become at least partially radiative the effects of clumpy ejecta should
no longer play a dominant role in the
source's emission and evolution. Clumps
with both high density and velocity will
have deposited their energy in the SN environment after shell penetration, while
lower density and/or velocity clumps will
have been ablated or evaporated in the
SNR interior, adding their energy to that
driving the diffuse shell.   The radiative shell of optical filaments surrounding
such objects has traditionally been interpreted as the result of the shock overrunning randomly occuring clouds in the
SN environment. This interpretation no
longer seems to be viable.

Individual cases that have been studied in detail have shown the existence of
cold, unshocked shells and cavity walls
which must pre-date the SN event. These
structures are illustrated in figure 1,
where the warm shock-heated dust emission from IC 443 centered at $\alpha, \delta =$
$6^h 14^m, 22.^{\circ}5$ has a dense unshocked shell
of cool dust immediately to the west
and south, and in figure 2, where much
of the western and northern rim of the
Cygnus Loop is seen to be abutting a
region of higher density traced by the
cool dust distribution. (The observations illustrated in these figures are discussed in Braun and Strom 1986a,b.)
In IC 443 the shock has already been
propagating within the high density gas
sufficiently long that shock-accelerated
molecular and atomic gas (e.g. Braun
and Strom 1986a) are seen in profusion.
In the Cygnus Loop on the other hand,
this interaction is only beginning, as evidenced by the high cavity wall densities
($n_H \geq 100 cm^{-3}$) but non-detection of
shocked CO and only faint accelerated
HI (Strom, priv.com.). The kinetic energy implied by the gas mass and velocity within a shell geometry in these two
cases, $E_k^{ISM} = 0.55 \times 10^{50}$ erg corre-

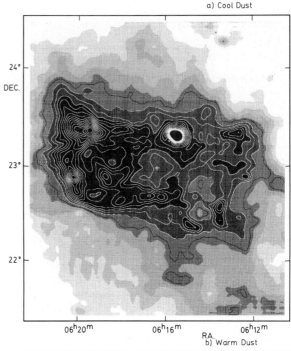

a) Cool Dust

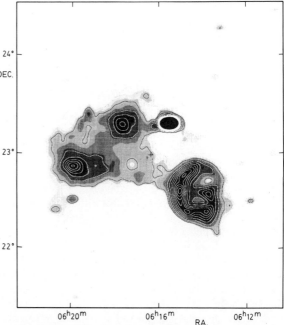

b) Warm Dust

Fig. 1a and b. Cool and warm dust emission components
within the IC 443 complex derived from images in the four
IRAS bands by a three-component spectral decomposition. The
shock-heated dust of the IC 443 SNR is centered near $\alpha, \delta =$
$6^h 14^m, 22.^{\circ}5$. Note the cool high-density shell immediately
outside the shocked gas to the south and west.

sponds to a total energy $E_o = 2 \times 10^{50}$ erg assuming the 28 % lower limit to the kinetic energy applicable to (post-) Sedov evolution. While few SNR have been studied carefully enough to detect unshocked shells, shock-accelerated HI has been found in every evolved SNR studied to date with sufficient sensitivity, IC 443, VRO 42.05.01, CTB 109, G78.2+2.1 (Braun and Strom 1986c) and OA184 (Routledge et.al. 1986). Consideration of the conditions required to detect accelerated HI implies columns of $n_H \cdot t \geq 3 \cdot 10^5 \mathrm{cm}^{-3} \mathrm{yr}$ shocked at 100 km s^{-1} (c.f. Braun and Strom 1986a). Since accelerated HI is found over a significant fraction of the shell, but yet a hard X-ray interior (detected so far in IC 443, the Cygnus Loop, CTB 109 and G78.2+2.1) indicates much higher shock velocities in the recent past, a shell geometry for this high density gas is strongly implied.

A further indication for the common occurrence of a pre-existing cavity is given by consideration of the implied initial energies of SNR obtained by various means. As indicated above, total energy estimates based on IR determined masses and source dynamics for Tycho, a type I SNR, and Cas A, presumably a type Ib or II SNR, are quite well determined, $E_o \leq 3 \times 10^{50}$. Similar total energies, $E_o = 2 \times 10^{50}$ erg, are implied above by considering the kinetic energy within a shell geometry for IC 443 and the Cygnus Loop, the evolved remnants of apparently massive SN. The analysis of eight X-ray detected SNR in the LMC by Long and Helfand (1979) assuming a Sedov description of the X-ray emitting gas results in a consistent mean initial energy estimate $E_o = 2 \times 10^{50}$ erg. This consistency disappears when considering the initial energies implied by assuming a Sedov description of the pressure driving the shock into the radiatively cooling gas observed optically (e.g. Dopita 1979; Blair, Kirshner and Chevalier 1981). Implied energies derived from this method

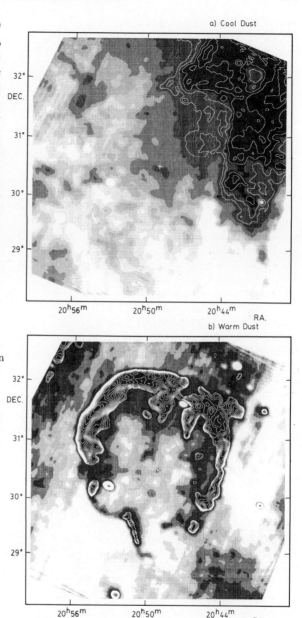

a) Cool Dust

b) Warm Dust

Fig. 2a and b. Cool and warm dust emission components in the Cygnus Loop region derived from images in the four IRAS bands by a three-component spectral decomposition. The shock-heated dust is very well delineated in the warm dust component. Note the cool dense cavity wall which wraps neatly around much of the western and northern rim of the SNR.

increase monotonically with SNR diameter between a few times 10^{49} erg at 5 pc and 10^{52} erg at 100 pc. It seems that the epoch of radiative cooling in SNR is accompanied by a radical departure from adiabatic shock propagation (like the transition from inner cavity to cavity wall shocks) rather than merely a perturbation resulting from randomly distributed clouds within a more tenuous substrate.

More General Considerations: It has too often been the case that supernova remnants were considered in isolation. While the remnants of type I SN may have the oppportunity of expanding into a relatively pristine environment (and as a consequence rapidly dissipate their energy in an unobtrusive manner after the first few hundred years of clumpy ejecta interception), the remnants of type II and probably type Ib SN must almost certainly encounter an environment which has been heavily processed by both the progenitor and possibly other association members. Such a scenario is graphically illustrated in figure 3 (from Braun and Strom 1986a), where the shock from the IC 443 SN is seen to propagate not only in the cavity of the precursor, but into the adjoining cavity surrounding at least one well-identified fellow association member.

The recent work of Heiles (1987) gives a refreshing analysis of the ISM that takes into account the spatial and temporal distribution of SN I and SN II. The major difficulty that this analysis has in confronting observations is the high predicted value of the porosity parameter, implying an inordinately high filling factor for the hot ionized medium (HIM).

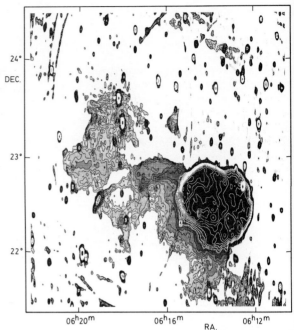

Fig. 3. Radio continuum emission from the IC 443 complex at 327 MHz. Thermal radio emission is confined to the diffuse region between $\alpha = 6^h18^m - 6^h20^m$, $\delta = 22.^{\circ}5 - 23.^{\circ}5$. Non-thermal emission is seen from the bright shell centered near $\alpha, \delta = 6^h14^m$, $22.^{\circ}5$ with radius $\sim 0.^{\circ}4$, as well as the fainter interconnected shell centered near $\alpha, \delta = 6^h16^m$, $22.^{\circ}2$ with radius $\sim 0.^{\circ}8$ containing the B0.5 V star HD44139.

As Heiles suggests, the most straightforward way out of this dilemma is to adopt an SN energy smaller than the oft-quoted but poorly established 10^{51} erg he assumes. In light of the preceding discussion this seems to be amply justified, and adopting $E_o = 2 \times 10^{50}$ erg gives HIM filling factors which are in reasonable agreement with those observed.

Ultimately, a better understanding of the makeup and energetic balance of a normal galaxy should emerge from the comprehensive study of our nearby neighbours M31 and M33. Both galaxies are currently being subjected to such an extensive observational assault that a much clearer view of the complex interplay of processes which mold their media should soon be forthcoming.

References:

Braun, R. and Strom, R.G. 1986a, *Astr. Ap.*, **164**, 193—207

Braun, R. and Strom, R.G. 1986b, *Astr. Ap.*, **164**, 208—217

Braun, R. and Strom, R.G. 1986c, *Astr. Ap. Suppl.* **63**, 345—401

Braun, R. 1987, *Astr. Ap.*, **171**, 233—251

Braun, R., Gull, S.F. and Perley R.A. 1987, *Nature* **327**,395—398

Blair, W.P., Kirshner, R.P. and Chevalier, R.A. 1981, *Ap. J.*, **247**, 879—893

Dopita, M.A. 1979, *Ap. J. Suppl.* **40**, 455—474

Dwek, E. 1987, in prep.

Heiles, C. 1987, *Ap. J.*, **315**, 555—566

Long, K.S. and Helfand, D.J. 1979, *Ap. J. (Letters)*, **234**, L77—L81

McKee, C.F. 1983, in *Supernova Remnants and their X-ray Emission*, eds. Danziger, I.J. and Gorenstein, P., Reidel, Dordrecht, 87—97

Routledge, D., Landecker, T.L. and Vaneldik, J.F. 1986, *M.N.R.A.S.*, **221**, 809—821

A MODEL OF SNR EVOLUTION FOR AN O-STAR IN A CLOUDY ISM

Peter Shull Jr.
Dept. of Physics, Oklahoma State Univ., Stillwater, OK 74078

John Dyson and Franz Kahn
Dept. of Astronomy, The University, Manchester M13 9PL

Abstract: We present an analytical model of SNR evolution
in a cloudy interstellar medium for a single progenitor
star of spectral type 05 V. The model begins with the
progenitor on the zero-age main sequence, includes the
effects of the star's wind and ionizing photons, and ends
with the SNR's assimilation by the ISM. We assume that the
ISM consists of atomic clouds, molecular clouds, and a hot
intercloud phase. The type of SNR that results bears a
strong resemblance to N63A in the Large Magellanic Cloud.

Introduction: For many years, it has been desirable to interpret
observations of supernova remnants in terms of models of SNR evolution.
The Sedov-Taylor model of the 1950s, for example, was the first widely
used paradigm. This model assumes the explosion occurs in a medium with
a constant or power-law density distribution. Disagreements between the
predictions and the observations, particularly when it came to time
scales and geometry, led to the realization that satisfactory models of
SNR evolution must include more realistic assumptions about the nature
of the ambient interstellar medium at the moment of the SN explosion.

Clearly, the state of the ambient ISM at the time of the SN
explosion depends not only on its condition at the birth of the SN
progenitor star, but also on the effects that this star has on the ISM
during the star's lifetime.

This inspired the development of an analytical model of SNR
evolution for the case of a single B-star progenitor by Shull, Dyson,
Kahn and West [1]. The most interesting prediction of this model is
that the SN progenitor should be surrounded at the time of its explosion
by a neutral shell of density $n \doteq 10^2$ cm^{-3}. This bubble results from
the expansion of hot, ionized gas around the star beyond the star's
Strömgren radius. The result is the creation of an SNR with a
filamentary, optical shell such as N49 or the Cygnus Loop.

Assumptions: As in the B-star model we assume a three-phase ISM in
approximate pressure equilibrium at 10^{-12} dyn cm^{-2} with solar
composition [2]. The intercloud medium has a temperature $T \doteq 10^6$ K, and
a density $n \doteq 10^{-2}$ cm^{-3}. Atomic clouds have $T \doteq 100$ K, $n \doteq 20$ cm^{-3}, and
radii of a few pc, while molecular clouds have $T \doteq 10$ K, $n \doteq 10^3 - 10^6$
cm^{-3}, and radii less than 10 pc. There are about 10 atomic clouds in
each spherical volume of radius 10 pc, and the clouds are isothermal,
self-gravitating spheres.

As a rather extreme example, we elected to model a star of spectral type 05 V. Such a star has a mass of about 50 $M_\odot$, radius = 14 $R_\odot$, main-sequence lifetime of 10^6 yr, and an ionizing photon luminosity of 5 x 10^{49} s^{-1}. Additionally, the stellar wind has a terminal velocity of 3000 km s^{-1} and a mass-loss rate of 3 x 10^{-6} $M_\odot$ yr^{-1} [3,4,5].

Pre-SN Evolution on the Main Sequence: The interesting differences between the current model and the model for the B-star are due to the O-star's more intense ionizing flux, energetically important wind, significant mass-loss effects, and shorter main-sequence lifetime. For reasons of brevity, we will present only the predictions of the model, and refer the reader to reference [6] for the details.

The ionizing photon flux will drive an R-type ionization front far into the surrounding medium. In a uniform medium of density n cm^{-3}, its ultimate radius would be 110 n$^{-2/3}$ pc, which would be attained in a timescale of 10^4/n yr.

As the Strömgren sphere is forming, the ionizing flux also evaporates the atomic clouds. Due to their lower temperatures, the molecular clouds will essentially not evaporate [7]. We neglect thermal conduction. Gas is injected into the intercloud medium at the rate of n = 2 x 10^{-6} cm^{-3} yr^{-1}. Clouds must have radii larger than 2 pc to survive until the end of the main-sequence phase. Due to the shortness of the main-sequence lifetime, the gas will not have time to evenly distribute itself, and densities will range from 10^{-2} to 1 cm^{-3}.

This gas will cool to 10^4 K within less than a main-sequence lifetime, have a pressure at most equal to the ambient value, and therefore not flow away from the star to form a neutral shell as it does in our B-star model.

The stellar wind simultaneously plays an important role in this process because, over the star's lifetime, the density of the injected wind's energy, when contained within radii of about 120 or fewer parsecs, is greater than or equal to the pressure of the ambient ISM.

The wind freely expands from the stellar surface out to a radius of approximately 1 pc. This radius is reached in 300 yr. Thereafter, the swept-up intercloud gas, whose density is steadily increasing, controls the wind's dynamics. Because of the low densities involved, this process involves no significant radiative losses. Out to a radius of about 6 pc, the wind radius r is proportional to $t^{3/5}$, where t is the elapsed time. Thereafter, the effect of the increasing density becomes noticeable, and r $\propto t^{2/5}$ until the stagnation radius of 21 pc is reached. At this point, the ram pressure of the wind is counterbalanced by the pressure of the ambient ISM.

Pre-SN Evolution on the Giant Branch: While the star is on the supergiant branch for a time T_{SG} roughly equalling a tenth of its MS lifetime, its ionizing photon flux will become negligible, and the mass-loss rate will increase by a factor of 10.

As the photon flux weakens, recombination of the gas within the Strömgren sphere occurs on timescales of $10^5/n$ yr, which is comparable to T_{SG} for the denser regions beyond 20 pc. Therefore, the pressure drops and the inner and outer boundaries of the sphere respectively move outward and inward. Since this in both cases involves the displacement of denser gas by lower-density gas, Rayleigh-Taylor instabilities may result at both boundaries. The density gradients characterizing these two new boundaries have scale heights of $c_s*T_{SG} \doteq 10$ pc.

The increased mass-loss rate enhances the motion of the inner boundary as it moves to seek a new equilibrium position.

The Circumstellar Region at the Time of Explosion: Within 20-30 pc of the star is a volume of ionized, unrecombined wind with $T \doteq 10^6$ K and n $\ll 1$ cm^{-3}. Beyond this, out to a radius of about 100 pc, is a region of neutral, recombined gas at $T \doteq 10^3$ K with n ranging up to 1 cm^{-3}. There are density gradients with scale heights of about 10 pc at both boundaries of this region. Furthermore, molecular clouds will stud both these volumes.

The Explosion and its Aftermath: We assume that the 0-star will explode by releasing 10^{51} erg of energy and ejecting about 10 $M_\odot$ of material at speeds of 2×10^4 km s^{-1}.

The ejecta propagate freely throughout the wind zone, the wind's mass being less than 0.2 $M_\odot$. The shock driven into the wind accelerates down the r^{-2} density gradient. The ejecta reach the inner boundary of the recombined region in approximately 10^3 yr.

A jumble of weak transmitted and reflected shocks are generated by the wind shock at the irregular inner boundary. When the ejecta arrive, the main shock driven into the recombined region will have a speed of only 200 $(E_0/10^{51}$ erg$)^{1/2}$ $(1$ cm$^{-3}/n)^{1/2}$ km s^{-1}. This shock will sweep up 10 $M_\odot$ of material by the time it has penetrated less than 0.1 pc into this region, and thus produce a reverse shock.

Due to the muffling effect of the wind cavity, the transmitted shock wave's adiabatic evolution can be described by a modified Sedov solution in which the epoch t of the shock when it is at radius r is given by $t \propto r^{7/2} [1-(r_0/r)^3]^{1/2}$. The latter factor indicates the muffling effect not present in the standard solution.

Whenever the blast wave encounters any remaining atomic or molecular clouds, optically radiative shocks will be driven into them. These clouds will be the only significant sources of optical emission from the SNR.

Comparison of the Model's Predictions to N63A: This model predicts that only hard X-rays and infrared radiation from shocked dust will be emitted while the ejecta traverse the wind zone, but at low intensity. Then, readily detectable X-rays and UV should emanate from the recombination zone, and soft X-rays from the reverse-shocked ejecta.

Optical emission will originate from any shocked clouds, but there will be no dense, neutral shell, in contrast to the case for a B-star. These events will be followed by the classical snowplow and radiative phases of the Sedov model, but on shortened timescales due to the wind cavity.

   The SNR N63A in the LMC strikingly resembles these predictions [8]. This SNR is in a group of O and B stars. Assuming that the progenitor was more massive than the stars remaining in the association, then the progenitor was at least of spectral type O9–O7. The remnant has center-filled X-ray and radio morphologies with radii of about 10 pc. Infrared emission has also been detected by the IRAS satellite. There is no optical shell whatever. The only optical emission comes from two small shocked clouds located between the SN and the Earth.

Acknowledgments: This research was supported by an NSF EPSCoR grant, an OSU Dean's Incentive Grant, and the Science and Engineering Research Council.

[1] Shull, P., Jr., Dyson, J. E., Kahn, F. D., and West, K. A. 1985, M.N.R.A.S., 212, 799.
[2] Spitzer, L., Jr. 1978, Physical Processes in the Interstellar Medium (New York: Wiley), p. 227.
[3] Garmany, C. D., et al. 1981, Ap. J., 250, 660.
[4] Panagia, N. 1973, A. J., 78, 929.
[5] Lamers, H. J. G. L. M. 1981, Ap. J., 245, 593.
[6] Shull, P., Jr., Dyson, J. E., and Kahn, F. D. 1988, Ap. J., submitted.
[7] Kahn, F. D. 1969, Physica, 41, 172.
[8] Shull, P., Jr. 1983, Ap. J., 275, 592.

RADIO EMISSION FROM YOUNG SUPERNOVA REMNANTS: EFFECTS OF AN INHOMOGENEOUS CIRCUMSTELLAR MEDIUM

John R. Dickel
Astronomy Department, University of Illinois
Urbana, IL

Jean A. Eilek
Physics Department, New Mexico Tech.
Socorro, NM

Eric M. Jones
Los Alamos National Laboratory
Los Alamos, NM

Abstract: The evolution of young supernova remnants has been modeled using a 1-dimensional hydrodynamic code. Turbulent dynamo amplification of magnetic fields and both turbulent and shock acceleration of relativistic electrons have been included macroscopically to produce synchotron radiation. The observed radio morphology cannot be reproduced by expansion into a homogeneous medium; it appears that many small cloudlets must be present in the circumstellar material.

Introduction: The evolution of a supernova remnant (SNR) as it expands into the local surroundings is a fundamentally straightforward hydrodynamic process (cf. Chevalier 1982). The ejected stellar material initially accelerates down an external density gradient and then decelerates when it reaches a constant-density circumstellar medium. This creates a double shock structure: an inner shock where the ejectum decelerates, a region of hot shocked ejectum, a contact surface, a region of shocked circumstellar material, and finally, an outer shock. This structure forms within the first few days of the remnant's life. When it has swept up about eight times the ejected mass, the remnant enters the blast wave phase modeled by Sedov (1959).

The radio luminosity, however, does not arise directly from the hydrodynamics. The luminosity traces processes which are essentially side effects: the amplification of magnetic fields in turbulent regions and the acceleration of relativistic electrons at shocks and in the turbulent regions. A shock will compress any tangential magnetic field and further accelerate any already relativistic electrons. The interface between the ejectum and the circumstellar medium is Rayleigh-Taylor unstable and will develop turbulence which will amplify the magnetic field and further accelerate the relativistic electrons.

To investigate the conditions in the SNR and the resultant radio emission we have used a hydrodynamic code to follow physical variables

and have modeled macroscopically both shock and turbulent energization
of fields and particles.

Model: We start with the 1-dimensional explicit Lagrangian hydrodynamic code developed by Jones and colleagues (Jones, Smith, and Straka
1981).   An artificial viscosity term is used to handle shock discontinuities.   The code follows the development of turbulence in the
remnant by evaluating the Rayleigh-Taylor stability conditions everywhere and by using a 1-dimensional method to trace the dynamical evolution of the resultant "fingers" in unstable regions.   Because the
fingers rarely move more than a few times their own diameter during
the length of the pre-Sedov phase, we need not consider the complication of fully developed turbulence.   We need only describe the growth
of the fingers at their source and their subsequent propagation
through the Lagrangian mesh used to calculate the mean flow.   In particular, at each boundary we describe the moving fingers by inward-
and outward-directed eddy velocities and corresponding eddy masses.

     Relativistic electrons can be accelerated by MHD turbulence which
should occur in the unstable regions even in the early stages of
development described herein.   The very small fraction of the turbulent energy transferred to relativistic electrons is kept track of
macroscopically by noting that the rate of transfer is proportional
to the rate of energy turnover in the turbulence (e.g. Eilek and
Henriksen 1984).   This depends upon the eddy size which is taken to
be the local density scale height.   Already relativistic electrons
can also be accelerated at shocks in a first order Fermi process (e.g.
Axford et al. 1982).   This is handled numerically by converting at
each shock a small fraction of the upstream kinetic energy to relativistic electrons.   The particle spectrum is not followed in the calculations but we note that shocks can apparently form power law spectra
whereas turbulent acceleration will require longer time scales to
develop a feedback mechanism to form a power law.

     The turbulent motion will also amplify any existing magnetic
field in the flow as the local averaged Lorentz force acts to drive
currents and generate new magnetic field energy (e.g. Moffatt 1979).
Again only a small fraction of the energy is transferred to fields but
this effect is most important in producing the radio morphology of the
SNR because the synchrotron radiation depends upon a power of the
magnetic field strength.

     As the remnant expands, flux freezing in the highly conductive
medium will redirect the field.   Although most of it remains disor-
dered on a small scale, a stretching of a small part of the field
along the radial direction of eddy motion at the unstable interface
can create the net (radial) orientation of the magnetic fields and
small fractional polarization observed in young SNRs (e.g. Matsui et
al. 1984).

Results: The expansion of such a model SNR into a homogeneous circumstellar medium can produce an apparent shell with some central bright-

ness, the appropriate polarization characteristics, and comfortably low efficiencies. For a 1.4 solar mass star and a circumstellar density of 1 particle cm^{-3}, the efficiencies for conversion of turbulent and shock energies to relativistic electrons are less than 1%; less than 5% of the turbulent energy goes into field amplification and 3% of the field is stretched radially.

Two important parameters are not reproduced by this model, however. These are brightness fluctuations of 20-30% between clumps within the shell and also the observed thickness of the shell which is typically 1/4 of the radius; the homogeneous model gives only about 0.1 to 0.15. To overcome these deficiencies, models with a clumpy circumstellar medium were constructed. By trial and error it was found that clumps with a peak density of about 2 cm^{-3}, halfwidths of about 10^{17} cm, and random spacings with a mean value of 5 x 10^{17} cm could reproduce the observations. Sample results are shown in the

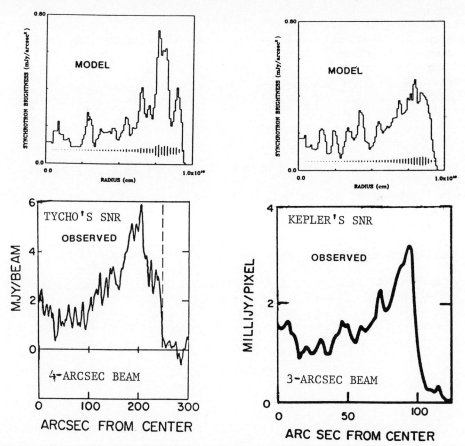

Figure: Observed and predicted radio synchrotron emission at a time of 400 years after explosion. The vertical vectors near the bottom represent the polarized power in a direction corresponding to a radial magnetic field.

figure. Because the model is only 1-dimensional but the observed slices represent an integral along the line of sight, we have made 4 model runs with different random spacings of the clumps. These were treated as wedges and randomly placed inside a semicircle. Summation through the semi-circle produced an effective slice through the remnant for comparison with the observations. Two different random summations are shown.

The good match between the model calculations and the data has shown that the expansion of the ejected material into a clumpy circum-stellar medium with the resultant unstable interfaces can produce the observed structure of young SNRs. The turbulent amplification of mag-netic fields plus the turbulent and shock acceleration of relativistic particles necessary to produce the synchrotron radiation require only a small fraction of the energy present and are dynamically unimportant.

References:

Axford, W. I., Leer, E. and McKenzie, J. R. 1982 Astron. Ap., 111, 117.

Chevalier, R. A. 1982 Ap. J., 258, 790.

Eilek, J. A. and Henriksen, R. N. 1984 Ap. J., 277, 820.

Gull, S. F. 1973 MNRAS, 161, 47.

Jones, E. M., Smith, B. W. and Straka, W. C. 1981 Ap. J., 249, 185.

Matsui, Y., Long, K. S., Dickel, J. R. and Griesen, E. W. Ap. J., 287, 295.

Moffatt, H. K. 1979 Magnetic Field Generations in Electrically Con-ducting Fluids, (Cambridge: Cambridge Univ. Press).

Sedov, L. I. 1959 Similarity and Dimensional Methods in Mechanics (New York: Academic Press).

HI ASSOCIATED WITH CAS A

W. M. Goss, NRAO, Socorro, New Mexico
P.M.W. Kalberla, Radio Astronomy Institute, Bonn
    Federal Republic of Germany
U.J. Schwarz, Kapteyn Astronomical Institute, Groningen,
    The Netherlands

Abstract: A small HI absorption feature has been
detected in front of Cas A at a velocity of -66 km/s.
This HI feature can probably be associated with a
recombined high density QSF.

Introduction: Using the 100 m telescope with a resolution of 9 arc
min, Mebold and Hills (1975) observed a weak absorption line at -65
km/s in the direction of Cas A. These authors suggested that the low
opacity line (tau=0.004) might result from a small cloud covering only
a part of Cas A and which could be physically associated with Cas A.
If this were the case, then the distance estimates of ~ 3 kpc based on
the HI absorption velocities need not be revised.

In November 1979, Cas A was mapped with the WSRT in a 2 x 12^h
period. The number of interferometers was 18 and the synthesized beam
was 27 x 32 arc sec (RA x Dec.) with grating rings at radii of 10 x 12
arc min. Since the size of Cas A is ~5 arc min, the grating rings
fall outside the source. The shortest spacing is 36 m and this
corresponds to a fringe spacing of 20 arc min. The velocity range
from -105 to 23 km/s was covered with a velocity resolution of 0.62
km/s. The rms noise in the line channels is 150 mJy/beam and is
limited by the precision of the video calibration. In Fig. 1 the
continuum map is shown; the total flux density is 2114 Jy.

Opacity maps in HI were calculated using a continuum cut-off
of 3.5 Jy/beam (5 % of the peak) in the clean map. In the Perseus arm
feature at -48 km/s, the optical depths are saturated over a great

Figure 1: WSRT 1420 MHz
continuum map of Cas A with a
resolution of 27 x 32 arc
sec. The contour units are
0.75 , 7.75 , 14.75 , 21.75
.... 70.75 Jy/beam. The peak
in the map is 74.4 Jy/beam.
The four crosses represent
various objects near the HI
feature ("HI"). The three
QSF R33,34 and 36 are
indicated as well as the
prominent radio non-thermal
knot near R36 ("NT"). The
HPBW is indicated in the upper
left hand corner. One Jy/beam
is a brightness temperature of
714 K.

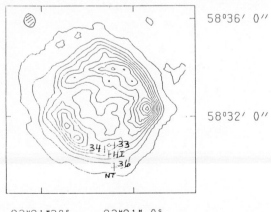

58°36' 0"

58°32' 0"

23ʰ21ᵐ30ˢ   23ʰ21ᵐ 0ˢ

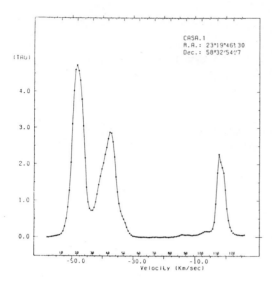

Figure 2: The average optical
depth   profile   over   Cas   A
including the local feature
(-1km/s) and the two prominent
Perseus arm features at -37 and
-48 km/s.   The cut-off in the
continuum was 3.5 Jy/beam and
in the -48 km/s feature an
optical depth cut-off of 5 was
applied.   127   of   the   255
channels   are   shown.   Velocity
is with respect to the lsr.

Figure 3 : A channel map in
opacity at a velocity of -36.9
km/s in the higher velocity
Perseus arm feature.   The max
and min tau's are 4.8 and
1.4.   The grey scale range is
1 to 5.

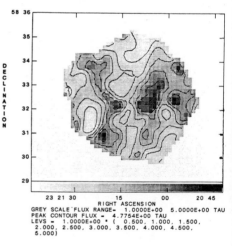

GREY SCALE FLUX RANGE- 1.0000E+00 5.0000E+00 TAU
PEAK CONTOUR FLUX = 4.7754E+00 TAU
LEVS = 1.0000E+00 * (0.500, 1.000, 1.500,
2.000, 2.500, 3.000, 3.500, 4.000, 4.500,
5.000)

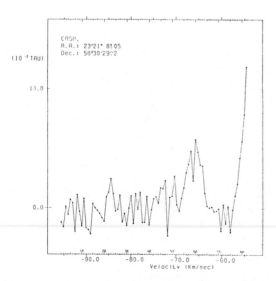

Figure 4 : A profile at the
position of the peak of the HI
feature at -66 km/s.   The
beginning of the -48 km/s
Perseus arm feature is seen at
the extreme right hand edge.
The continuum intensity at
this position is 22.8 Jy/beam.

deal of the source and an opacity cut-off at tau=5 was used.   In Fig.
2 the average opacity profile over the source is shown.   The two
prominent Perseus arm features are at velocities of -37 and -48 km/s
and the local absorption is at -1 km/s.   In Fig. 3 a sample opacity
channel map at a velocity of -36.9 km/s is shown.   These results will
be discussed in a later paper.

The -66 km/s feature: In the integrated opacity profile, the -66 km/s
line was not detected with a 2 sigma upper limit of 0.006 (the line
detected by Mebold and Hills was 0.004).   However at this velocity a
point source of HI absorption was detected at RA (1950)= 23^h 21^m 08.1^s
Dec. (1950) = 58° 30' 29" (2 sigma error in position is 10 arc sec).
The position is indicated in Fig. 1.   In Fig. 4 the profile at this
position is shown.   The upper limit for the angular size is 25 arc
sec.   The fitted parameters are: tau = 0.05 ± 0.01, integral over
velocity = 0.19 ± 0.03 tau km/s, velocity = -66.0 ± 0.2 km/s and full-
width at half maximum = 3.6 ± 0.7 km/s.   The velocity is in good
agreement with the previous 100 m observations and the optical depth
is an order of magnitude larger due to beam dilution.

The lower limit to the column density (since the source is
unresolved) is 2×10^{19} $(T_0/50)$ cm^{-2}, where T_0 is the spin temperature.
There is, of course, no independent information on the value of T_0.
Since the HI is probably combined shocked gas (see discussion), the T_0
could be much warmer and values in the range 100-1000 K are posssible.
The derived density is >20 $(T_0/50)$ cm^{-3} and the HI mass is < 0.01
$(T_0/50)$ solar masses.

The location of the HI feature is within 30 arc sec of three
prominent quasi-stationary flocculi (QSF) discussed by van den Bergh
and Kamper (vdBK, 1985).   These QSF are R33, R34 and R36 and are
indicated in Fig. 1 (the position of R34 is taken from their plate 4
and not from their Table 2).   R36 = A is one of the five QSF discussed
by van den Bergh and Kamper (1983) which lies outside the main
supernova shell.   It is noteworthy in two respects: 1) this QSF turned
on in the mid to late 1960's and 2) it may be associated with a knot
in the non-thermal radio emission (eg Bell, 1977).   The radio emission
position from Bell is also shown slightly south of the QSF R36 in Fig.
1.   As vdBK suggest: "Thus, we have been able to observe the
appearance of a new QSF, presumably as the expanding shock reaches the
pre-existing circumstellar or interstellar material."   Given the
small fraction of the surface of Cas A that is covered by QSF's (a
total of about 40, with a typical size of 2-3 arc sec), the proximity
of the HI feature to the three QSF does seem significant.

The HI feature is unlikely to be a standard interstellar HI
feature due to its small angular size and unusual velocity.   As Mebold
and Hills (1975) remark, there are no obvious emission counterparts
near Cas A at this velocity.   If the feature had shown more or less
continuous coverage over Cas A, then the kinematic distance to Cas A
would have to be revised.

The most likely interpretation for this feature is that it represents some HI which is physically associated with Cas A and is observed in projection. It is detectable due to the unusually bright continuum background of Cas A.

Discussion: If we make the assumption that the HI feature can be identified with a recombined high density QSF, then the true velocity may be estimated. Following McKee and Cowie (MC, 1975), we have assumed that the HI feature is moving radially outward with respect to the center of the SNR and that it lies in the region of the shocked interstellar (or circumstellar medium) between the blast wave and the inner surface of diffuse ejecta (Braun, 1987). These clouds will likely be rapidly ablated after they have collided with the ejecta itself (MC). The radial velocity of the HI feature (at a projected distance of 119 arc sec or 1.67 pc) is uncertain since the systemic velocity of Cas A is unknown. A reasonable estimate is -52 (±4) km/s (see Fig. 2) and thus an approximate value for the radial velocity is abs [-66 + 52] = 14 (±4) km/s. The total velocity is, of course, much higher since the HI feature is projected against the outer boundary of Cas A. The location of the HI feature near R 33-34 corresponds with the position 2 described by Braun (1987), where the radius of the outer blast wave is 1.76 pc (or 125 arc sec) and the surface of shocked diffuse ejecta is at 1.62 pc. If we assume that the young HI feature is physically located close to the blast wave itself, then a lower limit to the angle between the velocity vector and the line of sight can be estimated. This angle is about 70 degrees and thus the lower limit to the velocity is $\sim 14/\cos(70) = 44$ km/s. Naturally if the true location of the HI is closer to the inner shock the true velocity will be much higher.

The density of the HI feature must be higher than an average QSF in order that the object can cool faster (MC) and thus can have recombined to HI. If the density is higher than the average QSF, then the velocity of the shocked cloud will be lower since $v_c \propto n_c^{-0.5}$ (MC), where v_c and n_c are the velocity and density in the shocked cloud. The approximate lifetime of a QSF is the time it takes for the shocked diffuse ejecta to reach the material first shocked by the blast wave. Using the current velocity of the reverse shock wave at position 2 of 1700 km/s (Braun, 1987) and the distance between the two shocks of 0.14 pc, the lifetime, t, is about 80 years. This age is comparable with the estimate suggested by MC. This HI may be related to the shocked HI observed in a number of older SNR (Braun and Strom, 1986).

We can attempt to estimate the expected column density of HI and compare this with the limits derived in section 2. The density in the shocked QSF can be expressed using equation 6 of MC which can be written as: $n_c v_c^2 = 3 n_0 v_b^2$, where n_0 is the pre-shock density and v_b is the blast velocity. The factor 3 is the over-pressure in the shocked cloud for the maximum contrast case. At position 2 Braun finds $n_0 = 1.8$ cm^{-3} and $v_b = 2800$ km/s. The lower limit from the observations for v_c is 44 km/s. If the velocity were as high as 150

km/s (MC) the object would still be a conventional QSF and would not
have recombined.   Thus, we will assume that an estimate for v_c is
~100km/s.   In this case the value of n_c is 4×10^3 cm^{-3}.   The implied
column density, N_{HI}, would be $n_c L$, where L is the extent of the
recombined QSF along the line of sight.   An estimate for L is $v_c t$.
Thus L would be about 2.5×10^{16} cm or 0.6 arc sec.   This is comparable
with the size of the smallest QSF's.   The predicted N_{HI} is thus
10×10^{19} cm^{-2} which is consistent with the observed value of N_{HI}
$> 2 \times 10^{19}$ $(T_0 /50)$ cm^{-2}.

Acknowledgements: We gratefully acknowledge many useful discussions
and suggestions from R. Braun.   The Westerbork Synthesis Radio
Telescope is operated by the Netherlands Foundation for Radio
Astronomy (SRZM).   SRZM is supported by the Netherlands Organization
for the Advancement of Pure Research (Z.W.O.).   NRAO is operated by
Associated Universities Inc. under contract to the National Science
Foundation.

References:
Bell, A. R.:   1977, Monthly Notices Roy. Astron. Soc. _179_, 573.
Braun, R.:   1987, Astron. Astrophys. _171_, 233.
Braun, R., Strom, R.G.:   1986, Astron. Astrophys. Suppl. _63_, 345.
McKee, C. F., Cowie, L.L.:   1975, Astrophys. J. _195_, 715.
Mebold, U., Hills, D.L.:   1975, Astron. Astrophys. _42_, 187.
van den Bergh, S., Kamper, K. W.:   1983, Astrophys. J. _268_, 129.
van den Bergh, S., Kamper, K. W.:   1985, Astrophys. J. _293_, 537
   (vdBK).

THE INTERACTION OF THE SUPERNOVA REMNANT VRO 42.05.01 WITH ITS HI ENVIRONMENT

T.L. Landecker*, S. Pineault**, D. Routledge[+] and
J.F. Vaneldik[+]
*Dominion Radio Astrophysical Observatory, Penticton, B.C.
**Département de physique, Université Laval, Québec, P.Q.
+Electrical Engineering Dept., University of Alberta,
 Edmonton, Alberta

Abstract: VRO 42.05.01 (G166.0+4.3) is a SNR which has
broken into and re-energized an old interstellar cavity.
Observations of HI in a field containing the SNR show
features associated with it. An expanding shell of mass
~40 $M_\odot$ is associated with the semi-circular shell of the
SNR. The cavity, which the SNR has re-energized, is seen
in the HI. A cloud of HI of mass ~25 $M_\odot$ has apparently
been hit by the shock and has been accelerated by the
post-shock flow.

In the expansion of the
supernova remnant (SNR) VRO
42.05.01 the blast wave appears
to have encountered an abrupt
density discontinuity in the
interstellar medium (ISM).
This hypothesis was suggested
by the radio continuum
appearance of the SNR seen in
Figure 1 (Landecker et al. 1982
- Paper 1). In Papers 2 and 3
(Pineault et al. 1985, 1987)
we proposed that the blast wave
has broken into and
re-energized an old
interstellar cavity, itself the
product of previous supernovae
or stellar winds.

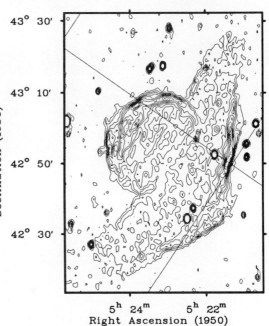

Figure 1: 1.4 GHz continuum map of
VRO 42.05.01 (Paper 1)

HI line observations
made with the DRAO Synthesis
Telescope (Roger et al. 1973)
are presented in part in
Figure 2. Angular and
velocity resolution as
observed were 1 x 1.4 arcmin and 2.6 km s^{-1}, but these quantities have
been altered for display to 4 x 4 arcmin and 2 km s^{-1}. The
observations cover a 2.2° square field of which the central half is
displayed, and encompass the entire range of galactic rotational
velocities in this direction. Figure 3 displays the average observed
spectrum in the 2.2° field; the range of velocities displayed in
Figure 2 is indicated.

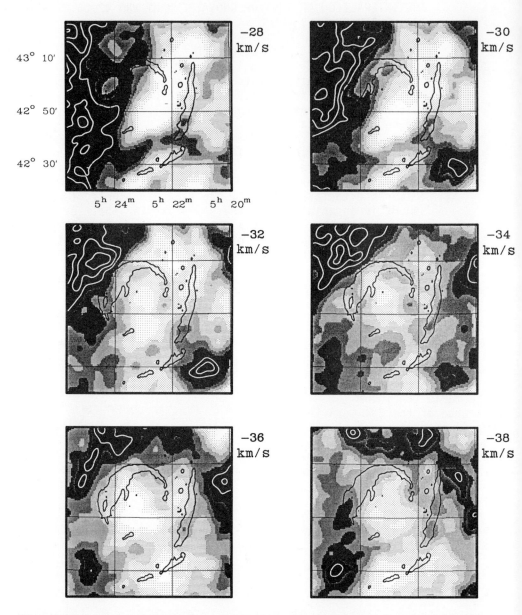

Figure 2: HI in the field containing VRO 42.05.01. HI emission is
represented by the grayscale; the increment between shades of gray is
2K in brightness temperature. Black represents bright emission and
white contours extend the grayscale. A constant equal to the average
emission in the field has been subtracted from each map; this does not
affect the appearance of the HI features. The lsr velocity is indicated beside each map. Angular resolution is 4 x 4 arcmin. A single
black contour indicates the 1.4 GHz continuum emission of the SNR.

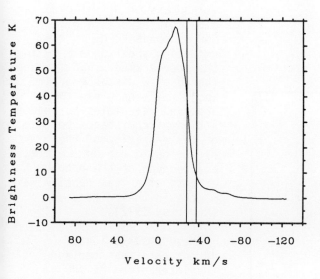

In this paper we simply point out some HI features related to the SNR. In describing them we use the term "shell" to denote the nearly circular component of diameter 31 arcmin in the east of the continuum object, and "wing" to refer to the western limb-brightened emission of extent ~1° (see Figure 1).

Figure 3: The average HI spectrum in the 2.2° field. The two vertical lines indicate the range of velocities displayed in the maps of Figure 2.

The semi-circular continuum shell appears to be completed by an HI feature, seen most clearly at -38 km s⁻¹. This HI shell can be seen from -34 to -40 km s⁻¹. Its diameter decreases as velocity becomes more positive, suggesting an expanding HI shell. A fit to the data yields (i) systemic velocity -40 km s⁻¹ (ii) expansion velocity 10 km s⁻¹ and (iii) radius 15 arcmin. This shell is confused with other HI between -30 and -40 km s⁻¹. The shell is incomplete: - an approaching component, expected between -40 and -50 km s⁻¹, is not seen.

Evidence of the cavity which the SNR has re-energized is found in the HI maps at -28 and -30 km s⁻¹, where it is seen as an elongated hole coincident with the interior of the wing. The northern part of the wing is clearly outlined by HI at -34, -36 and -38 km s⁻¹. The southern extremity of the wing appears to be a region of considerable confusion in the HI.

At velocities from -42 to -52 km s⁻¹ (not shown in Figure 2) we find HI that has apparently been hit by the SNR shock. At -42 and -44 km s⁻¹ we see an HI cloud which lies over the deep minimum at -28 km s⁻¹ which we identify with the old cavity. With increasingly negative velocity the intensity peak of the cloud moves radially outward from the centre of the SNR towards the edge of the wing and splits in two. Figure 4 shows the location of HI features associated with the cloud at -44 and -52 km s⁻¹. We interpret this as evidence that the cloud has been overtaken by the shock, and is now being accelerated by the post-shock flow (McKee and Cowie, 1975). The gas is apparently being ablated by the flow in the manner predicted by Nittman et al. (1982).

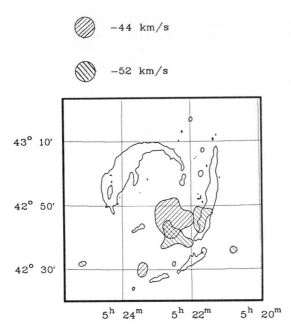

Figure 4: Shocked HI features. We interpret the -52 km s^{-1} features as gas ablated from the cloud at -44 km s^{-1} by the flow of post-shock gas.

The mass of HI which we can plausibly associate with the SNR is about 40 M$_\odot$ in the expanding shell and 25 M$_\odot$ in the accelerated cloud feature (based on a distance of 4 kpc). We cannot use the observed velocities to deduce a kinematic distance because of the proximity of the direction to the anticentre, and because of the probable effects of spiral density wave shocks in this direction (Roberts, 1972).

These observations demonstrate that our model of the interaction of VRO 42.05.01 with the ISM is substantially correct. We have been able to demonstrate for the first time that a SNR has re-energized an old interstellar cavity as envisaged by Cox and Smith (1974).

S.P., D.R., and J.F.V. were supported by grants from the Natural Sciences and Engineering Research Council. The DRAO Synthesis Telescope is operated by the National Research Council of Canada as a national facility.

References

Cox, D.P. and Smith, B.W., 1974, Ap. J. Letters 189, L105
Landecker, T.L., Pineault, S., Routledge, D. and Vaneldik, J.F., 1982, Ap. J. Lett. 261, L41 (Paper 1)
McKee, C.F. and Cowie, L.L. 1975, Ap. J. 195, 715
Nittman, J., Falle, S.A.E.G. and Gaskell, P.H. 1982, Mon. Not. Roy. Astron. Soc. 201, 833
Pineault, S., Pritchet, C.J., Landecker, T.L., Routledge, D. and Vaneldik, J.F., 1985, Astron. Astrophys. 151, 52 (Paper 2)
Pineault, S., Landecker, T.L. and Routledge, D. 1987, Ap. J. 315, 580 (Paper 3)
Roberts, W.W. 1972, Ap. J. 173, 259
Roger, R.S., Costain, C.H., Lacey, J.D., Landecker, T.L. and Bowers, F.K., 1973, Proc. IEEE 61, 1270

HI AND CO OBSERVATIONS TOWARDS THE SNR PUPPIS A

G.M. Dubner[1] and E.M. Arnal[2] [3]
[1] Instituto de Astronomia y Fisica del Espacio, Buenos
Aires, Argentina
[2] Instituto Argentino de Radioastronomia, Villa Elisa,
Argentina
[3] Observatorio Astronomico de La Plata, La Plata, Argentina

Abstract: The presence of interstellar clouds along the
northern and eastern edges of Puppis A is revealed by our
HI and CO observations.

Introduction:

The wealth of observational data in the X-ray, optical and radio
domains available for Puppis A, can be understood as the result of the
SN blast wave interacting with nearby clouds placed towards the
northern and eastern directions (e.g. Petre et al. 1982, Milne et al.
1983, Danziger 1983, etc.).

Based on the previous data, the presence of clouds with n_{HI} =
10 to 14 cm^{-3} was postulated. However, up to now, no study of the gas
distribution in the environs of the supernova has been undertaken in
order to confirm this contention.

Here, we report HI and CO observations carried out in the
direction of the Puppis A SNR. Our data disclose the presence of
clouds along both the eastern and northern borders of the
radio remnant. These features were also detected in the CO J:1-0
transition.

Observations and Results:

Neutral Hydrogen: The HI observations have been performed with
the 30 m radiotelescope of the Instituto Argentino de Radioastronomia.
The instrumental parameters are: Beam size: 30', velocity resolution:
2.1 km/s, system temperature: 85 K, rms noise: 0.1 K.

Figure 1 (left panel) displays the HI column density distribution
(in units of 10^{19} cm^{-2}) for three different radial velocities
(throughout this paper all radial velocities are referred to the LSR).
THe lowest contour of the 408 MHz emission (Green 1971) is included in
all the maps as a dashed line.

Carbon Monoxide: The CO data were obtained using the 1.2 m
Columbia Southern Millimeter Wave Facility at Cerro Tololo (Chile), in
two observing runs during 1985. The instrumental parameters can be
summarized as follows:
Beam size: 8', velocity resolution: 0.2 km/s, system temperature: 195 K
(excluding atmospherical radiation losses), rms noise: 0.2 K.

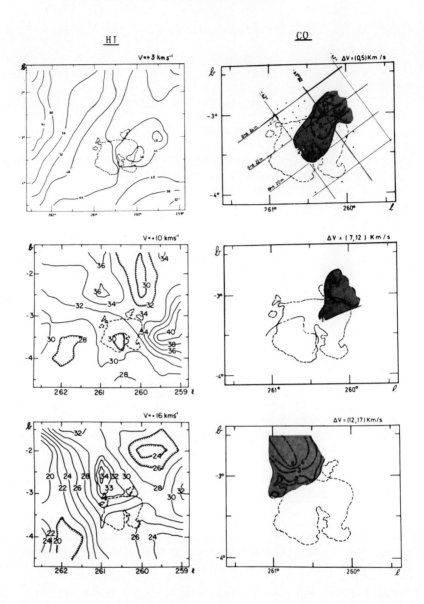

Figure 1: Contours of HI column density in units of 10^{19} cm^{-2} (left panels) and of CO integrated intensity over 5 km/s ranges in radial velocity (right panels).

The right panel of Figure 1, depicts the maps of CO integrated intensity, evaluated over the three velocity ranges (indicated in the upper right corner of each map) where the molecular emission is detected. The dots in the first map mark the observed positions, and a RA-DEC (1950) grid is also shown. Unfortunately, the CO features were only partially mapped due to a very tight observing schedule. As a consequence, only lower limits for the total masses can be derived.

Table 1

	VELOCITY RANGE (km/s)	HI	CO
NORTHERN FEATURES	N1 (0-5)	$(1,b)_{CENT}$= (259.7,-3.2) $V_{CENT} \cong$ 3 km/s M^*_{HI} = (747 ± 110) $M_{\odot}$ n_{HI} = (1.05 ± o.20)cm^{-3}	V_{CENT} =(3.2 ± 0.4) km/s V_{FWHM} = 1.2 to 2.6 km/s $N^{**}_{H_2}$(max)= 1.8 x 10^{21}m cm^{-2} M_{H_2} ⩾ 3810 $M_{\odot}$ n_{H_2} = 30 m cm^{-3}
NORTHERN FEATURES	N2 (7-12)	$(1,b)_{CENT}$= (258.0,-2.0) $V_{CENT} \cong$ 10 km/s M_{HI} = (6170 ± 920) $M_{\odot}$ n_{HI} =(1.32 ±0.25) cm^{-3}	V_{CENT} =(10.0 ± 0.1) km/s V_{FWHM} =(2.5 ± 0.7) km/s N_{H_2}(max)= 7.8 x 10^{20}m cm^{-2} M_{H_2} ⩾ 2850 $M_{\odot}$ n_{H_2} = 12 m cm^{-3}
EASTERN FEATURE	(12-18)	$(1,b)_{CENT}$= (261.0,-2.5) $V_{CENT} \cong$ 15 km/s M_{HI} = (3565 ± 530) $M_{\odot}$ n_{HI} = (12 ± 2.5)cm^{-3}	V_{CENT} =(13.8 ± 0.7) km/s V_{FWHM} =(4.67 ± 1.15) km/s N_{H_2}(max)= 1.2 x 10^{21}m cm^{-2} M_{H_2} ⩾ 3430 $M_{\odot}$ n_{H_2} = 13 m cm^{-3}

(*) Assuming d=2 kpc
(**)N_{H_2} = 2.3 x 10^{20} $\int T_A$ d ν (Murphy, Cohen & May,1986)

Conclusions: The most relevant results from our observations are: i) neutral hydrogen emission at three different radial velocities, namely: v=+3, +10 and +15 km/s, respectively, has been observed possibly associated with Puppis A; ii) every HI feature has a counterpart, in the same velocity interval, in the molecular gas.

The observed and derived parameters for the three features are summarized in Table 1.

Eastern feature: The existence of a dense HI cloud at v=+15 km/s, adjacent to the eastern edge of Puppis A, and at the same kinematical distance as the remnant, is clearly demonstrated. To the extent of our observations the spatial distribution of the CO emission within this velocity range, correlates with the HI cloud.

The coincidence between the derived HI volume density (n_{HI} = 12 cm^{-3}) and that predicted from X-ray data (n_{HI} = 10 to 14 cm^{-3}), is remarkable.

Northern features: Two different concentrations are observed both
in HI and CO lines in direction to the northern edge of Puppis A. In
both cases, the cold gas clouds are seen in projection against part of
the radio remnant.

Unlike the eastern feature, for the northern clouds the derived
kinematical distances do not agree with that of Puppis A, and the
volume densities derived for these concentrations are much lower than
the predicted ones. However, bearing in mind the effects of
interaction with surrounding clouds evidenced in other wavelength
ranges, we feel that a chance alignment of these clouds with the SNR is
highly improbable, and that they are actually associated with Puppis A.
Indeed, the HI map of a very large field containing the SNR Puppis A
(Dubner 1987, in preparation) shows clear evidence of the SN evolving
in the periphery of a large swept-up shell (centered around $\ell=260°$,
b=-2°). Therefore, the kinematics of the ambient gas interacting with
Puppis A might be reflecting non-circular motions.

Acknowledgements:

The authors wish to thank the technical staff of the IAR for their
support. One of us (E.M.A.) is grateful to Drs. R. Cohen and P.
Thaddeus for allocating observing time with the Columbia Southern
Millimeter Wave Facility at Cerro Tololo (Chile). The invaluable help
of F. Aviles and J. Montani is also acknowledged. J. Fernandez and M.
Pintos are acknowledged for the art work. Travel funds were provided
by the CONICET, Argentina. Both authors are members of the Carrera del
Investigador Cientifico of the Consejo Nacional de Investigaciones
Cientificas y Técnicas (CONICET), Argentina.

References

Danziger, I.J.: 1983, IAU Symp. 101 "Supernovae and their X-ray
     emission", ed. J. Danziger and P. Gorenstein, D. Reidel Pub. Co.,
     p. 193.
Green, A.J.: 1971, Aust. J. Phys. 24, 773.
Milne, D.K., Goss, W.M. and Danziger, I.J.: 1983, M.N.R.A.S. 204, 237.
Murphy, D.C., Cohen, R. and May, J.: 1986, A.A. 167, 234.
Petre, R., Canizares, C.R., Kriss, G.A. and Winkler, P.F.: 1982,
     Ap.J. 258, 22.

G18.95-1.1, A COMPOSITE SUPERNOVA REMNANT INTERACTING WITH THE AMBIENT INTERSTELLAR MEDIUM

E. Fürst, W. Reich, E. Hummel
Max-Planck-Institut für Radioastronomie
Auf dem Hügel 69, D-5300 Bonn 1, F.R.G.

Y. Sofue
Department of Astronomy, Faculty of Science
University of Tokyo, Bunkyo-ku, Tokyo 113, Japan

Abstract: New radio continuum and spectral line observations of the Galactic radio source G18.95-1.1 are reported.
The distance to G18.95-1.1 is 2 kpc as derived from HI-21
cm spectral line observations. These data also indicate an
interaction with the interstellar medium. The radio continuum
observations classify G18.95-1.1 as a composite supernova
remnant.

The Galactic radio source G18.95-1.1 was found to be nonthermal by
Fürst et al. (1985). The radio spectrum of its integrated flux density
between 57.5 MHz and 5 GHz is now well established ($\alpha = -0.28 \pm 0.05$,
$S_\nu \sim \nu^\alpha$, Odegard, 1986). From radio continuum maps (see Fig. 1 in the
paper by Fürst et al., 1985) it is apparent that G18.95-1.1 is composed
of two components. A large-scale diffuse emission (Fig. 1a), which contains 79% of the total flux density, peaks near the geometrical centre of

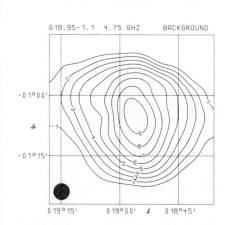
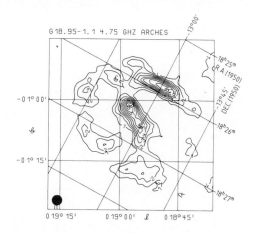

Figure 1: (a) Contour plot of the diffuse, centrally peaked component of
G18.95-1.1 at 4.75 GHz. Contour steps are in 20 mJy/2!45 beam. The map
is convolved to a HPBW of 4!2 as is indicated by the hatched circle.
(b) Contour plot of the arc component of G18.95-1.1 at 4.75 GHz. Contour
steps are 20 mJy/2!45 beam. The HPBW is 2!45 as is indicated by the
hatched circle.

the source. Superposed are various arc-like features (Fig. 1b) containing
21% of the total flux density. The most prominent arc features are shown
in Fig. 2 at an angular resolution of 19"x14" (p.a. = 0°) at 1.5 and 4.9
GHz. The observations were made with the Very Large Array (VLA) in its

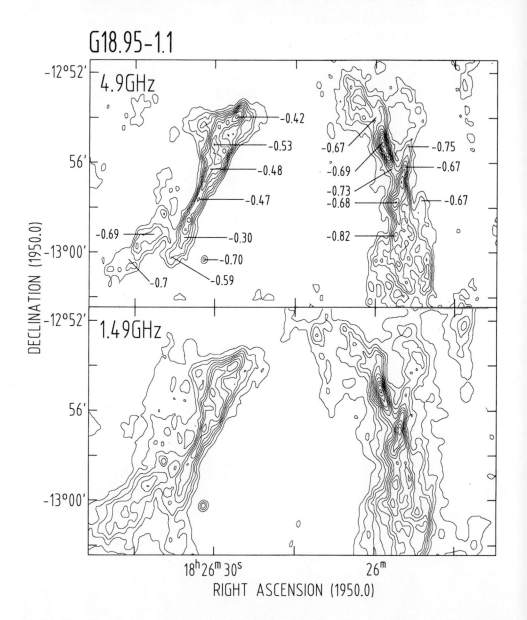

Figure 2: Contour plot of a part of G18.95−1.1 at 4.9 GHz (upper panel,
contour steps 0.5 mJy/beam) and 1.49 GHz (lower panel, contour steps
1 mJy/beam) obtained with the Very Large Array (C and D configuration).
Both maps are corrected for the primary beam attenuation. The HPBW is
19"x14". In the upper panel some spectral indices α ($S_\nu \sim \nu^\alpha$) are quoted.

D and C configurations. We also derived the spectral index between the two frequencies at some selected positions. These spectral indices have to be considered as lower limits (the spectra are probably flatter): due to missing short spacings we miss slightly more structure at 4.9 GHz. Taking also into account single-dish observations the spectral index of the central feature of G18.95-1.1 is $\alpha \approx -0.4 \pm 0.2$, i.e. similar to the spectral index of the integrated emission. On the other hand, all other arc-like structures visible in Fig. 1b have relatively steep spectra ($\alpha = -0.65 \pm 0.2$). This spectral index is typical for shell-type supernova remnants (SNRs). These steep spectral arcs and the flat spectrum, large-scale component classify G18.95-1.1 as a composite SNR (Weiler, 1983). Composite SNRs are expected to be powered by central activity. The flat spectrum central feature may be the signature of the outflow of relativistic electrons from this central activity feeding the large-scale component. The relativistic electrons are confined by the SNR-shock interacting with the interstellar medium. The steep spectrum arcs represent these shocks.

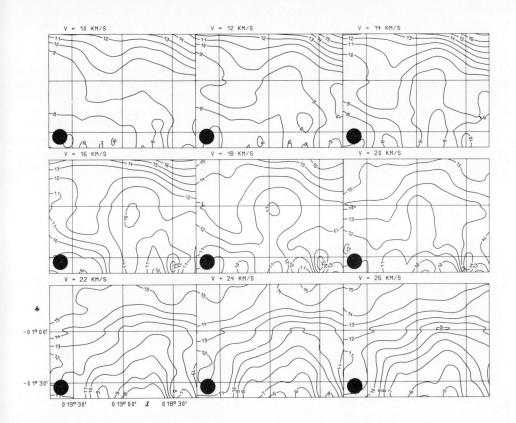

Figure 3: Contour plots of the HI-21 cm line antenna temperature obtained with the Effelsberg 100-m telescope. The HPBW is 9' as is indicated by the hatched circles. Plots are shown at radial velocities between 10 km s^{-1} and 26 km s^{-1} at an interval of 2 km s^{-1}. Contour steps are 5 K T_A beginning at 2 K T_A (contour 1).

The interaction of G18.95-1.1 with the interstellar medium is supported by HI-21 cm line observations made with the 100-m telescope by Braunsfurth and Rohlfs (1984). In Fig. 3 we show channel maps between the velocity v_{LSR} = 10 km s^{-1} and v_{LSR} = 26 km s^{-1}. A strong depression in the HI-antenna temperature is visible close to the geometrical centre of G18.95-1.1 at a velocity of v_{LSR} = 18 km s^{-1}. The discussion of the depression by Fürst et al. (1987) shows that it is most likely explained by missing HI-gas across the SNR. The association of the depression with G18.95-1.1 sets limits for the distance to the SNR. Common rotation models give 2 kpc or 15 kpc on the far side of the Galaxy. It can be shown that the large distance is unprobable: G18.95-1.1 would be much more energetic than the Crab nebula. Whence, the distance is 2 kpc (Sagittarius arm). The missing column density across G18.95-1.1 at the centre of the HI-depression is $\langle \Delta N \rangle$ = $9 \cdot 10^{19}$ cm^{-2}, which scales to N = 2 cm^{-3} at a distance of 2 kpc (linear size of the SNR $\approx$ 20 pc). Therefore, G18.95-1.1 is probably expanding into an ambient medium of density N $\gtrsim$ 2 cm^{-3}. The low angular resolution of the 21 cm line data obtained with the 100-m telescope is not sufficient to investigate the interaction conclusively. In particular, the question whether the depression is caused by the SNR ejecta/shock or by the stellar wind of the progenitor star cannot be answered, although the low velocity dispersion of the depression ($\Delta v \approx$ 8 km s^{-1}) argues in favour of the latter. More observations in HI and in molecular lines with better angular resolution are needed to study G18.95-1.1 as one of the rare cases, where a clear interaction of an SNR with the ambient interstellar medium is visible.

A detailed analysis of the new observations is in preparation (Fürst et al., 1987).

References
Braunsfurth, E., Rohlfs, K.: 1984, Astron. Astrophys. **57**, 189
Fürst, E., Reich, W., Reich, P., Sofue, Y., Handa, T.: 1985, Nature **314**, 720
Fürst, E. et al.: 1987, in preparation
Odegard, N.: 1986, Astron. J. **92**, 1372
Weiler, K.W.: 1983, The Observatory, 1054th edition

TWO REMARKABLY THIN CO FILAMENTS TOWARD THE SUPERNOVA REMNANT G109.1-1.0

K. Tatematsu, Y. Fukui, and T. Iwata
Department of Astrophysics, Faculty of Science,
Nagoya University, Chikusa-ku, Nagoya 464, Japan

M. Nakano
Department of Earth Science, Faculty of Education,
Oita University, Oita 870-11, Japan

Abstract: We observed the semicircular supernova remnant G109.1-1.0 in the $J = 1-0$ transition of CO with the Nobeyama 45-m radio telescope. It is found that two remarkably thin molecular filaments delineate the inner boundary of the X-ray jet feature in this remnant. These filaments seem to have experienced evaporation due to the hot gas in the remnant.

Introduction: The semicircular supernova remnant G109.1-1.0 was discovered at X-ray wavelengths (Gregory and Fahlman, 1980). There exists an X-ray pulsar at the center of the remnant (Fahlman and Gregory, 1981, 1983; Koyama et. al., 1987). At radio wavelengths this remnant contains radio arcs inside of the semicircular shell, but there is no feature corresponding to the X-ray pulsar or the X-ray jet (Downes, 1983; Sofue et al., 1983; Gregory et al., 1983; Hughes et al., 1984). Gregory and Fahlman (1981, 1983) proposed a precessing jet model analogous to SS433 based on the presence of the X-ray pulsar, the X-ray jet, and the radio arcs. Millimeter-wavelength studies using CO and/or less abundant ^{13}CO have revealed that there is a molecular cloud on the west of the remnant (Israel, 1980; Heydari-Malayeri et al., 1981; Tatematsu et al., 1985, 1987). Our previous study (Tatematsu et al., 1987) has shown that the main molecular cloud has prevented the isotropic expansion of G109.1-1.0 causing its semicircular shape. They have also revealed that the armlike molecular ridge ("CO arm" hereafter) extending from the main molecular cloud shows an apparent anticorrelation with the X-ray jet feature. We have investigated the structure and dynamics of the CO arm, and the interface between the main cloud and G109.1-1.0 using the Nobeyama 45-m radio telescope.

Observations: Observations were carried out in the $J = 1-0$ transition of CO at 115.271204 GHz in December 1986 and April 1987. The Nobeyama 45-m radio telescope had a half-power beamwidth of 17" and a beam efficiency of 0.45 at this frequency. The cooled Schottky diode receiver was employed and provided a system temperature of 600-700 K (SSB). The observations were made with 60" spacing in general, whereas the observations of the CO arm were carried out with 30" spacing. All spectra were obtained in the position switching

Fig. 1: The gray-scale map of CO (J = 1-0) intensity integrated over the radial velocity range V(LSR) = (-57, -43) km/s. The X-ray map of G109.1-1.0 (Gregory et al., 1983) is reproduced as white lines with a resolution of ∼ 3'. Contours of the X-ray emission are drawn every 20 from 20 to 200, every 50 from 200 to 500, and every 100 from 500 to 1700, in an arbitrary intensity unit.

Fig. 2: The gray-scale map of CO (J = 1-0) intensity integrated over the range V(LSR) = (-51, -49) km/s. The X-ray map of G109.1-1.0 (Gregory et al., 1983) is also reproduced as white lines.

Fig. 3: The same as Fig. 2 but for the range V(LSR) = (-49, -47) km/s.

mode and linear baselines were subtracted from them.

Results and discussion: The gray-scale map of CO intensity integrated from −57 to −43 km/s in V(LSR) is shown in Fig. 1. The X-ray map of G109.1-1.0 (Gregory et al., 1983) is reproduced as white lines with a resolution of ∼ 3'. To investigate the dynamics of the molecular cloud, we also show a gray-scale CO map for the range V(LSR) = (−51, −49) km/s in Fig. 2, and that for the range V(LSR) = (−49, −47) km/s in Fig. 3. The present observations give much more detail of spatial structure of the molecular cloud than previous lower-resolution studies. Figs. 2 and 3 clearly show that the CO arm consists of two filaments. One of them seen in Fig. 2 is a curled filament, and the other seen in Fig. 3 is a fairly straight filament. The west part of the straight filament is also seen in the range V(LSR) = (−47, −45) km/s. These two filaments emanate from a common root at the main molecular cloud and reach a common top of the CO peak. They have a width of ∼ 2 pc and a length of ∼ 18 pc, at a distance of 4.1 kpc (Sofue et al., 1983). It is remarkable that these filaments seem to delineate the inner boundary of the X-ray jet. In Fig. 1, there are small molecular clouds just outside of the X-ray jet feature.

To evaluate the mass of the two filaments, we reuse our CO and ^{13}CO data obtained with the Nagoya 4-m radio telescope (Tatematsu et al., 1987). We use the LTE method, and adopt Dickman's (1978) conversion factor from ^{13}CO column density to H_2 column density. The mean molecular weight per H_2 molecule is taken to be 2.76 m_H. The 4-m radio telescope could not resolve the two filaments. Therefore, we simply attribute the mass in the ranges V(LSR) = (−51, −49) km/s and V(LSR) = (−49, −45) km/s to the curled and straight filaments, and obtain masses, 4×10^2 M$_\odot$ and 1.6×10^3 M$_\odot$, respectively.

The interface between the main molecular cloud and G109.1-1.0 is clearly seen in Figs. 1 and 2, but the main molecular cloud is not conspicuous in Fig. 3. In the range V(LSR) = (−53, −51) km/s, the main molecular cloud extends toward the boundary of G109.1-1.0, but there is little CO emission of the CO arm. The main molecular cloud seems to be very clumpy or spongy: it may consist of many filamentary clouds. There is no significant enhancement in the CO intensity along the interface between the main molecular cloud and G109.1-1.0, however. The semicircular shape of G109.1-1.0 is not due to absorption by the molecular cloud, but a result of the anisotropic expansion taking place near the molecular cloud (Tatematsu et al., 1987). Expansions of supernova remnants near molecular clouds were studied also on the basis of hydrodynamical calculations by Tenorio-Tagle et al. (1985). Their result for a supernova remnant which has exploded slightly outside of the molecular cloud wall is similar to the case of G109.1-1.0, although parameters adopted by them are not exactly equal to these of G109.1-1.0.

In Fig. 2, there is an intense molecular ridge in the northern part of the main molecular cloud, and it is connected with the straight filament. The CO intensity decreases steeply just at the boundary of the remnant. This suggests that the straight filament has been partially evaporated due to the hot plasma of the remnant. The remarkable thin shape of the straight filament may be explained as a result of evaporation. The curled filament may have experienced evaporation, because the common top and root of these two filaments suggest the closeness between them. They seem to form a molecular loop surviving evaporation within the remnant. The appearance of the X-ray jet feature has been explained in terms of the presence of the CO arm as discussed in Sect. 4.2 of Tatematsu et al. (1987).

References:

Dickman, R.L. 1978, Astrophys. J. Suppl. 37, 407.

Downes, A. 1983, Monthly Notices Roy. Astron. Soc. 203, 695.

Fahlman, G.G. and Gregory, P.C. 1981, Nature 293, 202.

Fahlman, G.G. and Gregory, P.C. 1983, in Supernova Remnants and Their X-Ray Emission, IAU Symp. 101, eds. J. Danziger and P. Gorenstein (Reidel, Dordrecht), p. 445.

Gregory, P.C., Braun, R., Fahlman, G.G., and Gull, S.F. 1983, in Supernova Remnants and Their X-Ray Emission, IAU Symp. 101, eds. J. Danziger and P. Gorenstein (Reidel, Dordrecht), p. 437.

Gregory, P.C. and Fahlman, G.G. 1980, Nature 278, 805.

Gregory, P.C. and Fahlman, G.G. 1981, Vistas Astron. 25, 119.

Gregory, P.C. and Fahlman, G.G. 1983, in Supernova Remnants and Their X-Ray Emission, IAU Symp. 101, eds. J. Danziger and P. Gorenstein (Reidel, Dordrecht), p. 429.

Heydari-Malayeri, M., Kahane, C., and Lucas, R. 1981, Nature 293, 549.

Hughes, V.A., Harten, R.H., Costain, C.H., Nelson, L.A., and Viner, M.R. 1984, Astrophys. J. 283, 147.

Israel, F.P. 1980, Astron. J. 85, 1612.

Koyama, K., Hoshi, R., and Nagase, F. 1987, submitted to Publ. Astron. Soc. Japan.

Sofue, Y., Takahara, F., and Hirabayashi, H. 1983, Publ. Astron. Soc. Japan 35, 447.

Tatematsu, K., Nakano, M., Yoshida, S., Wiramihardja, S.D., and Kogure, T. 1985, Publ. Astron. Soc. Japan 37, 345.

Tatematsu, K., Fukui, Y., Nakano, M., Kogure, T., Ogawa, H., and Kawabata, K. 1987, Astron. Astrophys. in press.

Tenorio-Tagle, G., Bodenheimer, P., and Yorke, H.W. 1985, Astron. Astrophys. 145, 70.

Weiler, K.W. 1985, in The Crab Nebula and Related Supernova Remnants (eds. M.C. Kafatos and R.B.C. Henry) p. 227 (Cambridge University Press, Cambridge).

A CO (J = 1-0) SURVEY OF FIVE SUPERNOVA REMNANTS
AT ℓ = 70° - 110°

Y. Fukui and K. Tatematsu
Department of Astrophysics, Faculty of Science,
Nagoya University, Chikusa-ku, Nagoya 464, Japan

Abstract: A program of CO observations of five selected supernova remnants is in progress on the 4-m radio telescope at Nagoya. Here, we report observations of two supernova remnants, G78.2+2.1 (the γ Cygni SNR) and HB21. In these two remnants we have obtained evidence for the interaction between the supernova remnants and molecular gas.

Introduction: Since the middle 1970's indications of the interaction of supernova remnants (SNRs) with molecular gas, the densest component of the interstellar medium, have been searched for by several groups. Some of the previous observations of SNRs are difficult to interpret because of confusion due to unrelated clouds: e.g., W44 (Wootten, 1977; Dame, 1983), W28 (Wootten, 1981), and W50 (Huang et al., 1983). According to our re-examination of the published data, it seems that there are only two clear examples of interactions, namely IC443 and the Cygnus Loop. Discovery of shocked molecular gas having a wide linewidth in IC443 provides an unequivocal indication of the interaction (DeNoyer and Frerking, 1981), but it remains a unique example until now in spite of a few searches for broad molecular emission indicating shocks. On the other hand, in the Cygnus Loop, two small (~2 pc) molecular clouds on the western boundary of the remnant cause a distortion of the shell and an enhancement of the optical emission. In order to increase the number of the unambiguous examples of interacting SNRs, we have started to observe molecular clouds toward SNRs mainly using the Nagoya 4-m radio telescope. We compare the distribution of molecular clouds with X-ray and/or radio images of SNRs and search for morphological indications of interaction. Five SNRs were selected on the basis of three criteria: (a) the SNR is located within the galactic longitude range ℓ = 70° - 110° where confusion with unrelated clouds is not severe, (b) the SNR is relatively close, within ~5 kpc of the Sun, and (c) CO emission is detected toward the SNR area by the Columbia survey with the 1.2-m radio telescope (Huang, 1985; Cong, 1977; Israel, 1980). In this paper observations of G78.2+2.1 (the γ Cygni SNR) and HB21 with the 4-m radio telescope are reported. G78.2+2.1 is a circular SNR with a diameter of 62' and has a radio bright region in the southeast (Higgs et al., 1977). HB21 is a non-circular SNR with a size of ~1.5°. The distance to G78.2+2.1 is 1.6 kpc and that to HB21 is 1.1 kpc (Milne, 1979). Observations of another SNR, G109.1-1.0, are reported in Tatematsu et al. (1987a and b).

Observations: Observations were carried out with the Nagoya 4-m radio telescope of CO and ^{13}CO (J = 1-0) spectra between November 1986 and May 1987. This telescope has a half-power beamwidth of 2'.7 at 110

GHz.   A cooled Schottky diode receiver was employed, providing a
receiver temperature of 200-250 K (DSB).   Most of the spectra were
obtained in the frequency switching mode, whereas some of them for
G78.2+2.1 were obtained with position switching.   We first observed
areas covering the SNRs fully with coarse spacings, 8'-15', and then
observed in CO the areas which show a hint of interaction with spacings
of 3'. ^{13}CO observations were conducted at selected points.   The
intensity is calibrated by using the chopper-wheel method, corrected
for beam efficiency, and expressed as the radiation temperature T_R.

Results and discussion:
(1) G78.2+2.1 --- There is a molecular cloud on the southeast side of
the remnant.   This cloud has different radial velocity components and
the most conspicuous features which are likely associated with the SNR
are two clumps as shown in Figures 1 and 2.   These maps are close-ups
of the southeastern quadrant of the SNR.   These two clumps exist just
on the northeast and southwest sides of the elongated radio bright
region respectively.   The southern clump is stronger than the northern
one.   Other than these clumps there are a few molecular clouds lying
within the SNR shell including a filament crossing the remnant, but
they show only weak CO emission.   Close association between the two
clumps and the elongated radio bright region suggests that dynamical
interaction with these clumps has compressed the magnetic field of
G78.2+2.1 and enhanced the radio intensity of this region.
Furthermore, comparison with the soft X-ray image obtained with the
Einstein Observatory (Higgs et al., 1983) reveals that the southern
molecular clump shows an anticorrelation with the X-ray intensity
(Figure 3).   The most natural interpretation is that the southern clump

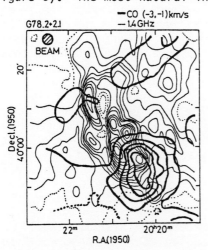
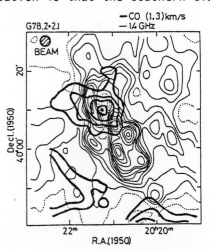

Fig. 1: The CO (J = 1-0) intensity
integrated in the radial velocity
range V(LSR) = (-3, -1) km/s around
the southeastern part of G78.2+2.1.
The contour interval is 6.7 K km/s.
The 1.4-GHz map of the remnant
(Higgs et al., 1977) is reproduced
as thin lines.

Fig. 2: The same as Fig. 1 but for
V(LSR) = (1, 3) km/s with a lowest
contour level of 6.7 K km/s and a
contour interval of 3.3 K km/s.

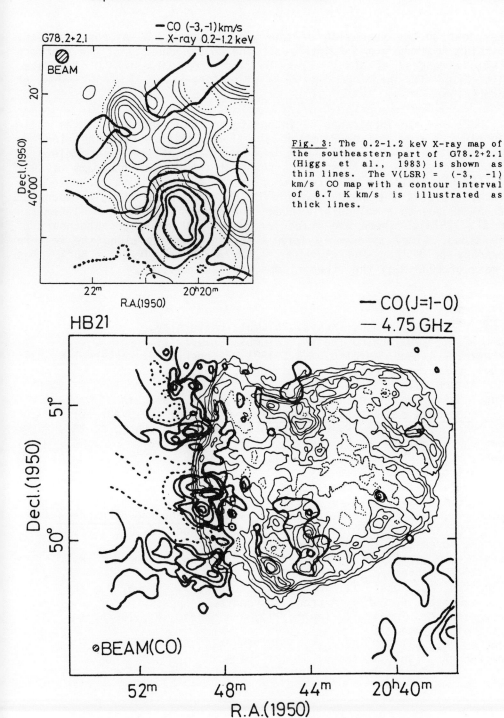

Fig. 3: The 0.2-1.2 keV X-ray map of the southeastern part of G78.2+2.1 (Higgs et al., 1983) is shown as thin lines. The V(LSR) = (-3, -1) km/s CO map with a contour interval of 6.7 K km/s is illustrated as thick lines.

Fig. 4: The integrated CO (J = 1-0) map around HB21 with a contour interval of 8 K km/s. The 4.75-GHz map of the remnant (Reich et al., 1983) is reproduced as thin lines.

lies just on the near side of the remnant shell and absorbs the X-ray emission from the remnant interior. The column density of the southern clump is deduced as $N(H_2) = 1 \times 10^{22}$ cm^{-2} from CO and ^{13}CO line data. This value is large enough to explain the X-ray map in terms of absorption.

(2) HB21 --- The integrated CO intensity map is superposed on the radio map of HB21 obtained by Reich et al. (1983) (Figure 4). The cloud on the east of the remnant is the westmost part of the giant molecular cloud Kh141. The nearly straight shape of the eastern boundary of HB21 can be well explained as a result of the interaction with the molecular cloud. This is the second example of the global anisotropic expansion of an SNR due to a molecular cloud; the other is G109.1-1.0 (Tatematsu et al. 1987a). There are molecular clumps on the western boundary of the cloud. They apparently form a string of beads along the eastern boundary of HB21. The clump at R.A. = 20^h49^m, Decl. = 50°15' has a mass of 2000 $M_\odot$; the other clumps are less massive.

References:
Cong, H.I., 1977, Ph.D. Thesis, Columbia University.
Dame, T.M., 1983, Ph.D. Thesis, Columbia University.
DeNoyer, L.K. and Frerking, M.A., 1981, Astrophys. J. Letters 246, L37.
Higgs, L.A., Landecker, T.L., and Roger, R.S., 1977,
    Astron. J. 82, 718.
Higgs, L.A., Landecker, T.L., and Seward, F.D., 1983, in
    Supernova Remnants and Their X-Ray Emission, IAU Symp. 101,
    eds. J. Danziger, P. Gorenstein (Reidel, Dordrecht), p. 281.
Huang, Y.-L., 1985, Ph.D. Thesis, Columbia University.
Huang, Y.-L., Dame, T.M., and Thaddeus, P., 1983,
    Astrophys. J. 272, 609.
Israel, F.P., 1980, Astron. J. 85, 1612.
Milne, D.K., 1979, Aust. J. Phys. 32, 83.
Reich, W., Fürst, E., and Sieber, W., 1983, in Supernova Remnants
    and Their X-Ray Emission, IAU Symp. 101, eds. J. Danziger,
    P. Gorenstein (Reidel, Dordrecht), p. 377.
Scoville, N.Z., Irvine, W.M., Wannier, P.G., and Predmore, C.R.,
    1977, Astrophys. J. 216, 320.
Tatematsu, K., Fukui, Y., Nakano, M., Kogure, T., Ogawa, H., and
    Kawabata, K. 1987a, Astron. Astrophys. in press.
Tatematsu, K., Fukui, Y., Iawata, T., and Nakano, M., 1987b, in
    this volume.
Wootten, H.A., 19877, Astrophys. J. 216, 440.
Wootten, A., 1981, Astrophys. J. 245, 105.

OBSERVATIONS OF "DENTS" IN THE RADIO SHELLS OF SNRs
EXPANDING INTO DENSE CLOUDS

T. Velusamy
Radio Astronomy Centre
Tata Institute of Fundamental Research
P.O. Box 8, Udhagamandalam (Ooty) 643 001, India

Abstract: High resolution (~36"x90") radio maps of SNRs
W28 and W44 observed with the OSRT at 327 MHz are
presented. The distortions or dents in their shell
structure are associated with the parts of the shells
expanding into dense molecular clouds in their vicinity
and provide direct observational evidence for interaction
between them.

Introduction: The evolution of the radio brightness and the shell
structure of supernova remnants (SNRs) strongly depends on the ambient
density distribution in the surrounding interstellar medium (ISM).
Thus, the differences in the structures of SNRs, particularly in the
older remnants, are indeed the result of the interaction of the
expanding shock wave with the surrounding interstellar gas. The
propagation of a shock wave in a cloudy interstellar medium and its
effects on the radio emission from SNRs have been well discussed, for
example, by McKee and Cowie (1975) and Blandford and Cowie (1982).
Recently Wootten (1977; 1982) mapped two dense molecular clouds
observed in the CO in the vicinity of the SNRs W44 and W28 and
concluded that they are interacting with the remnants. In this paper,
we report high resolution radio observations of the shell structure in
these two remnants and interpret the distortions observed in the radio
shell in terms of possible interaction of the shock wave with these
dense clouds.

Observations and results: The SNRs W44 and W28 were observed for a
3x10 hr period each, with the 4km Ooty Synthesis Radio Telescope (OSRT)
at 327 MHz during 1985 and 1987 respectively. The configuration of
OSRT was so chosen as to have a wider field of view of 90' in the
north-south direction and 170' in the east-west direction (Swarup 1984;
Sukumar 1986). OSRT sections were sampled every 25λ in order to
minimise the north-south grating effects in maps, particularly for W44.
The shortest spacings used for mapping are 50λ and 100λ for W28 and W44
respectively. The data were calibrated and CLEAN maps were obtained
using the standard procedure for synthesis radio telescopes.

    In Figures 1 and 2 are shown the OSRT contour maps of the
brightness distribution in W28 and W44 at 327 MHz obtained with
synthesised beams of 96"x36" at 6° PA and 87"x40" at 12° PA
respectively. The overall shell structure of both W44 and W28 are
quite consistent with the low resolution maps, for example at 10.6 GHz
with a 2.8 beam (Kundu and Velusamy 1972). Unlike the high frequency
maps, in the 327 MHz OSRT map the nonthermal remnant shell is seen more
clearly without confusion from the thermal sources in the region. The

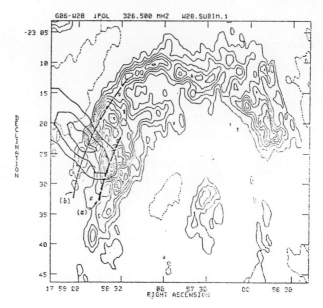

Fig.1   OSRT map of W28 at 327 MHz. Synthesised beam is 96"x36" at
        6°PA. The contour interval 100 mJy/beam. The thick contours
        are CO column density (from Wootten 1981). For arcs 'a' & 'b'
        see text.

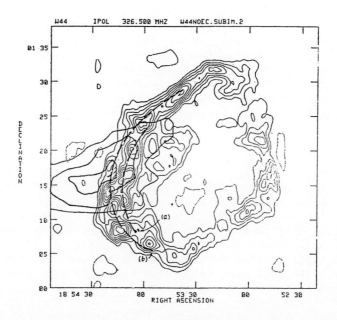

Fig.2   OSRT map of W44 at 327 MHz. Synthesised beam is 87"x40" at
        12°PA.The contour interval 120 mJy/beam. Thick contours are
        CO column density (from Wootten 1977). For arcs 'a' & 'b' see
        text.

high resolution maps in Figures 1 and 2 show more details along the shells. W44 has a fairly thin radio shell over most parts, except in the east. In W28 the brightest emission is seen along the eastern part of the shell. The shell in W28 is thick and it is possible that it is a superposition of several thin shells.

Discussion: In Figures 1 and 2, we have also plotted the contours of CO column density of the molecular clouds in the vicinity of W28 and W44 observed by Wootten (1977, 1982). From the positional coincidence in both SNRs, parts of the eastern shell appear to be interacting with the molecular clouds. The densities in these clouds are high, ~1000 cm^{-3} or greater. Kinematic considerations suggest that these clouds are at the same distances as the SNRs and are interacting with the remnants (Wootten 1977; 1982). Such interaction will be observable in the structure of the radio shell of the remnant. It is interesting to note that both the remnants show enhanced emission peaks just behind the densest region of the clouds. Further conspicuous distortions in the shell structure are also seen in the parts of the shell over the cloud region.

In Figure 3 is shown a simple sketch of the expansion of a remnant shell into a dense cloud. If we consider the pressure balance between the cloud shock and the ISM shock, the velocity of the shock in the cloud is less than that in the ISM and is given by $V_{c\ell}^2 n_{c\ell} \sim \beta V_0^2 n_0$, where n_0 and $n_{c\ell}$ are the densities in the ISM and in the cloud respectively, and β is of the order of unity (McKee and Cowie, 1975). The radio emission in SNRs originates in a shell of swept up interstellar gas behind the shock wave, due to the increased density of relativistic electrons and the amplification of the magnetic field by compression and/or by Fermi acceleration. Thus the radio shell

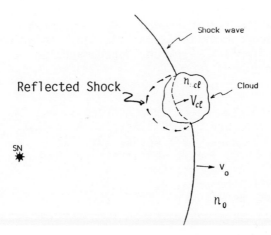

Fig. 3 Sketch of a shock wave colliding with a dense cloud.

expanding into the cloud will be decelerated more and will show an inward displacement with respect to the parts of the shell expanding into the less dense ambient medium. From the observed geometry of the shell it may be possible to obtain $n_0/n_{c\ell} \sim (V_{c\ell}/V_0)^2$. However detailed modelling may be required taking into consideration the magnetic field, the adiabatic or isothermal shocks, the energy losses and the fact that the densities in these clouds are very high ($\geqslant 1000$ cm^{-3}).

Figures 1 and 2 depict the geometry of the shell as observed (arcs 'a') and that for a shell expanding into a homogeneous ambient medium (arcs 'b'). In W28, while the radio shell north of declination $= -23°25'$ expands "freely", the southern part of the shell seems to be "stopped" by the dense cloud. Since little emission is seen beyond arc 'a' over the cloud region, the cloud is likely to extend along the line of sight at least as much as the extent of the remnant shell. In the case of W44 the apparent flattening of the arc 'a' is quite consistent with the interaction between the shell and the cloud as indicated in Figure 3. However, the significant fraction of strong emission along the arcs 'a' and 'b' suggests that only a small portion of the shell along the line of sight is interacting with the cloud. It is also interesting that there is considerable emission near the remnant centre closely following the lowest contour of CO density.

OSRT radio maps thus provide direct observational evidence for the interaction between the remnants and dense clouds near W44 and W28. As the occurrence of dense clouds in the vicinity of SNRs is likely to be common, the distorted shell structures seen in most SNRs may be the result of similar interactions between the remnant shells and dense clouds.

Acknowledgements: I thank Dr. V.R. Venugopal for discussions. The OSRT data were processed using the National Image Processing Facility for Astronomy, which is financed by the Department of Science & Technology, Government of India, at the Radio Astronomy Centre, Tata Institute of Fundamental Research, Udhagamandalam.

References:

Blandford, R.D. and Cowie, L. 1982 Astrophys. J., 260, 625
Kundu, M.R. and Velusamy, T. 1972 Astron. Astrophys. 20, 237
McKee, C.F. and Cowie, L. 1975 Ap. J. 195, 715
Sukumar, S. 1986 Ph.D. Thesis, University of Bombay
Swarup, G. 1984 J. Astrophys. Astron. 5, 139
Wootten, H.A. 1977 Astrophys. J. 216, 440
Wootten, H.A. 1981 Astrophys. J. 245, 105

RADIO STUDIES OF SUPERNOVA REMNANTS: PATTERNS AND STATISTICS

J.L.Caswell
CSIRO, Division of Radiophysics, Epping, NSW,
Australia, 2121.

Abstract: Most supernova remnants (SNRs) in our galaxy have been discovered from their radio emission and for the majority this remains the only means of studying them. In this review the impact of new radio observations is discussed.
    The increased detail of recent radio maps reveals some common patterns among SNRs, despite their generally diverse appearance. The patterns can give us clues to both the intrinsic properties of the supernova and the influence of the interstellar medium. With a fuller understanding of individual remnants, there are better prospects for meaningful interpretation of statistical studies.

1. INTRODUCTION

    My objective is to discuss recent radio observations and to assess what they tell us about the supernova and what they reveal about interactions with the interstellar medium (ISM).
    From the earliest studies of radio supernova remnants (SNRs) it was clear that the majority were shells (appearing as bright rings in projection on the sky), the Crab nebula being a notable exception. However, the quality of the maps used in early analyses (e.g. Milne,1970; Downes,1971; Clark and Caswell,1976) ranged from those showing considerable detail to others where only a crude flux density, angular size, and spectrum were available. With improved maps it has become possible to recognise more examples of centrally concentrated remnants resembling the Crab nebula (e.g. Caswell, 1979; Weiler,1983; Wilson,1983) and also 'composite' remnants which exhibit a flat-spectrum (Crab-like) central feature, enveloped by a steeper-spectrum shell (e.g. Clark et al.,1975; Helfand et al.,1986). A comparison with the Crab nebula suggests that centrally concentrated emission is excited by an embedded neutron star (pulsar). In contrast, emission from the shells of SNRs originates near the shock front where ejecta interact with the swept up interstellar medium (or circumstellar material); a fraction of the kinetic energy is converted to synchrotron radio emission by the amplification of magnetic fields and the acceleration of electrons to relativistic energies, probably by turbulence.
    The increased detail seen in the latest radio maps reveals marked deviations from simple shells, and also hitherto unrecognised patterns. Spectral line radio maps of nearby HI and CO also have been used to study

the impact of the expanding shell on the surrounding interstellar medium. I will deal first with these observations and then turn to investigations of a statistical nature - how useful are they and what do they tell us? Finally I will look ahead to the problems that will need to be tackled next.

2. RADIO CONTINUUM OBSERVATIONS

New observations show continued steady improvements with respect to:

a) Higher angular resolution for bright small-diameter remnants.

b) Increased sensitivity to large-diameter low surface-brightness remnants.

d) Modest improvements in the high-frequency (Mezger at al.,1986) and low-frequency (Odegaard,1986a,b) extremes at which maps are attainable without excessive loss of sensitivity or resolution respectively.

e) Improved dynamic range in maps - permitting much better recognition of faint detail. This has been assisted by new data presentations using grey scales and colour.

f) Improved algorithms (Masson,1986) facilitating proper motion measurements for additional remnants, and ultimately contributing age estimates for more remnants (see 5.1).

2.1 New measurements of known remnants. In some cases, new maps of remnants that previously had indeterminate morphology now enable us to classify them as shells, Crab-like or composite. The following examples illustrate typical results flowing from the new measurements.

a) VLA maps of Cas A with unprecedented resolution (Braun et al.,1987) permit the study of the miniscule changes that occur on a timescale of just a few years.

b) Wide-field mapping with the DRAO synthesis telescope at low frequency (408MHz) permits maps of low-brightness, large-diameter remnants in confused regions such as HB3 (Landecker et al.,1987).

c) Single-dish mapping at high frequency has led to the recognition of more Crab-like remnants (Reich et al.,1984). Basket-weaving mapping techniques yield good sensitivity over the field covered by very large SNRs such as S147 (Furst and Reich,1986).

d) Many of the southern galactic SNRs have now been surveyed by the MOST, which provides a good combination of resolution and sensitivity; references to the bulk of these observations are given by Milne et al.(1985).

2.2 Newly discovered remnants. There is a slow trickle of newly discovered remnants - both shell and Crab-like. The brighter fairly large-diameter shells have mostly been found already, but detailed comparisons of high-resolution radio maps over a wide frequency range, and comparisons with IR data should allow the trickle of new ones to continue (Haslam and

Osborne,1987). Extension of sensitive radio surveys to larger distances from the galactic plane is also paying dividends, as reported by Reich and Furst (1987,this colloquium). The wide fields achieved from single-dish basket-weaving (Bonn) or from synthesis telescopes (at DRAO) now make it feasible to search thoroughly for new remnants at large displacements from the galactic plane, where the expected low yield has previously discouraged such searches. New high-latitude remnants will be especially valuable in providing a sample in an environment less complex than close to the galactic plane (see the discussion of 'gradients' and 'barrels' in 3.1 and polarization in 3.2).

Centrally concentrated (Crab-like) remnants, with quite flat spectra have previously been discovered in a rather haphazard way. However, the recombination line survey by Caswell and Haynes (1987) searched sources in the southern galactic plane down to a peak brightness of ~1.3 Jy (at 5 GHz, with a 4' arc beam) and revealed very few remnants, suggesting that at least above this brightness level there are very few to be found. Reich et al.(1985) argue that G54.09+0.26 is a new, albeit weak, SNR in the Crab-like category, but much of the data, including that from IRAS, seem compatible with an HII region interpretation.

Dedicated searches for faint small-diameter remnants have begun with limited success (e.g.Turtle and Mills,1984; Green and Gull,1984;1986). This is a developing problem area that I will return to later, in Section 6.

The remainder of this review will concentrate on shells and composites, to the exclusion of centrally bright SNRs, which show little evidence of interaction with the ISM.

3. PATTERNS

I now turn to some of the more subtle <u>patterns</u> that can be discerned. In the radio continuum at least five principal patterns have been claimed in at least some remnants. Polarization and spectral line observations also reveal patterns which I will mention briefly.

3.1 Continuum.

a) Gradients. Shell SNRs tend to be brighter on the side closest to the galactic plane (Caswell,1977). Despite individual counter-examples, new observations (subsequent to the quantitative analysis by Caswell and Lerche,1979) have reinforced this conclusion. This can be seen qualitatively from the new data summarized in Table 1 which comprises roughly equal numbers of new remnants and previously known ones (with improved observations). The sample is restricted to SNRs estimated to lie more than 50 pc from the plane. Remnants that are brighter towards the plane are in the majority.

Table 1. Brightness gradients from new SNR observations.

SNR	Brighter towards plane?	References
G9.8+0.6	Yes	Caswell (1983)
G18.9–1.1*	?	Furst et al.(1985)
G30.7+1.0	No	Reich et al.(1986)
G33.2–0.6	Yes	Reich (1982)
G39.2–0.3	Yes	Caswell et al.(1982)
G40.5–0.5	Yes	Downes et al.(1980)
G41.1–0.3	Yes	Caswell et al.(1982)
G65.2+5.7	Yes	Reich et al.(1979)
G94.0+1.0	Yes	Landecker et al.(1985)
G340.4+0.4	Yes	Caswell et al.(1983)
G340.6+0.3	Yes	Caswell et al.(1983)
G359.1–0.5	?	Reich and Furst (1984)

*Probably a composite remnant - see papers presented at this colloquium by Furst et al. and by Barnes and Turtle.

A new analysis is now overdue and could be improved by treating separately the region of the Galaxy within the solar circle (where the disc is relatively thin and flat) and the outer region, which flares in thickness and is warped and may lead to less clearcut conclusions.

b) Barrels. In many SNRs the basically ring appearance (or inferred 3-dimensional shell structure) is modified in a regular manner to show two opposing arcs. Figure 1 illustrates an example (and is also an example of a remnant with increased brightness towards the galactic plane).

Figure 1. The SNR G340.6+0.3; it shows two opposing arcs of emission and is also brighter towards the galactic plane. FIRST map, taken from Caswell et al.(1983).

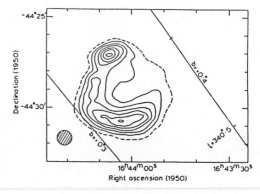

As in the example of Figure 1, there is often just one axis of mirror symmetry and it is found to pass through the regions of very low emission. The inferred 3-D structure is then a barrel, with emission from the staves but

not the end caps. The phenomenon is common, and two interpretations
have been briefly explored so far:

First by Kesteven and Caswell (1987), a proposal that the cylindrical
symmetry is defined in the outburst, because the ejecta has a broadly toroidal
distribution.

Secondly by Roger et al. (1987), that the magnetic field plays an
important role, with the cylinder axis along the prevailing magnetic field.

Each interpretation can draw some support from other arguments,
and it is possible that both mechanisms are operative. They clearly have
considerable implications for not only the SN explosion, but also its
interaction with circumstellar material, and subsequently with the ISM.

c) Double loops.   Manchester (1987) has argued that the double
loops seen in some remnants represent a distinct phenomenon which is
present in many or even most remnants. The loops are regarded as
enhancements of the shell emission in two annular zones. I think of this as
a 'fresco' or 'graffiti' model (launched appropriately in Venice at IAU
Symposium 101), insofar as the perturbations colour the appearance but are
thought not to be dynamically important. The annuli might originate from a
bi-conical flow from a pulsar. The most convincing example is G320.4–1.2, in
which the presence of a pulsar prompted the interpretation. In some other
examples, such as IC443, there are viable alternative explanations suggested
by Braun and Strom (1986a); Green (1986b); and Mufson et al.(1986).

d) Scalloped boundaries.   This variety of deviation from a
roughly circular boundary is exemplified by IC443 (Braun and Strom,1986a)
and OA184 (Routledge et al.,1986). Braun and Strom suggest that we are
observing inter-connected spherical sub-shells, corresponding to stellar-wind
driven bubbles which have pre-processed the ISM in the vicinity of the SNR
before its outburst. This notion has been invoked to account for several
other problems associated with SNR evolution, and we will return to it in
Section 5. Two important implications are that the SNR in such an
environment can rapidly expand to a large size before significant
deceleration sets in, and that the kinetic energy inferred under these
circumstances is much less than if the adiabatic expansion phase is reached
at much smaller radius.

e) Jets.   This term is intended to be descriptive and not to imply
a particular physical process. The jet, unlike the barrel, is a quite rare
phenomenon at the sensitivity level usually achieved. A quite remarkable
example is shown in Figure 2 (see Kesteven et al.,1987); the jet is narrow and
well-collimated and extends beyond the shell boundary for a further distance
at least equal to the shell radius.

A more complex example is shown by Roger et al.(1985), in which a jet expands to a large plume and is exceedingly difficult to account for; a satisfactory interpetation has not yet been devised, but the large plume may be evidence of a pre-existing low-density cavity in the ISM.

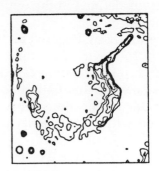

Figure 2. This radio source, G315.8-0.0, is believed to be an SNR which possesses a remarkable highly collimated jet.

A search for more jets is important but we must be wary of likely misclassifications: IC443 might readily (from Green's 1986b map) have been regarded as a jet, and the apparent jets in some other remnants seem likely to be superposed extragalactic sources; (eg SN1006 - Reynolds and Gilmore,1986; and G54.5-0.3 - Caswell,1985).

In summary, it should be stressed that many of these patterns are not mutually exclusive. Gradients and barrels often co-exist; scalloping can hamper the recognition of barrels and can coexist with gradients. However some features <u>are</u> the subject of competing, mutually exclusive descriptions, e.g. IC443 (as remarked earlier in the context of the double loop interpretation).

3.2 Polarization. Polarization observations can be used to map the magnetic fields in SNRs. Milne (1987) reviews the current state of this work and has briefly summarized it at the present meeting. This reinforces the conclusion (first suggested many years ago) that young shell SNRs exhibit predominantly radial fields whereas old SNRs at high latitudes show fields tangential to the periphery. Old remnants near the galactic plane show a confused picture.

For comments on the situation regarding Faraday rotation and depolarization I refer to Milne's review.

3.3 Radio spectral lines. These observations have the potential for revealing the interaction of the SNR ejecta with the ISM. Over the past few years such observations have progressed greatly with the increased use of synthesis telescopes, and more efficient image-processing packages. Nonetheless, the interpretation of the mass of data is a daunting task, and looks likely to be a major bottleneck to further advances.

Studies of HI with the DRAO synthesis telescope have been made by

Landecker et al. (1982) and Routledge et al. (1986) for example; likewise, Braun and Strom (1986b) have used the WSRT to study HI near four SNRs (see also Braun and Strom, 1986a). In several instances the observed patterns have been interpreted in terms of bubbles and thus these features are an important key to understanding interactions with the ISM.

Some of the early associations claimed for SNRs and CO seemed inconclusive but the evidence in the newer observations, such as those in the vicinity of G109.1-1.0 (Tatematsu et al.,1986; and at this meeting), seems quite compelling. The extensive studies by Huang and Thaddeus (1986) and Fukui and Tatematsu (this meeting) show probable associations for many more remnants. In some cases the systemic velocity of the CO clouds provides a useful kinematic distance for the SNR.

Because knowledge of SNR distances is so important I will briefly remark on the use of HI absorption measurements to determine kinematic distances - a technique that has been crucial for the quite young and bright SNRs that are too far away to be studied optically. Doubt was cast on these distances following marked revisions to those of 3C58 (Green and Gull, 1982) and Tycho's SNR (Albinson et al., 1986). However, such pessimism seems to be overstated. Many of the distances were determined with the Parkes interferometer more than a decade ago (Caswell et al.,1975); recent confirmatory measurements have become available with the VLA for several of these such as G29.7-0.3 (Becker and Helfand, 1984) and G21.5-0.9 (Davelaar et al.,1986); they are in very good agreement.

Within a few years we will be able to make improved measurements on the more southerly objects using the Australia Telescope.

Meanwhile, confidence in the present distance scale is provided by comparison with the Magellanic Clouds. Mills et al. (1984) show that the distance scale for the galactic SNRs (determined predominantly from HI absorption distances) is quite compatible with that of the Magellanic Cloud remnants; this is on the assumption that the distribution of remnants on the Σ - D plot is essentially similar for the two galaxies.

4. REMNANTS WITH DIFFICULT OR UNUSUAL INTERPRETATIONS

From time to time it is important to assess which remnants are really difficult to fit into a 'typical' mould. They may be so atypical as to tell us nothing about SNRs (especially if some of them ultimately prove not to be SNRs!) or they may, by exhibiting very extreme features, be a vital clue to understanding similar properties at a much lower level in all the others. My selection of remnants that may readily be misunderstood is as follows:

G166.0+4.3   The observations and discussion by Pineault et al.(1985; 1987) suggest that the two halves of this remnant appear to have evolved in quite different environments. The suggested breakout from a warm medium of intermediate density into a hot, low-density interstellar cavity provides a satisfactory expanation. Such a morphology is rare but the interpretation may serve to explain some other remnants not yet studied so exhaustively.

G69.0+2.7 (CTB80)   For years (e.g. Salter et al.,1983) this has been the most baffling of SNR morphologies. However it would now be instructive to consider it in the light of the data on G166.0+4.3 and IC443; alternatively, Manchester claims it can be fitted into his double-loop model, and the new results presented at this meeting by Strom may also allow construction of a viable model.

G292.0+1.8 The study of this source by Braun et al. (1986) is a warning that perhaps some shell remnants may masquerade as Crabs.

G65.7+1.2 (DA495) may be an old Crab, despite some affinity with shell SNRs (Landecker and Caswell, 1983); thus some Crabs may masquerade as shells.

G263.9–3.3 (Vela) may be a simple shell rather than a composite (Milne and Manchester, 1986).

G39.7–2.0 (W50), despite containing SS433, may be essentially a typical shell (Downes et al.,1986)

G5.4-1.2 may settle into the shell category, albeit with some unusual features (Caswell et al., 1987). This source emphasises that we need a combination of high sensitivity over a quite large field size (to see the weak eastern arc), and high resolution (to recognize that the neck is a distinct feature rather than just the blending of two nearby structures).

The radio nebula surrounding Cir X-1 (Haynes et al. 1987) suggests that there may exist other similar nebulae which are not necessarily SNRs.

At the other extreme, Kepler's SNR is undoubtedly an SNR, but a novel interpretation for its appearance is suggested by Bandiera (1987).

5. STATISTICS

We might expect that the interaction of SNRs with the ISM could best be determined from detailed studies of individual remnants, but important clues can also be gleaned from the ensemble of SNRs and their statistics. Unfortunately, the interpretations are not straightforward and two major areas of controversy lie in the Σ - D (surface brightness - diameter) relation and the N - D (number-diameter) relation. Some interpretations draw on both Σ - D and N - D results but it is convenient to discuss them separately.

   5.1 Σ - D relation. If there proves to be a reasonably good correlation between radio brightness (a parameter essentially independent of distance) and diameter, then this can be used to derive distances for remnants when the brightness and angular size are measured but the distance is not known. Green (1984) has been generally pessimistic about the use of the Σ - D diagram and Allakhverdiyev et al. (1983) have disputed any Σ - D - z relation. Mills (1983), from a study of the Magellanic Cloud remnants, remarks on the large scatter and suggested that perhaps a constant luminosity independent of diameter (corresponding to $\Sigma \propto D^{-2}$) fits the data; however, in the more detailed analysis of Mills et al. (1984) the steeper relation $\Sigma \propto D^{-3}$ is noted as better representing the Magellanic Cloud data, after rejection of three unresolved and two very faint remnants.

   On the positive side, Huang and Thaddeus (1985) argue that there is indeed a tight Σ - D relation for SNRs in our galaxy provided that one limits the analysis to the Type II SNRs associated with molecular clouds (as traced by CO). Other SNRs may be subluminous relative to this and thus the relationship derived is the upper boundary to the population distribution in the Σ - D plane. Berkhuijsen (1986) has investigated a combined Σ - D diagram for the Galaxy, other galaxies, and even prompt emission from very young supernovae. She argues that there is a well-defined upper boundary to the distribution.

   Thus despite much criticism, the Σ - D diagram does provide a useful indication of likely distances for newly discovered SNRs with no other distance determination, and provides a yardstick to gauge whether some individual SNRs are subluminous.

   Physical interpretation is much more contentious. Clearly, expanding SNRs increase in diameter as they evolve; it is less clear whether the brightness monotonically decreases, although, since the large-diameter SNRs are of low brightness, they must eventually fade as they evolve.

   A simple explanation of the Σ - D evolution was suggested by Duric and Seaquist (1986) and is a more detailed development of earlier models. It provides a satisfactory conventional explanation on the assumption that remnants evolve as $\Sigma \propto D^{-3}$.

   In contrast, Berkhuijsen (1987) has argued that comparison of X-ray and radio brightnesses suggests a quite different radio evolution, in which SNRs expand at constant brightness (corresponding to luminosity increasing as D^2) and then fade rapidly. Mills et al. (1984) also suggest rapid fading eventually. However, if the large galactic loops are assumed to be old SNRs, the fading cannot be very rapid. Returning to the suggested early evolution:

this could occur if the SN progenitor has generated a low-density cavity with the radio emission not reaching a peak until the SN ejecta have reached the boundary of the cavity. This interpretation has recently been favoured on other grounds but is controversial as to the origin and size of the cavity. Two possible origins are (1) the stellar wind of the progenitor, and (2) the HII region formed by its uv ionizing radiation. Berkhuijsen discounts bubbles from stellar winds, because the expected sizes are too large.

A detailed 'cavity' model for the LMC remnant N132D has been given by Hughes (1987; earlier suggestions along these lines were made by Chevalier, 1984). Hughes concludes that, for this particular remnant at least, the stellar wind bubble is, again, unsatisfactory, whereas a cavity formed by the HII region yields good correspondence with the data.

Additional support for an early phase of rapid expansion comes from G320.4-1.2 if the pulsar near its centre is indeed the core of the SN, and if the spin-down age is a valid measure, and if the distance (implying a quite large diameter) is reliable. In this case one could argue that the ejecta rapidly expanded to almost its present diameter, then was rapidly decelerated and the radio shell became bright. The non-detection of proper motion of the optical filaments (van den Bergh and Kamper, 1984) is only a slight problem, since the filaments could be regarded as dense regions of the ISM not appreciably accelerated by the shock.

If low-density cavities surrounding supernova progenitors are a general phenomenon, then the implications are considerable. Even for a small cavity size of 10 pc diameter, the cavity would be relevant to SNRs with ages up to 500 yr (assuming ejecta velocity of 10000 km/s). Not until this point would there be rapid deceleration; the subsequent evolution might then proceed roughly according to the Sedov relation. One consequence of this assumption is that very young ($\leq$ 500 yr) small-diameter SNRs might usually be undetectable since they would be in a free expansion phase with perhaps no significant radio emission.

On the other hand, in some remnants (Tycho, SN1006 - both believed to be Type I) deceleration and subsequent Sedov-like evolution has occurred at quite small diameters, requiring that any cavity in these instances be quite small. With radio maps at several epochs, sophisticated comparisons (Strom et al.,1982; Tan and Gull, 1985; Masson, 1986) should allow a determination of the expansion rate in many more SNRs.

5.2 N-D relation. The number (N) of SNRs smaller than a given diameter (D) is contentious both with respect to the data and the interpretation. If the diameter is a good measure of the age, then clearly the integral count of all SNRs up to a given age yields the rate of occurrence of supernovae. In addition, if a linear relationship between log N and log D is observed, it may be of use to infer the expansion law, D as a function of age.

Early work on SNRs in our galaxy suggested $N \propto D^{2.5}$, as expected in the adiabatic phase (Sedov relation, $D \propto t^{0.4}$). For many galactic remnants the value of D is poorly known, and has been derived indirectly from Σ. Mills et al. (1984; also Mathewson et al., 1983; Mills, 1983) showed that for the Magellanic Clouds, $N \propto D^{1.2}$, rather similar to expectations for free expansion. At face value this might suggest that the remnants we observe are indeed in the free expansion phase. In disagreement with this, there have been several alternative interpretations suggested (Green, 1984; Fusco-Femiano and Preite-Martinez, 1984; Hughes et al.,1984; and Berkhuijsen, 1987) but in every case the crucial argument is that the sample is not complete to a given age - firstly because it is not complete to even a given diameter and secondly because there is a scatter in the diameter-age relation. The latter could arise from a scatter in the energy of the ejecta, or a scatter in the ISM density, but another simple effect may be largely responsible: that the duration of the free-expansion phase is variable from source to source, causing some SNRs to become visible at quite small diameters and others not until a much larger diameter is reached. This might severely restrict the range of D over which a meaningful slope is obtainable; at still larger D there remains a completeness problem.

Note that in the simple model of Caswell and Lerche (1979), the scatter in the diameter-age relation can be corrected for because a simple analytic dependence of D and Σ on z is assumed; however, it is still necessary to assume completeness to a given diameter.

Some of the earliest suggestions that radio SNRs might be in the free-expansion phase came from Higdon and Lingenfelter (1980) and from Srinivasan and Dwarakanath (1982). Overall, it would seem fair to say that the free expansion interpretation has now met with considerable resistance and has decelerated! However, it has had the important effect of alerting us to the likelihood that even if the radio emitting remnants are not now in this phase, they may nonetheless have been in a state of free expansion for longer than hitherto realised.

6. THE FUTURE

Here I will summarize progress and problems that we expect to result from new radio observations:

a) More HI and CO studies are needed around remnants; these will help with distance estimates, and assist in understanding the interactions with the ISM.

b) More ages are needed; a contribution to this could come from proper motion studies of radio maps, which reveal current expansion rates.

c) More reliable measurements of the change of intensity with time. These might resolve disputes over the evolution of the radio brightness.

d) Better polarization data are needed to map magnetic fields.

   e) Recognition of large-diameter faint SNRs.   We hope to assess whether the galactic loops are indeed just old SNRs, and whether merging of old SNRs can wholly account for the non-thermal galactic disk.

   f) Recognition of small-diameter young remnants.

The last of these may be a much bigger problem than previously realised. Green (1984) has argued that appreciable numbers of young, small-diameter remnants are present in the Galaxy, as yet undetected. Others have likewise expressed optimism that many more will be found. However the controversy over the source G70.68+1.20 highlights a new problem. Green (1985) showed that the source had a shell structure and Reich et al. (1985) argued that it was non-thermal and (therefore) a supernova remnant. Green (1986a) presented new data intended to demonstrate the thermal nature of the source; but his measurement at 151 MHz reinforces the argument that it is non-thermal since the flux density (0.78 Jy) combined with the small size (20"arc) implies a brightness temperature in excess of 100000 K (much too high for thermal radiation from an HII region or planetary nebula). But if we accept the source as a non-thermal radio shell, the existence of the optical nebula counterpart noted by Green (1986a) is then difficult to understand.

The lack of obscuration suggests that this optical nebula is nearby (≤ 5 kpc) in which case the 20" diameter corresponds to 0.5 pc. At an expansion velocity of 10000 km/s, this would be attained in only 25 years and it is inconceivable that such a recent SN in an unobscured region should not have been noticed (since even if the outburst were in daytime it should still have been noticeable several months later as a night-time object). Even an expansion velocity as low as 2500 km/s would not allow an age greater than 100 yr.

   So do we have a new category of radio source, mimicking young SNRs? If so, other sources in this category could mislead us considerably if they lie in a direction where obscuration masks any optical clues. Two other objects invite comparison. First, the non-thermal radio emission from GK Persei = Nova Persei 1901 (Reynolds and Chevalier, 1984) suggests that nova events, despite their lower energy, might sometimes mimic supernovae in their radio emission; the new observations by Seaquist et al. (this meeting), which reveal a clear shell structure for the nova remnant, emphasise that this may be a very real problem. Secondly, the radio emission from η Car (Retallack,1983; Jones,1985), which is optically a strange variable and not readily classified, suggests that this too might be a variety of object that sometimes mimics weak young SNRs. Note that the radio intensity from η Car is 1 Jy, implying a luminosity (at 2.7 kpc) comparable to that of G70.68+1.20 , but it has not been established whether it is thermal or non-thermal; the emission from Nova Persei is clearly non-thermal but it is much weaker than G70.68+1.20.

   New optical observations towards G70.68+1.20 are urgently needed to resolve this puzzle.

This seems to be an appropriate place to consider the relevance of SN1987a in the LMC. Barring spectacular breakthroughs in medical science, those of us here today are unlikely to be embarrassed by our predictions for the future development of the remnant. However the evolution over the next few years may well tell us much about any pre-existing circumstellar shell and the ISM; our problem then will be to assess just how typical is SN1987a and whether its location in the LMC precludes its usefulness as a model for galactic SNRs.

7. CONCLUSIONS
I will end on a positive note by describing a working framework in which to fit the present observations. It is an edifice that may need shoring up and perhaps eventually ripping down, but it is habitable at present.

The Crab-like remnants have an embedded neutron star. They display a large range in their intrinsic luminosity and show little evidence of interaction with the ISM. Where they are encased in shell remnants, some systematic differences in properties should be looked for, but could be masked by an intrinsically large scatter.

The shell remnants are interacting with either circumstellar material or the ISM; in the case of composite remnants, the shell component should probably be regarded in the same way as isolated shells and not greatly affected by the inner Crab-like component.

Following a supernova outburst there may be a period of several hundred years before a strongly radio emitting shell builds up; this period of near-free expansion will depend on the size of the cavity around the SN and may be quite variable from one SN to another. Indeed G166.0+4.3, and perhaps IC443 suggest that sometimes the cavity is quite large and of complex shape, leading to gross departures from a spherical shell. However, such gross distortions seem rare. Overall, it seems unlikely that hot, low-density cavities dominate the ISM (Heiles, 1987; Kulkarni and Heiles,1987).

By the time that we detect SNRs by virtue of their radio emission, it seems clear that the free expansion phase is over. This is indicated by the optical velocities, where available, and by the clear interaction of old remnants with HI and CO in their vicinity.

When radio emission is seen, the typical corresponding values of Σ and D seen on the Σ - D diagram do provide some indication of D (and hence distance) if only Σ can be measured. Since the detected radio shell itself suggests that interaction with the interstellar medium is already taking place, a z dependence of Σ and D is not surprising. The shells are further distorted (both in shape and in their non-uniform brightness) by a tendency to a barrel structure.

The problem of Type I and Type II SN is as acute as ever, with no certain correspondence with distinctive radio morphologies, despite the general acceptance that Tycho, Kepler, and SN1006 were Type I, and that Cas A (and perhaps the Crab nebula) may have been Type II.

The rate of occurrence of supernovae in our galaxy remains uncertain. Existing estimates may refer to only the brightest supernovae, while subluminous supernovae (subenergetic with respect to the kinetic energy of the ejecta, and with subluminous radio emission) may occur more often; this is a problem likely to grow in importance in the coming years.

Since the emphasis of this review is on the observations and their immediate interpretation, I will put them in perspective with a quotation from the 1958 book 'Structure and Evolution of the Stars' by Martin Schwarzschild: -   'Pillars rather than crutches are the observations on which we base our theories'.

Acknowledgement I am grateful to all my colleagues at both CSIRO and DRAO for their enthusiastic cooperation in our SNR studies.

REFERENCES

Albinson,J.S.,Tuffs,R.J.,Swinbank,E.,and Gull,S.F.,(1986) Mon.Not.R.astr.Soc., 219,427-439

Allakhverdiyev,A.O.,Guseinov,O.H.,Kasumov,F.K.,and Yusimov,I.M.,(1983) Astrophys and Space Sci.,97,287-302

Bandiera,R.,(1987), Astrophys.J., in press

Becker,R.H.,and Helfand,D.J.,(1984),Astrophys.J.,283,154-157

Berkhuijsen,E.M.,(1986),Astron.Astrophys.,166,257-270

Berkhuijsen,E.M.,(1987),Astron.Astrophys., in press

Braun,R.,and Strom,R.G.,(1986a) Astron.Astrophys.,164,193-207

Braun,R.,and Strom,R.G.,(1986b) Astron.Astrophys.Suppl., 63,345-401

Braun,R.,Goss,W.M.,Caswell,J.L.,and Roger,R.S.,(1986) Astron.Astrophys.,162,259-264

Braun,R.,Gull,S.F.,and Perley,R.A.,(1987) Nature,327,395-398

Caswell,J.L.,Murray,J.D.,Roger,R.S.,Cole,D.J.,and Cooke,D.J.,(1975) Astron.Astrophys.,45,239-258

Caswell,J.L.,(1977) Proc.Astron.Soc.Australia,3,130-131

Caswell,J.L.,and Lerche,I.,(1979) Mon.Not.R.astr.Soc., 187,201-216

Caswell,J.L.,(1979) Mon.Not.R.astr.Soc.,187,431-439

Caswell,J.L.,Haynes,R.F.,Milne,D.K.,and Wellington,K.J., (1982) Mon.Not.R.astr.Soc.,200,1143-1151

Caswell,J.L.,Haynes,R.F.,Milne,D.K.,and Wellington,K.J., (1983) Mon.Not.R.astr.Soc.,203,595-601

Caswell,J.L.,(1983) Mon.Not.R.astr.Soc.,204,833-836

Caswell,J.L.,(1985) Astron.J.,90,1224-1230

Caswell,J.L.,Kesteven,M.J.,Komesaroff,M.M.,Haynes,R.F.,MilneD.K.,Stewart,R.T.,and
      Wilson,S.G.,(1987) Mon.Not.R.astr.Soc., $\underline{225}$,329-334
Caswell,J.L.,and Haynes,R.F.,(1987) Astron.Astrophys., $\underline{171}$,261-276
Chevalier,R.A.,(1984) in 'Gas in the Interstellar Medium' pp128-138, workshop at
      Rutherford Appleton Laboratory, ed. P.M.Gondhalekar
Clark,D.H., and Caswell,J.L.,(1976) Mon.Not.R.astr.Soc., $\underline{174}$,267-305
Clark,D.H.,Green,A.J.,and Caswell,J.L.,(1975) Aust.J.Phys.Astrophys.Suppl.no.37,75-86
Davelaar,J.,Smith,A.,and Becker,R.H.,(1986) Astrophys.J.Letts,$\underline{300}$,L59-L62
Downes,D.,(1971) Astron.J.,$\underline{76}$,305-316
Downes,A.J.B.,Pauls,T.,and Salter,C.J.,(1980) Astron.Astrophys.,$\underline{92}$,47-50
Downes,A.J.B.,Pauls,T.,and Salter,C.J.,(1986) Mon.Not.R.astr.Soc.,$\underline{218}$,393-407
Duric,N.,and Seaquist,E.R.,(1986) Astrophys.J.,$\underline{301}$,308-311
Furst,E.,Reich,W.,Reich,P.,Sofue,Y.,and Handa,T.,(1985) Nature,$\underline{314}$,720-721
Furst,E., and Reich,W.,(1986) Astron.Astrophys.,$\underline{163}$,185-193
Fusco-Femiano,R.,and Preite-Martinez,A.,(1984) Astrophys.J.,$\underline{281}$,593-599
Green,D.A.,and Gull,S.F.,(1982) Nature,$\underline{299}$,606-608
Green,D.A.,and Gull,S.F.,(1984) Nature,$\underline{312}$,527-529
Green,D.A.,and Gull,S.F.,(1986) Nature,$\underline{320}$,42-43
Green,D.A.,(1984) Mon.Not.R.astr.Soc.,$\underline{209}$,449-478
Green,D.A.,(1985) Mon.Not.R.astr.Soc.,$\underline{216}$,691-700
Green,D.A.,(1986a) Mon.Not.R.astr.Soc.,$\underline{219}$,39p-43p
Green,D.A.,(1986b) Mon.Not.R.astr.Soc.,$\underline{221}$,473-482
Haslam,C.G.T.,and Osborne,J.L.,(1987) Nature,$\underline{327}$,211-214
Haynes,R.F.,Komesaroff,M.M.,Little,A.G.,Jauncey,D.L.,Caswell,J.L.,Milne,D.K.,
      Kesteven,M.J.,Wellington,K.J.,and Preston,R.A.,(1986) Nature,$\underline{324}$,233-235
Heiles,C.(1987) Astrophys.J.,$\underline{315}$,555-566
Helfand,D.J.,and Becker,R.H.,(1987) Astrophys.J,$\underline{314}$,203-214
Helfand,D.J.,Becker,R.H.,Lockman,F.J.,and Velusamy,T.
      (1986),Bull.Amer.Astr.Soc.,$\underline{18}$,1052
Higdon,J.C.,and Lingenfelter,R.E.,(1980) Astrophys.J.,$\underline{239}$,867-872
Huang,Y.-L.,and Thaddeus,P.,(1985) Astrophys.J.Letts.,$\underline{295}$,L13-L16
Huang,Y.-L.,and Thaddeus,P.,(1986) Astrophys.J.,$\underline{309}$,804-821
Hughes,J.P.,Helfand,D.J.,and Kahn,S.M.,(1984) Astrophys.J.Letts.,$\underline{281}$,L25-L28
Hughes,J.P.,(1987) Astrophys.J.,$\underline{314}$,103-110
Jones,P.A.,(1985) Mon.Not.R.astr.Soc.,$\underline{216}$,613-621
Kesteven,M.J.,and Caswell,J.L.,(1987) Astron.Astrophys.,in press
Kesteven,M.J.,Caswell,J.L.,Roger,R.S.,Milne,D.K.,Haynes,R.F.,and
      Wellington,K.J.(1987) in IAU Symp 125, 'The Origin and Evolution of
      Neutron Stars', in press
Kulkarni,S.R.,and Heiles,C.,(1987) in 'Galactic and Extragalactic Astronomy'
      eds.K.I.Kellermann and G.L.Verschuur, in press
Landecker,T.L.,Roger,R.S.,and Dewdney,P.E.,(1982) Astron.J.,$\underline{87}$,1379-1389
Landecker,T.L.,and Caswell,J.L.,(1983) Astron.J.,$\underline{88}$,1810-1815
Landecker,T.L.,Higgs,L.A.,and Roger,R.S.,(1985) Astron.J.,$\underline{90}$,1082-1093
Landecker,T.L.,Vaneldik,J.F.,Dewdney,P.E.,and Routledge,D., (1987) Astron.J.,in press
Manchester,R.N.,(1987) Astron.Astrophys.,$\underline{171}$,205-215
Masson,C.R.,(1986) Astrophys.J.Letts,$\underline{302}$,L27-L30

Mezger,P.G.,Tuffs,R.J.,Chini,R.,Kreysa,E.,and Gemund,H.-P.,(1986)
        Astron.Astrophys.,167,145-150
Mathewson,D.S.,Ford,V.L.,Dopita,M.A.,Tuohy,I.R.,Long,K.S.,and Helfand,D.J.,(1983)
        Astrophys.J.Suppl.Ser.,51,345-355
Mills,B.Y.,(1983) in IAU Symp.101,'Supernova Remnants and their X-Ray Emission'
        eds.J.Danziger and P.Gorestein,p551-558
Mills,B.Y.,Turtle,A.J.,Little,A.G.,and Durdin,J.M.,(1984) Aust.J.Phys.,37,321-357
Milne,D.K.,(1970) Aust.J.Phys.,23,425-444
Milne,D.K.,Caswell,J.L.,Haynes,R.F.,Kesteven,M.J.,Wellington,K.J.,Roger,R.S.,and
        Bunton,J.D.,(1985) Proc.Astron.Soc.Aust.,6,78-89
Milne,D.K.,and Manchester,R.N.,(1986) Astron.Astrophys., 167,117-119
Milne,D.K.,(1987) Aust.J.Phys.,in press
Mufson,S.L.,McCollough,M.L.,Dickel,J.R.,Petre,R.,White,R.,and
        Chevalier,R.A.,(1986) Astron.J.,92,1349-1357
Odegaard,N.,(1986a) Astron.J.,92,1372-1380
Odegaard,N.,(1986b) Astrophys.J.,301,813-824
Pineault,S.,Pritchet,C.J.,Landecker,T.L.,Routledge,D., and Vaneldik,J.F.,(1985)
        Astron.Astrophys.,151,52-60
Pineault,S.,Landecker,T.L.,and Routledge,D.,(1987) Astrophys.J.,315,580-587
Reich,W.,Berkhuijsen,E.M.,and Sofue,Y.,(1979) Astron.Astrophys.,72,270-276
Reich,W.,(1982) Astron.Astrophys.,106,314-316
Reich,W.,and Furst,E.,(1984) Astron.Astrophys.Suppl.Ser.,57,165-167
Reich,W.,Furst,E.,and Sofue,Y.,(1984) Astron.Astrophys.,133,L4-L7
Reich,W.,Furst,E.,Altenhoff,W.J.,Reich,P.,and Junkes,N.,(1985)
        Astron.Astrophys.,151,L10-L12
Reich,W.,Furst,E.,Reich,P.,Sofue,Y.,and Handa,T.,(1986) Astron.Astrophys.,
        155,185-192
Retallack,D.S.,(1983) Mon.Not.R.astr.Soc.,204,669-674
Reynolds,S.P.,and Chevalier,R.A.,(1984) Astrophys.J.Letts,281,L33-L35
Reynolds,S.P.,and Gilmore,D.,(1986),Astron.J,92,1138-1144
Roger,R.S.,Milne,D.K.,Kesteven,M.J.,Haynes,R.F.,and Wellington,K.J.,(1985)
        Nature,316,44-46
Roger,R.S.,Milne,D.K.,Kesteven,M.J.,Wellington,K.J.,and Haynes,R.F.,(1987)
        Astrophys.J.,in press
Routledge,D.,Landecker,T.L.,and Vaneldik,J.F.,(1986) Mon.Not.R.astr.Soc.,221,809-821
Salter,C.J.,Mantovani,F.,and Tomasi,P.,(1983) in IAU Symposium 101,'Supernova
        Remnants and their X-Ray Emission' eds.J.Danziger and P.Gorenstein,
        pp343-346
Srinivasan,G.,and Dwarakanath,K.S.,(1982) J.Astrophys.Astron.,3,351-361
Strom,R.G.,Goss,W.M.,and Shaver,P.A.,(1982) Mon.Not.R.astr.Soc.,200,473-487
Tan,S.M.,and Gull,S.F.,(1985) Mon.Not.R.astr.Soc.,216,949-970
Tatematsu,K.,Fukui,Y.,Nakano,M.,Kogure,T.,Ogawa,H.,and Kawabata,K.,(1987)
        Astron.Astrophys.,in press
Turtle,A.J.,and Mills,B.Y.,(1984) Proc.Astron.Soc.Aust.,5,537-540
van den Bergh,S. and Kamper,K.W.,(1984) Astrophys.J.Letts., 280,L51-L54
Weiler,K.W.,(1986) Observatory,103,85-106
Wilson,A.S.,(1986) Observatory,103,72-84

EFFECT OF AMBIENT DENSITY ON STATISTICAL PROPERTIES OF SNRs: Σ-D RELATION, N-D RELATION, ENERGY CONTENT

Elly M. Berkhuijsen
Max-Planck-Institut für Radioastronomie
Auf dem Hügel 69, 5300 Bonn 1, F.R.G.

Abstract: Variations in ambient density can largely account for: slope and extent of the observed Σ-D relations, the slope of about +1 of the cumulative N-D relation, and the increase of E_0 with D as derived from optical line data. E_{min} increases during the evolution; the behaviour of E_X remains uncertain.

Introduction

A sample of 84 SNRs at <u>known</u> distances in the Galaxy, the Magellanic Clouds, M31 and M33 observed at radio wavelengths formed the basis for a comparison of data at 1 GHz with X-ray data and [SII]-line data (Table 1 in Paper I = Berkhuijsen, 1986). A close correlation between the surface brightnesses in radio (Σ_R) and X-rays (Σ_X) was found, and a study was made of various SNR-properties as a function of apparent ambient density n_0 as derived from the X-ray data. A review of significant relationships is given in Table 1, some of which are discussed below. Details may be found in Papers I, II (Berkhuijsen, 1987) and III (Berkhuijsen, 1988).

Σ-D relations

Evolutionary models of X-ray SNRs (Fusco-Femiano and Preite-Martinez, 1984) indicate that during the adiabatic phase Σ_X is fairly constant until it drops sharply at the beginning of the radiative phase. Similarly, since $\Sigma_R \propto \Sigma_X^{0.7}$, radio remnants may evolve at nearly <u>constant</u> Σ_R through the observed strip in the Σ_R-D_R diagram until they reach the maximum observable diameter at the beginning of the radiative phase. This conclusion is confirmed by the increase of the minimum energy in relativistic particles and magnetic fields with radius after correction to $n_0 = 1$ cm^{-3} (Fig. 1, Table 1), i.e. $E_{min}(n_0=1) \propto R_R(n_0=1)^{2.7\pm0.3}$, which implies $\Sigma_R(n_0=1) \propto R_R(n_0=1)^{0.5\pm0.5}$.

The slopes of the Σ-D relations observed in radio continuum and X-rays then may <u>not</u> reflect evolutionary tracks. Instead, the slopes of the distributions and part of the variations in Σ and D can be explained by the dependence of Σ and D on apparent ambient density n_0. It is remarkable that data of SNR-candidates in M82 and radio supernovae are consistent with this picture (Paper I).

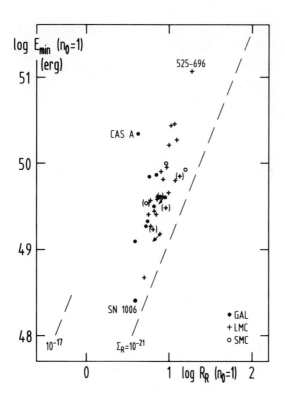

Table 1. Review of significant[a] correlations for shell-type SNRs[b]
Orthogonal (o) and normal (n) regression lines

Y	X	#pp.	o/n	Slope (Y on X)	Zero point	Correl. coeff.
$\log \Sigma_R$	$\log \Sigma_X$	37	o	0.58±0.07	-39.03∓2.49	0.77±0.06
$\log \Sigma_R$	$\log n_0$	33	o	1.13±0.17	-19.34±0.11	0.75±0.08
$\log D_R$	$\log n_0$	33	o	-0.37±0.04	1.19±0.02	0.84±0.05
$\log E_{min}$ -SN1006, 525-696	$\log n_0$	32	o	-0.54±0.12	49.69±0.03	0.62±0.11
$\log E_{min}(n_0=1)$	$\log R_R(n_0=1)$	33	n	2.72±0.31	47.22±0.28	0.84±0.05
$\log T_S$[c]	$\log R_X(n_0=1)$	23	n	-1.57±0.33	1.62∓0.28	0.71±0.11
$\log E_X$[c]	$\log R_X(n_0=1)$	23	n	1.31±0.36	49.21∓0.31	0.61±0.13

Units: Σ_R in W Hz^{-1} m^{-2} sr^{-1}, Σ_X in erg s^{-1} pc^{-2}, n_0 in cm^{-3}, D_R and $R_R(n_0=1)$ in pc, $E_{min}(n_0=1)$ and E_X in erg, T_S in keV.

a) level of significance P < 0.0027; b) without Cas A;
c) for 14 objects with known T_S

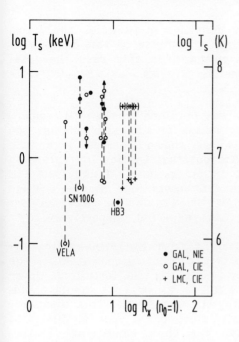

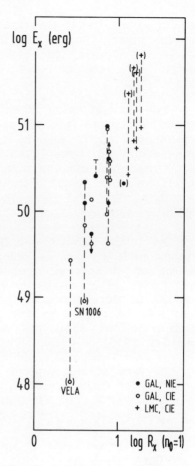

Fig. 2: Dependence of T_S on $R_X(n_o=1)$. T_S has been deduced from X-ray spectra by means of models of collisional ionization in equilibrium (CIE) or not in equilibrium (NIE). References are given in Paper III.

Fig. 3: Dependence of E_X on $R_X(n_o=1)$ for SNRs with known T_S.

N-D relation

The observed exponent of about +1 of the cumulative N-D relations could result from a random distribution of diameters, which may be largely ascribed to variations in n_o. After correction to $n_o = 1$ cm^{-3} the exponent is considerably larger than +1. For a total sample of 74 radio diameters the exponent of 2.5±0.7 suggests adiabatic expansion of the remnants in the sample. However, the statistical significance does not permit definite conclusions on their average expansion law (Paper II).

Energy content

The total energy in relativistic particles and magnetic fields, $E_{tot}(n_o=1)$, would be higher than $E_{min}(n_o=1)$ if the energy in particles were very different from that in magnetic fields. Indeed, in Kepler's SNR

(Matsui et al., 1984) the ratio of energy densities in particles and fields is much higher than in the ISM suggesting a decrease of this ratio with $R_R(n_0=1)^{-1.5}$ during the evolution of a remnant. In this case the total energy in particles and fields possibly varies as $E_{tot}(n_0=1) \propto R_R(n_0=1)^{2.0}$.

The thermal energy seen in X-rays was obtained from shock temperatures T_S derived from X-ray spectra by means of adiabatic models. The dependences of T_S and E_X on $R_X(n_0=1)$ are shown in Figs. 2 and 3. Gronenschild and Mewe (1982) have pointed out that the low T_S-values in young remnants and the high T_S-values in old remnants found from CIE models may not apply. Disregarding such CIE points Figs. 2 and 3 yield $T_S \propto R_R(n_0=1)^{-1.6\pm.3}$ and $E_X \propto R_R(n_0=1)^{1.3\pm.4}$ (Table 1), whereas for adiabatic remnants exponents of -3 and 0 are expected, respectively.

The initial kinetic energy E_0 as derived from [SII]-line ratios increases with R^3, but is independent of $R(n_0=1)$ in agreement with the assumption of adiabatic expansion. This result may be fortuitous, however, since for adiabatic remnants $N_e(SII) \propto R(n_0=1)^{-3}$ should hold, whereas no dependence of $N_e(SII)$ on $R(n_0=1)$ was found. This disagreement may be caused by the inclusion of values of $N_e(SII)$ for filaments interior to the shell not representative for shell conditions (Preite-Martinez, 1985).

The lack of direct proof for adiabatic expansion casts doubt on the reliability of estimates of E_X and E_0 obtained above. An estimate of the mean value of $E_{tot}(n_0=1)$ indicates that a mean value of $E_0 \simeq 2 \ 10^{51}$ erg might be consistent with the data (Paper III).

References
Berkhuijsen, E.M.: 1986, Astron. Astrophys. **166**, 257 (Paper I)
Berkhuijsen, E.M.: 1987, Astron. Astrophys., in press (Paper II)
Berkhuijsen, E.M.: 1988, Astron. Astrophys., in preparation (Paper III)
Fusco-Femiano, R., Preite-Martinez, A.: 1984, Astrophys. J. **281**, 593
Gronenschild, E.H.B.M., Mewe, R.: 1982, Astron. Astrophys. Suppl. Ser. **48**, 305
Matsui, Y., Long, K.S., Dickel, J.R., Greisen, E.W.: 1984, Astrophys. J. **287**, 295
Preite-Martinez, A.: 1985, Proceedings of Workshop on "Model Nebulae", Meudon, July 1985

VLA OBSERVATIONS OF SNRs IN M33

N. Duric
Institute for Astrophysics,
Department of Physics and Astronomy,
University of New Mexico, Albuquerque, NM , U.S.A.

Abstract: The results of a VLA study of the radio continuum properties of supernova remnants in the nearby spiral galaxy M33 are presented. It is shown that a relationship exists between the radio surface brightnesses and the diameters of the 13 detected supernova remnants. The significance of the relation and the constraints it imposes on models of SNR evolution are discussed.

Introduction: The evolution of supernova remnants (SNRs) and their interaction with the interstellar medium is a subject of growing debate, fuelled by both theory and observation. The traditional view holds that SNRs undergo distinct phases of evolution, namely the free expansion, adiabatic and isothermal stages (Woltjer, 1972; Chevalier, 1974; Spitzer, 1978). This view has been supported by radio continuum observations and the establishment of the $\Sigma - D$ relation (Clark and Caswell, 1976; Huang and Thaddeus, 1986) which suggests that the radio emission of an SNR evolves systematically with time. Furthermore, Duric and Seaquist (1986) suggest that, on theoretical grounds, the $\Sigma - D$ relation is a direct consequence of the adiabatic phase of SNR evolution. The opposing viewpoint contends that SNRs undergo free expansion into a rarefied medium, possibly created by the SN precursor wind, and that this phase is terminated by isothermal deceleration caused by the interaction with a dense medium (possibly the material swept up by the wind). In this scenario, the adiabatic phase is not an important factor in the evolution of SNRs (eg Braun, 1985). This view has been supported by observations of SNRs in the LMC (Mathewson et al 1983; Mills et al 1984) in which the cumulative number vs diameter $(N(< D) - D)$ relation was found to have the shallow slope characteristic of free expansion.

The difficulty of determining distances to SNRs in our galaxy makes studies of other galaxies, such as the LMC, appealing. However, in order to shed light on SNR evolution in our galaxy, it is important to study galaxies where the SNRs and their environments are more representative of those in our galaxy. Although not a perfect match, M33 is a good candidate because of its proximity and face-on orientation. Optical studies of SNRs in M33, for example, have led to the establishment of $N(< D) - D$ relation that is consistent with free expansion (Blair and Kirshner, 1985). In this paper we wish to describe a radio study of SNRs in M33 with the aim of establishing the presence or absence of a $\Sigma - D$ relation. As noted above, the $\Sigma - D$ relation can be used to test for the adiabatic phase of SNR evolution. Its presence or absence should therefore help shed some light on how SNRs evolve in M33.

Observations: The SNRs in M33 were observed in the process of surveying the 6cm radio continuum emission of this galaxy. The observations were carried out with the VLA in the C configuration. A total of 19 primary beam fields were needed to map the entire galaxy with integration times ranging from 20 to 90 minutes per field. This allowed us to look for radio emission from all the SNRs referred to in Blair and Kirshner's (1985) paper. The data were 'naturally weighted' in order to maximize the signal to noise ratio and to improve our detection limits. The resulting *rms* noise levels range from 0.03 mJy/beam to 0.1 mJy/beam. Most of the observed SNRs were in fields having a limiting noise level of 0.03 mJy/beam. The resolution ranged from 4.5″ to 5″ which at the assumed distance of 650 kpc corresponds to an average linear resolution of $\approx$ 15 pc.

The radio SNRs were identified by positional coincidence using the optical positions listed by DGD (D'Odorico, Goss and Dopita, 1982). Positional coincidence to 1-2″ was achieved for all but the largest remnants. Integrated flux densities and their uncertainties were calculated using areal integration routines in the AIPS (Astronomical Image Processing System, developed by NRAO) package. Of the 20 bona fide SNRs in DGD's list we have detected 13 at the 3σ level or better. Three other SNRs were confused by emission from HII regions and 4 were clearly undetected. The average uncertainty in the integrated flux measurements was $\approx$ 0.15 mJy.

The $\Sigma - D$ Relation: Using the flux measurements described above and SNR diameters from optical measurements (D'Odorico, Dopita and Benvenuti, 1980; Sabbadin, 1979) we have constructed a plot of $\log \Sigma$ vs $\log D$ to determine if the SNRs in M33 form a $\Sigma - D$ relation. In order to compare with the galactic relation, the 5 GHz fluxes were transformed to 1 GHz by assuming a spectral index of -0.5. The resulting $\log \Sigma - \log D$ plot is shown in the figure below.

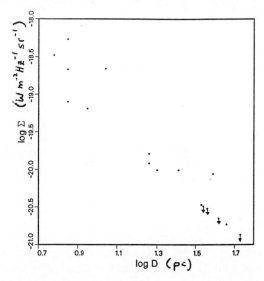

A least squares fit to the plotted points results in the following relation:

$$\log \Sigma = -2.5 \log D - 16.5.$$

The uncertainties in the fitted slope and intercept were $\pm$ 0.2 and 0.3 respectively.

The corresponding relation for the galaxy (Clark and Caswell, 1976) has the form:

$$\log \Sigma = -3.0 \log D - 14.8.$$

The slope of the relation is somewhat flatter than that of the galactic relation. The accuracy of the fit appears to be good enough to suggest that the slope is significantly greater than 2. A slope of 2 corresponds to no systematic evolution of the radio emission. The extent to which selection effects may be producing this result is discussed below. An unusual aspect of the above relation is the low value of the intercept relative to the galactic relation. This suggests that the SNRs in M33 are at least an order of magnitude fainter, at a given D, than those in our galaxy. Possible causes for this difference are discussed below.

Selection Effects: Before a proper interpretation of the above relation can be made it is necessary to consider possible selection effects and observing biases that may affect it. The optically selected sample may be biased against larger remnants because of their lower surface brightnesses and confusion by background objects such as HII regions. This tends to favour the optically brighter remnants. As suggested by Blair and Kirshner (1985) SNRs with $D >\approx 25$ pc are probably undersampled in DGD's list. The detection threshold of the radio observations is such that for resolved remnants, it is independent of their diameters. This favours the radio detection of higher surface brightness remnants for $D >\approx 15$ pc. Very small remnants become inconspicuous as they approach the optical seeing limit and their diameters tend to be overestimated. As a result, the surface brightnesses of these remnants are underestimated. The net result of the selection effects at both very small and very large D is to flatten the $\Sigma - D$ relation. A competing effect occurs when larger remnants are partially resolved so that only portions of their fluxes are detected which results in underestimates of their surface brightnesses. This would tend to significantly steepen the $\Sigma - D$ relation. This problem was avoided in our observations by using integration areas at least as large as the optical areas of the remnants so that any additional uncertainty in the flux densities were incorporated into the error estimates. It is thus unlikely that selection effects are producing the observed steep slope (in fact, they tend to flatten it) and we conclude that the observed $\Sigma - D$ relation signifies a significant departure from the non-evolutionary case ($\Sigma \propto D^{-2}$).

Discussion: The results of the above analysis suggest that a $\Sigma - D$ relation may indeed hold for M33 SNRs. According to Duric and Seaquist (1986) the presence of such a relation is consistent with adiabatic evolution. However, the scatter in the relation is greater than that for the galactic remnants. How can this be reconciled with the fact that M33 SNRs are free of distance uncertainties and that they should therefore form a tighter relation? Furthermore, how can this result be reconciled with Blair and Kirshner's (1985) results that argue against simple adiabatic evolution of SNRs in M33? The answer may lie in Blair and Kirshner's (1985) own discussion of selection effects in their $N(< D) - D$ relation and the possibility that simple adiabatic evolution may be modified by supernova precursor differences and an inhomogeneous interstellar medium. If the interstellar medium of M33 is more clumpy than in our galaxy, varying degrees of SNR interaction with their environment will result for SNRs of the same age and they will achieve different diameters and radio emissivities. Our data support this possibility. The significant difference in the intercepts between the galactic and M33 relations argues for a systematic difference in SNR properties between our galaxy and M33. It suggests that the ISM of M33 has a lower average density (and therefore less interaction between remnant and environment) or that supernovae in M33 have systematically lower blast energies. Owing to M33's lower overall mass and possible evolutionary differences, neither scenario can be ruled out. Hughes, Helfand and Kahn (1984) showed that, for the LMC, a flat $N(D) - D$ relation is consistent with adiabatic evolution if there is sufficient scatter in the interstellar gas densities and/or blast energies. Blair and Kirshner (1985) present the same possibility to explain their flat $N(D) - D$ relation for M33. We propose that these same factors may be the cause of the intrinsic scatter in our $\Sigma - D$ relation. If that is the case, adiabatic evolution is consistent with both, our analysis and Blair and Kirshner's optical analysis for M33.

We caution against over-interpretation of our $\Sigma - D$ relation which, after all, is based on only 13 detections. However, the effect of the selection effects described above and the fact that all non-detections are for large remnants ($D > 35$ pc) suggest that the true slope may actually be steeper than measured. As poor as the relation may be in terms of scatter, the data does appear to favour the existence of a $\Sigma - D$ relation.

Conclusions: A VLA, 6 cm study of the radio continuum properties of 13 SNRs in M33 has led to the establishment of a $\Sigma - D$ relation. The relation appears to have a flatter slope than the galactic relation. The much lower intercept of the M33 relation and the large scatter in the data point to significant differences in the blast energies and/or environmental gas densities among the remnants and systematic differences between these factors in M33 and the galaxy. It is proposed that the $\Sigma - D$ relation described here and Blair and Kirshner's (1985) optically derived $N(D) - D$ relation are consistent with adiabatic evolution of SNRs in M33. 20cm VLA observations are planned to measure flux densities for the confused and undetected SNRs thereby increasing the total sample and improving the statistics. More optical searches and identifications of SNRs in M33 would be invaluable in extending the study and verifying the results described above.

References:
Blair, W.P. and Kirshner, R.P. 1985, _Ap. J._**289**, 582
Braun, R. 1985, _PhD Thesis_, University of Leiden
Chevalier, R.A. 1974, _Ap. J._**188**, 501
Clark, D.H. and Caswell, J.L. 1976, _M.N.R.A.S._**174**, 267
D'Odorico, S., Dopita, M.A. and Benvenuti, P. 1980, _Ap. J._**236**, 628
D'Odorico, S., Goss, W.M. and Dopita, M.A. 1982, _M.N.R.A.S._**198**, 1059
Duric, N. and Seaquist, E.R. 1986, _Ap. J._**301**, 308
Huang, Y.-L. and Thaddeus, P. 1985, _Ap. J. (Letters)_**295**, L13
Hughes, J.P., Helfand, D.J. and Kahn, S.M. 1984, _Ap. J. (Letters)_**281**, L25
Mathewson, D.L. et al 1983, _Ap. J. Supp._**51**, 345
Mills, B.Y. et al 1984, _Austr. J. Phys._ **37**, 321
Spitzer, L. 1978, _Physical Processes in the Interstellar Medium_ (New York: Wiley), p.255
Woltjer, L. 1972, _Ann. Rev. Astr. Ap._**10**, 129

NEW SUPERNOVA REMNANTS FROM DEEP RADIO CONTINUUM SURVEYS

W. Reich, E. Fürst, P. Reich and N. Junkes
Max-Planck-Institut für Radioastronomie
Auf dem Hügel 69, D-5300 Bonn 1, F.R.G.

Abstract: Based on radio continuum surveys of the Galactic plane at wavelengths of 21 cm and 11 cm we have so far identified about 32 new supernova remnants in the area $357°.4 \leqslant \ell \leqslant 76°$, $|b| \leqslant 5°$. This increases the number of known objects in this field by about 68%. Most of them are in the Galactic latitude range $|b| > 0°.5$. Some implications are discussed.

Search for new Supernova Remnants

Deep radio continuum surveys with the Effelsberg 100-m telescope at 21 cm and 11 cm wavelength have been used to search for previously undetected supernova remnants (SNRs). The first part of the 11 cm (2.695 GHz) survey covers the area $357°.4 \leqslant \ell \leqslant 76°$, $|b| \leqslant 1°.5$ (Reich et al., 1984b). This survey has an angular resolution of $4'.27$ (HPBW) and a sensitivity of 50 mK T_B (20 mJy/beam area) and has been extended for the area $|b| \leqslant 5°$. Linear polarization is recorded along with total power (Junkes, 1985; Junkes et al., 1987). The 21 cm survey (HPBW $9'.4$, sensitivity ≈ 100 mK T_B or 50 mJy/beam area) covers at least Galactic latitudes $|b| \leqslant 4°$ (Reich et al., in prep.). The section $92° \leqslant \ell \leqslant 162°$ has been already published (Kallas and Reich, 1980). Both surveys will finally cover the entire Galactic plane visible from Effelsberg.

We have used three different methods to identify SNRs. For several objects we have carried out additional observations at 6 cm wavelength with the 100-m telescope (including linear polarization) and at 3 cm wavelength with the Nobeyama 45-m telescope (Reich et al., 1986; Fürst et al., 1987a). The identification is mainly based on the radio spectral index α.

Particularly in case of plerionic-type SNRs (radio spectrum similar to that of optically thin HII-regions) this identification relies also on the analysis of the linear polarization data. Analysing the polarization data of the 11 cm survey we have found several previously unknown SNRs (Junkes, 1985; Junkes et al., in prep.). This method is limited to regions with a small amount of thermal gas along the line of sight, i.e. to regions of low depolarization.

As a third method we have compared radio and infrared data. Fürst et al. (1987b) have used the 11 cm survey and the IRAS 60 μm survey (IRAS, Beichmann et al., 1985). They have found a ratio R of the infrared to radio emission of typically $R \lesssim 10$ for SNRs and of $R \approx 1000$ for HII-regions. This contrast in R is used to identify sources even in regions of high confusion. A systematic search for SNRs is in progress.

Table 1: Preliminary list of newly identified Supernova Remnants based on the Effelsberg 11 cm survey

Source	S(1 GHz) (Jy)	Size (arcmin)	$\Sigma_{1GHz}\cdot10^{-21}$ (Watt/m² Hz sr)	α $S_{\sim}\nu^{\alpha}$	Type	Ref.
G357.7+0.3	10.5	24	2.7	0.4	S	1
G358.4-1.9	12.5	40x36	1.3	0.5?	S	*
G359.0-0.9	23.0	23	6.5	0.5	S	2
G359.1-0.5	15.0	24	3.9	0.4	S	1
G 4.2-3.5	3.2	28	0.6	0.6?	S	*
G 5.2-2.6	2.6	18	1.2	0.6?	S	*
G 5.9+3.1	3.3	20	1.2	0.4	S	*
G 6.1+1.2	4.0	30x26	0.8	0.3?	P	3,4
G 6.4+4.0	1.3	31	0.2	0.4?	S	*
G 8.7-5.0	4.4	26	1.0	0.3	S	*
G 15.1-1.6	5.5	30x24	1.2	0.8?	S	*
G 16.8-1.1	4.9	30x24	1.0	0.2	U	5
G 17.4-2.3	4.8	24	1.3	0.8?	S	*
G 17.8-2.6	4.0	24	1.1	0.3?	S	*
G 18.9-1.1	37.0	33	5.1	0.3	C	6,7
G 24.7+0.6	20.0	30x15	6.7	0.2	P	8
G 27.8+0.6	30.0	50x30	3.0	0.3	P	8
G 30.7+1.0	5.0	24x18	1.7	0.4	S	5
G 30.7-2.0	0.5	16	0.3	0.7?	U	*
G 31.5-0.6	1.5	18	0.7	0.4	S	9
G 33.2-0.6	4.5	18	2.1	0.7	S	10
G 36.6-0.7	---	>25	---	0.9	U	9
G 36.6+2.6	0.7	17x13	0.5	0.5?	S	*
G 42.8+0.6	2.0	24	0.5	0.5	S	9
G 43.9+1.6	8.6	~60	0.4	0.0?	P	*
G 45.7-0.4	3.8	22	1.2	0.4	S	9
G 54.1+0.3	0.5	1.5	33.4	0.1	P	11
G 59.8+1.2	1.6	20x16	0.8	0.5	U	3,4
G 68.6-1.2	0.7	28x25	0.2	0.0?	U	3,4
G 69.7+1.0	1.6	16	0.9	0.8	S	3,4
G 70.7+1-2	1.1	0.3	1840.	0.6	S	11,12
G 73.9+0.9	9.0	22	2.8	0.3	S	5
G179.0+2.6	7.0	70	0.2	0.4	S	13

? = very uncertain

Type: S = shell, P = plerion, C = combined (plerion+shell), U = unclear morphology

References: * = this paper; 1 = Reich and Fürst, 1984; 2 = Reich et al. 1987b; 3 = Junkes, 1985; 4 = Junkes et al., in prep.; 5 = Reich et al. 1986; 6 = Fürst et al., 1985; 7 = Fürst et al., this volume; 8 = Reich et al., 1984a; 9 = Fürst et al., 1987a; 10 = Reich, 1982; 11 = Reich et al., 1985; 12 = Reich et al., 1987a; 13 = Fürst and Reich, 1986

List of new SNRs

The actual list with some data of SNRs based on the application of the identification methods described above is given in Table 1. The systematic search for SNRs was limited to $l \leqslant 76°$, because of still incomplete observations outside this longitude range. At present one object, G179.0+2.6, was found at $l > 76°$. An example of two new shell-type SNRs at high Galactic latitudes is shown in Figure 1. Some spectral index data of the very weak high latitude objects listed in Table 1 are uncertain, because they are based on 21 cm and 11 cm data only. Consequently the derived flux and surface brightness at 1 GHz are also uncertain.

It is apparent from Table 1 that all except two objects have diameters $\geqslant 15'$. For objects of smaller diameter the distinction between SNRs and extragalactic sources requires observations with high angular resolution to obtain their morphological structure. These observations are available only for a very limited number of sources. The two entries in Table 1, G54.1+0.3 and G70.68+1.2, have been observed by Green (1985) and proved to be Galactic.

G70.68+1.20 is of particular interest. This shell-type object (size ≈20") is seen towards a high density molecular cloud of a diameter of ≈2' (Reich et al., 1987a). All available data indicate that the SNR exploded inside the molecular cloud, which has a kinematic distance of 5.5 kpc.

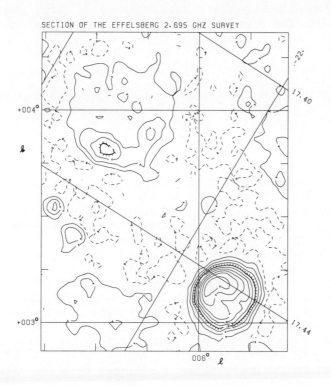

SECTION OF THE EFFELSBERG 2.695 GHZ SURVEY

Figure Caption: Contour map at 11 cm wavelength of the two shell-type SNRs G5.9+3.1 and G6.4+4.0. The contours are 50 mk T_B or 20 mJy/beam area apart. The angular resolution is 4'27.

The size of the SNR is ~0.5 pc and it is still in free expansion. For an average expansion velocity of 2000 km s^{-1} its age is about 135 years. This means it is the youngest Galactic SNR known at present.

Some implications

In the area $357°.4 \leqslant \ell \leqslant 76°$, $|b| \leqslant 5°$ the total number of SNRs reported in the literature and/or listed in Table 1 is 79. At $|b| \leqslant 0°.5$ the number of SNRs is 33 including 4 objects from Table 1, while for $0°.5 \leqslant |b| \leqslant 5°$ the total number is 46 including 28 new sources. The detection limit of the 11 cm survey corresponds to $\Sigma_{1GHz} = 2 \cdot 10^{-22}$ (Watt/m^2 Hz sr) ($\alpha = -0.5$, $S_\nu \sim \nu^\alpha$). While at $|b| \leqslant 0°.5$ numerous objects may be still undetected due to the high confusion with sources and background emission, the list of objects at $|b| > 0°.5$ seems fairly complete close to the detection limit except for small diameter sources ($\leqslant 12'$).

No distances are available for most of the objects, so statistical results fully depend on the validity of the application of $\Sigma-D$ relations. We have applied the $\Sigma-D$ relation given by Milne (1979): $\Sigma_{1GHz} = 2.88 \cdot 10^{-14}$ D^{-4} exp($-|z|/54$), to derive diameters and we calculated the cumulative count $N(D)$. For SNRs in the adiabatic phase a dependence of $N(<D) \sim D^{5/2}$ is expected. For objects of $|b| \leqslant 0°.5$ this dependence is found for diameters up to ~30 pc, and for objects at $|b| \geqslant 0°.5$ up to ~40 pc. This difference is probably due to the higher resolution at lower Galactic latitudes, which hampers the detection of large SNRs with low surface brightness. The diameter of ≈ 40 pc may, therefore, be taken as a lower limit, up to which the adiabatic phase controls the expansion.

References

Fürst, E., Reich, W., Reich, P., Sofue, Y., Handa, T.: 1985, Nature **314**, 720
Fürst, E., Reich, W.: 1986, Astron. Astrophys. **154**, 303
Fürst, E., Handa, T., Reich, W., Reich, P., Sofue, Y.: 1987a, Astron. Astrophys. Suppl. (in press)
Fürst, E., Reich, W., Sofue, Y.: 1987b, Astron. Astrophys. (in press)
Green, D.A.: 1985, Monthly Notices Roy. Astron. Soc. **216**, 691
IRAS Explanatory Supplement, 1985, eds. C.A. Beichmann, G. Neugebauer, H.J. Habing, P.E. Clegg and T.J. Chester
Kallas, E., Reich, W.: 1980, Astron. Astrophys. Suppl. **42**, 227
Junkes, N., Fürst, E., Reich, W.: 1987, Astron. Astrophys. Suppl. (in press)
Junkes, N., 1985, Diploma Thesis, Bonn University
Milne, D.K.: 1979, Aust. J. Phys. **32**, 83
Reich, W.: 1982, Astron. Astrophys. **106**, 314
Reich, W., Fürst, E.: 1984, Astron. Astrophys. Suppl. **57**, 165
Reich, W., Fürst, E., Sofue, Y.: 1984a, Astron. Astrophys. **133**, L4
Reich, W., Fürst, E., Steffen, P., Reif, K., Haslam, C.G.T.: 1984b, Astron. Astrophys. Suppl. **58**, 197
Reich, W., Fürst, E., Altenhoff, W.J., Reich, P., Junkes, N.: 1985, Astron. Astrophys. **151**, L10
Reich, W., Fürst, E., Reich, P., Sofue, Y., Handa, T.: 1986, Astron. Astrophys. **155**, 185
Reich, W., Junkes, N., Fürst, E.: 1987a, Astron. Astrophys. (submitted)
Reich, W., Sofue, Y., Fürst, E.: 1987b, Publ. Astron. Soc. Japan (submitted)

APERTURE SYNTHESIS OBSERVATIONS OF THE SUPERNOVA REMNANT G73.9+0.9

P. Chastenay and S. Pineault
Université Laval, Québec, Canada

ABSTRACT: Radio continuum observations, obtained with the DRAO Synthesis Telescope, are presented of the new supernova remnant (SNR) G73.9+0.9. Our map at 1420 MHz shows indications of spatially resolved knots of emission in the brightest part of the remnant. The 408 MHz map, although of lower resolution, shows the same general morphology. The spectral index α ($S_\nu \propto \nu^{-\alpha}$) between 1420 and 408 MHz is about 0.5, a value typical for shell type SNRs. The morphology however is more suggestive of a filled centre SNR.

I. INTRODUCTION

G73.9+0.9 is a new SNR recently identified by Reich *et al.* (1986). Their best resolution map (HPBW = 2.4 arcmin) is a 4750 MHz map obtained with the 100-m Effelsberg telescope.

Here we report observations obtained with the DRAO Synthesis Telescope (Roger *et al.* 1973) at 1420 and 408 MHz. The resolution of the telescope (arcmin EW x NS) at the declination of the source is 1.0 x 1.7 and 3.5 x 5.9 respectively. Our observations included all spacings from 13 to 605 m. The resultant maps were corrected for the polar diagram of the 9 m paraboloids. Data for spacings shorter than 13 m were derived from the measurements of Kennedy (1975) at 1420 MHz and Haslam *et al.* (1982) at 408 MHz and were added to the Synthesis Telescope maps. The rms noise near the map centre is approximately 0.4 K at 1420 MHz and 5 K at 408 MHz. The sensitivity, particularly at the lower frequency, has been severely affected by the limited dynamic range of the instrument and the near proximity of Cygnus A.

II. RESULTS

Figure 1 shows the part of our map at 1420 MHz which includes the SNR. Although the radio emission from this object is rather smooth, there are nonetheless indications of spatially resolved knots of emission near the centre of the remnant. The object is elongated with the long axis nearly perpendicular to lines of constant galactic latitude. Reich *et al.* (1986) note that the northernmost point source appearing within the outline of the remnant coincides with an IRAS source.

Figure 2 shows the SNR and the area surrounding it to the north-east. A number of interesting sources can be seen, the properties of which are summarized

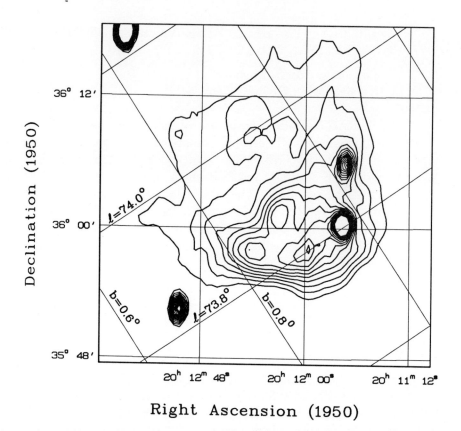

Right Ascension (1950)

Figure 1. DRAO 1420 MHz continuum map of G73.9+0.9. Resolution is 1 x 1.7 arcmin. Contour steps are 1 K in brightness temperature. The first contour is at 9 K.

in Table 1. The spectral index α ($S_\nu \propto \nu^{-\alpha}$) between 1420 and 408 MHz is derived from our two continuum maps. For G73.9+0.9, we obtain α on the order of 0.5. Although this value is typical for shell type SNRs, the morphology of the object is however more suggestive of a filled centre SNR.

The object labelled B on Figure 2 may be of particular interest. Its morphology, consisting of a central source surrounded by two lobes of extended emission, is reminiscent of a double radio source. It is very likely that this object is indeed a distant extragalactic radio source consisting of a flat spectrum central component and steep spectrum lobes. One would thus expect the composite spectral index for the entire source to have an intermediate value and this is indeed what we determine from our maps. However, because the spectral index is close to the one derived for the SNR, one could speculate on the possibility that it and G73.9+0.9 are in fact two parts of a single remnant. We note that object A, which lies between G73.9+0.9 and object B, has a spectral index of 1.0.

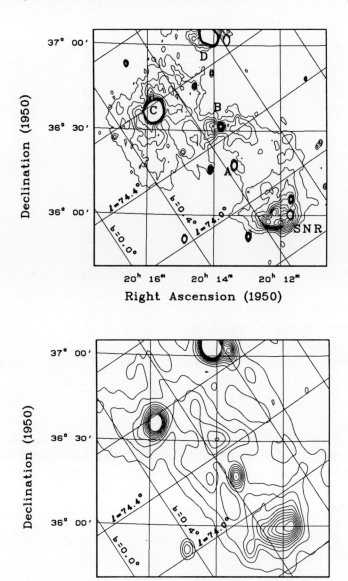

Figure 2. DRAO continuum maps of G73.9+0.9 and of its immediate vicinity. Top: 1420 MHz map with resolution 1 x 1.7 arcmin. Contour steps are 1 K in brightness temperature. The first contour is at 8 K. Bottom: 408 MHz map with resolution 3.5 x 5.9 arcmin. Contour steps are 20 K in brightness temperature. The first contour is at 160 K.

Table 1

Properties of Sources near G73.9+0.9

Spectral Index	Object	Identification	Notes
0.5	SNR	- - -	(1)
0.6	SNR	- - -	(2)
1.0	A	- - -	(2)
0.6-0.7	B	- - -	(3)
0.1	C	S 104	(2)
0.4	D	CTB 87	(2)

Notes to Table:
(1) From integrated flux
(2) From TT-plot
(3) From TT-plot; derived value of spectral index very sensitive to area used

We thank the staff of the DRAO for help with the data reduction. The DRAO is a national facility operated by the Herzberg Institute of Astrophysics. This work was supported by NSERC.

REFERENCES

Haslam, C.G.T., Salter, C.J., Stoffel, H., and Wilson, W.E.: 1982, *Astron. Astrophys. Suppl.*, **47**, 1

Kennedy, J.E.: 1975, Ph. D. Thesis, York University, Toronto

Reich, W., Furst, E., Reich, P, Sofue, Y., and Handa, T.: 1986, *Astron. Astrophys.*, **155**, 185-192

Roger, R.S., Costain, C.H., Lacey, J.D., Landecker, T.L., and Bowers, F.K.: 1973, *Proc. IEEE*, **61**, 1270

Weiler, K.W.: 1984, in *Supernova Remnants and their X-ray emission*, IAU Symp. No. **101**, eds. I. Danziger, P. Gorenstein, D. Reidel, Dordrecht, p. 299

HIGH RESOLUTION RADIO OBSERVATIONS OF THE SNR G160.9+2.6 (HB9)

D.A. Leahy* and R.S. Roger**
* Dept. of Physics, University of Calgary, Canada
** Dominion Radio Astrophysical Observatory, National
   Research Council of Canada, Penticton

Abstract
G160.9+2.6 (HB9) is a supernova remnant of large angular
diameter and low radio surface brightness. We report new
observations of the continuum emission from HB9 at 408 MHz
and 1420 MHz with angular resolutions of 3.5' and 1.0',
respectively, which reveal significant filamentary
structure not previously detected. The 1420 MHz field
covers only the central and eastern parts of HB9. The
408-1420 MHz spectral index ($S \propto \nu^{\alpha}$) of regions common to
both maps is $\alpha = -0.68$, with no significant spatial
variation. The radio filamentary structure closely follows
the optical structure. X-ray emission from HB9 is more
centrally concentrated than the radio or optical emission.
The radio, optical, and particularly, the X-ray surface
brightness are all diminished in the northern and northwestern portions of the remnant, in directions
approximately coincident with an extensive molecular
cloud detected in CO.

Observations

The observations of HB9 were made with the DRAO synthesis
telescope during the period Dec. 1986 to Feb. 1987. The telescope
(Roger et al., 1973) receives left-hand circularly polarized radiation
in the HI spectral band and in a 15 MHz continuum band at 1420 MHz
which excludes the HI band. The telescope simultaneously receives
right-hand circularly polarized radiation in a continuum band at 408
MHz (Veidt et al., 1985). The complex visibilities were calibrated and
Fourier transformed to produce maps at 408 and 1420 MHz, which were
subsequently cleaned. To produce a final 408 MHz map with accurate
large scale structure, visibilities near the center of the uv plane
were added from the single dish survey of Haslam et al., (1982) and
combined with the synthesis data. For the 1420 MHz map, visibilities
near the center of the uv plane were derived from the Effelsberg survey
(Kallas and Reich, 1980).

Results

a) Radio Maps –

Figure 1 shows the 408 MHz map of HB9 with the full resolution of
3.5' (E-W) by 4.8' (N-S) and a contour interval of 12 K. HB9 is
apparent as the 2-degree diameter extended source with a marked outer
boundary except in the northwest sector. The 1420 MHz map of Figure 2
has been convolved to a resolution of 2' x 2', and has a contour
interval of 0.25 K. Diffuse emission and several filamentary arcs are
seen in the interior and along the edges of HB9. A number of

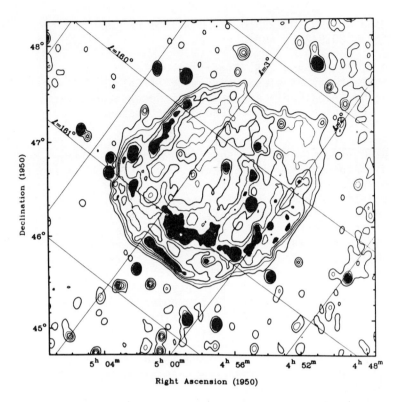

Figure 1. 408 MHz map of HB9. 3.5' (E-W) x 4.8' (N-S) resolution,
contour interval 12K.

unresolved (probably extragalactic) radio sources also appear in these
maps. The results of analysis of these will be presented elsewhere.

b)   Spectral Index-
     The 1420 MHz map was convolved to a resolution of 3.5' x 4.8' in
order to directly compare the brightness temperatures with those from
the 408 MHz map. Pixels within the area common to the two maps were
compared. The brightness temperature of a pixel at 408 MHz versus its
temperature at 1420 MHz was plotted (a T-T plot) for pixels with
temperatures greater than 1 K in the 1420 MHz map and in the interior
of HB9 (excluding any bright point sources). The slope in the T-T plot
is equivalent to a spectral index of -0.68. Within errors this same
index was derived from T-T plots for different subareas of the interior
of HB9.

c)   Comparison with Other Wavelengths-
     The 2.7 GHz map of Reich et al., (1983) agrees closely with the
408 MHz structure. A low resolution (~30 arcminute) map of HB9 at 34.5
MHz has been presented by Dwarakanath et al., (1982). They also
summarize flux density measurements, with care to correct all data to a
common scale, and find a good fit with a power law spectrum with a

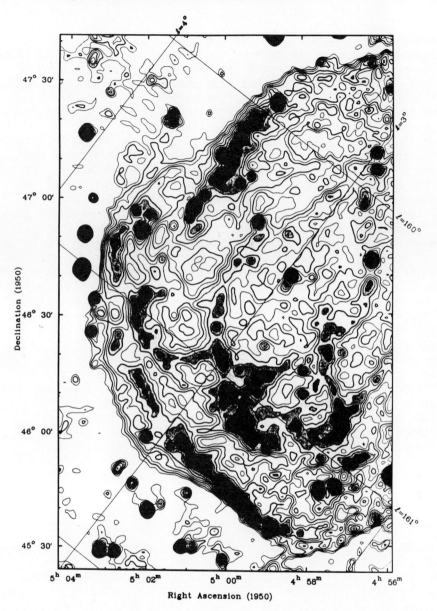

Figure 2. 1420 MHz map of HB9. 2' x 2' resolution, contour interval 0.25K.

spectral index from 34.5 to 2700 MHz of -0.58 ± 0.06. They also see no significant variation of 34.5-1400 MHz spectral index across the SNR. Willis (1973), in contrast, finds a spectral index of -0.44 with a possible steepening above 1.3 GHz. Our data, which lies on either side of the reported break at 1.3 GHz by Willis (1973), shows no evidence for a change of spectral index and is consistent with the result of Dwarakanath et al., (1982).

An optical photograph (van den Bergh et al., 1973) reveals diffuse and filamentary emission throughout the central and southern portions of the remnant, being brightest along the southeastern rim. A comparison of the 408 and 1420 MHz maps with the optical photograph shows that the optical and radio filamentary structure agrees in position to the limits of the map resolutions (i.e., 3.5 and 1.0 arcminutes for 408 and 1420 MHz, respectively).

The Einstein IPC X-ray map (Leahy, 1987) shows HB9 to be brighter in the center than along the rim and also to have a region in the northwest part of the radio maps which is not seen in X-rays. The CO map of Huang and Thaddeus (1986), for radial velocities -12 to -30 km/s, shows that a molecular cloud overlaps the northern and northwestern boundary of HB9. This cloud has not distorted the shape of the boundary of HB9 in radio and optical, but is roughly associated with an area of diminished brightness. A comparison of CO and X-ray images shows that the edge of the molecular cloud coincides with the outer boundary of the X-ray emission. Thus, it appears that the X-ray brightness has been reduced by absorption, and that the cloud may lie on the near side of, or may be interacting with, HB9. Comparison of the CO map with the IRAS maps shows 60 and 100 micron emission coincidental with brighter CO emission from the molecular cloud.

Summary

The radio and optical structure are seen to agree closely in detail indicating that both radio and optical emission are associated with shocks propagating into sheets or filaments of material inside HB9. For the large scale appearance, the radio and optical are significantly brighter in the southern regions of the remnant. The X-ray structure (Leahy, 1987) shows the same tendency even more strongly. This may be caused by the presence of a molecular cloud along the line of sight to, or interacting with, the northwest portion of HB9.

References

Dwarakanath, K., Shevgankar, R., and Sastry, C.V. 1982. J. Astrophys. Astr. 3, 207.
Haslam, C.G.T., Salter, C.J., Stoffel, H., and Wilson, W.E. 1982. Astron. Astr. Suppl., 47, 1.
Huang, Y.L. and Thaddeus, P. 1986. Ap. J., 309, 804.
Kallas, E. and Reich, W. 1980, Astron. Astrophys. Suppl. Ser. 42, 227.
Leahy, D.A. 1987. to appear in Ap. J.
Roger, R.S., Costain, C.H., Lacey, J.D., Landecker, T.L., and Bowers, F.K. 1973. Proc. I.E.E.E., 61, 1270.
Reich, W., Fürst, E., and Seiber, W. 1983. IAU Symposium No. 101 "Supernova Remnants and Their X-ray Emission", J. Danziger, P. Gorenstein (eds.), (Reidel, Dordrecht), p. 377
van den Bergh, S., Marscher, A., Terzian, Y. 1973. Ap. J. Suppl. 26, 19.
Veidt, B.G., Landecker, T.L., Vaneldik, J.F., Dewdney, P.E., and Routledge, D. 1985. Radio science, 20, 237.
Willis, A.G. 1973. Astron. Astrophys. 26, 237.

RADIO CONTINUUM OBSERVATIONS OF SUPERNOVA REMNANTS
AT 22 GHz

C.E. Tateyama, N.S.P. Sabalisck and Z. Abraham
INPE: Instituto de Pesquisas Espaciais, MCT.
C.P. 515, 12.200 Sao José dos Campos, SP, Brazil

Abstract: We describe in this work the results of high
frequency radio continuum observations of 96 supernova
remnants (SNRs) obtained at the Itapetinga Observatory
at 22 GHz. We have detected 15 SNRs. Most of the
strongest SNRs detected were previously classified in terms
of their spectral and morphological properties as filled-
center type (Crab-like). Some of the others have doubtful
classifications.

Introduction: Our knowledge of SNRs is based primarily on radio
observations. Their non-thermal spectra made them objects of intense
investigation at low frequencies. There is, however, a lack of high
frequency observations. To obtain information on their spectral
indices at high frequencies we have made radio continuum observations
at 22 GHz. This is the highest frequency survey of SNRs done up to the
present time.

Observations: The SNRs were observed during November 1985 at 22 GHz
using the 13.7 m radome-enclosed Itapetinga radio telescope. The
transmission of the radome was 0.77 and the beamwidth 4.2 arcmin. The
main beam efficiency was 63%. The receiver was a K-band mixer with a
1 GHz d.s.b. providing a system temperature of about 1000K. We
operated the receiver in the total power mode. We used a rectangular
horn sensitive to the horizontal E vector, i.e. we detected only one
polarization component; the total intensity was obtained by assuming
that the linear polarization of the sources was negligible. The use of
a room-temperature load as a calibrator provided a value for the source
temperature free from the effects of atmospheric attenuation (Ulich and
Hass, 1976; Abraham et al., 1984).   Virgo A was the absolute
calibration source, with a flux density of 21.5 Jy at 22 GHz (Janssen
et al., 1974).
    The observations consisted of scans through each source in right
ascension and declination.   Each scan lasted 20 seconds; each
observation consisted of the average of 30 scans (10 minute
integration), preceded by a calibration. A baseline was fitted to the
points at the beginning and at the end of the scan to eliminate the sky
temperature. The amplitude of the scan was chosen according to the
sources; 1° for compact sources and 2° for the extended ones. The
minimum detectable surface brightness was about 3×10^{-21} Wm^{-2} Hz^{-1}
sr^{-1}.
    We observed all the SNRs from the catalogue by Green (1984) with
declination $\delta<12°$, and angular sizes less than 2°.

Results: Table I is the list of the 15 SNRs detected at 22 GHz with the Itapetinga radio telescope.

TABLE I

Galactic source number	Size (')	S_{22} (Jy)	Σ_{22} (W m^{-2} Hz^{-1} Sr^{-1})	Peak flux 22GHz (Jy)	Peak flux 5GHz (Jy)	Map* refer ence	$\alpha_{408,5}$	$\alpha_{5,22}$
G 0.0+0.0	3.0	11.5	0.19 E-18	55.0	116.0	4	-	-0.5
G11.2-0.3	4.2	3.9	0.33 E-19	2.5	5.9	1	-0.56	-0.58
G21.5-0.9	1.2	5.1	0.53 E-18	5.9	5.9	3	0.00	0.00
G31.9+0.0	4.8	4.2	0.27 E-19	1.3	3.5	3	-0.71	-0.67
G41.1-0.3	3.6	4.2	0.32 E-19	1.5	3.3	2	-0.49	-0.53
G43.3-0.2	4.2	8.0	0.68 E-19	7.8	11.2	1	-0.47	-0.24
G290.1-0.8	12.6	12.4	0.12 E-19	1.5	2.5	3	-0.55	-0.35
G291.1-0.1	10.0	5.5	0.82 E-20	2.2	3.8	1	-0.35	-0.37
G292.0+1.8	5.4	4.2	0.21 E-19	5.2	6.0	1	-0.41	-0.10
G298.5-0.3	3.7	1.8	0.19 E-19	1.0	1.6	1	-0.36	-0.32
G328.4+0.2	4.0	7.7	0.72 E-19	5.9	8.8	1	-0.24	-0.27
G337.0-0.1	7.6	4.0	0.10 E-19	2.6	5.5	1	-0.47	-0.50
G338.5+0.1	12.4	9.9	0.96 E-20	3.0	5.4	1	-0.33	-0.40
G349.7+0.2	1.7	4.4	0.23 E-18	4.6	8.6	2	-0.49	-0.42
G357.7-0.1	5.2	9.8	0.54 E-19	4.6	5.8	2	-0.43	-0.16

* 1 Goss and Shaver (1970)
  2 Caswell, Clark and Crawford (1975)
  3 Haynes, Caswell and Simons (1979)
  4 Whiteoak,J.B. and Gardner,F.F. (1973)

Column 1 gives the galactic identification number for each SNR; column 2, the angular size (θ) obtained from the catalogue of Clark and Caswell (1976); column 3, the integrated flux density S(22) extrapolated to 22 GHz from data at 0.408 GHz and 5 GHz (Clark and Caswell, 1976); column 4, the corresponding surface brightness at 22 GHz, calculated from the relation:

$$\Sigma(22) = 1.505 \times 10^{-19} \frac{S(22)}{\theta^2} \ \text{W m}^{-2} \ \text{Hz}^{-1} \ \text{Sr}^{-1}$$

Column 5 is the peak flux density equivalent to a point source at 22 GHz; column 6, the peak flux density at 5 GHz obtained from the references listed in column 7; column 8, the spectral index between 0.408 GHz and 5 GHz derived by Clark and Caswell (1976) and column 9 gives the spectral index calculated from the flux densities at 5 GHz and 22 GHz; both observations have the same angular resolution.

The SNRs G4.5 + 6.8, G34.7-0.4, G39.2 - 0.3, G348.5 + 0.1 & G348.7 + 0.3 were not detected at 22 GHz although they had an expected surface brightness higher than the minimum detectable value. For some of them the expected flux density was up to ten times higher than the upper limit obtained. These supernova remnants must present a break in the spectra somewhere between 5 GHz and 22 GHz.

Discussion: Three sources, G357.7 - 0.1, G292.0 + 1.8 and G43.3 - 0.2 show a spectral index derived from the 5 GHz data ($\alpha_{5,22}$) which is flatter than the index obtained from 0.408 GHz and 5 GHz ($\alpha_{408,5}$). This flattening may be due to a change in the spectral index across the supernova, since $\alpha_{5,22}$ was calculated at the center of the remnant (integrated over a $4'$ beam) and the $\alpha_{408,5}$ index was calculated from the integrated flux of the remnant. G43.3 - 0.2 is recognized as a composite type of SNR. The radio morphology is shell type while in X-ray it shows centrally peaked emission without limb-brightening. This region was mapped at 22 GHz (Sabalisck et al., 1987) and it is partially resolved with the 4.2' beam. The strong molecular cloud (G43.2 - 0.0) may be contributing to the emission. G357.7 - 0.1 and G292.0 + 1.8 are classified as filled-center SNRs, however X-ray observations of G292.0 + 1.8 do not show the normal morphology of a centrally localized source. The other SNR, G357.7 - 0.1 shows a peculiar morphology and was even classified as a new class of non-thermal object by Becker and Helfand (1985).

The sources G328.4 + 0.2, G349.7 + 0.2 and G21.5 - 0.9 are filled center SNRs. There is no reported detection of X-rays for G328.4 + 0.2. The SNR G349.7 + 0.2 has a spectral index consistent with the typical value of a shell-type SNR, but it does not have the ring-like appearance which is characteristic of an SNR with a prominent cavity in the central region.

The SNRs G11.2 - 0.3, G31.9 + 0.0 and G290.1 - 0.8 are weak sources at 22 GHz. They are classified as shell-type SNRs and their spectral indices are consistent with objects of this type. The source G290.1 - 0.8 shows a flattening in the spectral index $\alpha_{5,22}$.

There are no identifications of the sources G41.1 - 0.3, G298.5 - 0.3, G337.0 - 0.1 and G338.5 + 0.1 as either filled center or shell type. They are weak sources at 22 GHz. The SNR G337.0 - 0.1 and G338.5 + 0.1 are in the vicinity of strong HII regions and the determination of the peak flux is difficult because of the uncertainty in the determination of the base line.

References:

Abraham, Z., Medeiros, J.R. and Kaufmann, P.: 1984, Astron. J., 39, 200
Becker, R.H. and Helfand, D.J.: 1985, Nature, 313, 115-118
Caswell, J.L., Clark, D.H. and Crawford, D.F.: 1975, Aust. J. Phys.
     Astrophys. Suppl., 37, 39-56.

Clark, D.H. and Caswell, J.L.: 1976, Mon. Not. R. Astr. Soc., 174,
    267-305
Goss, W.M. and Shaver, P.A.: 1970, Aust. J. Phys. Astrophys. Suppl.,
    14, 1
Green, D.A.: 1984, Mon. Not. R. Astr. Soc., 209, 449-478
Haynes, R.F., Caswell, J.L. and Simons, L.W.: 1979, Aust. J. Phys.
    Astrophys. Suppl., 48, 1-30
Janssen, M.A., Golden, L.M. and Welch, W.J.: 1974, Astron. Astrophys.,
    33, 373
Sabaliīsck, N.S.P., Tateyama, C.E. and Abraham, Z.: 1987, in preparation
Shaver, P.A. and Goss, W.M.: 1970, Aust. J. Phys. Astrophys. Suppl.,
    14, 77
Ulich, B.L. and Hass, R.W.: 1976, Astrophys. J., 30, 247
Whiteoak, J.B. and Gardner, F.F.: 1973, Astrophys. J. Letters, 218,
    L103

SHOCK WAVES, PARTICLE ACCELERATION AND NON-THERMAL EMISSION

R. D. Blandford
130-33 Caltech
Pasadena, California 91125, U.S.A.

Abstract: Some recent developments in the theory of particle acceleration at supernova shock fronts are reviewed and the confrontation of this theory with measurements of galactic cosmic rays and observations of supernova remnants is discussed. Supernova shock waves are able to account for the energetics, spectrum and composition of galactic cosmic rays, though it remains difficult to understand acceleration of $\sim 10^5$ GeV particles. Recent developments in the analysis of interplanetary shock waves and in the numerical simulation of quasi-parallel shocks are encouraging. Interpretations of different categories of remnants are reviewed and a speculative interpretation of the optical companion to SN1987a is discussed.

Introduction: The primary interaction of an expanding supernova remnant with the interstellar medium is mediated by the bounding shock front. For most astronomers this shock front can be treated as a discontinuity—a sort of Dedekind cut separating the unshocked gas from higher entropy shocked gas in thermal equilibrium at some temperature T whose subsequent evolution is to be modelled. This is not the view of a plasma astrophysicist for whom high Mach number collisionless shock waves possess structure that has to be understood prior to analysing the downstream flow. In this talk I shall discuss this structure.

The interstellar medium has several distinct components. The substrate is the thermal plasma and, as is well known, a strong shock wave moving with speed $V_s = 1000V_{s8}$km s^{-1} will, according to the Rankine-Hugoniot conditions, quadruple its density and increase its temperature to $T = 1.4 \times 10^7 V_{s8}^2$K. However, this temperature is really only a measure of the rms thermal ion speed. There is every expectation that a collisionless shock will not transmit electrons and ions with the same temperature and indeed a Maxwellian distribution function is not guaranteed. Suprathermal tails of electrons and ions are created at shocks and can persist in the face of Coulomb collisions, probably bolstered by wave damping. This may invalidate existing analyses of optical and X-ray line strengths which generally assume a Maxwellian electron distribution and often at the same temperature as the ions. (Aschenbach, Kirshner, this volume). The interaction of dust grains with shock fronts has important implications for the IR emission, which may be the dominant radiative loss from an expanding remnant (Dwek, this volume).

However, in this talk I shall be mostly concerned with the interaction of cosmic rays and magnetic fields with shocks. In the following two sections, I shall summarise what is generally understood about non- thermal processes in supernova remnants. I shall then describe some more recent developments in the study of collisionless shocks. Finally, I shall return to the interpretation of observations of supernova remnants and suggest some specific investigations which should now be practical. Recent reviews of this and associated topics can be found in Blandford (1982) Drury (1983), Kennel, Edmiston and Hada (1985), Blandford and Eichler, (1987).

Acceleration of Galactic Cosmic Rays: Galactic cosmic rays are observed with kinetic energy T from 1 to 10^{11} GeV. The differential number spectrum from $\sim 5 - 10^5$ GeV is a power law with logarithmic slope ~ 2.7 for the primary particles (e.g.p, C, O) and slope ~ 3.1 for the secondaries (e.g.Li, Be, B) created by spallation in the interstellar medium. The secondary spectrum tells us that ~ 5 GeV primary particles traverse a grammage $\lambda_e \sim 7$ g cm^{-2} which declines $\propto T^{-0.4}$ at higher energy. From this we infer that the source spectrum has logarithmic slope ~ 2.3 and that the cosmic ray energy density (dominated by $\sim$ GeV particles) is $u_{CR} \sim 10^{-12}$ erg cm^{-3}. As we know the mean grammage through the galactic disk, $\lambda_d \sim 2$ mg cm^{-2} (Cox, this volume), the local flux leaving the galaxy, $\sim \lambda_d u_{CR} c / \lambda_e$ can be computed. Integrating over the disk gives a galactic cosmic ray power of $\sim 3 \times 10^{40}$ erg s^{-1}, consistent with an independent determination based on the γ-ray background. This is 3 percent of the fiducial supernova energy (10^{51} erg) times the fiducial supernova rate (30 yr)$^{-1}$. With the possible exception of spiral arms, supernovae are the principal heat source for the interstellar medium. They must therefore be efficient particle accelerators. The elemental and isotopic abundances of cosmic rays, (Simpson, 1983) although differing somewhat from solar composition (especially in the under-abundance of hydrogen) show sufficient similarity to those of solar cosmic rays that it is suspected that in both instances, the particles are injected from a hot ($\sim 10^6$ K) coronal gas (Breneman and Stone 1985, Meyer 1985). Electrons are conspicuously underabundant relative to protons (by a factor ~ 30 at the same kinetic energy).

    These and other properties of Galactic cosmic rays are broadly consistent with the theory of particle acceleration by the first order Fermi process at a shock front. In this mechanism, the background plasma is idealised as a uniformly moving fluid approaching the shock with speed u_- and leaving it with speed $u_+ = u_-/r$ (in the frame of the shock), where $r = 4$ for a strong shock. The fluid convects elastic scatterers (in practice Alfvén waves) which can scatter high energy particles. Cosmic rays which travel much faster than the shock (with speed v) can cross the shock $\sim v/u$ times. As the scatterers are approaching each other with speed $\sim 3u_-/4$, a typical particle will gain energy by an amount $\sim u/v$ per shock crossing, giving a mean fractional energy increase for the transmitted particles of order unity. As this is a Fermi process, the distribution of particle energies is a power law. However, unlike with most Fermi processes, the slope of this power law is simply fixed by the kinematics. Specifically, we find that the momentum space distribution function, $f(p) \propto p^{\frac{-3r}{r-1}}$. For relativistic particles incident upon a strong shock, the transmitted energy distribution function is $dN(T)/dT \propto p^2 f(p) \propto T^{-2}$. If we allow particles to be freely injected at the shock front and admit some small inefficiency in the acceleration rate, then we see that this proceess naturally accounts for the source spectrum of galactic cosmic rays. If we further notice that protons (and electrons) have smaller Larmor radii at a given energy than heavier nuclei and will therefore be less readily injected into the acceleration mechanism, then we can also account for the observed abundances (Eichler and Hainebach, 1981).

Magnetic Field: Magnetic field is amplified at a plane adiabatic shock. If the angle between the shock normal and the field direction ahead of the shock is designated θ_{BN}, then the strength of the post-shock field is given by $B_+ = (\cos^2 \theta_{BN} + r^2 \sin^2 \theta_{BN})^{\frac{1}{2}} B_-$ in terms of the field strength ahead of the shock, B_-. Shocks with $\theta_{BN} \lesssim 45°$ are called "quasi-parallel" and those with $\theta_{BN} \gtrsim 45°$ are "quasi-perpendicular". For a general field orientation and standard interstellar field strength $\sim 4\mu G$, these amplifications are quite inadequate to account for the large field strengths inferred to be present in young supernova remnants. For example in Cas A, a lower bound on the ambient field strength within the remnant of $80\mu G$ can be derived from the reported absence of γ-rays (Cowsik and Sarkar, 1980). What seems quite reasonable dynamically (Gull 1973) and is quite consistent with the emissivity distribution (Braun, Gull and Perley, 1987) is that most of the field amplification (i.e., stretching of the field lines) occurs at the interface between the ejecta and the shocked interstellar medium. This makes the outer shock wave rather difficult to locate. In Tycho's supernova remnant, the volume emissivity and hence the implied amplification is much smaller and so the shock is more prominent (Bell, 1979).

In an older remnant like IC443, much of the shocked gas can cool on the expansion timescale and will therefore be crushed by the large post-shock gas pressure. This will in turn accelerate the trapped electrons and compress the magnetic field giving a substantially enhanced volume emissivity from a small fraction of the volume (e.g., Blandford and Cowie 1982).

The polarisation observed from supernova remnants is roughly consistent with these interpretations. In the young remnants, the fields are usually predominantly radial and presumably caused by the strong radial velocity gradients associated with the ejecta. However, in the older remnants, the transverse expansion and the cloud crushing will both tend to accentuate the tangential component of the field as also appears to be generally true. These observations parallel similar trends present in extragalactic jets (e.g., Bridle and Perley, 1984).

Planetary Bow Shocks and Interplanetary Shock Waves: Before dealing in more detail with supernova blast waves, it is instructive to look at collisionless shocks from three differing perspectives. The first is that of a space physicist. The nearest shock is the earth's bow shock which stands off from the earth at ~ 10 earth radii. Conditions in the interplanetary medium ($\rho \sim 10^{-23}$gcm^{-3}, $B \sim 30\mu G$, $u \sim 400$km s^{-1} and $T \sim 10^5$K are similar to those typically associated with supernova remnants when they interact with the interstellar medium. However, there is one respect in which this shock is crucially different from interstellar shocks and this is that its size is quite small, typically 10^{10}cm, only 30 times the Larmor radius of a 10 keV proton whereas a supernova blast wave expands out as far as ~ 30pc. Nevertheless spacecraft observations are able to demonstrate that when the shock is quasi-perpendicular, the actual shock transition, as measured by the thermal ion distribution function or the magnetic field for example, is quite thin—typically a few thermal ion Larmor radii. By contrast, quasi-parallel shocks are quite thick and difficult to localise. A variety of wave modes can be detected in the upstream region, large amplitude low frequency MHD waves, whistlers, Langmuir waves, ion acoustic turbulence, in addition to supra-thermal ion and electron distribution functions. Flybys of the Halley (Sagdeev, et al. 1987) and Giacobinni-Zinner comets and the outer planets have detected their bow shocks. As the solar wind is cooler in the outer solar system, the Mach numbers of the shock tend to be larger than at 1AU, (up

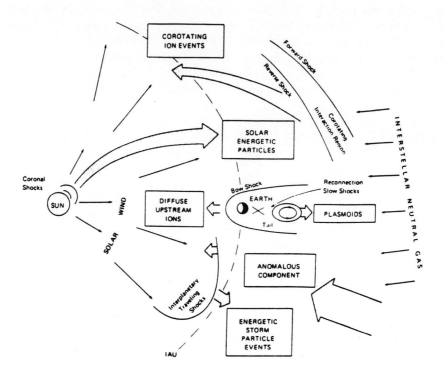

Figure 1. Various particle acceleration sites within the heliosphere (adapted from Scholer 1984).

to $M \sim 20$ in the case of Jupiter).

The travelling interplanetary shock waves, being several AU in radius, are easier to relate to interstellar shocks. Mach numbers in excess of 5 have been reported and again we find that strong quasi-parallel shocks are efficient at accelerating suprathermal protons. The shocks are observed to exhibit several scale lengths associated with the individual components, (Figure 2). A simple but powerful adaptation of the supernova remnant theory of shock Fermi acceleration (Lee 1982) is mostly encouragingly consistent with the detailed observations (e.g.Kennel et al. 1985). In particular, the relationship between distribution function slope q and shock compression r has been verified, as has deceleration of the background fluid ahead of the shock by backstreaming ions. Unfortunately, interplanetary shock waves are too small to accelerate relativistic particles.

Numerical Simulations of Collisionless Shocks: Another way to try to understand the structure of collisionless shocks is to simulate them on a computer. To date most work has been carried out on perpendicular or nearly perpendicular shocks (e.g., Leroy et al. 1982). This is obviously an easier proposition numerically than parallel shocks, because the post shock thermal ions cannot migrate more than a few ion Larmor radii upstream before being convected back into the shock. The magnetic structure seems to be well

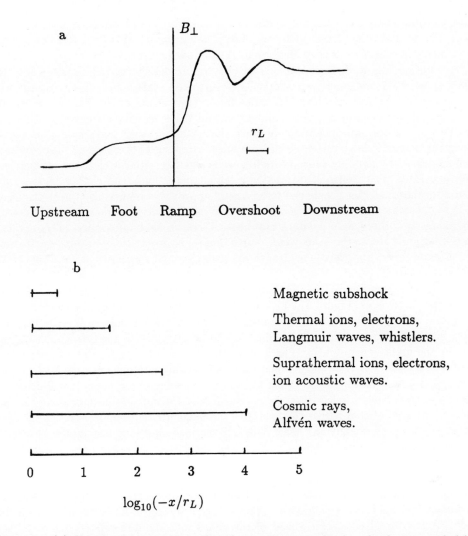

Figure 2. (a) Sketch of magnetic profile of a quasi-perpendicular shock as revealed by spacecraft observations and numerical simulations. The length r_L is the gyro radius of a proton moving with the shock speed in the downstream magnetic field. The thermal ions and electrons are thermalised in a few r_L. (b) Structure of a quasi-parallel shock as inferred from spacecraft observations and theoretical considerations. The scale lengths of the various components ahead of the shock are indicated in units of r_L. Of course, this is only schematic and the details are sensitive to the parameters M, β, θ_{BN}.

established (and is in fair agreement with bow shock observations). Quasi-perpendicular shocks have a precursor "foot" created by the reflected ions. This is followed by a ramp where the field increases rapidly to overshoot its asymptotic downstream value and this is in turn followed by a region in which the field undergoes oscillations about its asymptotic value. The reflected ions have $T_\perp \gg T_\parallel$ (with respect to the magnetic field direction.)

This distribution is unstable and ought to excite lower hybrid waves, especially in the foot. These, and other wave modes can heat the electrons. However, the *downstream* electron temperature is usually lower than the ion temperature.

Rather less work has been carried out on parallel shocks (e.g. Mandt and Tan 1985, Quest 1987 and references therein). Nevertheless, this is broadly consistent with the theory of Fermi acceleration at a shock front. Simulations of this type are restricted to temporal evolution in one space dimension and three velocity dimensions. This may not be too bad an approximation, because the fastest growing wave modes propagate along the magnetic field, although the non-linear coupling of these waves may not be so well modelled. In fact only the ions are followed (together with the electromagnetic fields they generate). The electrons are treated as an adiabatic fluid with charge density equal to that of the ions. This is believed to be a good approximation because the electrons are so light in comparison with the ions. However, in making this simplification, the electrons are expressly forbidden to conduct any heat. Electron heat conduction is observed to be quite significant in the solar wind. Furthermore, electrostatic variations on scales of the plasma period and the Debye length are averaged and assumed not to influence the shock structure.

In the simulations, the ions are fired at a reflecting wall and a stand off shock is allowed to develop (Figure 3). This shock takes several tens of gyro periods to build up and is many Larmor radii in thickness. In the high Mach number ($M \sim 5$) shocks of most interest to us, the incident beam of cold ions is coupled to the post-shock thermal ions by the firehose instability. (The firehose instability is the plasma physics version of the fluid instability that develops when water flows along a sinuous flexible hose and the centrifugal force exceeds the restoring tension in the walls of the hose. It will grow in a plasma when the particle pressure, or more generally the momentum flux along the field exceeds the sum of that across the field and the magnetic tension, $B^2/4\pi$. The firehose instability is non-resonant and essentially all the incoming ions can interact with the magnetic field. Its importance in collisionless shock structure was first recognised by Parker (1961), Kovner (1961), and Kennel and Sagdeev (1967).)

The firehose instability grows so rapidly that in the simulation, the instability criterion is only marginally satisfied. Hydromagnetic waves of non-linear amplitude are sustained just behind the shock and the ions are typically decelerated in ~ 3 Larmor radii. A few ions backscatter ahead of the shock into the undecelerated flow. These suprathermal particles are able to excite Alfvén modes resonantly (as we describe below) and these same waves are carried into the shock by the background fluid because it is moving faster than the Alfvén speed. In addition they are responsible for eventually reversing the motion of the backscattered ions. These ions are the particles that may be injected into the Fermi process. Unfortunately, it is not possible to simulate the entire quasi-parallel shock, because present-day computers still have inadequate speed and memory to encompass all the length scales involved.

Astrophysical Shock Waves: An astrophysicist has yet a third perspective on the general problem of shock structure. As we have emphasised, if cosmic rays are accelerated at supernova shock fronts, then the acceleration has to be pretty efficient. This implies that the cosmic rays are not strictly test particles and may have a strong influence on the shock compression. Furthermore, the simple fact that particles escape the galaxy means

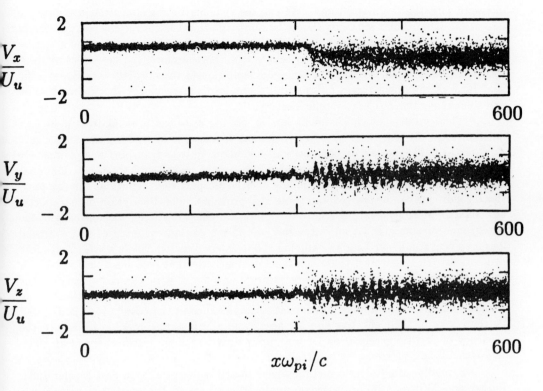

Figure 3. Simulation of a parallel shock by Quest (1987). The shock has an Alfvén Mach number of 5. Distance is measured in units of the ion inertial length (equal to ~ 0.8 thermal ion gyro radii downstream). Note the oscillations in the downstream transverse bulk velocity identifiable as a firehose instability. Also note the backstreaming suprathermal ions ahead of the shock ($V_X > 0$).

that they must be replenished—roughly a thousand times over the age of the galaxy. This process of particle injection, which dictates the abundances in observed cosmic rays, has been a mystery ever since Fermi made his original proposal. There cannot be a pool of subrelativistic though suprathermal particles waiting in the interstellar medium to be overtaken by a shock front and accelerated to GeV energy because they would lose energy so quickly in Coulomb collisions with the ambient plasma that more power would have to be devoted to sustaining this population than would be needed to accelerate the cosmic rays in the first place. Fresh cosmic rays are presumably created at the shock front out of the pool of back streaming suprathermal ions. In this case acceleration can follow immediately after injection so that Coulomb losses are negligible.

These two features of astrophysical shocks were emphasised by Eichler (1979) who also proposed that they might be causally related in the sense that the pressure of the high energy cosmic rays might decelerate the background thermal plasma just enough to control the injection of suprathermal particles. The key point is that the high energy particles should diffuse further ahead of the shock as their gyro radii are proportional to their momenta and so they can act on the incoming fluid before it interacts with the thermal plasma. We can see that high energy particles are possibly important dynamically at strong shocks by observing that the cosmic ray energy density $u_{CR} = \int 4\pi p^2 dp f_+(p)(p^2+m^2c^2)^{\frac{1}{2}}c$ diverges logarithmically as the upper cut off increases when the compression ratio $r = 4$ so that $q = 4$. Furthermore, the net effect of these relativistic particles on the Rankine-Hugoniot conditions is to reduce the effective specific heat ratio of the downstream plasma and therefore to increase the shock compression and to enhance the acceleration efficiency.

This problem of shock mediation by cosmic rays has been addressed in four different ways. Firstly, there have been attempts to finesse it by treating the cosmic rays as a second fluid ignoring its composition, but allowing it to transport heat diffusively (e.g., Drury and Völk, 1981). This probably can only make sense when the dominant cosmic ray pressure is provided by mildly relativistic particles. Solutions for the post-shock conditions can be derived and these may provide a valid description of real shocks, although, as we shall see, they ignore the important effect of Alfvén wave heating of the background plasma. (For a more specific description of the scattering of suprathermal ions involving non-linear Landau damping of oppositely directed Alfvén waves, see Galeev et al. 1986). In particular, these solutions demonstrate that it is possible for a shock to exist in which all the entropy generation is associated with the cosmic rays. The background plasma can be simply compressed adiabatically and not pass through an abrupt subshock discontinuity. The relative importance of the cosmic rays is dictated by the choice of their effective specific heat ratio and it is not possible to deduce the efficiency of particle acceleration from a purely fluid treatment (Achterberg et al. 1984, Heavens, 1984).

Secondly, Monte Carlo methods have been used (e.g. Ellison and Eichler 1984, Ellison and Möbius 1987). In these computations, an *ad hoc* scattering operator is introduced which is supposed to act on the suprathermal particles as well as the cosmic rays. The reaction of the scattering particles on the pre-shock gas is included self-consistently. What is quite impressive about this approach is that it can provide a good fit to the measured particle spectra for H, He, and C+N+O at the earth's bow shock, which verifies that it is the particle rigidity which effectively controls the composition of the accelerated particles. The form of scattering assumed has little formal justification at present, but the agreement with observations suggest that it is the gyro rather than the Debye length which is important.

Thirdly, kinetic models that solve the convection-diffusion equation for the cosmic ray distribution function $f(p)$ have been computed (e.g., Achterberg 1985, Bell 1987, Falle and Giddings 1987). Here the degree of cosmic ray dominance in the shock is controlled by the low energy source function of the cosmic rays usually parametrised as some small fraction of the incident thermal particle flux. It turns out to be difficult numerically to accommodate a diffusion coefficient that increases with energy as rapidly as expected (roughly linearly). However, the particle spectra are noticeably concave reflecting the larger shock compression and more efficient particle acceleration experienced by the higher energy particles that are able to stream further ahead of the shock.

Finally Eichler (1985) has derived an analytic model of shock mediation incorporating magnetic pressure and wave generation. This approach is able to reproduce the results of Monte Carlo simulations.

So we see with all four approaches, that although the computed shock models may be valid descriptions of real astrophysical shocks, we cannot claim to understand shocks properly until we understand the transport of the suprathermal ions. It is here that the numerical simulations offer great hope, especially if they are successful in reproducing the spacecraft measurements of interplanetary shock waves.

Wave-Particle Interactions: The scattering of high energy particles is believed to be effected by circularly polarised Alfvén waves propagating parallel to the magnetic field. The scattering is resonant with condition

$$\frac{\Omega_g}{\gamma} = k_\| v_\| - \omega \simeq k_\| v_\| \tag{1}$$

where Ω_g is the non-relativistic gyro frequency and the subscript $\|$ designates the component resolved along the magnetic field direction. The particle speed is usually much larger than the Alfvén speed $\omega/k_\|$, and so the resonance condition is equivalent to requiring that the wavelength be equal to the ion gyro radius. The scattering rate can be simply estimated by observing that each time a particle orbits the field its pitch angle ϕ will change by an amount $(\delta B/B)_{res}$ where res refers to the the resonant field amplitude. These changes in pitch angle are essentially stochastic and so a particle will random walk in ϕ so that its pitch angle will change by of order 1 radian in a time $\sim \nu_c^{-1}$, where the collision frequency ν_C is given by

$$\nu_c \sim \left(\frac{\Omega_g}{\gamma}\right)\left(\frac{\delta B}{B}\right)^2_{res} \tag{2}$$

(One consequence of the relatively small Alfvén speed v_a is that the ratio of the energy change in the particle to the momentum change is $\sim v_A \ll v$. This implies that the waves are far more efficient at scattering than they are at particle acceleration, which in fact also requires that waves propagating in anti-parallel directions be present.) If we ignore the much discussed but probably illusory difficulty in scattering through 90°, then this is also an estimate of the time for a particle to reverse its direction along the field. We can now estimate the spatial diffusion coefficient

$$D \sim \frac{v^2}{3\nu} \propto \gamma v^2 (B/\delta B)^2 \tag{3}$$

roughly proportional to the particle kinetic energy. This explains why it is generally assumed that higher energy particles stream further ahead of the shock than lower energy particles. In the limiting case, when $\delta B \sim B$ the particles undergo Bohm diffusion, i.e., they random walk of order a gyroradius every gyro period. Numerical simulations by Zachary (1987), verify that large amplitude magnetic fluctuations are created when the cosmic ray density is high and that there is no problem in scattering through 90°.

Long wavelength Alfvén waves must exist in the undisturbed medium where they can inhibit the escape of cosmic rays as we discussed above. However, larger amplitude waves are required and these can be generated by the particles themselves. The linear growth rate can be computed. Imagine that we have an Alfvén wave propagating along the field. Transform into the wave frame where the disturbance is purely magnetostatic.

Now suppose that resonant cosmic rays (of density n_{CR}) are streaming through this frame in the opposite direction to the background plasma with a mean drift speed v_d. There will be a resonant current density of magnitude $\delta j_{res} \sim n_{CR} e v_d (\delta B/B)_{res}$ and an associated force density acting on the wave of $(\delta F)_{res} \sim (\delta j)_{res}(\delta B)_{res}/c$. This force does no work in the wave frame, but in the plasma frame, it increases the energy of the waves at a rate per unit volume $\sim (\delta F)_{res} v_A$ as long as $v_d > v_A$. The growth rate of the instability is therefore

$$\Gamma \sim \frac{n_{CR} e v_d}{c \rho^{1/2}} \tag{4}$$

We can now use these results to estimate the maximum energy to which a supernova blast wave can accelerate cosmic rays. Two criteria must be satisfied. Firstly, the diffusion length of the particles ahead of the shock must be shorter than the radius of curvature of the shock front, R. This implies that $D/u_- \lesssim R$ or

$$E \lesssim 10^4 (\delta B/B)^2 \, \text{GeV} \tag{5}$$

Secondly, scattering waves must be able to grow to non-linear amplitude by the time the background plasma has been convected into the shock. Setting $v_d \sim u_-$ the condition $\Gamma^{-1} \lesssim R/u_-$ gives a lower estimate of the maximum particle energy,

$$E \lesssim \left(\frac{u_{CR}(> E)}{10^{-12} \text{ergs}^{-1}} \right) \left(\frac{R}{1pc} \right) \left(\frac{\rho}{10^{-24} \text{gcm}^{-3}} \right)^{-1/2} \sim 10^3 \, \text{GeV} \tag{6}$$

(e.g., Blandford and Ostriker, 1978; Fedorenko and Fleischman, 1987). In our view, condition (5) is more reliable because quantitative study of interplanetary shocks reveals that the spectrum of hydromagnetic turbulence is not well-described by quasi-linear theory (Kennel et al. 1985). Either way, it seems to be very difficult to accelerate protons up to the "knee" in the cosmic ray spectrum at 10^5 GeV.

It is possible that the problem of accelerating to high energy is illusory. The source spectrum can only be inferred directly up to ~ 100 GeV, and if the grammage traversed by the particles λ_e continues to decline up to 10^5 GeV, then the expected particle anisotropy, $\sim \lambda_d/\lambda_e \sim 0.01$ might exceed the observed value. We know that additional components dominate the cosmic ray spectrum at energies $\gtrsim 10^5$ GeV and that there is the possibility that Fe (which can be accelerated by SNR) is also present above 10^4. It seems, to this reviewer, entirely reasonable that supernova shock waves accelerate protons up to 10^4 GeV with non-linear Alfvén waves and that high energy particles are a mixture of heavier nuclei and protons accelerated at larger scale shocks (e.g., a galactic wind termination shock, Jokipii and Morfill 1985).

In fact, conditions for acceleration may be even more stringent than those described because if a shock is efficient at particle acceleration, then the waves must be rapidly damped To see this, it suffices to evaluate the work by the cosmic ray pressure gradient on the Alfvén waves.

$$U_A = \int \frac{dP_{CR}}{dx} v_A \frac{dx}{u_-} = P_{CR}/M_{A-} = \left(\frac{P_{CR}}{\rho_- u_-^2} \right) M_A \left(\frac{B^2}{4\pi} \right) \tag{7}$$

So, if cosmic ray pressure comprises a fraction $\geq M_A^{-1}$ of the total momentum flux, then the waves must be driven non-linear and are probably damped. The energy probably ends up heating the background medium. A critical question to ask of the X-ray observations is: "must the post-shock electron temperature exceed the Rankine-Hugoniot value $(3\mu_P m_P u_-^2/16k)$ divided by M_A?" If it must, then we can conclude that a subshock is present, and the shock is not totally mediated by cosmic ray pressure.

A somewhat different concern involves the origin of the scattering waves *behind* the shock. Short wavelength hydromagnetic waves are presumably created in abundance by the firehose instability. However, larger wavelength waves, resonant with cosmic rays, may have to be transmitted through the shock. Unfortunately they may be rapidly damped downstream at a general shock with $\theta_{BN} \sim 60°$ because a field parallel mode will be transmitted as an oblique mode subjected to transit time (magnetic Landau) damping. Calculation of the damping rate (Achterberg and Blandford, 1986) suggests that relativistic cosmic rays can be backscattered by transmitted waves but that acceleration of lower energy particles requires wave generation behind the shock.

Electron Acceleration: A topic of more direct interest to the radio astronomers than ion acceleration is electron acceleration. Unfortunately, this is much harder to discuss because it accounts for only a small fraction of the energy available. As is well known, electrons comprise only 3 per cent of the cosmic ray flux at a given kinetic energy. In fact it is one of the ironies of the subject that supernova remnants, which appeared historically to be such spectacularly powerful emitters, are actually quite under-luminous. Cas A radiates roughly 10^{-4} times the maximal synchrotron power of a source of the same energy density. Extragalactic radio sources are often supposed to be radiating the maximum power for their pressure.

It is not hard to understand why shock acceleration should be prejudiced against electrons. The electron gyro radius is smaller than that of a proton of similar energy by a factor ~ 0.02. It is therefore much more difficult to inject electrons into the Fermi mechanism. Indeed, some authors have proposed that electrons are accelerated by a quite separate mechanism. For example, the back streaming reflected ions are able to radiate lower hybrid waves which will be preferentially Landau damped by hot electrons which can in turn be accelerated to relativistic energy (e.g. Galeev *et al.* 1987). Alternatively, the weaker shocks that must surely be present in supernova remnants, can be responsible for second order Fermi acceleration. A third, and quite popular possibility is that a spectrum of hydromagnetic waves be established in the remnant and that this be damped resonantly by the relativistic particles. However, all of these alternative mechanisms require some fine tuning of the effective acceleration and escape times in order to account for the relatively narrow range of observed electron synchrotron radiation spectral indices.

By contrast, suprathermal electrons are produced copiously at interplanetary shocks and even if we are a long off accounting quantitatively for their density, it is surely simplest to imagine that similar electron ejection occurs at supernova blast waves.

A quite different problem associated with electron acceleration is posed by the plerionic remnants like the Crab Nebula. In this case, a central pulsar is believed to lose its rotational kinetic energy through a relativistic electron-positron wind. This wind will shock at a radius where its momentum flux becomes comparable with the ambient remnant pressure. The shock will be relativistic, and so the mean post shock energy per particle will also be relativistic. Plerions are generally observed to have flat radio spectral indices and this could be a signature of a relativistic shock. Indeed, it is possible to account for the IR to γ-ray spectrum from the Crab Nebula by assuming that a power law distribution of positrons and electrons is accelerated with the same mean energy per particle as in the pulsar wind (Kennel and Coroniti, 1984). However, we must ask if the post-shock distribution should be a power law created by the sort of Fermi process we have discussed for non-relativistic shocks. So far most attention has been devoted to the theoretical problem of Fermi acceleration at a mildly relativisitic shock. It is somewhat discouraging that Monte Carlo computations by Kirk and Schneider (1987) indicate that the transmitted particle spectrum steepens as the particle energy is increased.

Observations of electron synchrotron radiation are also very important because they can locate the shock wave and thence constrain the dynamics of the expanding remnant. However, even here there are many puzzles. In the case of Tycho's remnant and that of SN 1006AD (Reynolds, this volume), bright circumferential arcs are seen which are believed to be the shock wave seen tangentially. If the external medium is uniform and its mean compression after being passed by the shock is $4k, (k \sim 2)$ then the shell of shocked gas should occupy a fraction $\gtrsim 1/12k$ of the radius. Some arcs seem to be thinner than this. Furthermore, the polarisation is believed to signify a radial field. A resolution of both problems is possible if the freshly accelerated relativistic electrons are observed through their scattering hydrodynamic turbulence at the shock front.

The outer shocks are not seen in the case of Cas A and the Crab Nebula, although composite supernova remnants (Helfand and Becker, 1987) tell us that they may have quite large radii compared with the brightest emitting regions. Radio maps that can limit the volume emissivity behind an outer bow shock are very important for testing theories of particle acceleration.

Acceleration Efficiency at Quasi-perpendicular Shocks: The empirical evidence from the solar system is that supra-thermal particles can propagate freely upstream in quasi-parallel but not quasi-perpendicular shocks, which ought then to be less efficient particle acclerators. An ingenious argument due Edmiston, Kennel and Eichler (1982), (e.g. Galeev et al. 1987), assumes that post-shock particles have a Maxwellian distribution with temperature given by the Rankine-Hugoniot conditions and estimates the fraction of these particles that have sufficient energy to escape upstream away from the shock front. This is large in the case of the quasi-parallel shocks and of course, vanishingly small for perpendicular shocks. Injection should therefore be much easier at parallel shocks.

This argument has been turned on its head by Jokipii (1987) who points out that once particles achieve high enough energy to diffuse freely, they will be accelerated more efficiently at perpendicular shocks because the effective diffusion coefficient away from the shock front will be reduced by a factor $\cos^2 \theta_{BN}$. If a random distribution of the angle θ_{BN} is established along the shock front, then it might be possible to increase the maximum energy to which particles are accelerated. If we consider the limiting case of a perpendicular shock, we find that each time a particle encounters the shock, it must cross it $\sim v/u$ times before it is transmitted downstream. However it acquires an order unity increase in energy in the process. (Strictly, the adiabatic invariant $p_\perp^2/B$ is conserved.) A particle being accelerated at a curved shock front can therefore gain energy in steps $O(1)$ when the shock is locally perpendicular, and this allows particles to continue to increase their energy until the radius of curvature exceeds a few Larmor radii. This may allow the maximum energy accelerated to rise as high as 10^5 GeV. These ideas may also be of relevance to the so-called "barrel" supernova remnants discussed here by Caswell.

Shock Multiplicity: One issue of concern to cosmic ray physicists, which radio observations may elucidate, is the number of shock waves that accelerate a given cosmic ray before it leaves the galaxy. As discussed above, the difference in the primary and secondary cosmic ray spectra already implies that energetic cosmic rays are not continuously accelerated in the interstellar medium. However, low energy particles may interact several times with weak shocks, and there is some evidence from the detailed abundance ratios that this is actually occurring (Blandford and Ostriker 1980, Wandel this volume).

As has been discussed here many times, a supernova remnant is typically highly inhomogeneous and should contain many weak and some strong secondary shocks. It should be possible to use radio astronomical observations of nearby remnants to quantify this. Furthermore the incidence of weak shocks is strongly influenced by the maximum radius to which a remnant can expand before cooling, which again may be determined observationally. An additional complication, highlighted by McCray's talk here (cf. also Pineault, Landecker, and Routledge, 1987) is that many supernovae may explode in superbubbles of tenuous gas and may lead to efficient acceleration from poorly visible shock waves like that presumed to surround the Crab nebula.

Cosmic Ray Radiative Shocks: We are familiar with the idea of a radiative shock. When the gas is sufficiently dense, the post shock flow can radiate away its internal energy on the expansion timescale, through free-free and line emission. The gas will compress to maintain pressure equilibrium and therefore cool even faster, ending up in a dense shell at a temperature $\sim 10^4$ K where cooling is ineffective. Something similar can occur with cosmic rays. Suppose that we have a high Mach number shock that is able to accelerate cosmic rays with high efficiency. As we have discussed, the highest energy particles will be able to stream ahead of the shock if their Larmor radii are large enough. Furthermore, the pressure in the highest energy cosmic rays may account for most of the post-shock momentum flux because the relativistic particles can increase the overall shock compression to $r > 4$ and the slope of the high energy momentum space distribution function to $q < 4$. For this reason, particle acceleration may become a runaway process and high Mach number shocks may convert most of the incident bulk kinetic energy flux into a high energy cosmic ray precursor (Eichler 1984).

Now the evidence is against this actually occurring in the supernova shock waves that accelerate most of the Galactic cosmic rays, because the inferred source spectrum has a high energy slope with $q \sim 4.2$. However, it could be that most cosmic rays are accelerated by non-radiative, low Mach number shocks, and that only the highest Mach number shocks are cosmic ray radiative and the particles accelerated in these shocks are subsequently overtaken by the blast wave and ultimately lose energy in the expansion.

In this connection, it is of interest to speculate about the "mystery spot" recently discovered by speckle interferometry close to SN1987a. This feature has a luminosity of $\sim 3 \times 10^{40}$ erg s^{-1}, only ~ 15 times fainter than the supernova itself and also somewhat redder, which are both claimed to rule out a scattering explanation. Excepting the neutrinos, the major reservoir of energy is the expanding blast wave and so it is possible that a fraction of this can be converted into non-thermal emission via a cosmic ray precursor. If we take an extreme view then all of the energy flux in the frame of the shock will be converted into sufficiently energetic relativistic particles to escape. The mass flux in the wind is estimated to be $\dot{M}_w \sim 10^{-5} M_\odot \mathrm{yr}^{-1}$ from the observations of the non-thermal radio source three days after the explosion (Manchester, private communication), the wind speed is presumably $v_w \sim 500 \mathrm{km\ s^{-1}}$ and the shock speed is at least $v_s = 30,000 \mathrm{km\ s^{-1}}$. The total cosmic ray precursor luminosity would then be

$$L_{CR} \sim \frac{\dot{M}_w v_s^3}{2 v_w} \sim 2 \times 10^{41} \mathrm{ergs^{-1}} \tag{8}$$

roughly 7 times the luminosity of the spot (if it radiates isotropically). Avoidance of the Razin effect in the radio source requires that the field strength in the wind exceed $\sim 10^{-2}$ G and this implies that the escape energy is $\sim 3 \times 10^4 \mathrm{GeV}$, independent of radius.

We must account for the directionality of the emission. If the progenitor star is rotating, then both the mass flux and the magnetic field strength ought to be strongest at the equator. This means that the high energy particles (presumably protons, as electrons will cool rapidly by inverse Compton scattering the supernova light) should escape preferentially along the rotation axes. However, assuming that the streaming velocity is $\sim c/2$, consistent with the location of the spot, then Doppler beaming should enhance one polar jet relative to the other (and may also reduce the overall power requirements somewhat). We would then observe one jet and if most of the dissipation were at the end, as in the powerful extragalactic radio sources, the jet would appear to be a single spot on one side of the supernova (cf. Rees, 1987).

It is more difficult to account for the spectrum. The natural radiation process is synchrotron radiation by relativistic electrons accelerated at the end of the jet in a locally amplified magnetic field; electron energies $\sim 30 \mathrm{GeV}$ and a field strength $\sim 0.1 \mathrm{G}$ will suffice. However the absence of radio emission and the reported steepness of the optical spectrum, which is inconsistent with any pure synchrotron model (Phinney, private communication) are severe problems for this model.

Conclusions: The theory of particle acceleration at a shock front seems to be able to account for most of the observed features of galactic cosmic rays either qualitatively or semi-quantitatively. The spectrum, energetics and overall composition have natural and convincing explanations. The major difficulty with the theory is that it is difficult to account for the observed smoothness of the spectrum up to $\sim 10^5 \mathrm{GeV}$ when supernova shock waves find it difficult to accelerate beyond $\sim 10^4 \mathrm{GeV}$. Spacecraft observations of interplanetary shocks are providing invaluable empirical information on the mechanics of particle transport and wave generation and, in particular, verify that small scale, non-relativistic particle acceleration is actually occurring. Numerical simulations, now that they are starting to address the problems of quasi-parallel shocks, are equally encouraging and seem to verify the conjecture that the incident ions create scattering waves non-resonantly via the firehose instability and that these ions create scattering Alfvén waves upstream which backscatter them and inject them into the acceleration mechanism. The major uncertainty in the theory, and this unfortunately affects our ability to interpret the best diagnostics we have, *i.e.* the radio and X-ray observations, lies in quantifying the electron temperature and suprathermal particle injection rate in different types of shock.

What has become clear in recent years is that the problem of collisionless shock structure and the theory of cosmic ray origin can no longer be considered in isolation. What is equally true is that supernova remants occupy a pivotal position between the interplanetary shocks and the far more energetic activity associated with active galactic nuclei and that a reliable understanding of interstellar particle acceleration is a pre-requisite to unravelling the mysteries of quasars. It is hoped that future research will emphasise these linkages.

Acknowledgements: I thank D. Eichler, R. Jokipii, S. Phinney, S. Reynolds and especially C. Kennel for advice and helpful suggestions. I acknowledge support by the National Science Foundation under grant AST86-15325.

References

Achterberg, A, Blandford, R. D., and Periwal, V.,1984. *Astr. Astrophys.,* **132**, 97.

Achterberg, A.,1984.*Radiation in Plasmas,* , (ed. B. McNamara) World Scientific Publishing Co., Singapore. p3.

Achterberg, A., and Blandford, R. D.,1986. *Mon. Not. R. astr. Soc.,* **218**, 551.

Bell, A. R.,1979. *Mon. Not. R. astr. Soc.,* **182**, 443.

Bell, A. R.,1987. *Mon. Not. R. astr. Soc.,* **225**, 615.

Blandford, R. D., and Ostriker, J. P.,1978.*ApJLett,* **221**, L29.

Blandford, R. D. and Cowie, L. L.,1982. *Astrophys. J.,* **260**, 625.

Blandford, R. D.,1982.*Supernovae and Supernova Remnants,* , ed. Rees, M. J. and Stoneham, R., Dordrecht, Reidel, Holland.

Blandford, R. D. and Eichler, D.,1987.*Physics Reports,* , in press.

Braun, R., Gull, S. F. and Perley, R.,1987. *Nature,* **327**, 395.

Breneman, H., and Stone, E. C.,1985.*ApJLett,* **299**, L57.

Bridle, A. H. and Perley, R.,1984. *Ann. Rev. Astr. Astrophys.,* **22**, 319.

Cowsik, R., and Sarkar, S.,1980. *Mon. Not. R. astr. Soc.,* **191**, 855.

Drury, L. O'C. and Völk, H.,1981. *Astrophys. J.,* **248**, 344.

Drury, L. O'C,1983.*Rep. Prog. Phys.,* **46**, 973.

Edmiston, J. P., Kennel, C. F. and Eichler, D.,1982.*Geophys. Res. Lett.,* **9**, 531.

Eichler, D.,1979. *Astrophys. J.,* **229**, 419.

Eichler, D. and Hainebach, K.,1981.*Phys. Rev. Lett.,* **47**, 1560.

Eichler, D.,1984. *Astrophys. J.,* **277**, 429.

Eichler, D.,1985. *Astrophys. J.,* **294**, 40.

Ellison, D. C. and Eichler, D.,1984. *Astrophys. J.,* **286**, 691.

Ellison, D. C. and Möbius,1987. *Astrophys. J.,* , in press.

Falle, S.A.E.G., and Giddings, J. R.,1987. *Mon. Not. R. astr. Soc.,* **225**, 399.

Fedorenko, V. N., and Fleischman, G. D.,1987.*Sov. Astron. Lett.,* , in press.

Galeev, A. A., Sagdeev, R. Z., and Shapiro, V. D.,1986.*Proc. Workshop on Plasma Astrophysics,,* **ESA SP-251**, Nordwijk, Holland.

Heavens, A. F.,1984. *Mon. Not. R. astr. Soc.,* **210**, 813.

Helfand, D. J., and Becker, R. H.,1987. *Astrophys. J.,* , in press.

Jokipii, J. R., and Morfill, G.,1985. *Astrophys. J. (Letters)* , **290**, L1.

Jokipii, J. R.,1987. *Astrophys. J.,* **313**, 842.

Kennel, C. F., Sagdeev, R. Z.,1967.*J. Geophys. Res,* **72**, 3303.

Kennel, C. F. and Coroniti, F. V.,1984. *Astrophys. J.,* **283**, 710.

Kennel, C. F., Edmiston, M. and Hada, T.,1985.*J. Geophys. Res,* **90**, A1.

Kirk, J. G., and Schneider, P.,1987. *Astrophys. J.,* , in press.

Kovner, M. S.,1961.*Sov. Phys. JETP,* **13**, 369.

Lee, M. A.,1982.*J. Geophys. Res,* **87**, 5063.

Leroy, M. M., Winske, D., Goodrich, C. C., Wu, C .S., and Papadopoulos, K.,1982.*J. Geophys. Res,* **87**, 5081.

Mandt, M. E., and Kan, J. R.,1985.*J. Geophys. Res,* **90**, 115.

Meyer, J.-P.,1985. *Astrophys. J. Suppl.,* **57**, 173.

Parker, E. N.,1961.*J. Nuclear Energy,* **C2**, 146.

Pineault, S., Landecker, T. L., and Routledge, D.,1987. *Astrophys. J.,* **315**, 580.

Quest, K. B.,1987.*J. Geophys. Res,* , in press.

Rees, M. J.,1987. *Nature,* , in press.

Sagdeev, R. Z., Galeev, A. A., Shevchenko, V. and Shapiro, V.I.,1987.*preprint,* , .

Scholer, M.,1984.*Adv. Sp. Res.,* **4**, 419.

Simpson, J. A.,1983.*Ann. Rev. Nucl. Part. Sci.,* **33**, 323.

Zachary, A.,1987.*Unpublished thesis,* , Univ. of California, Berkeley.

SUPERNOVA REMNANTS AND THE ISM:
CONSTRAINTS FROM COSMIC-RAY ACCELERATION

Amri Wandel
Center of Space Science and Astrophysics
Stanford University, ERL
Stanford, CA 94305 USA

Abstract.   Supernova remnants can reaccelerate cosmic rays and modify their distribution during the cosmic ray propagation in the galaxy. Cosmic ray observations (in particular the boron-to-carbon data) strongly limit the permitted amount of reacceleration , which is used to set an upper limit on the expansion of supernova remnants , and a lower limit on the effective density of the ISM swept up by supernova shocks. The constraint depends on the theory of cosmic ray propagation: the standard Leaky Box model requires a high effective density, $> 1\text{cm}^{-3}$, and is probably inconsistent with the present picture of the ISM. Modifying the Leaky Box model to include a moderate amount of weak-shock reacceleration , a self consistent solution is found, where the effective density in this solution is ≈ 0.1 cm^{-3}, which implies efficient evaporation of the warm ISM component by young supernova remnants , during most of their supersonic expansion.

I. INTRODUCTION

If supernova remnants occupy a large fraction of the galaxy (McKee and Ostiker 1977, hereafter MO) additional acceleration by weaker shocks of supernova remnants , distributed over cosmic ray residence time in the galaxy is unavoidable. Detailed model calculations of cosmic ray acceleration by supernova shocks (Blandford and Ostriker 1980; Frannson and Epstein 1980) show that the resulting spectrum is consistent with the observed cosmic ray spectrum. Distributed acceleration cannot be a major source of cosmic ray energy, because it leads to a positive age-energy correlation, which is rejected by the observations. Even a relatively small amount of reacceleration can alter the cosmic ray spectrum significantly,in particular the energy dependence of the secondary-to-primary cosmic ray ratio (e.g. boron/carbon, hereafter B/C). It has been shown (e.g. Eichler 1980; Cowsik 1986) that observations of this ratio strongly constrain the acceptable amout of reacceleration . Wandel *et. al.* (1987, hereafter WELST) developed a model in which reacceleration of cosmic rays is treated self consistently, showing that a moderate amount of reacceleration by weak shocks (Mach number< 2) is consistent with cosmic ray observations. We show that the constraint on reacceleration derived from B/C ratio leads to constraints on the expansion of supernova remnants in the interstellar medium, and on the effective density of the ISM, felt by supernova remnants .

II. ACCELERATION OF COSMIC RAYS BY SUPERNOVA REMNANTS

Shock acceleration transforms a group of upstream particles with momentum p_0 into a (differential) power law distribution $J(p) \propto p^{-q}$ with an exponent (shock index)
$q = (2M^2 + 2)/(M^2 - 1)$, $\gamma = 5/3$ where M is the Mach number of the shock.

Consider supernova remnants with a shock index q, corresponding to a Mach number $M_q = u_q/c_s$, where u_q is the shock velocity and c_s is the sound speed in the interstellar medium. In the initial adiabatic expansion phase, supernova remnants (and consequently the probability to encounter them) are small, and their contribution to reacceleration is negligible. In the homologous expansion (Sedov-Taylor) phase the radius of the shock expanding in a uniform medium of density ρ, is given by (e.g. Spitzer 1968) $r = (2.1E/\rho)^{1/5}t^{2/5} = 12.7n^{-1/5}t_4^{2/5}$ pc, where $\rho = 1.3 m_H n$, $t = 10^4 t_4$ yr is the remnant's age, and E is the explosion energy; here and in the following we assume $E = 10^{51}$ erg (for different values our results scale with E in the same manner as they do with n^{-1}). In terms of the shock velocity u we have $r_q = 36\, n^{-1/3}u_{q,2}^{-2/3}$ pc, and $t_q = 1.3 \times 10^5\, n^{-1/3}u_{q,2}^{-5/3}$ yr, where $u_q = 10^2 u_{q,2}$ km s^{-1}. Expansion in a cloudy medium is modified by cloud evaporation (MO). In the early, evaporation dominated phase, the expansion behaves as $r \propto t^{3/5}$, due to the change in the effective density. However, as discussed in III, when evaporation is efficient, we may use the homogeneous medium expression, with a higher effective density. Finally, when the internal pressure drops to that of the ambient medium, the remnant stops expanding, reaching a radius (Shull 1987) $r_{snr} \approx 50 E_{51}^{11/35} n^{-13/35}$ pc.

In the standard Leaky Box model (e.g. Omes and Protheroe 1983) cosmic rays escape from the galaxy after having traversed a column of interstellar matter (path length), which at a few GeV/n has the mean value of $\sim 9 - 10$ g cm^{-2}. The residence time of cosmic rays in the galaxy estimated from the relative abundance of the unstable nucleus ^{10}Be, produced by spallation is $\tau_{cr} \sim 10^7$yr at a few GeV/n (Garcia Munoz et. al. 1981).

In the presence of a reacceleration process which transforms a given initial distribution into a power law distribution, p^{-q}, the steady state differential distribution of cosmic rays can be described by the equation (cf. WELST)

$$J_0(p) - (R+S)J(p) = B\left[J(p) - (q-1)\int_{p_0}^{p} \frac{dp'}{p'}\left(\frac{p'}{p}\right)^q J(p')\right]. \tag{1}$$

where R, B, and S are the rates [1] of escape, reacceleration and spallation, respectively, J_0 is the cosmic ray source spectrum, and p_0 is a cutoff due to ionization losses at low energies. The secondary distribution satisfies a similar equation, with the source term replaced by the primary distribution.

The B/C data from the HEAO-3 experiment (Engelmann et. al. 1983), strongly limit the allowed amount of reacceleration . In the frame of the standard leaky box (SLB) model ($R_0 = 0.11$g^{-1}cm^2), this limit is very stringent (eq. [2.b]). If reacceleration is taken into account self consistently, WELST find a less stringent constraint. Their best fit of the B/C data compared to the constraint from the SLB with the same shock index gives (see Fig. 1)

$$B \approx (0.2 \pm 0.05)\ \text{g}^{-1}\text{cm}^2 \qquad R_0 = 0.2;\ q = 4\ \ (WELST), \tag{2.a}$$

$$B < 0.03\ \text{g}^{-1}\text{cm}^2 \qquad R_0 = 0.11;\ q = 4\ \ (SLB). \tag{2.b}$$

At the energies under consideration the Larmor radius and the mean free path to pitch angle scattering, λ, of cosmic rays are small compared to the scale height of the galactic disk, h_d, and the diffusion approximation may be used. Over a time $t \gg \lambda/c$, an average cosmic ray covers a volume of radius $r_{cr}(t) \approx (Dt)^{1/2}$, where $D = \lambda c/3$ is the diffusion coefficient. During their lifetime in the galaxy cosmic rays at a few GeV/n travel through a column of R_0^{-1}, which, assuming a mean density $n_d = 1$cm^{-3} implies a residence time in the galactic disk of $\tau_d = (1.3\, m_H n_d R_0 c)^{-1} \approx 5 \times 10^5 R_0^{-1}$yr. An average cosmic ray propagates a linear distance of $d_{cr} = r_{cr}(\tau_{cr})$ covering a portion of the disk of volume $V_{cr} = 2\pi h_d d_{cr}^2$. Assuming supernovae are distributed uniformly in the galactic disk [2] the number of supernovae within this volume is $N_{sn} = \pi d_{cr}^2 \tau_d \sigma$, where $\sigma = 10^{-10}\sigma_{-10}$ supernovae pc^{-2}yr^{-1} is the surface rate of supernovae in the galaxy ($\sigma_{-10} = 0.6$ corresponds to a total rate of one supernova every 20 years within a disk radius of 15kpc).

---

[1] These rates are expressed in inverse path length (gram^{-1}); in particular B is the inverse path-length between shocks, so that B/R is the average number of shocks encountered during a cosmic ray residence in the galaxy.

[2] Type II supernovae are clustered in space and in time, but the clustering effect may increase the reacceleration rate over the rate produced by a uniform distribution, since presumably cosmic rays are initially accelerated by supernova shocks, and the clustering effect increases the probability to encounter a shock after leaving the parent supernova .

A cosmic ray particle will encounter a supernova remnant of radius r_q within a time t_q if it happens to be within a sphere of radius $r_q + r_{cr}(t_q)$ about the supernova . Since cosmic rays are assumed to be homogeneously distributed in the volume V_{cr}, the probability for such an event is given by the ratio between the volume defined by this sphere and V_{cr}, $P_{cr}(q) = 2[r_q + r_{cr}(t_q)]^3/3d_{cr}^2 h_d$. The average number of shocks encountered by a cosmic ray particle during its residence time in the disk is given by $N_{sn}P_{cr}$. From the definition of the reacceleration and escape rates, this number is also given by the ratio B/R. Comparing the two expressions we have

$$B(q) = N_{sn}P_{cr}(q)R_0 = 1.5\sigma_{-10}h_{d,2}^{-1}r_{q,2}^3\left[1 + \frac{d_{cr}}{r_q}\left(\frac{t_q}{\tau_{cr}}\right)^{1/2}\right]^3$$

$$= 0.07\ \sigma_{-10}h_{d,2}^{-1}n^{-1}C_q^3 u_{q,2}^{-2}\ , \tag{3}$$

where $C_q = 1 + 0.3d_{cr,2}\tau_{cr,7}^{-1/2}(nr_{q,2})^{1/4} = 1 + 0.3d_{cr,2}\tau_{cr,7}^{-1/2}n^{1/6}u_{q,2}^{-1/6}$, B is in units of $g^{-1}cm^2$, and subindices denote obvious units. (As P_{cr} is strongly peaked towards higher values of q, we have assumed the reacceleration due to supernova remnants with $q' \le q$ is dominated by the largest ones, having $q' \approx q$).

III. IMPLICATIONS

Eqs. (2) and (3) yield a direct constraint on the expansion of supernova remnants . One can, however, use this result to constrain the effective density felt by the expanding remnant. The magnetosonic sound speed in the hot component [3] (density n_0) of the ISM ($P/k = 3 \times 10^3$ and a magnetic field of $3\mu G$) is $c_s = (\gamma P_{gas}/\rho + v_A^2)^{1/2} \approx 150(n_0/3 \times 10^{-3}cm^{-3})^{-1/2}km\ s^{-1}$. A shock index of $q = 4$ ($M = 1.7$) then corresponds to $u = 250km\ s^{-1}$.

If the effective density were close to the density of the hot component, the reacceleration rate is much too large; e.g. for $n = 3 \times 10^{-3}cm^{-3}$ eq. (3) yields $B > 2\ g^{-1}cm^2$, which, as seen from Fig. 1, grossly violates the B/C data. However, cloud evaporation in the early expansion phases gives a higher effective density. Solving for n eq. (3) gives

$$\frac{n}{(1 + 0.8n^{1/6})^3} \approx 0.015\ B^{-1}c_{s,2}^{-2}, \tag{4}$$

where $c_{s,2}$ is in units of $100km\ s^{-1}$ and we have taken $\sigma_{-10} = 0.6$, $h_{d,2} = \tau_{cr,7} = 1$ and $d_{cr} = 300pc$. (Note that eqs. (3) and (4) are only weakly dependent on the parameters of the cosmic ray model, τ_{cr} and d_{cr}). Combining this with the constraints on the reacceleration rate derived from the B/C data, eqs. (6.a,b), we get $n \approx 0.1cm^{-3}$ (WELST) and $n > 0.8cm^{-3}$ (SLB). The respective limits on the radius are $r(q = 4) \approx 40pc$ and $r < 20pc$.

Assuming once more a uniform supernova distribution, the porosity of hot supernova cavities in the ISM is

$$Q = \left(\frac{\sigma}{2h_{sn}}\right)\left(\frac{4\pi}{3}r_{snr}^3\right)\left(\frac{r_{snr}}{v_{rms}}\right) \approx 0.2\ E_{51}^{1.26}n^{-1.5} \tag{5}$$

where h_{sn}(taken here as 300pc) is the scale height of the supernova distribution and $v_{rms} \approx 10km\ s^{-1}$ is the random velocity of the interstellar gas. The filling factor of hot coronal gas is $f \approx Q/(1+Q)$. For an effective density $n \approx 0.1$ (WELST), f is close to unity and our choice of the sound speed is justified. However, in the SLB scenario $f < 0.25$, and the shock is effectively expanding in a much denser medium ($n_0 \approx 0.1cm^{-3}$), corresponding to a lower sound speed, which gives a yet higher limit on n (Wandel 1987). However, since the effective density cannot be larger than the average density in the disk, $1cm^{-3}$, this case may probably be ruled out, making the SLB inconsistent

---

[3] The density of the hot phase determines the sound speed as long as its filling factor is of order unity; the effective density, n, may however be much higher.

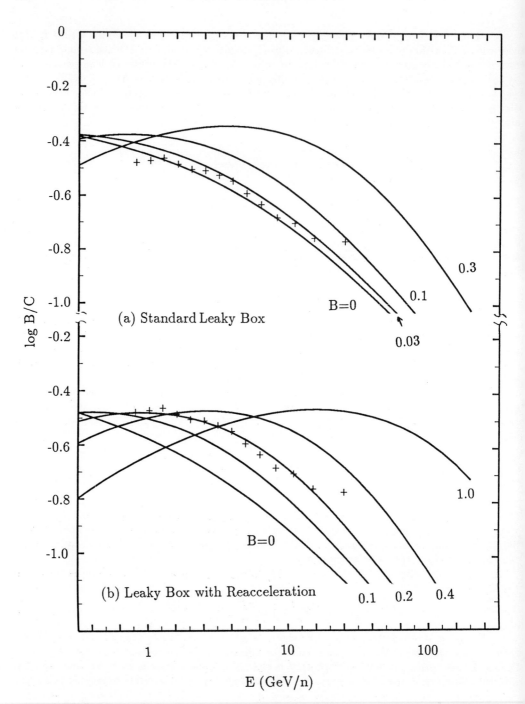

Fig 1 (a) The secondary-to-primary ratio (normalized to the Boron/carbon ratio) calculated in the standard Leaky Box model (a), ($R_0 = 0.11$, $\alpha = 0.6$ and various amounts of reacceleration ($B=0$, 0.03, 0.1, 0.3), and in the modified Leaky Box model (WELST) (b), with an increased escape rate, $R_0 = 0.2$ and shock acceleration index $q = 4$. Curves correspond to no reacceleration ($B = 0$), $B=0.1, 0.2, 0.4$ and 1. Data points are from Engelmann (1983).

with reacceleration by supernova remnants . In the MO model the early adiabatic expansion of supernova remnants is dominated by evaporation of clouds inside the remnant, rather than by sweeping up of the hot intercloud medium by the shock front. The expansion in this phase is given by $r \approx 12(E_{51}/n_h)^{-1/5}t_4^{2/5}$ pc, where n_h is the gas density inside the remnant. As long as the evaporation of clouds is efficient, n_h is much higher than the density of the hot intercloud medium, n_0. At later times the evaporation time becomes larger than the expansion time and the density in the gas inside the remnant approaches n_0. The transition is given by $n_h/n_0 = 1 + (r/r_{es})^{-5/3}$, and occurs at the characteristic radius $r_{es} \approx 10\ (E_{51}/a_{pc})^{2/5}f_{cl}^{1/5}n_0^{-3/5}$ pc, where a_{pc} and f_{cl} are the clouds' radius (in parsec) and filling factor, respectively. As long as the evaporation is fast enough to ensure $n_h \approx n_{av}$, we have the familiar homologous expansion with an effective density $n = n_{av}$ which is the approximation used in this work). This is the case for

$$r < r_{es}(n/n_0)^{-3/5} \approx 10\ (E_{51}/a_{pc})^{2/5}f_{cl}^{1/5}n^{-3/5}\ \text{pc}. \tag{6}$$

When n_h drops, the expansion proceeds at a faster rate, $r \propto t^\eta$, $\eta = 3/5$. Numerical calculations (Cowie, McKee and Ostriker 1981) show that the expansion is actually intermediate between these two extremes.

For $n = 0.1\text{cm}^{-3}$ and warm clouds, eq. (6) gives $r < 30$pc, which is roughly consistent with the constraint $(r(q = 4) \approx 40$pc) derived from the reacceleration argument in the WELST model.

The SLB model, however, requires $n > 1\text{cm}^{-3}$, which must invoke evaporation of the cold clouds too, for which eq. (11) yields $r < 6$pc. Since this is much less than the reacceleration constraint $(\sim 20$pc), evaporation cannot keep the required high effective density long enough, and supernova remnants will expand too fast, producing too much reacceleration .

REFERENCES

Axford, W.I. 1981, 17th Int. Conf. Cosmic Rays (Paris).

Blandford,R.D.and Ostriker, J.P. 1980, *Ap. J.*, **237**, 793.

Cowie, L.L, McKee, C.F., and Ostriker, J.P. 1981, *Ap. J.*, **247**, 908.

Cowsik, R. 1986, *Astr. Ap.*, **155**, 344.

Eichler, D.S. 1980, *Ap. J.*, **237**, 809.

Engelmann , J.J., *et. al.* 1983, *Proc. 18th Internat. Cosmic Ray Conf.* (Bangalore), **2**, 17.

Fransson, C. and Epstein, R.I. 1980, *Ap. J.*, **242**, 411.

Garcia-Munoz, M., Simpson, J.A. and Wefel, J.P. 1981, *Proc. 17th Internat. Cosmic Ray Conf.* (Paris) **2**, 72.

McKee, C.F., and Ostriker, J.P. 1977 (MO), *Ap. J.*, **218**,, 148 .

Ormes, J. and Protheroe, R.J. 1983,*Proc. 18th ICRC* (Bangalore), **2**, 221.

Shull, J.M. 1987, in *Interstellar Processes*, ed. D. Hollenbach and H. Thornson, Reidel.

Spitzer, L. 1968, in *Diffuse Matter in Space* p.199, John Wiley, New York.

Wandel, A., Eichler, D.S., Letaw, J.R., Silberberg, R., and Tsao, C.H. 1987 (WELST), *Ap. J.*, **316**, 676.

Wandel, A. 1987, in preparation.

SMALL-SCALE STRUCTURE IN YOUNG SUPERNOVA REMNANTS

Stephen P. Reynolds
Department of Physics, North Carolina State University
Raleigh, NC 27695-8202 USA

Abstract: Recent VLA observations of the shell supernova remnant SN 1006 AD (Reynolds and Gilmore 1986) and the Crablike remnant 3C 58 (SN 1181 AD?; Reynolds and Aller 1987) show features at high resolution that contain information on details of particle acceleration and transport in the remnants. Thin arcs at the edge of SN 1006 require time-variable particle acceleration and/or magnetic field amplification. Filaments in 3C 58 probably result from interaction of pulsar-generated relativistic fluid with filaments of thermal gas formed early in the remnant's life by cooling or dynamical instabilities. Their sharp edges imply efficient scattering by Alfvén waves; as much as 1% of the large-scale magnetic energy density may be in magnetic turbulence on length scales of 10^{11} cm.

Introduction. The wealth of information present in a high-resolution radio image of a supernova remnant can be translated into quantitative information about the behavior of the relativistic electrons and magnetic field responsible for the emission. The observation of structure on scales of order 0.1 pc or less begins to constrain the diffusive properties of the post-shock turbulent medium. Very small-scale features place limits on the diffusion coefficient and effective mean free path of electrons; from this information, inferences can be drawn about the degree of disorder of the magnetic field. The origin of such features can pose significant problems for simple pictures of shock acceleration of electrons. In Crablike supernova remnants, the detailed nature of the nonthermal filamentation that dominates the appearance of 3C 58 and the Crab Nebula itself can provide clues to the nature and origin of the filaments; these questions bear on the evolutionary history of the remnant as well as on the characteristics of the turbulent plasma.

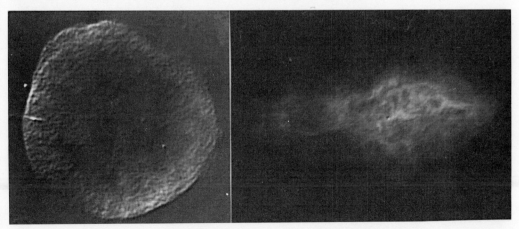

Fig. 1(a): 1.4 GHz image of SN 1006.   Fig. 1(b): 1.4 GHz image of 3C 58.

SN 1006. Figure 1(a) shows the 1.4 GHz total-intensity image of SN 1006 (Reynolds and Gilmore 1986; RG) with a resolution of $16'' \times 20''$. Contrast is set to enhance bright structure. The edge of the remnant is very sharp over much of its circumference; profiles show that it is unresolved. This suggests that the edge of the radio emission marks the blast wave, not the unstable contact discontinuity between shocked interstellar medium and supernova ejecta, since in the latter case one would expect a highly irregular appearance such as that of Cas A. However, the sharp edge in the eastern and southern parts of the remnant appears to be a narrow filament, dropping off as sharply behind (toward the remnant center) as in front. The eastern filament describes an arc of very nearly constant curvature and can be traced over almost $90°$. The southern edge appears to show two parallel linear structures, again unresolved in the radial direction. These narrow filaments raise significant difficulties in modeling.

I assume a distance of 1.7 kpc to SN 1006, a recent upward revision (Kirshner, Winkler, and Chevalier 1987). Then $20''$ corresponds to 0.17 pc. The profiles of RG indicate that the first minimum behind the eastern and southern filaments is at a level of about 70% of peak. This decrease over less than two beams or 0.3 pc cannot result from any smooth, steady, spherically symmetric process unless unacceptably huge gradients are invoked. To see this, assume that SN 1006 is in the adiabatic or Sedov phase of evolution, as indicated by the proper motion of the optical filament (Hesser and van den Bergh 1981). Then a parcel of gas currently at a relative position $\rho = r/r_{\text{shock}}$ was shocked at a time and radius calculable from the Sedov relations (Sedov 1959). The observed filament width of less than $40''$ out of a total radius of $15'$ provides a conservative lower limit to the value of ρ beyond which the emissivity must be larger, since any curvature would increase the projected width of this region. For $\rho = 0.96$, the Sedov relations imply that the material was shocked some 300 yr ago; for $\rho = 0.98$, the material was shocked less than 170 yr ago. Thus if the particle acceleration and magnetic field amplification/compression efficiencies have been constant over more than the last 300 yr, a wider feature than observed would be produced.

A quantitative estimate of the required gradients can be made assuming that very near the edge at least, the intensity varies as a power of radius. Then the rise in intensity I_ν by 1.4 over a fractional increase in projected radius ρ of 4% or less requires that I_ν rise as at least ρ^8. Of course, this is completely inconsistent with the *average* profile of emission dropping reasonably toward the center; a crude average from one of the RG profiles suggests an increase more slow than ρ^3. Since the lines of sight corresponding to the filament peak and the adjacent interior minimum are so close together, the paths through the remnant are not greatly different, and the ratio in I_ν is roughly a ratio in mean emissivities. Thus the increase of a factor of about 1.4 would require an increase of the electron energy density by about the same factor, or an increase in the magnetic field strength by a factor 1.25 (since $j_\nu \propto B^{1-\alpha}$ and $\alpha \cong -0.6$ for SN 1006; Milne 1971). Now this must happen over a radial interval of less than 0.3 pc, but coherently over almost a quarter of the remnant circumference. The hint of two parallel filaments along the southern edge suggests that whatever might create this "burst" of emissivity might act repeatedly.

Another possibility exists for altering the synchrotron emissivity: the orientation of the mean magnetic field may change systematically with location. The simplest example is an exactly radial field. In order to investigate this effect quantitatively, I calculated theoretical profiles for such a geometry under several assumptions about the radius dependence of the magnetic field strength and relativistic particle density. If the latter quantities are roughly constant, the projection of the radial field produces a profile that peaks at 71% of the total radius, with a relative width at 70% of peak of 55% of the total radius. Adding a steep power-law increase to the magnetic field narrows the peak and moves it outwards, but even with the field growing roughly as ρ^4, the width at 70% of peak is still 23%. This kind of simple geometry cannot account for the filaments; only a perturbation extensive in the tangential direction but very narrow in the radial, changing the field orientation by several tens of degrees, could produce the observed structures.

RG proposed that the filaments represented sheets with properties like those of the band across the remnant interior, but seen edge-on. While this is possible, a simple restriction of emissivity to a sector (an "orange slice" roughly perpendicular to the line of sight) cannot produce a narrow enough feature. Profiles of a constant-emissivity, radial field model so restricted have 70% widths of 21% even for a sector with opening angle only 20°. The emissivity must therefore decrease radially behind the edge, at a rate too great to have been constant over the remnant lifetime. The particle acceleration or magnetic-field enhancement rate must vary sharply with time.

Further information on the nature of the turbulent medium itself may be derived from the observation that the features remain narrow, implying that the diffusion of radio-emitting electrons is not rapid. Assume that the emissivity increase that produces the edge filaments was instantaneous. Then the observed limit on the filament widths is a limit on the diffusion coefficient for relativistic electrons. A one-dimensional delta-function of particles in a medium of diffusion coefficient $D = v\lambda/3$, where λ is the mean free path and v the particle velocity, will diffuse in time t into a Gaussian profile with dispersion $\sigma = (2Dt)^{1/2}$. Assuming that the observed width ℓ of the SN 1006 filaments is entirely due to electron diffusion from a plane of zero thickness, one obtains $\lambda < (3/2)(\ell^2/ct)$ or $\lambda < 1.3 \times 10^{16}(\ell/0.3 \text{ pc})^2(t/100 \text{ yr})^{-1}$ cm. While not yet a very strong limit, this length scale is already smaller than that on which the magnetic field can be completely tangled, or no significant polarization at all would be observed. If this scattering is due to resonant Alfvén waves with energy density R times that in the large-scale magnetic field, the mean free path is roughly $(2/\pi)r_L R^{-1}$ where r_L is the electron Larmor radius (Wentzel 1974). For radio-emitting electrons in SN 1006, $r_L \sim 10^{11}$ cm in the equipartition magnetic field of 3×10^{-5} Gauss, implying that at least 10^{-5} of the large-scale magnetic field energy is in magnetic turbulence on length scales of about 10^{11} cm. Higher-resolution observations could substantially improve this limit.

3C 58. The filamentary structure in 3C 58 must represent a totally different physical phenomenon. The bright filaments occur predominantly in the remnant interior. The relativistic particles and magnetic field responsible for the synchrotron emission presumably come from a central pulsar according to the conventional assumption for Crablike supernova remnants. While thermal material probably dominates the energy density in shell supernova remnants, the amount of thermal material in 3C 58 is unknown, and possibly quite small, as judged from the faintness of optical features (van den Bergh 1978).

The VLA image of Figure 1(b) (Reynolds and Aller 1987: 2″ resolution at 1.4 GHz), emphasizes the bright filamentary skeleton; the faint envelope is less apparent at this contrast. A few conspicuously long, narrow filaments can be seen, but most are fairly broad (widths 10″ − 30″). However, the edges of filaments are sometimes unresolved and frequently narrower than 10″ from valley to peak. Filament contrasts are comparable to those for the edge filaments of SN 1006; factors of 1.5 to 2 are typical for the ratio of a peak to the adjacent valley. If filaments are cylindrical and in the plane of the sky, this implies an emissivity increase of a factor of 30 to 40 in a typical filament.

While the question of the origin of filaments in SN 1006 drew on the physics of shock acceleration, the origin of 3C 58's filaments must be looked for elsewhere. Spectral-index maps between 1.4 and 4.9 GHz (Reynolds and Aller 1987) show that there is no appreciable difference between the spectral index of bright filaments and that of the remnant mean. Thus filaments are neither locally shock-accelerated particles, nor compressed ambient cosmic-ray electrons; both sources would imply spectral indices $\alpha \lesssim -0.4$ $(S_\nu \propto \nu^\alpha)$. The time-scales on which the pulsar-generated relativistic fluid could penetrate the surrounding thermal supernova ejecta are enormously long, so the thermal and relativistic gases are not coextensive. Cooling or dynamical instabilities acting on the thermal gas are the most likely mechanism to produce thermal filaments, but such mechanisms must act early while the shock driven by the injected relativistic fluid is slow and post-shock cooling times are short (Reynolds 1987). The nonthermal filaments then result from pulsar-injected material interacting with the thermal obstructions, creating nonthermal sheaths. The difficulty with this picture remains the faintness of optical emission, though the thermal gas required for this nucleation picture (roughly $0.2M_\odot$, a tenth of that in the Crab Nebula) might, with some extinction, be difficult to detect.

Independently of the origin question, as with SN 1006 the observed sharpness of features constrains particle diffusion in 3C 58. For a distance to 3C 58 of 2.6 kpc (Green and Gull 1982), $2''$ corresponds to 8×10^{16} cm. The relations applied to SN 1006 above then give a limit on the mean free path of $\lambda < 1.2 \times 10^{13}(\ell/8 \times 10^{16}$ cm$)^2(t/806$ yr$)^{-1}$ cm in order that unresolved filament edges not be shallower than observed. For the equipartition field of 6×10^{-5} Gauss, the Larmor radii of radio-emitting electrons are of order 10^{11} cm; arguments like those presented above for SN 1006 imply that almost 1% of the large-scale magnetic field energy density is present in magnetic turbulence on the scale of r_L. This surprisingly strong result requires that the filaments be nearly as old as the remnant, as suggested above; if they were formed much later, or are transitory phenomena, the limit is weaker.

For both SN 1006 and 3C 58, radio polarimetry holds substantial promise for investigating the nature of particle diffusion. One should be able to learn about the degree of small-scale order in the field, and the geometry of large-scale order. Detailed theoretical modeling of profiles of features should further constrain the scattering lengths in the gas and, in SN 1006, can improve our understanding of shock acceleration.

REFERENCES

Green, D. A., and Gull, S. F. 1982, *Nature*, **299**, 606.

Hesser, J. E., and van den Bergh, S. 1981, *Ap. J.*, **251**, 549.

Kirshner, R. P., Winkler, P. F., and Chevalier, R. A. 1987, *Ap. J. (Letters)*, **315**, L135.

Milne, D. K. 1971, *Aust. J. Phys.*, **24**, 757.

Reynolds, S. P. 1987, *Ap. J.*, submitted.

Reynolds, S. P., and Aller, H. D. 1987, *Ap. J.*, submitted.

Reynolds, S. P., and Gilmore, D. M. 1986, *A.J.*, **92**, 1138 (RG).

Sedov, L. I. 1959, *Similarity and Dimensional Methods in Mechanics* (New York: Academic), p. 219.

van den Bergh, S. 1978, *Ap. J. (Letters)*, **220**, L9.

Wentzel, D. G. 1974, *Ann. Rev. Astr. Ap.*, **12**, 71.

A SEARCH FOR SHELLS AROUND CRABS

R.H. Becker, Physics Department
University of California, Davis
and Institute of Geophyiscs and Planetary Physics, LLNL

D.J. Helfand, Astronomy Department
Columbia University, New York, NY

Abstract

Prior to conducting a survey of the galactic plane at 327 MHz using the VLA[1], we have imaged four fields near galactic longitude of 20 degrees. Each image will cover a 2.5 degree field with ~1 arcmin resolution. The fields have been chosen to include the remnants G20.0-0.2, G21.5-0.9, and G24.7+0.6. The first two are isolated Crab-like objects, that is, there is no discernible associated shell. Since such shells have relatively steep spectra, images at 327 MHz will be more sensitive to their presence. The absence of a shell can constrain the density of the ISM in the vicinity of the SNR (Reynolds and Aller 1985 A.J. 90, 2312). Since ~50% of Crabs are naked, the implications can be extended to a significant fraction of the ISM.

Introduction

Several hundred years after exploding most supernova remnants(SNR) are observable as a result of the interaction between the expanding shell of ejecta and the interstellar medium (ISM). They appear as coincident shells of nonthermal radio emission and thermal x-ray emission. In stark contrast to this, the Crab Nebula is visible as a result of the acceleration of relativistic electrons (and positrons) by the Crab Pulsar, appearing as a source of nonthermal radio and x-ray emission with a filled-center brightness distribution. Over the last 10 years, approximately 15 additional sources similar to the Crab Nebula (Crab-like remnants) have been identified, representing ~10% of the known SNR (Helfand and Becker 1987). Of these Crab-like remnants, half show evidence of a surrounding shell and half do not. The absence of a shell in these sources is something of an embarrassment if you believe pulsars form in Type II SN which presumably yield a high mass of ejecta. Alternatively, the lack of a shell could be indicative of the local environment around the supernova. In this paper we present new 327 MHz radio observations of two Crab-like remnants, G21.5 - 0.9 and G20.0 - 0.2, and place upper limits on the presence of a surrounding radio shell. In addition, we place severe limits on any x-ray shell surrounding G21.5 - 0.9 using data from the Einstein Observatory.

[1]The National Radio Astronomy Observatory is operated by Associated Universities, Inc. under contract with the NSF.

Observations
        The VLA is in the process of installing 327 MHz receivers on
all of its telescopes. In early 1987, we carried out a pilot program
for an eventual galactic plane survey which covered four 2.5 degree
fields near l^{II} = 20°. At that time 13 telescopes were available at 327
MHz. Observations were taken in C and D arrays resulting in images with
~1 arcmin resolution. Both sets of observations were made in spectral
line mode utilizing 8 channels (7 narrow-band channels each ~400 KHz
wide and 1 broad band channel). This mode is useful for isolating
narrow-band interference. The seven narrow-band channels were
concatenated together and treated as a single data base from which
images were made. Images were obtained by first taking an FFT of the
visibility data and then by using a maximum entropy routine in AIPS
(VTESS) to deconvolve the array response function from the image.
        Two of the fields were centered near G21.5 - 0.9 and G20.0 -
0.2 respectively and the remnants were clearly seen in the images.
Generally the images contained 3-5 additional strong (>1 Jy) sources as
well as many weaker sources. The observed 327 MHz radio fluxes of G21.5
- 0.9 and G20.0 - 0.2 were measured to be 6.8 and 10.7 Jy respectively,
in reasonable agreement with their extrapolated spectra. The rms noise
in both maps is ~0.015 Jy/25 arcsec pixel. We concluded that a shell
around either source of surface brightness equal to or greater than the
rms noise would have been discernible in the images. No evidence for a
shell was apparent in either image and therefore we place an upper
limit of 7 x 10^{-21} Wm^{-2} Hz^{-1} Sr^{-1} at 327 MHz for shell emission around
G21.5 - 0.9 and G20.0 -0.2 out to a radius of 30 arcmin at which point
confusion from neighboring sources becomes a problem.
        We also re-examined the Einstein IPC (Giacconi et al. 1979)
observation of G21.5 - 0.9 which lasted 4700 s. This data, originally
published by Becker and Szymkowiak (1981), and taken in conjunction with
observations by the HRI and SSS, revealed that the x-ray emission from
G21.5 - 0.9 was nonthermal in origin as expected for a Crab-like SNR.
The IPC image can be used to search for any thermal x-rays from a
surrounding shell. In fact, there is no indication of an excess of
counts between 5-15 arcmin where shadowing from the support structures
becomes important. We conclude that the x-ray luminosity of a shell
around G21.5 - 0.9 in the IPC band (0.5 - 4 keV) is less than 1 x 10^{27} W
assuming a distance of ~5 kpc (Davalaar, Smith, and Becker 1986).

Discussion
        The VLA and Einstein observations of G21.5 - 0.9 and G20.0 -
0.2 serve to reinforce the dichotomy between Crab-like objects with
shells and those without shells. Our results set limits comparable to
these recently obtained for 3C58 (Green 1986). As Green pointed out for
3C58, the limits on x-ray luminosity are more compelling, being a full
two orders of magnitude lower than the measured values for Kepler,
Tycho, and SN1006. Green understated the importance of the limits on
the radio brightness of any shell emission by comparing the limits to
the average surface brightness of SN1006, the faintest young remnant
shell, and finding them comparable. But in reality, the shell, by its
very nature, will be enhanced by a factor of four or more over the

average surface brightness. Therefore, the limit on a radio shell
surrounding 3C58 is at least 6 times smaller than the observed shell in
SN1006, while limits for G21.5 - 0.9 and G20.0 - 0.2 are at least 3
times smaller.

Becker (1987) recently reviewed the current status of Crab-
like SNR and listed 15 galactic SNR split 8 to 7 between "naked" Crabs
and composite remnants respectively. That is to say that approximately
half of the galactic remnants known to have formed active pulsars show
no indication of an expanding shell of ejecta. Either the original SN
failed to eject a significant amount of material or the surrounding ISM
is of so low a density that the shock front is invisible.

It is of interest to speculate on the true ratio of "naked"
Crabs to composite SNR in our galaxy in so far as the selection effects
for the two classes are totally different. The composite SNR are
usually discovered in low frequency surveys because of their steep
spectra shells while "naked" Crabs are usually found by their absence of
recombination lines. As noted by Helfand and Becker (1987) "five of the
eight composite remnants first catalogued as a result of their bright
shells, have radio core components less luminous than the lowest
luminosity Crab-like objects known." Until recombination line surveys
(or polarization surveys) are extended to much weaker objects, many
"naked" Crabs will remain undiscovered, suggesting that the true ratio of
"naked" Crabs to composite is greater than the current observational
result of unity.

Alternatively, this ratio could be misleading if the Crab-like
cores in composite remnants are short-lived in comparison to the
shells. To test the hypothesis we compared the linear diameter of
composite shells to the diameters of all other remnants using size
estimates from Milne (1979) and Helfand and Becker (1987). The
comparison suffers from the small number of known composite remnants.
In any case, there are four composites with diameters of 20 - 30 pc and
one between 30 - 40 pc (Vela) suggesting that if the cores fade sooner
than their associated shells, it's not a lot sooner.

Becker (1987) and Helfand and Becker (1987) made a number of
other comparisons between "naked" Crabs, composites, and shell remnants
and also concluded that no clear distinctions existed between the class
of shell remnants and the shell components of composite remnants or
between "naked" Crabs and the Crab cores of composite remnants.

If the majority of SN which produce active pulsars do not
produce observable shells, the explanation is far from obvious. We
generally assume that pulsars are formed in Type II events from high
mass progenitors. These stars have the ability to alter their
environment through a strong stellar wind so that the SN, when it
occurs, will be within a stellar wind bubble (McKee, Van Buren, and
Lazareff 1984). In an extreme case of a Wolf-Rayet star which has a
mass loss rate of ~2 x 10^{-5} Mo yr^{-1}, a low density bubble could extend
out to ~30 pc resulting in a very weak shell. Of course, Cas A offers
one counterexample where a presumed massive star created a strong shell
and no observable evidence for a pulsar. This is particularly
interesting in light of the conclusion of Fesen and Becker (1987) that
the progenitor of Cas A was a Wolf-Rayet star.

In conclusion, the dichotomy between composite remnants and "naked" Crabs remains one of the two important questions to be addressed in understanding Crab-like remnants; the other being the seemingly low percentage of remnants which appear to exhibit Crab-like properties.

REFERENCES

Becker, R.H. and Szymkowiak, A.E. 1981 Ap.J. (Letters) 248,L23.
Becker, R.H. 1987 IAU Symp. 125 on the Origin and Evolution of
    Neutron Stars ed.   D.J. Helfand, in press.
Davelaar, J.A., Smith, A., and Becker, R.H. 1986 Ap.J. (Letters)
    300,L59.
Fesen, R.A., Becker, R.H., and Blair, U.P. 1987, Ap.J. 313, 378.
Giacconi, R. et. al. 1979 Ap.J. 230,540.
Green, D.A. 1986 MNRAS 218, 533.
Helfand, D.J. and Becker, R.H. 1987 Ap.J. 314,203.
McKee, C.F., Van Buren, D., and Lazareff, B. 1984, Ap.J. (Letters)
    278, L115.
Milne, D.K. 1979 Aust J. Phys. 32, 83.

A SHELL SNR ASSOCIATED WITH PSR1930+22?

D. Routledge and J.F. Vaneldik
University of Alberta, Edmonton, Canada

Abstract: Observations at 408 and 1420 MHz are reported which show a large (1.7 degree diameter) faint shell of nonthermal emission partially surrounding PSR1930+22. This may be the SNR associated with the pulsar, whose spin-down age is only 40000 years.

The lifetimes of radio-detectable SNRs are ~100 times shorter than those of rotationally-powered pulsars, making it difficult to confirm their association. The best candidate in the northern hemisphere is PSR1930+22, whose spin-down age $P/2\dot{P}$ is 40000 years[1]. In 1980, a sensitive 610 MHz synthesis survey[2] failed to reveal any SNR in the vicinity. In 1982, however, single-dish observations[3] at 1.4 GHz and 2.3 GHz showed what appeared to be a plerion of ~40' diameter near the pulsar position. We now report new higher resolution observations at 408 and 1420 MHz which show a much larger faint shell of nonthermal emission which partially surrounds the pulsar. This may be the SNR associated with PSR1930+22.

The observations were performed with the Synthesis Telescope of the Dominion Radio Astrophysical Observatory (DRAO)[4] [5]. Figure 1 shows the CLEANed maps. At 1420 MHz the u-v coverage has been restricted to produce the same resolution (3.5' x 9.2') in both maps for direct comparison of extended emission.

In each map, low-latitude galactic emission appears as a band of patchy brightness, and the pulsar is indicated by a cross. The discrete sources agree in position with those catalogued[2] at 610 MHz. These include the three bright sources near RA = 19H 29M, DEC = 22°30' which were identified[2] as being the true cause of extended emission reported in low-resolution surveys of this area.

The SNR is visible at both frequencies as a faint partial shell whose NW limb nearly coincides with the three bright sources mentioned above. A superposition of discrete sources does not account for the diffuse emission comprising the shell. The diameter of the 408 MHz shell is found by a least-squares technique to be 1.7°. The centre of the shell is 24' ± 4' from the pulsar position. Direct comparison of the two maps shows that the shell has spectral index $\alpha = 0.48 \pm 0.14$ and is therefore non-thermal. The 408 MHz surface brightness at this location is 7.9×10^{23} W/m^2/Hz/sr. In Figure 1, the shell limb is reasonably well defined, it is circular (standard deviation ~5%), and the central brightness is only half that of the shell. These are characteristics of SNRs. The relatively steep spectral index is consistent with the object being a shell SNR, rather than a plerion, while the diffuseness of the shell and its extreme faintness are indicators of old age. We emphasize its detection specifically as a

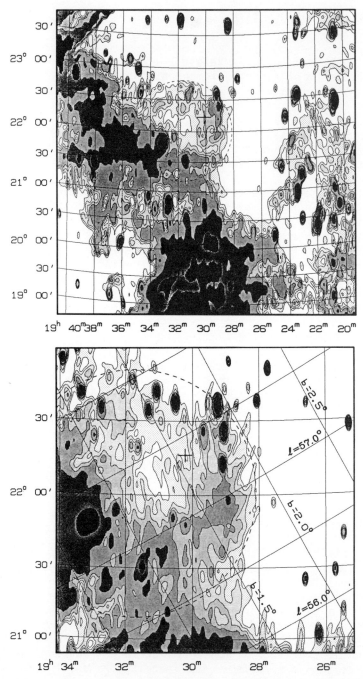

Figure 1: Field of PSR1930+22 at 408 MHz (top) and 1420 MHz (bottom). Cross indicates pulsar position and dashed circle outlines shell of SNR. 1420 MHz resolution degraded to match that at 408 MHz.

result of searching the neighbourhood of PSR1930+22.

The Huang-Thaddeus Σ-D relation[6] gives a distance for the SNR of 4.5 kpc $\pm$ 30%. The shell diameter is then 135 pc. Combining Manchester and Taylor's galactic electron density model[1] with the pulsar dispersion measure[7] gives a distance for PSR1930+22 of 6.6 kpc $\pm$30%. These distance estimates overlap. Other Σ-D relations[8] [9] give distances of 3.6 and 6.0 kpc, respectively. We conclude that since the pulsar lies well within the SNR boundary, there is positional agreement in three dimensions.

If the pulsar was formed at the centre of the SNR at a distance of 4.5 kpc, then the transverse velocity v_t of the pulsar is 750 km/s. This is larger than any v_t reported[10] in a set of 26 measured pulsar proper motions. (Three of these 26 pulsars had $v_t > 300$ km/s.) At the same time, PSR1930+22's z-distance of 93 pc falls well within the distribution of pulsar galactic z-distances, whose half-density height is 400 pc[10].

Assuming Sedov expansion with $E_0/n_0 = 5 \times 10^{51}$ erg cm^3 an SNR would require 290000 years to reach 135 pc diameter. This is several times the pulsar spin-down age. The age disparity is in the same sense as that between PSR1509 and the SNR G320.4-1.2, however[11].

Preliminary analysis of HI data from the DRAO synthesis survey shows that HI near $v_{LSR} \simeq 51$ km/s may be physically associated with the SNR. The HI appears to outline the western limb of the shell over ~30' in declination. Gas with this velocity is likely located near the subcentral point[12], so that the distance of the SNR would be ~5.4 kpc, in fair agreement with the pulsar dispersion measure distance[7].

The fact that PSR1930+22 appears to be younger than the SNR is consistent with "late pulsar turn-on" as suggested by Helfand and Becker[13]. Late turn-on would also obviate the high apparent pulsar proper motion mentioned earlier.

We greatly appreciate the help of the staff of the DRAO, particularly that of Dr. T.L. Landecker. The DRAO Synthesis Telescope is operated as a national facility by the Herzberg Institute of Astrophysics. This work was supported by NSERC operating grants.

References:

1) Manchester, R.N., & Taylor, J.H., Astr. J., 86, 1953-1973, 1981
2) Goss, W.M., & Morris, D.J., Astr. Astrophys. 1, 189-192, 1980
3) Gómez-Ganzález, J., and del Romero, A., Astr. Astrophys., 123, L5-L7, 1983
4) Roger, R.S., Costain, C., Lacey, J.D., Landecker, T.L., & Bowers, F., Proc. I.E.E.E., 61, 1270-1276, 1973
5) Veidt, B.G., Landecker, T.L., Vaneldik, J.F., Dewdney, P.E., & Routledge, D., Radio Science, 20, 1118-1128, 1985

6) Huang, Y.-L., & Thaddeus, P., Astrophys. J., 295, L13-L16, 1985
7) Hankins, T.H., Astrophys. J., 312, 276-277, 1987
8) Milne, D.K., Austr. J. Phys., 32, 83-92, 1979
9) Caswell, J.L., & Lerche, I., Mon. Not. R. Astr. Soc., 187, 201-216, 1979
10) Lyne, A.G., in Birth and Evolution of Massive Stars and Stellar Groups, W. Boland & H. van Woerden, eds., Reidel, 189-194, 1985
11) Manchester, R.N., Tuohy, I.R., & d'Amico, N., Astrophys. J., 262, L31-L33, 1982
12) Weaver, H.F., & Williams, D.R., Astr. Astrophys. Suppl., 17, 1, 1974
13) Helfand, D.J., and Becker, R.H., Nature, 307, 215-221, 1984

CTB80: A SNR WITH A NEUTRON-STAR DRIVEN COMPONENT

R.G. Strom
Netherlands Foundation for Radio Astronomy
Dwingeloo, The Netherlands

Abstract: Pulsar-like emission has been detected from the flat-spectrum component of the galactic radio source CTB80. The consequences for the source are discussed.

Introduction: Although it is widely accepted that neutron stars have their origin in supernovae, there are relatively few pulsars associated with supernova remnants. The best established cases are the Crab Nebula and Vela remnants, each of which has a central pulsar with a fairly high spin-down rate, consistent with the view that the neutron star still supplies substantial energy to the remnant (e.g. Manchester and Taylor, 1977). Searches for pulsar emission from other remnants have, however, failed to turn up additional candidates (Seiradakis and Graham, 1980; Manchester et al., 1983; Mohanty, 1983).

The hot surface of a neutron star can also be detected at X-ray wavelengths, and this strategy has turned up several candidates. Unresolved X-ray emission has been observed in the supernova remnants 3C58 (Becker et al., 1982), CTB80 (Wang and Seward, 1984) and RCW103 (Tuohy et al., 1983). All of these sources are weak, however, so it is impossible to say for certain whether their emission is pulsed (as in the case of PSR 0531+21, the Crab Nebular pulsar) or steady (as would be expected from a hot neutron star surface). In this paper I discuss the detection of an unresolved source in the unusual remnant CTB80 which has the characteristics of a radio pulsar. Clues to its interaction with the surroundings are found in morphological features, and they are discussed with reference to similar objects.

Observational results: CTB80 is an extended galactic radio source of unusual morphology (e.g., Angerhofer et al., 1981). Near its geometrical center, at the western edge of a bright radio plateau, lies a dominant flat spectrum component extended on scales of 0.5-1 arcmin (Strom et al., 1984), upon which my discussion will focus. The emission from this feature is unquestionably nonthermal (polarization was mapped in both of the papers just cited), but its spectrum is unusually flat: the flux density is virtually constant throughout the radio spectrum.

Maps made of this component at 2, 6 and 20 cm with the VLA show an almost unchanging structure, so the 20 cm brightness distribution can be taken as representative (Figure 1). The data were obtained in the A-configuration and have been CLEANed and restored with a 1.1 arcsec beam. The only significant difference between Figure 1 and the 2 and 6 cm maps is the presence of a point source near the southwestern apex of the triangularly-shaped plateau (arrow). Its nondetection at 6 cm dynamic pressure with the internal pressure of the hot spot. (In this and other calculations I will follow Blair et al., 1984, in assuming a distance to CTB80 of 2.5 kpc.)

If the neutron star is moving at a velocity v, its interaction with gas of density ρ produces a dynamic pressure ρv^2 which can be equated with p, the internal pressure of the hot spot. The usual energy equipartition assumption is made in estimating p for synchrotron emission, while ρ has been deduced from depolarization observed in adjacent regions. This gives $v \gtrsim 400$ km/s, a high value but by no means an unusual one for pulsars (Lyne et al., 1982). The neutron star would then be moving toward the hot spot (in the direction indicated by the arrow in Figure 1), and I will assume its velocity to be 500 km/s.

The fact that the hot spot has the same, flat, spectrum as the rest of the component suggests that it may represent a major center of particle acceleration. Its proximity to the neutron star is crucial in this respect, and I wish to point out its similarity to features in the Crab Nebula, in particular Wisp 1. The wisps near PSR 0531+21 have been studied by Scargle (1969) who suggests that they are compressional enhancements in the relativistic plasma. He has interpreted the activity and oscillatory motion of Wisp 1 in terms of a piston, feeding energy to the rest of the remnant. I suggest that Wisp 1 and the hot spot in CTB80 are regions of major particle acceleration, through which much of the energy released by each neutron star passes.

The physical properties of both the CTB80 hot spot and Wisp 1 are rather similar: size, 0.01-0.05 pc; minimum energy density, $0.5-1\times10^{-9}$ erg/cm^3; magnetic field directed along the feature's major axis; and elongated shape. But the most striking similarity is that in both cases the neutron star is moving toward the wisp-like feature. In CTB80 this is based on indirect arguments, but in the case of PSR 0531+21, the proper motion has actually been determined and found to correspond to a speed of 123 km/s (Wyckoff and Murray, 1977). In both objects the wisp/hot spot is elongated perpendicular to the line joining it with the neutron star: it has the morphology of a bow shock being pushed ahead of a fast moving object.

By analogy with the situation in the Crab Nebula, the energy which passes through the hot spot region to be deposited in the flat spectrum component must have originated in the rapidly rotating neutron star. Most of the energy radiated from the entire component appears as X-rays, with a luminosity of 4×10^{33} erg/s (Wang and Seward, 1984). Another estimate of the minimum rate at which energy is supplied to the component can be made from the parameters of the radio hot spot. It has a thickness of 0.01 pc or less so, for a speed of 500 km/s, it must be completely renewed in no more than 20 yr. This time scale combined with the minimum total energy (2×10^{42} erg) implies a rate of 3×10^{33} erg/s. Since much of the energy must pass through the hot spot without being radiated away, and some of it must produce expansion of the component, these values are lower limits.

establishes a fairly steep spectrum ($\alpha \lesssim -2$), and it is found to be
polarized at about the 30% level.

These facts - an unresolved, strongly polarized steep spectrum
source immersed in a flat spectrum nonthermal radio component which
appears to be part of a supernova remnant - are the signature usually
associated with a pulsar: radio emission from a neutron star. This view
is strengthened by the likelihood that the source also emits X-rays; the
radio and X-ray point source positions agree within the errors (Becker
et al., 1982). The existing radio data cannot be used to search for
pulsed emission. The fact that no pulsar has been found at the source's
position in existing surveys is, however, not inconsistent with what is
known about its flux density and spectrum. I will now consider the
consequences of this discovery, assuming that the object is indeed a
neutron star associated with CTB80.

Discussion: One of the most striking aspects of the neutron star is its
location: not centered on the bright plateau, but located near the
southwestern rim, just behind the brightest peak on the ridge. It has
been argued in the past (Strom et al., 1984) that there is evidence for
a general westward motion in this component of CTB80, and the eccentric
location of the neutron star would seem to underline this. Its proximity
to the hotspot suggests a causal relationship, such as an interface
between its atmosphere and the ambient medium. Following this line of
thought, we can estimate the speed of the neutron star by equating the

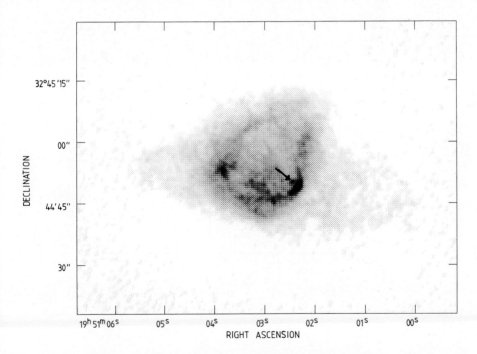

Fig. 1. 20 cm map of flat spectrum component, with point source (arrow)
(copyright University of Chicago Press)

We can now use these estimates in the usual way (e.g., Goldreich and Julian, 1969) to say something about the CTB80 neutron star. If its properties (mass, magnetic field strength, braking index) were precisely the same as those construed for PSR 0531+21, then the pulsar in CTB80 would be expected to have a period of 0.38 s, given the Crab Nebula's 5000 times greater luminosity and the 33 ms period of its pulsar. However, lower values for the magnetic field strength and perhaps mass of the CTB80 neutron star would shorten the period, possibly to under 0.1 s.

Conclusions: It is virtually certain that the flat spectrum component of CTB80 is driven by a neutron star which produces X-ray and radio emission. A search for pulsations is clearly desirable to tie down its period (which may be about 0.1 s) and determine the spin-down rate. The velocity I have estimated is sufficiently large that proper motion may be detectable in a few years.

Acknowledgements: I thank the director of NRAO for providing observing time and facilities at the VLA, and Dr. B.G. Clark for scheduling the observations. NRAO is operated by Associated Universities, Inc., under contract with the U.S. National Science Foundation. The Netherlands Foundation for Radio Astronomy is financed by the Organisation for the Advancement of Pure Research (Z.W.O.).

References:

Angerhofer, P.E., Strom, R.G., Velusamy, T., and Kundu, M.R. 1981, Astron. Astrophys. 94, 313.
Becker, R.H., Helfand, D.J., and Szymkowiak, A.E. 1982, Astrophys. J. 255, 557.
Blair, W.P., Kirshner, R.P., Fesen, R.A., and Gull, T.R. 1984, Astrophys. J. 282, 161.
Goldreich, P., and Julian, W.H. 1969, Astrophys. J. 157, 869.
Lyne, A.G., Anderson, B., and Salter, M.J. 1982, Mon. Not. R. astr. Soc. 201, 503.
Manchester, R.N., and Taylor, J.H. 1977, Pulsars (W.H. Freeman).
Manchester, R.N., Tuohy, I.R., and D'Amico, N. 1983, in Supernova Remnants and their X-ray Emission (eds. Danziger, J. and Gorenstein, P.) 495 (D. Reidel).
Mohanty, D.K. 1983, in Supernova Remnants and their X-ray Emission (eds. Danziger, J. and Gorenstein, P.) 503-504 (D. Reidel).
Scargle, J.D. 1969, Astrophys. J. 156, 401.
Seiradakis, J.H. and Graham, D.A. 1980, Astron. Astrophys. 85, 353-355.
Strom, R.G., Angerhofer, P.E. and Dickel, J.R. 1984, Astron. Astrophys. 139, 43.
Tuohy, I.R., Garmire, G.P., Manchester, R.N., and Dopita, M.A. 1983, Astrophys. J. 268, 778.
Wang, Z.R., and Seward, F.D. 1984, Astrophys. J. 285, 607.
Wyckoff, S., and Murray, C.A. 1977, Mon. Not. R. astr. Soc. 180, 717.

A NEWLY-RECOGNISED GALACTIC SUPERNOVA REMNANT WITH SHELL-TYPE AND FILLED-CENTRE FEATURES

Peter J. Barnes (Astronomy Department, University of Illinois, 1011 West Springfield Ave., Urbana, Illinois 61801-3000, USA) and

A. J. Turtle (School of Physics, University of Sydney, New South Wales 2006, Australia)

Introduction

While the number of galactic supernova remnants (SNRs) now known is fairly large (>150), the subset among these that are known to resemble the Crab Nebula is still distressingly small, about 15 or so (Green, 1984). Thus any object that can be unambiguously included in this exclusive club forms a valuable addition to our knowledge of this class. We report here observations of a newly recognised nonthermal galactic object, G18.94-1.06, having all the hallmarks of the classical shell-type SNRs, while also appearing to have a filled-centre component located inside the shell. Among the known Crab-like remnants, about one third show this dual nature (Green, 1984). This diagnosis of G18.94-1.06 is supported mainly by the variations in spectral index α ($S_\nu \propto \nu^\alpha$) across the source, as seen between the two observation frequencies, 408 MHz and 5.0 GHz.

Observations

In a study begun in 1980 and completed in 1985 (Barnes and Turtle, in preparation), several areas of the galactic plane were selected for detailed analysis using the results of two galactic plane continuum surveys: one at 408 MHz (Green, 1974), and the other at 5.0 GHz (Haynes et al., 1978; 1979). The telescopes, observations and data reduction procedures will be described more fully in the above report. In these selected areas of the plane, over 100 separate objects were studied, including many apparent HII regions, SNRs, and extragalactic sources. Most of the 22 apparent SNRs are new candidates. Very early in this investigation, it was realised that G18.94-1.06 is one of the best new SNR candidates in this group, based on its morphology, size, brightness and spectral index. Fig. 1 shows the SNR as it was observed at 408 MHz with the 1.6 km cross-type radiotelescope at the Molonglo Observatory, near Hoskinstown, New South Wales, while Fig. 2 depicts its appearance as seen at 5 GHz with the 64 m dish near Parkes, New South Wales.

Note that at both frequencies the SNR is perched on the uneven background of the galactic plane, which hinders the derivation of the source parameters. After subtracting a sloping baseplane from each map, we obtain the following integrated flux densities: 58 ± 9 Jy at 408 MHz, and 23 ± 6 Jy at 5 GHz. Thus the overall spectral index for the SNR between these two frequencies is α = -0.37 ± 0.17. This value is intermediate between those typical of shell-type SNRs ($\bar{\alpha}$ = -0.45; Clark and Caswell, 1976) and filled-centre SNRs ($\bar{\alpha}$ = -0.28; Caswell,

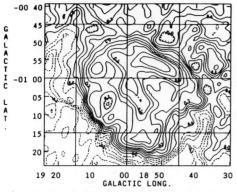

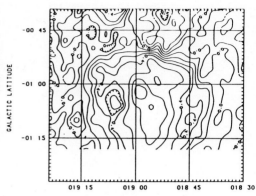

Figure 1: Map of G18.94 – 1.06 at 408 MHz <u>before</u> convolution and baseplane subtraction. Contour levels are at -0.4 -0.3, -0.2 (labelled), -0.15, -0.1, -0.05, 0.0 (labelled), 0.05, 0.1, 0.15, 0.2 (labelled), 0.3, 0.4, 0.5, 0.6 (labelled), 0.8, and 1.0 Jy/beam. Negative contours are shown as dashed lines. The angular resolution is 2.̈86 (R.A.) x 3.̈09 (Dec.), and so 1 Jy/beam = 272 K.

Figure 2: Map of G18.94 – 1.06 at 5 GHz <u>after</u> a sloping baseplane has been subtracted. Contour levels are at -0.4 (labelled), -0.3 (dashed), -0.2 (dashed), -0.1 (dashed), 0.0 (labelled), 0.1, 0.2, 0.3, 0.4 (labelled), 0.6, 0.8 and 1.0 Jy/beam. The angular resolution is 4.̈4 (R.A.) x 4.̈1 (Dec.), and so 1 Jy/beam = 0.750 K.

1979). Comparison with other observations amplifies this peculiarity: Altenhoff <u>et al.</u> (1970) measured flux densities at 1.4, 2.7 and 5.0 GHz of 42, 27 and 28 Jy respectively, while Fürst <u>et al.</u> (1985) obtained (±~10%) 32.9, 27.4, 23.8 and 14.6 Jy at 1.42, 2.695, 4.75 and 10 GHz respectively. Thus, while these data are consistent with our flux densities and spectral index, a more careful inspection of the spectrum (Fig. 3) suggests a flattening of the spectral index towards higher frequencies (i.e., above 1.4 GHz), although the measurement at 10 GHz (14.6 Jy), if confirmed, would suggest a re-steepening of the spectrum beyond 5 GHz, to $\alpha \sim$ -0.66.

Discussion

An explanation of this behaviour may be found in the detailed structure of the source as revealed in the maps (Figs. 1,2). G18.94-1.06 shows a definite, bright shell structure around most of its peri-meter at 408 MHz, and so the classification as a SNR would seem to be a good one. However, the appearance at 5 GHz, while showing some shell-like features, is dominated by a brighter patch near the middle of the SNR, and this Crab-like blob has its counterpart at 408 MHz. The object's spectrum can now be explained as follows. At lower frequencies the shell part of the remnant dominates the emission, and it gives a value for the spectral index typical of shell-type SNRs. At higher frequencies, however, the central, flatter-spectrum part of the SNR dominates over the fainter shell, thus giving rise to the observed spectrum.

Because such a clear combination of Crab-like and classical SNR features would make G18.94-1.06 a valuable contribution to the list of known SNRs, one would like to confirm the spectral index variations across it in a more quantitative manner. To this end, the 408 MHz map was convolved to the resolution of the 5 GHz map and a spectral index map was formed (Fig. 4). As was suspected, we see that the SNR has $\alpha \sim$ -0.5 to the north and east, typical of SNR shells, while towards its centre the spectral index drops to $\sim$ -0.25, close to the canonical

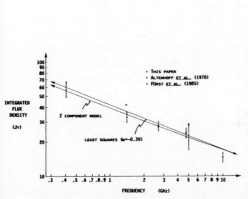

Figure 3:   The radio spectrum of G18.94-1.06

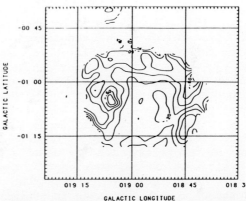

Figure 4: Spectral index map of G18.94 - 1.06 between the observation frequencies 408 MHz and 5 GHz. The spectral index has been calculated only where the signal in the convolved, subtracted 408 MHz map exceeded 30 mJy/beam, and where the signal in the subtracted 5 GHz map exceeded 50 mJy/beam. The gap at the south end of the map is due to lack of data at 5 GHz. The contour levels are in steps of 0.10 in spectral index, from -0.80 to -0.10; labelled contours are at -0.60 and -0.20.

filled-centre value. We may fit the integrated spectrum by a 2 compo-nent model, S_ν (Jy) = 33 $(\nu_{MHz}/408)^{-\frac{1}{2}}$ + 23 $(\nu_{MHz}/408)^{-\frac{1}{4}}$. This curve is shown in Fig. 3, along with a single component least squares fit.

Thus the case for G18.94-1.06 being a new member of the class of dual component SNRs appears to be a strong one. It could be confirmed further by detecting significant radio polarisation in the Crab-like component or the bright parts of the shell (a characteristic of both types of SNR; Weiler and Shaver, 1978) and/or by seeing the central blob as an X ray nebula. In fact, recent observations at 4.75 GHz have shown that both the central blob and the bright, northern edge of the shell are linearly polarised by up to 10% of the point-to-point emis-sion, although the integrated polarisation is only 2.5% at this fre-quency (Fürst et al., 1985). Hence the flatter spectrum in the middle cannot be explained by an unpolarised thermal contribution to the radio emission from an HII region.

PARAMETERS FOR G18.94-1.06

OBSERVED:

Major axis	35 arcmin
Minor axis	31 arcmin
Flux density S_{408}	58 Jy
Mean surface brightness Σ_{408}	0.80×10^{-20} W m^{-2}Hz^{-1}sr^{-1}

DERIVED:	CL	MTLD
Distance d	4.5 kpc	3.4 kpc
Diameter D	43 pc	33 pc
Height above plane z	-82 pc	-63 pc
Age t	11300 yr	2200 yr
Mean expansion velocity V	1900 km s^{-1}	7300 km s^{-1}

The table presents the observed and derived data for the SNR. The first column of derived parameters (labelled "CL") was obtained using the Σ-(D,z) and related relationships of Caswell and Lerche (1979). An alternative approach to deriving SNR parameters was presented by Mills et al. (1984): this approach yields the quantities in the second column (labelled "MTLD"). Note that both distance scales give roughly the same distances and dimensions for the SNR, but that the two age estimates are mutually incompatible. Either distance scale would place G18.94-1.06 somewhere between the Scutum and Sagittarius spiral arms of the galaxy, leading us to expect the SNR to be very faint or invisible at optical wavelengths, due to extinction.

As a final note, we see that at both frequencies the Crab-like component appears extended in roughly the direction of the long axis of the remnant (about 30° inclination to the galactic plane). The obvious question is then: is the central component still interacting with the shell? Fürst et al.'s (1985) interpretation, that this source represents not a SNR but a new class of object produced by an accreting binary system, is now seen to be unnecessary, as the low-frequency data especially show that G18.94-1.06 resembles a SNR very closely. Although their model for the centre might yet prove to be relevant, their misinterpretation of the SNR nature of G18.94-1.06 is seen to be the result of a lack of high-resolution low-frequency data. Even higher resolution observations of this fascinating radio source should prove to be enlightening.

We would like to thank Drs. R.F. Haynes, J.L. Caswell and L. Newton for making the Parkes 5 GHz galactic plane survey data available to us.

References

Altenhoff, W.J., Downes, D., Goad, L., Maxwell, A. & Rinehart, R. Astr. Astrophys. Suppl. (1970) 1, 319.

Caswell, J.L. Mon. Not. R. astr. Soc. (1979) 187, 431.

Caswell, J.L. & Lerche, I. Mon. Not. R. astr. Soc. (1979) 187, 201.

Clark, D.H. & Caswell, J.L. Mon. Not. R. astr. Soc. (1976) 174, 267.

Fürst, E., Reich, W., Reich, P., Sofue, Y. & Handa, T. Nature (1985) 314, 720.

Green, A.J. Astr. Astrophys. Suppl. (1974) 18, 267.

Green, D.A. Mon. Not. R. astr. Soc. (1984) 209, 449.

Haynes, R.F., Caswell, J.L. & Simons, L.W.J. Aust. J. Phys. Astrophys. Suppl. (1978) 45.

Haynes, R.F., Caswell, J.L. & Simons, L.W.J. Aust. J. Phys. Astrophys. Suppl. (1979) 48.

Mills, B.Y., Turtle, A.J., Little, A.G. & Durdin, J.M. Aust. J. Phys. (1984) 37, 321.

Weiler, K.W. & Shaver, P.A. Astr. Astrophys. (1978) 70, 389.

SNR POLARIZATION AND THE DIRECTION
OF THE MAGNETIC FIELD

D.K. Milne
Division of Radiophysics, CSIRO
PO Box 76 Epping, NSW 2121, Australia

Abstract: At the CSIRO Division of Radiophysics we
are currently engaged in a program to map polarization
in SNRs at 8.4 GHz. These results are compared with
earlier Parkes 5 GHz maps to deduce the direction of
magnetic field, Faraday rotation and depolarization.

Over the past decade or so linear polarization observations have been published for ~70 supernova remnants. We accept that this polarized radio emission is generated by the synchrotron process, which implies a magnetic field directed normal to the intrinsic electric vector. For a uniform magnetic field the degree of polarization is independent of frequency and is given in terms of the spectral index (α) by

$$P = (3-3\alpha)/(5-3\alpha), \qquad \text{i.e. } P = 0.7 \text{ for } \alpha = -0.5.$$

If the magnetic field is not uniform throughout the emission region the degree of polarization is reduced, and if the magnetic field is completely random it is zero. The observed polarization will be decreased even in the presence of uniform magnetic fields by differential Faraday rotation of emission from regions at different depths. The observed polarization may be further reduced by Faraday rotation across the telescope beam.

Faraday rotation of the polarization position angle is generally observed and, in summary, the position angle should vary proportionally to wavelength squared if the rotation is outside the emitting region and will follow a more complicated relationship if it is inside.

Measurements of the distribution of polarization, Faraday rotation, depolarization and the direction of the magnetic field could tell us much about the energy process and evolution of these objects.

At the CSIRO Division of Radiophysics we are mapping SNR polarization at 8.4 GHz (3' arc beam) and comparing these with the Parkes 5 GHz maps (Milne and Dickel, 1975) to obtain the distribution of rotation measure, depolarization and direction of the projected magnetic field over the remnants; the first results, for the remnants G291.0-0.1 and G7.7-3.7, have been published (Roger et al., 1986 and Milne et al., 1986 respectively). A further 10 fields are currently being investigated.

In Figure 1 we illustrate this work with a preliminary map of G316.3-0.0, which shows the projected directions of the magnetic field, deduced from 5 and 8.4 GHz measurements, superimposed on an 843 MHz Molonglo synthesis map (Milne et al., 1987).

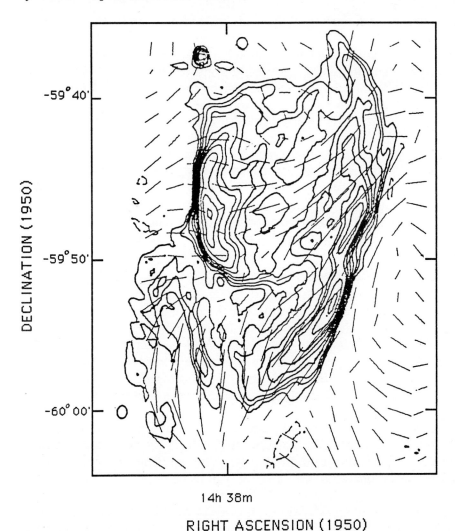

Fig. 1 - Direction of the projected magnetic field in the SNR G316.3-0.0 deduced from 5 and 8.4 GHz polarization. The magnitude of these vectors indicates the mean polarization intensities at these two frequencies. The vectors are shown superimposed on the 843 MHz Molonglo total power map (Milne et al., 1987).

At present there are 27 SNRs for which polarization observations have been published - sufficient to deduce the direction of the projected magnetic field over the remnant. These results have been collated to form "An Atlas of SNR Magnetic Fields" (Milne, 1987). In Table 1 we list these remnants, together with their ages, structural form and the suggested direction of the projected magnetic field.

Table 1   SNRs with observed magnetic field directions

Remnant	Age (yrs)*	Structural form	Magnetic field direction
G5.4-1.0	2500	Shell	Radial in northern arc
G6.4-0.1	6000	Shell	Tangential
G7.7-3.7	2000	Shell	Across shell
G18.8+0.3	4000	Shell	Radial
G21.5-0.9	500	Filled + shell(?)	Radial
G21.8-0.6	4000	Shell	?
G34.6-0.5	2500	Shell	Radial (?)
G74.0-8.6	12000	Shell + blow-out (?)	Tangential in blow-out
G89.0+4.7	7000	Shell?	
G93.3+6.9	1500	Shell	Tangential
G111.7-2.1	300	Shell	Radial
G120.1+1.4	413*	Shell	Radial
G127.1+0.5	18000	Shell	Tangential
G130.7+3.1	805*	Filled	Longitudinal
G184.6-5.8	932*	Filled	Radial on periphery
G189.1+2.9	2500	Shell	Radial
G260.4-3.4	3000?	Shell	Radial
G263.9-3.0	11000	Shell	?
G291.0-0.1	4000	Filled + faint shell	Directed along central bar
G296.5+10.0	5000	Shell	Tangential
G315.4-2.3	1800*	Shell	Radial
G316.3-0.0	700	Shell + blow-out (?)	Radial in shell
G320.4-1.2	1600	Shell	Radial
G326.3-1.8	3000	Faint shell + bright region	?
G327.6+14.0	980*	Shell	Radial
G327.4+0.4	6000	Shell(s)	?
G332.4+0.1	5000	Shell + blow-out (?) + jet	Tangential

* Age obtained from observation of SN outburst, otherwise calculated from the relationship of Caswell and Lerche (1979) using the data from Milne (1979).

The conclusions that can be drawn from our current knowledge can be summarized:

(a)   A general large-scale field is readily seen at low resolutions, but at higher resolutions this may be concealed by increased detail.

(b) Radial fields are more prevalent than tangential fields.

(c) Radial fields predominate in the young remnants. All objects for which we have a definite (and consequently young) age have a field that is radial around the periphery; this is as we would expect it to be, since firstly, the young SNR has not been broken up as much by interaction with the ISM, and secondly, a radial field in a spherical shell is seen as a radial field in all projections.

(d) Blowouts have possibly occurred in G74.0-8.6 (the Cygnus Loop), G316.3-0.0 (MSH 14-57) and in G332.4+0.1 (Kes 32). In these objects and perhaps others it appears that an otherwise spherical shell has been breached, dragging the magnetic field into the blowout at this point (e.g. the most southern part of G316.3-0.0 in Figure 1).

(e) In spite of the preference for alignment of the brightest parts of a remnant nearest the galactic plane (Caswell, 1977) there seems to be no evidence for any preferred orientation in the direction of magnetic field.

Finally, there does not appear to be an answer yet to the question whether polarization or the magnetic field direction exhibits any interaction with the ISM or with the interstellar magnetic field.

References

Caswell, J.L. (1977). Proc. Astron. Soc. Aust. 3, 130.

Caswell, J.L., and Lerche, I. (1979). Mon. Not. R. Astron. Soc. 187, 201.

Milne, D.K. (1979). Aust. J. Phys. 32, 83.

Milne, D.K. (1987). "An atlas of SNR magnetic fields", Aust. J. Phys. (in press).

Milne, D.K., and Dickel, J.R. (1975). Aust. J. Phys. 28, 209.

Milne, D.K., Roger, R.S., Kesteven, M.J., Haynes R.F., Wellington, K.J., and Stewart, R.T. (1986). Mon. Not. R. Astron. Soc. 223, 487.

Milne, D.K., Roger, R.S., Kesteven, M.J., Haynes R.F., and Wellington, K.J. (1987) (in preparation).

Roger, R.S., Milne, D.K., Caswell, J.L., and Little, A.G. (1986). Mon. Not. R. Astron. Soc. 219, 8.

THE ROTATION MEASURE OF BACKGROUND RADIO SOURCES SEEN THROUGH THE SUPERNOVA REMNANT OA184 (G166.2+2.5)

K.–T. Kim and P. P. Kronberg, University of Toronto, Canada
T. L. Landecker, Dominion Radio Astrophysical Observatory, Penticton, B.C., Canada

Abstract: Radio sources in the field of the extended SNR OA184 (G166.2+2.5) have been studied to determine the excess rotation measure (RM) arising from the SNR. Of a total of 32 radio sources observed with the VLA in the C configuration, eight are found to be polarized above 7σ. The sources seen through the SNR show significantly high RM in comparison to background sources. The excess RM due to the Faraday active plasma in the SNR is estimated to be 150 ± 20 rad m^{-2}, which corresponds to $< n_e B_{-6} >_{\rm rim} = (30\pm5)L_{10pc}^{-1}$. The sign of RMs of the sources within an area of about $2° \times 2°$ centred on the SNR shows a systematic longitudinal polarity change on either side of $l \approx 166°.2$. Although a larger sample is needed to justify this, we tentatively interpret this "flip" as due to the reversal of an irregular component of the galactic magnetic field on a scale of order 100 pc.

INTRODUCTION:

Polarization measurement using background sources as probes is known to be an efficient technique for investigation of the magnetic field structure of the intervening medium if a sufficient number of "Faraday" probes can be obtained. This technique has been used for galactic objects, such as the Gum nebula (Vallée and Bignell 1983), the Monogem region (MacLeod *et al.* 1984) and notably for the whole Galaxy (see Simard-Normandin and Kronberg 1980 and references therein) by examining the spatial correlations of the Faraday rotations of a large number of extragalactic sources. This technique should be applicable to supernova remnants (SNRs). Unfortunately SNRs are small; the average angular size of 125 SNRs listed by Milne (1979) is $35' \pm 5'$. Only a small number of background sources are seen through most SNRs and therefore this type of study has been previously applied only to very extended "bubbles".

The large angular extent of the SNR OA184 (G166.2+2.5; 76' in diameter) makes it well suited to a "Faraday Probe" experiment using polarized background sources. A previous DRAO map has revealed 32 background radio sources whose flux density at 1.4 GHz exceeds 25 mJy in the $2° \times 2°$ area centred on OA184 (Routledge, Landecker and Vaneldik 1986, RLV). Furthermore, OA184 lies towards the galactic anticentre and is above the galactic plane. This makes OA184 an ideal SNR to probe for rotation measure (RM) variations, since the general interstellar medium (ISM) component of RM is small at this longitude (*cf.* Simard-Normandin and Kronberg 1980), and is therefore less likely to overwhelm the contribution from the SNR shell itself. This paper describes the results of an attempt to study the magnetic field strength in the SNR OA184 by observing the excess of Faraday rotation measures of background radio sources in the field.

From *a priori* estimates of physical conditions in OA184 ($n_e \sim 1$, $B_{-6} \sim 1$, size(L)$\approx$100pc [RLV, revised diameter, Landecker, private communication 1987]), the RM produced in the SNR might easily be of the order of 100 $rad\ m^{-2}$. (B_{-6} is the magnetic field strength in microgauss, n_e is the electron density in cm^{-3} and L is the pathlength). This estimate of RM is well supported by the results given in Bignell and Vallée (1983).

Figure 1

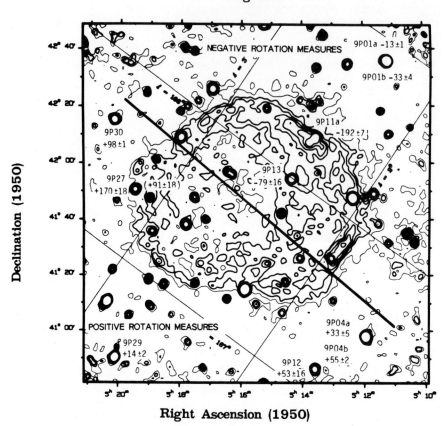

Right Ascension (1950)

Table 1   LIST OF FARADAY ROTATION MEASURES FOR THE 8 SOURCES
BEHIND THE SNR OA184

NAME (1)	RM ($rad\ m^{-2}$) (2)	IPA (3)	χ^2 (4)	OTHER RM (5)	χ^2 (6)	COMMENTS (7)
9P01a	−13±1	2±2	4.8×10^{-4}	NONE	—	
9P01b	−33±4	56±9	2.8×10^{-4}	NONE	—	
9P04a	+33±5	127±10	7.6×10^{-3}	−45±14 −855±12	6.6×10^{-2} 5.1×10^{-2}	
9P04b	+55±2	28±4	1.3×10^{-3}	NONE	—	
9P11a	−192±7	98±14	1.1×10^{-2}	NONE	—	RIM
9P12	+53±16	169±37	2.0×10^{-2}	+119±58	1.5×10^{-1}	*
9P13	−79±16	8±36	5.3×10^{-3}	+338±57	7.2×10^{-2}	+
9P27	+170±18	33±38	7.3×10^{-2}	+91±18	7.4×10^{-2}	RIM
9P29	+14±2	170±3	2.3×10^{-4}	NONE	—	
9P30	+98±1	21±1	4.4×10^{-7}	NONE	—	

Note: (1) Name of the source, (2),(3), Accepted value of Faraday Rotation Measure and Intrinsic Polarization Position Angle in degrees, (4) chi-square values of least-squares fit, (5) Other
acceptable RMs, (6) chi-square values for RMs in column (5), (7) Comments; RIM = source is
superimposed on the rim of the SNR, + Source seen through the SNR, * Ambiguous RM

OBSERVATIONS AND RESULTS:

Observations were made with the VLA* in the C configuration at 1385, 1465, 1515, 1665, 4835 and 4885 MHz. In Table 1, we list the RMs measured for the 8 sources in the field of SNR OA184 which had detectable polarization. These sources are marked on the map of OA184 in Figure 1, and the measured RMs are indicated.

An intriguing result evident in Figure 1 is that the signs of the RMs are opposite on either side of a line drawn at $l \approx 166.2°$. This separation of the signs of the RMs could occur by chance, and a series of computer simulations was therefore performed to investigate this. The probability that 8 sources with random RMs should by chance fall into two groups of 5 and 3 separated by a straight line (as shown in Table 1 and Figure 1) was calculated in the Monte Carlo manner. It proved to be $P(5:3) = 19.8\pm1.4\%$ after several thousand simulations. However, the observed distribution has greater significance than this because the double sources 9P01 and 9P04 both have two components showing RM of the same sign. Therefore a further probability p_p^2 was applied, where $p_p = \frac{1}{2}$ is the probability that the two components of a double source have the same sign of RM. Then the probability of chance occurrence of the observed distribution of RMs is

$$P_{2D}(5:3) = P(5:3) \times p_p^2 = 5\pm2\%. \tag{1}$$

This probability is by coincidence about the same as $P(6:4) = 3.6\pm0.6\%$.

For the cases of equal numbers of sources of each sign, the probabilities $P(2:2)$, $P(3:3)$, $P(4:4)$, $P(5:5)$, and $P(6:6)$ are estimated to be $81\pm3\%$, $40\pm2\%$, $16.5\pm1.3\%$, $4.0\pm0.6\%$, and $2.0\pm0.4\%$. This implies that the significance of the suggested polarity flip would be firmly tested with a small number of additional sources in the field. It is unfortunate that of 32 sources observed, only 8 showed measurable polarization.

DISCUSSION:

The average RM of the background sources outside the SNR is $< |RM| > \approx 40\pm13\ rad\ m^{-2}$ (for 7 sources), whereas that of the inside sample (for 3 sources) is $< |RM| > \approx 150\pm30\ rad\ m^{-2}$. Taking the mean separately according to the sign of the RM, we have $+50\pm15$ and -23 ± 10 $rad\ m^{-2}$ for each side of the SNR. Note that these are small number statistics.

It is particularly noteworthy that the two polarized sources lying in the radio continuum "rim" have large rotation measures. These are 9P11a $(-192\ rad\ m^{-2})$ and 9P27 $(+170\ rad\ m^{-2})$. This reinforces the sense of the statistics above, and suggests that we are detecting an excess RM due to the SNR.

The excess amount of RM from the rim of the SNR can be estimated as follows. Taking the mean of the two RMs, -192 and $170\ rad\ m^{-2}$, of the sources in the rim, and assuming that the pathlength intercepting the SNR is less than 10% of the diameter of OA184, that is about 10 pc, we have

$$n_e B_{-6} = 18\pm2L_{10pc}^{-1}(cos\theta)^{-1}, \tag{2}$$

where L_{10pc} is the pathlength in units of $10pc$ and θ is the angle between the line of sight and the magnetic field. By taking an average RM of the sources seen outside the SNR, the mean intrinsic RM of the background sources, $< |RM| >_{inc}$, is estimated to be $36\ rad\ m^{-1}$ and this amount is subtracted to estimate the net RM of the SNR. For an old SNR like OA184 the magnetic field configuration is dominated by a circumferential component (Milne 1987). Hence, we can assume that θ lies in a plane tangent to the rim of the SNR. As a good approximation, where the line of sight passes through the rim of the SNR, we can estimate that $< cos\theta >=0.54$ and we expect:

$$< n_e B_{-6} >_{rim} \approx (30\pm5)L_{10pc}^{-1}. \tag{3}$$

---

* The Very Large Array of the National Radio Astronomy Observatory is operated by Associated Universities Inc. under contract to the National Science Foundation.

This amount of $n_e B_{-6}$ is consistent with the value inferred from shock wave amplification of $n_e B$ in the ambient galactic medium.

Another way of estimating the magnetic field of the SNR is to use the equipartition argument. Using the flux density at 1.4 GHz of 9.0 Jy and $\alpha = -0.54$ (Routledge *et al.* 1987; Milne 1979), we have

$$B_{eq} = 4.1 \left(\frac{1+k}{\psi}\right)^{\frac{2}{7}} \text{ microgauss,} \qquad (4)$$

where ψ is the volume filling factor, k is the proton/electron energy density and 100 pc is used as the diameter of OA184. The edge-brightened morphology of OA184 suggests that most of the total radio flux is likely to originate from a thin shell where the interactions of the SNR with the ISM are intense. Assuming the thickness of the shell is of the order of 10% of the diameter of OA184, then the best estimate on the term in parentheses in eq (4), $\left(\frac{1+k}{\psi}\right)^{\frac{2}{7}}$, which is relatively insensitive to k, is about $4-6$ for $k = 1-100$. Substituting this estimate into eqs (4) and (3), then gives

$$B_{eq} \approx 16 \pm 8 \text{ microgauss,} \qquad (5)$$

$$< n_e >_{\text{rim}} \approx (1.9 \pm 1.2) L_{10pc}^{-1}. \qquad (6)$$

These estimates on B and n_e are in good agreement with those from other studies (*e.g.*, Milne 1987).

The longitudinal gradient in RM across OA184 is estimated to be about 300 $rad\ m^{-2}$/degree. It is interesting to compare this result with those from two previously studied SNRs. Both appear to show a polarity flip in RM; they are 1209-51/52 and Puppis A (Dickel and Milne 1976). The polarity flip appears not to be associated with the SNR itself, since background sources located well away from the centroid of the SNR also show this trend. Instead the SNR, as it expands, amplifies the ambient galactic magnetic field and enhances the contrast in magnetic field polarity already existing in the galactic field.

The location of OA184, especially the longitude, is of particular interest. The galactic longitude $l \approx 165°$ is perpendicular to the Orion and Perseus arms (Simard-Normandin and Kronberg 1980; Vallée and Bignell 1983), and the galactic RM arising from the *regular* component of galactic magnetic field threading the spiral arms should be small. Hence the RM distribution of the sources in the field of OA184 might reflect the spatial coherence of the *random* field component of the general interstellar magnetic field. Based on the data presented here, the scale size of the magnetic field reversal should be of the order of a degree, about 100 pc in linear dimension, to be consistent with the polarity flip shown in the map. However, of course, more data are needed to better specify the scale size and strength of the random magnetic field reversals.

REFERENCES:

Dickel, J. R., and Milne, D. K. 1976, *Aust. J. Phys.*, **29**, 435.

MacLeod, J. M., Vallée, J. P., and Broten, N. W. 1984, *Astron. Astrophys. Suppl. Ser.* **56**, 283.

Milne, D. K., 1979, *Aust. J. Phys.*, **32**, 83.

Milne, D. K. 1987, *Preprint.*

Routledge, D., Landecker, T. L., and Vaneldik, J. F., 1986, *M.N.R.A.S.*, **221**, 809. [RLV]

Simard-Normandin, M., and Kronberg, P. P., 1980, *Ap. J.*, **242**, 74.

Simard-Normandin, M. 1980, *Ph.D. Thesis* University of Toronto.

Vallée, J. P. and Bignell, R. C., 1983, *Ap. J.,* **272**, 131.

BARREL-SHAPED SUPERNOVA REMNANTS

M.J.Kesteven and J.L.Caswell
Division of Radiophysics, CSIRO
PO Box 76 Epping, NSW, Australia, 2121

Abstract: We argue that the majority of radio supernova remnants have a three-dimensional distribution of emissivity which is barrel shaped, with little emission from the end-caps. We examine some mechanisms which could produce this distribution.

Introduction. Shell-type supernova remnants tend to be roughly circular in outline, suggesting that their emissivity has spherical symmetry. Closer examination reveals systematic departures from circularity - these "defects" have been interpreted as the result of an interaction with irregularities in the interstellar medium. (Whiteoak and Gardner, 1968; Tenorio-Tagle et al, 1985). We offer a simpler explanation: the remnants have cylindrical symmetry.

Two definite barrels : G296.5+10.0 & G327.6+14.6

Maps of G296 and G327 (SN1006) obtained at 843 MHz with the Molonglo Observatory Synthesis Telescope (Mills, 1980) are shown in figure 1. There are three characteristics which are quite inconsistent with spherical symmetry, but are compatible with a barrel shape, (cylindrical symmetry):

* a single axis of mirror symmetry. (In effect, a two-arc appearance).
* a gradient in brightness along the the arcs.
* regions of low (or null) emission where the axis intersects the shell boundary.

These two remnants are at high galactic latitude where a uniform interstellar medium can be expected.

Are all remnants barrels?

Few remnants are as simple and symmetrical as the two examples discussed above. We argue that what is seen is in general a distorted version of the three characteristics: the underlying "barrel" modified by the surrounding medium. The observed brightness distribution depends also on on the orientation of the barrel axis relative to the line of sight. A barrel seen end-on is ring-shaped, whereas it will have a two-arc appearance when seen edge-on .

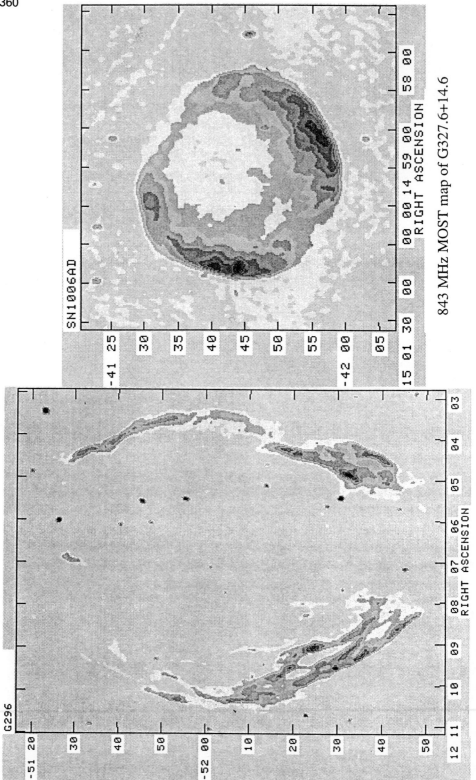

843 MHz MOST map of G327.6+14.6

843 MHz MOST map of G296.6+10.0

The expected proportions of each category in the population at large are given below, along with the observed proportions. The agreement is fair, suggesting that a large fraction of all remnants are barrel-shaped.

Supernova remnant morphologies: predicted and observed distributions.

Appearance	Probability of occurrence	
	Predicted	Observed
Ring-shaped	10%	6% (4)
Two arcs	70%	63% (44)
Intermediate	20%	31% (22)
		(70 remnants)

What causes barrels?

(a) Some axial order in the interstellar medium? A smooth density gradient, or a uniform magnetic field will impose some measure of cylindrical symmetry on a remnant. The deep nulls along the cylinder axis are difficult to explain, however.

(b). A cylindrical supernova outburst? G327 is a young remnant (981 years old); G296 is most likely old (Clark and Caswell, 1976). The barrel shape is thus unlikely to be the result of an evolutionary process. We suggest that the remnants are intrinsically barrel-shaped. This might be the result of an outburst concentrated to the equatorial plane. (Bodenheimer and Woosley, 1983).

Conclusions.

* Most remnants have a barrel-shaped distribution of emissivity;
* The cylindrical symmetry is likely to be intrinsic to the supernova outburst, and not imposed on the remnant by some interaction with an ordered structure in the interstellar medium.

References.

Bodenheimer,P and Woosley,S.E. Astrophys.J. (1983) 269, 281
Clark,D.H and Caswell,J.L. Mon.Not.Roy.Ast.Soc. (1976) 174, 267
Mills,B.Y. Proc.Astron.Soc.Aust. (1981) 4, 156
Tenorio-Tagle,G, Bodenheimer,P. and Yorke,H.W. Astron. and Astrophys. (1985) 145, 70
Whiteoak,J.B. and Gardner,F.F. Astrophys. J. (1968) 154, 807

INFRARED ANALYSIS OF SUPERNOVA REMNANTS

Eli Dwek
Laboratory for Astronomy and Solar Physics
NASA/Goddard Space Flight Center, Maryland

Abstract. Infrared observations of supernova remnants obtained
with the Infrared Astronomical Satellite provide new insights into the
dynamics and energetics of the remnants, and into their interaction
with the ambient interstellar medium. In most remnants the infrared
emission arises from dust that is collisionally heated by the X–ray
emitting gas. The infrared observations can therefore be used as a
diagnostic for the physical conditions of the shocked gas. In
particular, it is shown that all the prominent X-ray remnants in the
Galaxy and in the LMC cool mainly by dust grain collisions instead
of atomic processes.

1. Introduction

Theoretical calculations (Ostriker and Silk 1973; Silk and Burke 1974) suggest that at
temperatures above ~ 10^6 K, a dusty plasma should cool mainly at infrared (IR) wavelengths.
Supernova remnants (SNR) should therefore appear as prominent IR sources in the galaxy.
However, searches for IR emission from Tycho, Kepler, and Cas A (Wright et al. 1980), and the
Crab Nebula (Harvey et al. 1978), yielded essentially negative results. Searches for IR emission
from the optical knots in Cas A, which were suggested by Dwek and Werner (1981) as possible
sites of supernova condensate material, were more successful (Dinerstein et al. 1987). However, the
detected radiation has been attributed to IR line emission from ionic species present in the shocked
optical nebulosities of Cas A.

This situation has drastically changed with the success of the Infrared Astronomical Satellite
(IRAS) which in November 1983 completed an unbiased all-sky survey at 12, 25, 60, and 100 μm.
Details on the instrument, observing strategy, and mission objectives can be found in Neugebauer
et al. (1984), the first paper in a special Ap. J. Letters issue dedicated to the IRAS mission. In the
same issue, Marsden et al. (1984) reported the first detection of IR emission from dust in a SNR,
the Crab Nebula. Figure 1, taken from Mezger et al. (1986), shows the spectrum of the IR and radio
emission from the Crab. The Crab has a strong non-thermal radiation field that fills the nebula, and
the figure clearly shows the thermal IR emission just barely rearing its peak above this nebular
emission component. Dwek and Werner (1981) have shown that in the Crab the dust is most likely
to be radiatively heated by the ambient radiation field. The Crab nebula may therefore represent a
special class of SNR in which the IR emission represents only a minor fraction of the total
radiative output from the remnant.

The opposite is true of the emission of all other remnants that will be discussed in this review.
Figure 2 (based on Dwek et al. 1987a) depicts the power emitted by Cas A from radio to X-ray
wavelengths. The figure shows that most of the radiative output of the remnant is at IR
wavelengths. The dust is most likely of interstellar origin and collisionally heated by the shocked
gas.

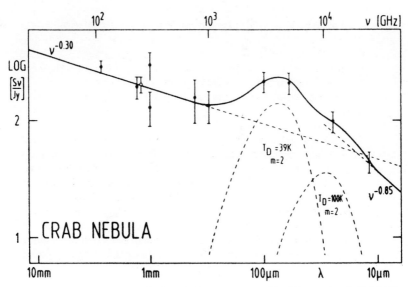

Figure 1 – *The spectrum of the Crab Nebula as presented by Mezger et al. 1986. The dust is radiatively–heated in the Crab, and its IR emission represents only a small fraction of the total radiative output from the nebula.*

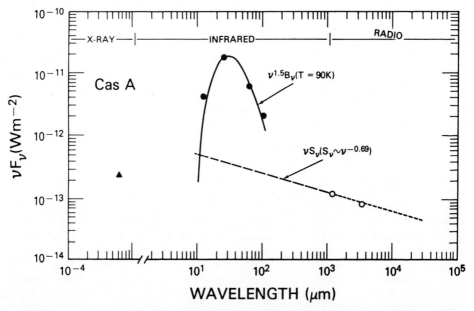

Figure 2 – *The power spectrum of Cas A. The dust is collisionally–heated in the remnant, and its IR radiation dominates the emission at all other wavelengths.*

The most extensive search for SNR in the IRAS skyflux plates is being conducted by Arendt (1987a). From a preliminary unbiased sample of 70 remnants, identified by their radio emission, 16 showed evidence for IR emission that is clearly associated with the remnant, and 24 sources showed evidence for IR emission, but its association with the remnants is uncertain. The remaining 30 regions showed no evidence for IR emission that could be attributed to the remnant. This investigation shows that the IR detection of SNR is mostly hampered by confusion with other IR objects near the galactic plane. The latter is suggested by preliminary findings that the average galactic lattitude of the remnants detected in the search is |b|=2.0°, whereas remnants with no detectable IR emission seem to be located closer to the plane with an average value of |b|=0.8°. However, some remnants elude detection even though they are high–lattitude objects, and are therefore expected to figure quite prominently in the IRAS skyflux plates. An example of such a remnant is SN 1006 (l; b = 327.5°; +14.6°) which is expanding into a low–density interstellar medium (ISM). This suggests that a minimum ISM density is required to observe the IR emission from shock heated dust (Dwek et al . 1987b; Braun 1986a).

In the remainder of this review I will emphasize remnants that are most prominent in the IR. Multiwavelengths analysis of these remnants have been presented in a series of papers: Tycho, Kepler, and Cas A (Braun 1987), Cas A (Dwek et al. 1987a); IC443 (Braun 1986b; Mufson et al. 1986; McCollough and Mufson 1987); Cygnus Loop (Braun 1986c); and G292.0+1.8 (Braun 1986a). For these remnants, the IRAS observations constitute the first observational evidence of collisionally heated dust in the ISM. The IR emission may therefore be used as a diagnostic for the plasma conditions in the remnant , and a new means for deriving various parameters that determine its evolution.

2. Infrared Emission Mechanism

The galaxy is transparent at wavelengths of the IRAS bands. Therefore, several dust components not associated with the remnant may contribute to the IR flux from the direction of the remnant. These include: 1) emission from interplanetary dust particles (zodiacal dust); 2) emission from interstellar dust; 3) infrared cirrus; and 4) any galactic or extragalactic extended or point sources. Any attempt to analyze the IR emission from the remnant must therefore be preceded by a separation of all these various emission components.

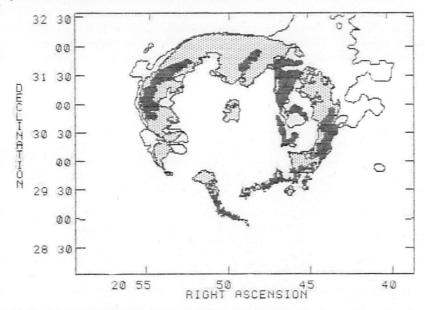

Figure 3 – *A composite image of the Cygnus Loop at 60 μm (outer contour lines), X–ray (light shaded area), and optical wavelengths (dark shaded areas).*

The main zodiacal emission component (described by Hauser et al. 1984) has a flat spatial distribution over the dimensions of all extended remnants, provided they are sufficiently far removed from the ecliptic plane. The contribution of this component to the IR can therefore be removed from the data by the subtraction of a flat or smoothly varying background. The same is true for the interstellar dust component whose emission,   a few degrees from the galactic plane,   can be described by a cosec law in galactic latitude. The infrared cirrus, a new component of IR emission discovered by the IRAS (Low et al. 1984), consists of dust heated by the interstellar radiation field, and exhibits filamentary structure on all scales. However, since its morphology and color temperature are significantly different from that of the shock–heated dust, its contribution can be separated from that of the remnant by decomposing the emission into its various spectral components. Finally, extended IR sources are easily distinguishable and separable from the remnant emission. A good example of a region confused by zodiacal light, cirrus, galactic emission , and IR point sources, where the IR emission was separated into its various spectral components, is the Cygnus Loop (Braun 1986c). Figure 3 is a composite image of the Cygnus Loop at 60 µm, X–ray, and optical wavelengths. The figure shows that the IR emission from dust arises from the same region that gives rise to the X–ray emission, except for regions where the IR emission is dominated by cirrus, and regions where the X–ray is obscured by foreground material (the 'carrot' is such a region). Enhanced IR emission arises from the shocked, optical line emitting filaments.

Figure 4 is a schematic diagram of a supernova remnant expanding into a cloudy ISM depicting various sites and sources of infrared emission. The expanding SN blast wave sweeps up interstellar dust (or circumstellar dust in young remnants) that is collisionally heated by the X-ray emitting gas. The blast wave may engulf a nebulosity or impinge upon a molecular cloud. Infrared lines from atomic or ionized species, continuum free-free emission, and thermal emission from dust are the main sources of IR emission expected from these nebulosities. Table 1 lists several lines that may be important in the IRAS bands. In young remnants, the pressure behind the forwardly expanding blast wave will drive a reverse shock through the cooling ejecta. If dust formed in this metal-rich material, this shock will radiate mostly at IR wavelengths. Finally, in centrally-filled, Crab-like remnants, the dust may be heated by the ambient non-thermal radiation field. The wide variety of IR emission mechanisms and sites discussed above demonstrates the usefulness of using IRAS data as a diagnostic tool for studying the evolution and interaction of supernova remnants with the ISM.

Table 1

Infrared Lines in the IRAS Bands

IRAS Band	Atomic/Ionic Specie
12 µm	S IV (10.5 µm), Ne II (12.8 µm)
25 µm	S III (18.7 µm), Fe III (22.9 µm), Fe II (26.0 µm)
60 µm	O III (51.8 µm), S I (56.3 µm), O I (63.2 µm), Si I (68.5 µm)
100 µm	O III (88.4 µm)

In most remnants (important exceptions to this statement will be discussed later on) the dominant source of IR emission is collisionally heated dust. Therefore, the study of the physical interaction between these dust particles and the hot gas can yield valuable information on plasma conditions and remnant parameters.

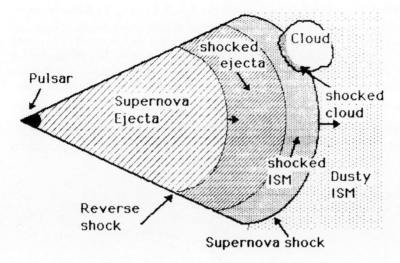

Figure 4 – *A schematic diagram of a supernova remnant expanding into an inhomogeneous, dusty medium.*

3. The Physics of Dust Particles Behind Strong Shocks

The IR emission from SNR is primarily determined by the initial conditions (i.e., the dust–to–gas mass ratio, and the grain composition and size distribution of the preshocked gas). After passing through the shock front, the dust is subjected to various physical processes (Burke and Silk 1974; Shull 1977; Draine and Salpeter 1979a, 1987b; Dwek 1987) that determine the efficiency at which the blast wave energy is converted to infrared emission. These include: 1) collisional charging of the dust; 2) grain destruction by sputtering or sublimation; and 3) collisional heating (mostly by electrons) of the dust grains.

The charge on the dust particle determines its effective collisional cross-section (Spitzer 1978), and how effectively it is coupled (via magnetic fields) to the shocked gas. In the absence of a magnetic field, a dust particle will move ballistically through the shock and spend a reduced fraction of its lifetime near the shock front. The calculations of Draine and Salpeter (1979a) show that at temperatures above $\sim 10^6$ K, secondary electron emission becomes important. Consequently, the grain charge is too small to affect its collisional heating rate with the ambient gas, but sufficiently large to effectively couple it to the shocked gas with moderate values of the magnetic field (Shull 1977).

Grain destruction by sputtering will shift the weight of any initial grain size distribution to smaller grain sizes, reducing the effective grain area available for heating. For gas temperatures above $\sim 10^6$ K, the dust lifetime against sputtering is approximately given by t (yr) $=10^6$ a(μm)/n(cm^{-3}) (Draine and Salpeter 1979a; Seab 1986), so that 0.1 μm dust particles will survive for about 10,000 yrs in a 10 cm^{-3} gas. This is longer than the shock crossing time in all the bright X-ray remnants considered here.

The dust particles are primarily heated by electronic collisions, thereby cooling the gas. Figure

5 (taken from Dwek 1987) depicts $\Lambda_d(T)$, the cooling function (in erg cm^3 s^{-1}) of a dusty plasma containing a cosmic abundance of interstellar graphite and silicate grains (Mathis, Rumpl, and Nordsieck 1977; hereafter MRN) whose grain size distribution has been extended to very small sizes (solid line). Thin lines in the figure correspond to various single–size distributions of silicate–graphite grains. At temperatures below ~2×10^7 K (the exact value depends on the grain size), all the electrons are stopped in the grain, and Λ_d increases as $T^{3/2}$. At higher temperatures, the dust particles become transparent to the incident electrons, and at $T > 3 \times 10^8$ K, Λ_d actually decreases with gas temperature. However, at these temperatures, the ionic contribution to Λ_d is still rising with gas temperature and equals that of the electrons, thus maintaining an approximately constant value of $\Lambda_d \approx 5 \times 10^{-21}$ erg cm^3 s^{-1} at $T > 2 \times 10^7$ K.

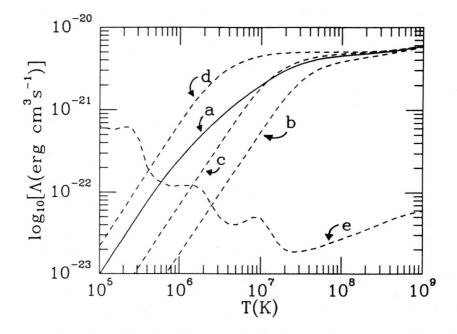

Figure 5 – *The cooling function of a dusty plasma via gas–grain collisions (curve a) is plotted as a function of gas temperature for an MRN interstellar dust model in which the grain size distribution has been extended to very small grain sizes. Curve (e) is the cooling function of the gas due to atomic processes. For more information see Dwek (1987).*

3. The Infrared Cooling of A Dusty Plasma

Also shown in the figure is $\Lambda(T)$, the cooling function of a plasma due to atomic (free-free, bound-free, and bound-bound) transitions (e.g. Raymond,Cox, and Smith 1976). The ratio between Λ_d and Λ is a measure of the relative importance of the cooling mechanisms they represent. From the figure we see that plasma cooling by gas-grain collisions is the dominant cooling mechanism in the shocked gas, exceeding the atomic cooling rate by about two orders of magnitude at temperatures above ~ 10^7 K.This fact may have a significant effect on the thermal and dynamical evolution of SNR. However, the dominance of IR cooling may not be realized in real astrophysical environments, and the ratio between the two cooling functions can deviate substantially from its theoretically predicted value. Reasons for these deviations may be: 1) the dust may be depleted compared with its average interstellar value in the shocked gas; 2) the dust may not be effectively coupled to the gas in the postshock region; 3) a significant fraction of the IR emission may originate from radiatively–heated dust that does not reside in the X-ray emitting gas; and 4) infrared lines may contribute significantly to the observed emission in the IRAS bands.

To compare the theory with the observations, it is useful to introduce an infrared-to-X-ray flux ratio (hereafter referred to as the IRX ratio) defined as

$$IRX = \Lambda_d(T) / \Lambda_{0.2 - 4.0}(T) \tag{1}$$

where $\Lambda_{0.2 - 4.0}(T)$ is the value of the atomic cooling function in the 0.2 – 4.0 keV energy interval. For a given dust-to-gas mass ratio and dust model, and for a plasma that is in ionization equilibrium, the IRX ratio is only a function of gas temperature, which is graphically shown in Figure 6 (solid line). To construct an 'observed' IRX ratio, Dwek et al. (1987b) carried out a comparison between the IR and X-ray cooling rates of nine selected remnants including Cas A, Tycho, Kepler, SN 1006, RCW 103, IC 443, Puppis A, and the Cygnus Loop. Two corrections need to be made to the observed X-ray fluxes to allow for an intercomparison between remnants: 1) all flux estimates from the remnants must be converted to the chosen common X-ray band; 2) observed fluxes must be corrected for interstellar photoelectric absorption along the line of sight to the remnant; and 3) the observed flux must be corrected for non–equilibrium ionization effects. Hamilton, Sarazin, and Chevalier (1983) showed that a shocked plasma may not achieve ionization equilibrium, and its X-ray emission will be significantly enhanced (by a factor η) compared to that expected from a remnant in ionization equilibrium. Therefore, for comparison with the theoretical curve (which assumes ionization equilibrium), the observed IRX ratio has to be *increased* by the factor η.

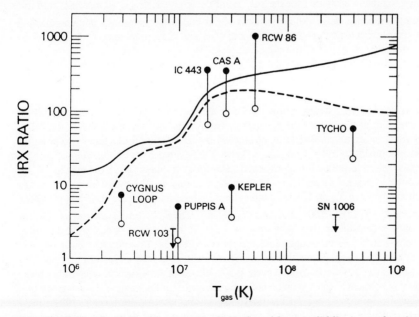

Figure 6 – *The IRX ratio (defined by equation 1) is plotted here (solid line) as a function of gas temperature. Open circles represent the observed IR to (0.2 – 4) keV flux ratio of various SNR, and the filled circles represent these flux ratios, corrected for non–equilibrium ionization effects in the remnant.*

Figure 6 shows the result of the comparison. The IRX ratio, as defined in eq. (1), is plotted as a solid line in the figure. The dashed line represents the ratio of the IR to total gas cooling. The data in the figure are the observed IRX ratios for the various remnants. The open and solid circles represent the data before and after the observations have been 'corrected' for non–equilibrium ionization effects. The figure shows that for all remnants (with the possible exception of SN 1006 and RCW 103), the IRX ratio is significantly larger than unity. *These results clearly demonstrate that IR emission, mainly attributed to gas–grain collisions, is the dominant cooling mechanism in these SNR over large periods of their evolutionary lifetime.* The same qualitative conclusion was reached by Graham et al. (1987; see also this volume), who compared the IR and X–ray emission of various SNR in the Large Magellanic Cloud. They found that on the average, remnant cooling by IR emission dominates the atomic cooling by a factor of ~ 10.

4. The Infrared Diagnostic of a Supernova Remnant

(a) Plasma Parameters

An important quantity that can be derived from the IR observations of a remnant is the temperature of the radiating dust. It is independent of any assumed distance to the remnant, and can be obtained by a simple fit to the IR spectrum without any major assumptions about the dust model. Figure 7 depicts the temperature, T_d, of 0.1 μm silicate and graphite grains that are collisionally heated by a plasma with an electron density of 1 cm^{-3}, as a function of electron temperature T. As in the behavior of $\Lambda_d(T)$, the dust temperature initially rises with T as all the energy of the impinging electrons is deposited in the dust. Above a given gas temperature (T$\approx 2 \times 10^7$ K), the electrons penetrate the grains and T_d reaches a plateau value of $\approx$ 58 K. This behavior of T_d versus T suggests that we should , for the purpose of the analysis, divide all remnants into two categories: young remnants, defined here as remnants in which the postshock temperature is $\geq$ 2×10^7 K (or expansion velocities above ~ 1200 km s^{-1}); and older remnants, characterized by postshock temperatures $\leq 2 \times 10^7$ K (or expansion velocities below ~ 1200 km s^{-1}). In young remnants T_d is essentially independent of the gas temperature, whereas in older remnants the dust temperature 'merely' constrains the allowable combinations of plasma densities and temperatures.

In young remnants, dust temperatures above or below 58 K can only be achieved by respectively increasing or decreasing the gas density from 1 cm^{-3}. As a result, *the IR emission from these objects is an excellent diagnostic of plasma density.* Figure 8 illustrates this point for the three historical remnants, Kepler, Cas A, and Tycho. The figure depicts the dust temperature (of an 0.1 μm graphite grain) as a function of plasma density for two values, T= 2×10^7 and 10^8 K, of gas temperature which bracket all possible values in these remnants. The proximity of the curves illustrates the insensitivity of T_d on T. Dust temperatures in these remnants are (Dwek et al. 1987a; Braun 1987; Arendt 1987b) ~ 100 , 85, and 80 K, respectively. The figure shows that the corresponding plasma densities in these remnants are ~ 20, 10, and 7 cm^{-3}. Uncertainties in these derived values are ~ 50%. Even though these results were derived for a specific grain size and composition, a careful analysis shows that in young remnants, the dust temperature is not very dependent on grain size or composition (Dwek 1987). Electron densities derived from X–ray observations are 0.4 – 4.6 cm^{-3} for Tycho (detailed explanations for the range of values given here are given by Seward, Gorenstein, and Tucker 1983); 6 cm^{-3} for Cas A (Murray et al. 1979); and 7 cm^{-3} for Kepler (Matsui et al. 1984). The agreement between the IR and X–ray derived densities is excellent for Cas A and Kepler, but quite poor for Tycho. This discrepancy should clearly be a subject for further investigation.

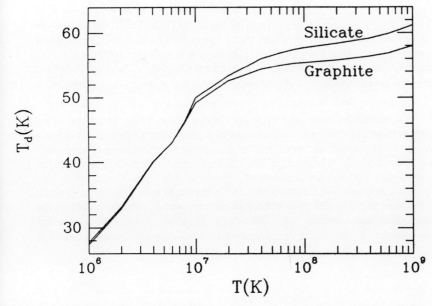

Figure 7 – *The temperature of 0.1 μm silicate and graphite grains is depicted as a function of gas temperature T, for a gas density of 1 cm^{-3}.*

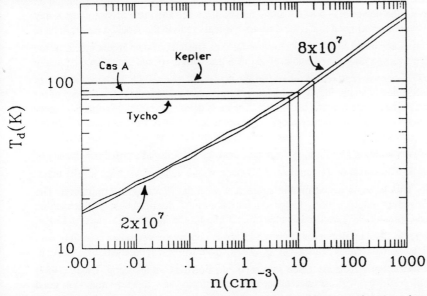

Figure 8 – *The temperature of 0.1 μm graphite grains is depicted here as a function of gas density n, for two values of the gas temperature bracketing the range of values expected in the remnants shown. The figure illustrates how gas densities of young remnants can be derived from their dust temperatures.*

For older remnants $T_d \sim n^\alpha T^\beta$, where (α, β) are (0.20,0.30) and (0.17,0.25) for 0.1 μm graphite and silicate grains, respectively. The above relation is of little use in practical applications since, unlike in young remnants, the dust temperature is strongly determined by the grain size. A more useful approach, therefore, is to use the planar shock models of Draine (1981) for the analysis of older remnants. In these models Draine presented the IR flux emitted by an MRN distribution of grain sizes for various gas densities, and shock velocities ranging from 200 to 1000 km s^{-1}. The peak of the IR emission (determined by the range of grain temperature behind the shock) is related to the gas density and shock velocity (or plasma temperature) and given by $\lambda_{peak}(\mu m) \approx 75$ $n_H{}^{-0.18} v_s{}^{-0.30}$, where n_H is the H–density of the unperturbed ISM, and v_s is the shock velocity in 1000 km s^{-1}. This relation has been used by Braun (1986a, 1986b, 1986c, and 1987) in his analysis of the various remnants. Two new developments took place since the publication of these models: the first was the realization that an interstellar dust component consisting of very small grains is prevalent in the ISM (e.g. Weiland et al. 1986; and references therein); and the second is that these very small dust particles are stochastically heated in a hot gas (Dwek 1986). The wavelength of peak IR emission will therefore shift, with respect to their position in Draine's models, in much the same way it shifted when very small particles were included in models for the IR cirrus clouds (Draine and Anderson 1985; Weiland et al. 1986). Since various remnant parameters are very sensitive to the value of λ_{peak}, one must carefully consider the uncertainties in the derived parameters that result from adopting the above relation.

(b) X-ray Emitting Mass

From the infrared observations one can derive the mass of dust neccesary to account for the IR emission. The dust mass is directly proportional to the IR luminosity, and therefore dependent on dust temperature (to the power of ~ 5 to 6) and remnant distance. The mass of swept–up X–ray emitting gas can then be determined if Z_d, the dust–to–gas mass ratio in the shocked gas, is known.

If the IR emission is attributed to swept–up interstellar dust (instead of dust in the ejecta of the remnant) then the average undepleted value of Z_d in a gas with a cosmic abundance of heavy elements is ≈ 0.0075. However, the dust present in the postshock gas will be depleted relative to the unperturbed ISM due to sputtering. This depletion factor can be estimated if the density of the postshock gas and the age of the remnant are known. The mass determined from this analysis can be compared to the mass of swept–up interstellar gas. For example, in Cas A the dust mass was found to be ~ 0.004 $M_\odot$ (Dwek et al. 1987a). The postshock depletion factor is ~ 0.64, giving a value of 0.8 $M_\odot$ for the X–ray emitting gas. From our previous analysis we deduced a postshock density of ~ 8 cm^{-3} in Cas A. The radius of the remnant is 1.7 pc, giving a value of 1.4 $M_\odot$ for the mass of swept–up gas. This value is in reasonable agreement with the previous determination. The difference may, however, imply a filling factor < 1 for the 2 cm^{-3}–density phase of the medium around the remnant. Braun (1987) deduced a value of ~1.3 $M_\odot$ for the shocked ISM mass in Cas A. Similar values were obtained by Braun for Tycho and Kepler. These values compare favorably with the value of ~ 3.5 $M_\odot$ derived by Murray et al. (1979) from the X–ray analysis of Cas A, if non–equilibrium ionization effects are taken into account. For the Cygnus Loop, Dwek (1987) deduced a mass of ~ 70 $M_\odot$, with an uncertainty of about a factor of ~ 2. This value is in good agreement with the value of ~ 100 $M_\odot$ derived by Ku et al. (1984) from the X–ray analysis of the remnant.

(c) Remnant Energetics

From the dynamics of the remnant and the mass of swept–up interstellar gas, one can infer $E_k{}^{ISM}$, the kinetic energy of the shocked ISM. This quantity can be related to E_0, the energy of the explosion, if the evolutionary state of the remnant is known. For a remnant that has entered the adiabatic stage of its evolution, $E_k^{ISM} = 0.28\ E_0$ (Chevalier 1974). From an analysis of the dynamics of Cas A, Braun (1987) deduced that the kinetic energy of the swept–up ISM is 4.2×10^{49} erg, and that the remnant evolved beyond the adiabatic stage of its evolution. These two facts lead then to an initial energy E_0 of 1.5×10^{50} erg. Values of $E_0 = 1.4 \times 10^{50}$ and 1.8×10^{50} erg were derived for Kepler and Tycho, respectively. The major uncertainty in these values arises from the uncertainty in the evolutionary stage of the remnant. The analysis of Braun implies that the mass of diffuse ejecta in these remnants is $\sim 0.3 - 0.4\ M_\odot$. This may be problematic for Cas A, which is believed to be the result of the explosion of a massive star, unless most of the material was ejected in dynamically non–interacting clumps.

(d) IR Morphology of SNR

The medium into which a supernova is expanding is inhomogeneous. The presence of density gradients and clouds in the ambient medium affect the propagation of the supernova blast wave (McKee 1987), and will have an important influence on the IR appearance of a remnant. For example, a comparison of the IR and X–ray emission from RCW 86 shows a very good correlation between the two images in the soutwestern part of the remnant (see Figure 9). However, no IR emission is detected from the X–ray emitting gas in the east. This suggests that the remnant is expanding into a steep east–west density gradient, and that a minimum ISM density is required to detect the IR emission from the shocked gas. The same conclusion was reached by Pisarski, Helfand, and Kahn (1984) from their X–ray analysis of the remnant. Comparison of the IR and X–ray morphology of Cas A (Dwek et al. 1987a) suggests the presence of non–uniform extinction across the remnant. Furthermore, differences between the IR brightness in the northern and southern parts of the remnant suggest the presence of a density gradient between the front and back part of the remnant (Braun 1987). A comparison among the optical, X–ray, and IR emission from the Cygnus Loop (Figure 3) shows enhanced IR emission associated with the optical line emitting filaments encountered by the shock.

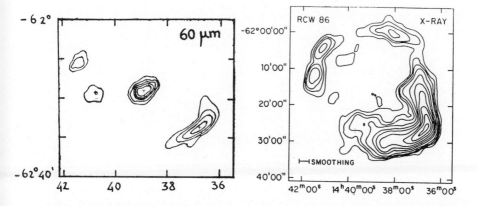

Figure 9 – *Comparison of the IR (Figure 9a) and Einstein X–ray image (Figure 6b) of RCW 86.*

An IR analysis of different regions in a remnant can put valuable constraints on the existence of various gas 'phases' and their filling factor in its interior. For example, the IR spectrum of Cas A is very well approximated by a single–temperature dust component with $T_d \sim 80 - 90$ K. This fact can be used to constrain the density contrast and the filling factor of the two phases responsible for the double hemisphere structure that was pointed out by Braun. A density contrast of ~ 10 will suggest that the dust temperature in the low–density phase is $\sim 50 - 60$ K. Even though the IR *luminosity* of the dust residing in this phase will be reduced by a factor of ~ 10, it may still contribute considerably to the 60 and 100 μm *flux* from the remnant, depending on its filling factor. Braun (1987) suggested a filling factor of ~ 0.9 for this low–density phase, which may be somewhat too high since it will result in a significant broadening of the IR spectrum from the remnant. A similar analysis of other young remnants as well as evolved remnants such as the Cygnus Loop or Puppis A will be useful in determining the morphological structure of the remnant's interior.

5. Dust Temperature Fluctuations in a Hot Gas

Figure 10 shows the observed IR emission from Cas A in the four IRAS bands. The dashed line in the figure depicts the IR spectrum expected from a standard MRN graphite–silicate mixture of interstellar dust that is swept up by the expanding SN blast wave (Dwek et al. 1987a). In the standard dust model, the grain size distribution is characterized by an $a^{-3.5}$ power law in grain sizes with an upper cutoff of $a_{min} \approx 0.25$ μm. The lower limit, a_{min}, on the grain size distribution is 0.005 μm for graphite particles, and 0.025 μm for silicates. In the calculation, the dust particles are collisionally–heated by a plasma with a density and temperature of n = 6 cm^{-3} and 4.6×10^7, respectively.

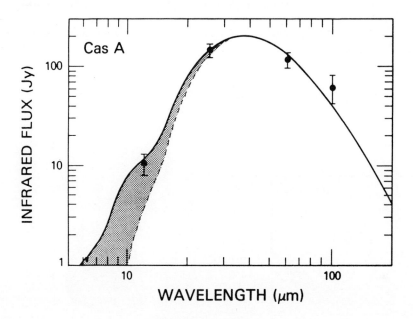

Figure 10 – *The IR spectrum from an MRN distribution of grain sizes (dashed line) is compared to the observed IRAS fluxes. The solid line represents the IR spectrum from an MRN interstellar dust model whose size distribution has been extended to 3 Å. The shaded area represents the excess IR flux resulting from the stochastic heating of the very small (<250 Å) grains in the size distribution.*

The figure shows that the standard MRN dust model gives a reasonable fit to the IRAS data at all wavelengths except at 12 μm. A better fit to the data (solid line) is obtained if the MRN size distribution is extended to very small dust particles (i.e. if $a_{min} = 3$ Å). The addition of small particles produces an excess of IR emission at short wavelengths (shaded area in the figure), compared to the standard size distribution of dust particles. The reason for the enhanced IR emssion at these wavelengths is that small dust particles are stochastically heated in a low–density plasma (Dwek 1986). Two effects occur when a dust particle becomes smaller than a certain critical size (which depends on the conditions of the ambient gas): 1) the heat capacity of the dust particle becomes very small, so that collisions with individual electrons can cause the dust temperature to surge to a value, T_{max}, well above its equilibrium value; and 2) the cooling time of the dust is significantly shorter than the time between electronic collisions, so that the dust cools to a temperature, T_{min}, well below its equilibrium temperature before it is hit by another electron. The result is that the dust temperature fluctuates between T_{min} and T_{max}, with most of the energy deposited in the collision radiated at T_{max}. Figure 11 is a schematic illustration of the effects of temperature fluctuations in a very small dust particle that is immersed in a hot gas. The small dust particles needed to explain the excess 12 μm emission in Cas A could not have been created by the erosion of larger grains in shock – the remnant is too young for this process to be significant. Very small grains must therefore be present in the ambient medium into which the remnant is expanding. The IRAS observations of Cas A therefore constitute the first observational evidence for the prevalence of these very small dust grains around SNR, as well as in the general ISM.

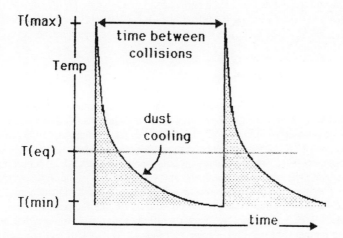

Figure 11 – *A graphic representation of the temporal behavior of a collisionally–heated, very small, dust particle. Following a single electronic collision, the temperature of a 20Å dust particle, embedded in a 10^7 K gas, will surge to a value of ~ 600 K. The equilibrium value of the dust temperature is ~ 56 K. The dust cools from 600 K with an e–folding time of ~ 0.4 sec, whereas the time between electronic collisions is ~ 4000 sec.*

6. Effect on Remnant Evolution

An important question is – will the IR cooling of the remnants accelerate their evolution compared with the standard evolutionary picture presented, for example, by Cox (1972)? The first evolutionary stage of the remnant is the free expansion phase, terminated when enough mass has been swept up to begin decelerating the shock. This stage is not affected by radiative processes behind the shock. The second stage of remnant evolution, the adiabatic phase, is terminated when about half of the thermal energy of the remnant has been radiated away. The end of this phase is marked by the formation of a dense shell behind the shock and the beginning of the "snowplow" phase of remnant evolution. Enhanced cooling resulting from gas-grain collisions may therefore significantly affect the duration of the adiabatic phase.

The importance of dust cooling depends on the density of the ISM, and can be qualitatively assessed in the following way: Consider a remnant with $E_0=10^{51}$ erg expanding into an ISM with $n_0=1$ cm^{-3}. The postshock temperature at the end of the adiabatic phase, T_c, is about 5×10^5 K, whereas IR emission dominates the cooling of the remnant when $T > 6 \times 10^5$ K (see Figure 4). Gas-grain collisions will therefore have only a moderate effect on the duration of the adiabatic phase, since most of the thermal energy is lost from the remnant when its volume is largest (i.e. around temperatures of $\sim 6 \times 10^5$ K when IR and atomic cooling are about equally important). At densities of $n_0=100$ cm^{-3}, T_c is about 2×10^6 K, so that most of the thermal energy is lost when IR emission dominates the cooling by a large factor. Calculations by Dwek (1981) show that the adiabatic phase is shortened by only $\sim 10\%$ in the former case, and by as much as $\sim 40\%$ in the latter case. At higher densities, IR cooling should be more pronounced. However, grain destruction becomes increasingly important as well, resulting in declining effect of gas-grain collisions on remnant evolution. All adiabatic remnants considered here expand into media with densities < 10 cm^{-3}, so that the IR emission will have a negligible effect on their general evolution. However, as suggested by Graham et al. (1987), the IR emission will affect the internal temperature structure of the remnant.

7. Future Prospects

The IRAS all sky survey opened a new wavelength window for the studies of SNR. An interesting question is whether SNR form a class of IR objects that is distinct from other known IR sources like HII regions or planetary nebulae. Such a distinct spectral 'signature' will prove valuable in the identification of SNR that may explode in molecular clouds (Shull 1980; Wheeler, Mazurek, and Sivaramakrishnan 1980). The preliminary investigation of Arendt (1987a) shows that the IR spectrum of an 'average' SNR peaks between the 60 and 100 μm bands of the IRAS, or that it is still rising at 100 μm. The spectrum of these remnants is therefore essentially identical to that of compact HII regions. Planetary Nebulae are hotter objects with IR spectra that peak between 25 and 60 μm (e.g. Pottash 1984). They are therefore distinctly different from the average SNR but look very similar to young remnants like Tycho, Kepler, and Cas A. The lack of any unique IR spectral signature and the confusion with IR sources near the galactic plane suggest that the most promising approach to discover new SNR is still to search for their radio emission (Caswell 1987).

However, even if the IR cannot currently be used to identify new remnants, it provides researchers with powerful means to study the dynamics and energetics of existing remnants, and their interaction with the ambient interstellar medium. A stepwise approach to such analysis would include the following: removal of zodiacal, galactic, and cirrus emission components from the IR image of the remnant; analysis of the relative contribution of IR lines and continuum radiation to the emission; comparison between the IR and X–ray emission to estimate the contribution of collisionally–heated dust to the emission; comparison between IR and optical, HI, and CO maps of the region to estimate the contribution of IR lines from shock–heated filaments or clouds; search for adjacent OB associations to estimate the relative contribution of radiatively–heated dust to the emission. This outline shows the intricate relation between studies at the various wavelengths at our disposal, a collection of which IR radiation is the newest member.

Acknowledgements

I thank Robert Braun for helpful conversations. Walter Rice provided the infrared images presented in the figures. Ted Gull provided the optical image of the Cygnus Loop digitized by Dan Klinglesmith, III. Fred Seward provided the X–ray image of the Cygnus Loop, and Ryszard Pisarski provided that of RCW 86.

References

Arendt, R. G. 1987a, this volume.
Arendt, R. G. 1987b, private communications.
Braun, R. 1986a, Astr. Ap., 162, 259.
_____. 1986b, Astr. Ap., 164, 193.
_____. 1986c, Astr. Ap., 164, 208.
_____. 1987, Astr. Ap., 171, 233.
Burke, J. R., and Silk, J. 1974, Ap. J., 190, 1.
Caswell, J. L. 1987, this volume.
Chevalier, R. A. 1974, Ap. J., 188, 501.
Cox, D. P. 1972, Ap. J., 178, 159.
Dinerstein, H. L., Lester, D. F., Rank, D. M., Werner, M. W., and Wooden, D. H. 1987,
        Ap. J., 312, 314.
Draine, B. T. 1981, Ap. J., 245, 880.
Draine, B. T., and Salpeter, E. E. 1979a, Ap. J., 231, 77.
Draine, B. T., and Salpeter, E. E. 1979b, Ap. J., 231, 438.
Draine, B. T., and Anderson, N. 1985, Ap. J., 292, 494.
Dwek, E. 1981, Ap. J., 247, 614.
Dwek, E., and Werner, M. W. 1981, Ap. J., 248, 138.
Dwek, E. 1986, Ap. J., 302, 363.
Dwek, E., Dinerstein, H. L., Gillett, F. C., Hauser, M. G., and Rice, W. L. 1987a,
        Ap. J., 315, 571.
Dwek, E. 1987, Ap. J., in press.
Dwek, E., Petre, R., Szymkowiak, A., and Rice, W. L. 1987b, Ap. J. (Letters), in press.
Graham, J. R., Evans, A., Albinson, J. S., Bode, M. F., and Meikle, W. P. S. 1987,
        Ap. J., 319, 000.
Hamilton, A. J. S., Sarazin, C. L., and Chevalier, R. A. 1983, Ap. J. Suppl.,
        51, 115 (HSC).
Harvey, P. M., Gatley, I., and Thronson, H. A. 1978, Pub. A. S. P., 90, 143.
Hauser, M. G., et al. 1984, Ap. J. (letters), 278, L15.
Ku, W. H.-M., Kahn, S. M., Pisarski, R. L., and Long, K. S. 1984,
        Ap. J., 278, 615.
Low, F. J., et al. 1984, Ap. J. (letters), 278, L19.
Marsden, P. L., Gillett, F. C., Jennings, R. E., Emerson, J. P., de Jong, T., Olnon, F. M. 1984,
        Ap. J. (Letters), 278, L29.
Mathis, J. S., Rumpl, W., and Nordsieck, K. H. 1977, Ap. J., 217, 425.
Matsui, Y., Long, K. S., Dickel, J. R., and Greisen, E. W. 1984, Ap. J., 287, 295.
McCollough, M. L., and Mufson, S. L. 1987, this volume.

McKee, C. F. 1987, this volume.

Mezger, P. G., Tuffs, R. J., Chini, R., Kreysa, E., and Gemund, H.–P. 1986,
        Astr. Ap., **167**, 145.

Mufson, S. L., McCollough, M. L., Dickel, J. R., Petre, R., White, R., and Chevalier, R. 1986,
        A. J., **92**, 1349.

Murray, S. S., Fabbiano, G., Fabian, A. C., Epstein, A., and Giacconi, R. 1979,
        Ap. J. (Letters), **234**, L69.

Neugebauer, G. et al. 1984, Ap. J. (Letters), **278**, L1.

Ostriker, J. P., and Silk, J. 1973, Ap. J. (Letters), **184**, L113.

Pisarski, R. L., Helfand, D. J., and Kahn, S. M. 1984, Ap. J., **277**, 710.

Pottash, S. R. 1984, Planetary Nebulae (Dordrecht: Reidel), page 302.

Raymond, J. C., Cox, D. P., and Smith, B. W. 1976, Ap. J., **204**, 290.

Seab, C. G. 1986, Interstellar Processes, eds. D. Hollenbach, and H. A.
        Thronson, Jr. (Dordrecht: Reidel).

Seward, F., Gorenstein, P., and Tucker, W. 1983, Ap. J., **266**, 287.

Shull, J. M. 1977, Ap. J., **215**, 805.

Shull, J. M. 1980, Ap. J., **237**, 769.

Silk, J., and Burke, J. R. 1974, Ap. J., **190**, 11.

Spitzer, L. 1978, Physical Processes in the Interstellar Medium (New York: Interscience), p. 199.

Weiland, J. L., Blitz, L., Dwek, E., Hauser, M. G., Magnani, L., and Rickard, J. L 1986,
        Ap. J. (Letters), **306**, L101.

Wheeler, J. C., Mazurek, T. J., and Sivaramakrishnan, A. 1980, Ap. J., **237**, 781.

Wright, E. L., Harper, D. A., Lowenstein, R. F., Keene, J., and Whitcomb, S. E. 1980,
        Ap. J. (Letters), **240**, L157.

AN INFRARED SURVEY OF GALACTIC SUPERNOVA REMNANTS

Richard G. Arendt
University of Illinois
349 Astronomy Bldg.
1011 W. Springfield
Urbana, IL 61801

Abstract: Presented here are preliminary results from a survey of supernova remnants (SNRs) in the data base collected by the Infrared Astronomical Satellite (IRAS). About one-third of the known galactic SNRs are visible in the IRAS data. Confusion with other sources in the galactic plane prohibits the detection of many remnants. The objects that are detected have similar spectral characteristics and temperatures, except that the three youngest remnants known, Tycho, Kepler, and Cassiopeia A, are distinctly warmer.

1. INTRODUCTION

Galactic supernova remnants are usually identified by non-thermal radio spectra, relatively strong polarization, and shell-like morphology. Many SNRs are also observed as x-ray sources. Remnants which are nearby or along paths of low extinction, can often be detected optically as well. However, until the recent work of IRAS, observations of SNRs in the far-infrared ($\lambda \geq 10\mu$) were extremely scarce. This omission in spectral coverage is significant. Most of the IR emission arises from collisionally heated dust. This energy loss may be a major fraction of the total luminosity and can have a significant effect on the evolution of many SNRs (e.g. see Dwek 1981). The presence of this dust is not directly observed in any other wavelength regime.

This paper presents preliminary results of a project to measure flux densities in the infrared for all known galactic SNRs. The full catalog is not yet complete, but the sample is large enough that several trends are apparent. Section 2 describes the process by which the flux densities are measured. Section 3 describes some of the general properties of SNRs in the infrared.

2. PROCEDURE

The IRAS satellite provides survey observations in four broad bands centered on 12, 25, 60 and 100μ. These data, in three different formats are being used for this project. The standard Sky Brightness Images (calibrated surface brightness maps with 2' pixels, see IRAS Explanatory Supplement 1985 for a complete description) are being used for most objects. For 37 SNRs, two-dimensional coadded fields of higher resolution (0.25 - 1.0 pixels) are also being used. These coadded fields use the data from all of the satellite's passes over the selected regions. These maps are processed to yield a flat background over the field. For most of the smallest SNRs (≤ 8' diameter), we also have one-dimensional co-added data. These data provide the highest resolution, and are suitable for objects which have a high degree of symmetry, or which are not well resolved by the IRAS satellite. These data are in the from of averaged slices across the source in each of the four IRAS bands.

Total flux densities are measured from the images by integrating over the area occupied by the remnant, and subtracting a background level determined by integration over a roughly equal area surrounding the remnant. These integrations are done over both circular and rectangular regions. Obvious bright and unrelated sources are excluded from the process. In cases where only an upper limit can be established, the background is chosen to be the lowest level in the region occupied by the SNR. In most cases there is good agreement between the results from the various data formats and the different integration techniques. The differences in the resulting flux densities are comparable to the variations which arise from alterations of the region over which the background is determined.

3. RESULTS

So far we have examined the regions of 70 supernova remnants. Currently the sample is slightly biased towards the brighter and more well known SNRs, and more noticeably towards objects near the galactic center. In 16 regions there is infrared emission clearly associated with the remnant; 30 regions show no infrared emission attributable to the SNRs; and in 24 regions there is emission which may be from the SNR, but the association is uncertain. Confusion with the complex background provided by the galactic plane appears to be a significant problem. The mean distance from the galactic plane for detectable SNRs is 2°0, while the mean distance for undetected SNRs is 0°8. However, there are several examples of SNRs at high galactic latitudes in apparently unconfused regions which show no IR emission.

In the most convincing detections the IR emission correlates fairly well with the optical and/or radio emission (e.g. Cygnus Loop, OA 184, RCW 86), or the IR emission correlates with that of shock excited molecules (e.g. IC 443). In several of the uncertain cases, the IR emission is from a relatively small clump which has a suggestive position in relation to the radio morphology (e.g. G 7.7-3.7, Kes 67, G 323.5+0.1).

In most cases where infrared emission is detected from a SNR, the IR luminosity is greater than both the x-ray luminosity and the radio luminosity. This is based on the objects in the present sample for which x-ray fluxes or luminosities are found in the literature. Examination of ~20 such SNRs indicates that only the Crab nebula clearly has its greatest luminosity in a wavelength regime other than the infrared. In SNRs for which only upper limits are obtained, the IR luminosity is not required to be less than the luminosity in other regimes.

The limited spectral information provided by the four IRAS bands is not of great assistance in discriminating SNRs from other infrared

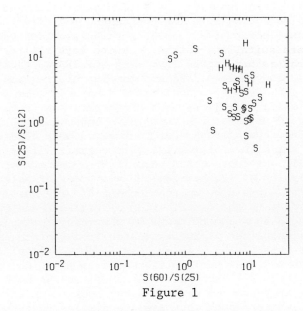

Figure 1

In this color-color diagram S designates an SNR, H designates a compact
HII region. There is some tendency for SNRs to appear lower than
compact HII regions on this diagram. This reflects somewhat greater
12μ emission (relative) in SNRs than in compact HII regions.

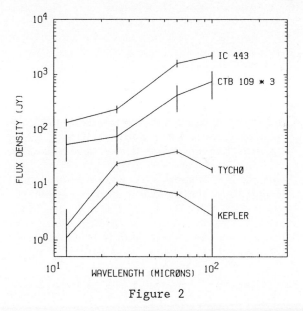

Figure 2

This figure illustrates the difference between the spectra of typical
old and young supernova remnants. The spectra of IC 443 and CTB 109
are representative of the spectra of older SNRs. Tycho, and Kepler are
typical of young SNRs.

sources.   Most SNRs are brightest in the 100μ band, although it is
usually easiest to detect them in the 60μ band.   The same is true for
compact HII regions (Chini, et al. 1986, Antonopoulou and Pottasch
1987), which have relative flux ratios that are very similar to those
of SNRs.   On a color-color diagram involving the longer wavelengths,
there is no distinction between SNRs and compact HII regions.   A color-
color diagram using the shorter wavelengths does show some separation
between the two classes of objects, but the separation is not clear
enough to use as a discriminator between SNRs and compact HII regions
(see figure 1).   The separation is due to the tendency of SNRs to have
slightly more 12μ emission (relatively) than compact HII regions.   If
this effect is real, one cause could be differing amounts of fine
structure line emission at 12μ between the two types of objects.
Another possible cause would be an enhancement of very small grains in
most SNRs with respect to the distribution of grains found in compact
HII regions.   Such an enhancement could be the result of sputtering
grains down to smaller sizes behind the shock fronts of the SNRs.
      The youngest SNRs (Tycho, Kepler, Cas A) show energy
distributions which are distinctly different from the older ones (see
figure 2).   These three all have peak flux densities between 25μ and
60μ.   The dust temperatures derived are about 100K (~40K warmer than
most other SNRs).   Objects such as these are similar in infrared
appearance to planetary nebulae (Pottasch, et al. 1984).

Thanks are extended to Dr. John R. Dickel for his valuable comments and
suggestions on the preparation of this paper.

References:
Antonopoulou, E., and Pottasch, S. R., 1987, Astr. Ap., 173, 108.
Chini, R., Kreysa, E., Mezger, P. G., and Gemünd, H.-P., 1986, Astr.
      Ap., 154, L8.
Dwek, E., 1981, Ap. J., 247, 614.
IRAS Catalogs and Atlases, Explanatory Supplement 1985, edited by
      Beichman, C. A., Neugebauer, G., Habing, H. J., Clegg, P. E., and
      Chester, T. J., (Washington D.C.: U. S. Government Printing
      Office.)
Pottasch, S. R., Baud, B., Beintema, D., Emerson, J., Habing, H. J.,
      Harris, S., Houck, J., Jennings, R., and Marsden, P., 1984, Astr.
      Ap., 138, 10.

IRAS OBSERVATIONS OF COLLISIONALLY HEATED DUST IN LARGE MAGELLANIC CLOUD SUPERNOVA REMNANTS

James R. Graham[1,2], A. Evans[3], J. S. Albinson[3], M. F. Bode[4] & W.P.S. Meikle[1]

[1]Imperial College, London, [2]Lawrence Berkeley Laboratory, University of California, [3]University of Keele, [4]Lancashire Polytechnic.

ABSTRACT

IRAS additional observations show that luminous (10^4-10^5 $L_\odot$) far-IR sources are associated with the Large Magellanic Cloud (LMC) supernova remnants N63A, N49, N49B, and N186D. Comparison of the IRAS and X-ray data shows that a substantial fraction of the IR emission from three of the SNRs can be accounted for by collisionally heated dust. The ratio of dust-grain cooling to total atomic cooling is ~10 in X-ray emitting gas (T~10^6 K). We show why dust cooling does not dominate, but probably speeds SNR evolution in an inhomogeneous interstellar medium.

INTRODUCTION

Dust grains may be an important coolant of astrophysical plasmas, because grains embedded in hot gas are heated by inelastic collisions with electrons and ions (Ostriker & Silk 1973). The energy deposited raises the grain temperature so this energy is radiated in the far-IR. Grain cooling should be important at temperatures characteristic of the X-ray emitting gas (~10^6K) in SNR's. Therefore, SNR's should be an ideal astrophysical laboratory where grain cooling can be investigated.

We observed LMC SNR's because they form a well studied sample at X-ray, optical, and radio wavelengths; they have sizes of the order of the IRAS apertures; they should be less confused than Galactic SNR's because we are not looking through a galactic disc; the LMC SNR's are all at the same distance (we adopt 55kpc).

RESULTS

Of 9 LMC SNR's for which IRAS AO data have been obtained, 4 are unambiguously associated with IR sources. The position of the IR source is shown on X-ray maps of the SNR's in Figure 1. The IR source is precisely coincident with the peak of the X-ray emission in the case of N63A, N49 and N49B. The IR source associated with N186D is offset from the X-ray peak by ~2' to the SW. There are upper limits for a further 3 SNR. The remaining 2 were too close to extended, bright sources to obtain useful limits.
The 8-120μm luminosity, the band III-IV temperature and the total luminosity, calculated assuming a black-body emission with a λ^{-1} grain emissivity law, for 55kpc, are presented in Table 1.

Table 1

SNR	T (K)	$L_{(8-120\mu m)}$ ($10^5 L_\odot$)	L_{tot} ($10^5 L_\odot$)	M_d/M_x	L_{IR}/L_x
N63A	30	1.1	1.8	0.03	12
N49	40	0.5	0.6	0.006	12
N49B	30	0.05	0.1	0.002	4
N186D	25	0.5	1.1	0.3	2100

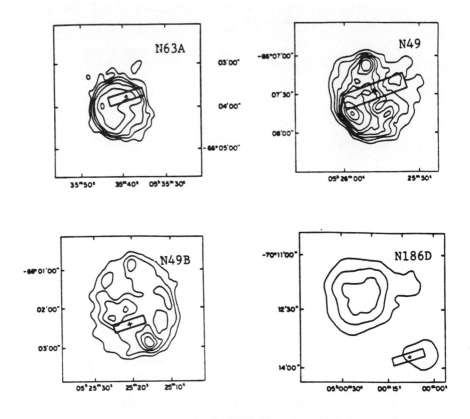

Figure 1
The mean IR position from IRAS is indicated by a cross along with the X-ray contours (reproduced with permission) from Mathewson et al. (1983). The positional uncertainty (1σ) is represented by the rectangle.

INTERPRETATION

The positional coincidence between the IR and the X-ray sources constitute *prima facie* evidence that the IR emission originates from the SNR; from LMC source counts the chance of detecting an unassociated source at 100μm is ~ 0.002.

The IR sources are thermal in nature. Although these SNR's are bright radio objects, extrapolation to 100μm indicates that the non-thermal contribution to the IRAS fluxes is 0.02-0.4%. A significant contribution by IR fine structure lines of [O III], [NII] and [NIII] can be ruled out, but, if any of these SNR's are interacting with dense molecular material, like IC443, then [O I] 63μm may contribute to the band III flux.

Galactic far-IR sources are usually due to dust grains re-radiating the light of luminous young stars. We have considered direct heating of dust by luminous young stars and by resonantly trapped Lyman α and been able to show that radiative heating cannot account for the IR luminosity associated with N63A, N49, and N49B, but not N186D (Graham et al. 1987). Consequently, a strong case can be made for investigating alternative energy sources in these remnants.

The respective masses of IR (M_d) and X-ray (M_x) emitting material can be calculated (Table 1). If the X-ray emitting gas heats the dust, then the ratio of these masses, in the absence of grain destruction, should just be the dust-to-gas ratio in front of the shock. The mass of hot gas for the LMC SNR's has been calculated from the X-ray luminosity and the shock radius assuming a Sedov blast wave and   emissivity in the *Einstein* energy band Λ_x = 1.5 x10^{-23} erg cm^3 s^{-1} appropriate for T= 6x10^6 K and LMC abundances (0.5 of Galactic; Lequeux et al. 1979, Raymond, et al. 1976). The mass of radiating dust M_d is calculated assuming mass absorption coefficient of κ=250 cm^2/g at 100μm (Gatley et al 1977; Harvey, Hoffmann, and Campbell 1979; Harvey, Campbell, and Hoffmann 1979; Harvey, Thronson, and Gatley 1979)

A dust-to-gas ratio of ~0.0015 is representative of the LMC. The measured dust-to-gas ratios for N49B is close to this value, the value for N49 is comparable to the Galactic value. In N63A the dust-to-gas ratio is substantially higher than either the typical LMC or Galactic values. The N186D IR source is clearly not due to dust heated by X-ray emitting gas because a large fraction of the SNR's luminosity is due to dust in cool gas (T«10^{6}K) which is unobservable by *Einstein*. This may explain why the N186D IR source does not coincide precisely with the peak of the X-ray emission.

It is important to emphasize that there are uncertainties in the determination of the dust-to-gas ratio because of uncertainties in κ, Λ_x, L_x . Nevertheless, the inferred dust-to-gas ratios for N63A, N49, and N49B are in reasonably good agreement with the hypothesis that this hot dust is embedded in, and heated by the X-ray emitting gas.

The ratio of IR luminosity to X-ray luminosity L_{IR}/L_x =Λ_{dust}/Λ_x if collisional heating is invoked. At 6x10^{6}K this ratio should be ≈30 for Galactic metallicity and dust-to-gas ratio. For a dust-to-gas ratio which is a factor of 4 lower, and a LMC metallicity which is lower by a factor of 2 we predict Λ_{dust}/Λ_x = 15.

The observed IR to X-ray luminosity ratio is presented in Table 1. To calculate this ratio we have used our estimate of the total IR luminosity and the 0.14-4.5 keV X-ray luminosity from Mathewson et al (1983). L_{IR} is probably known to ~ 25%. However, L_x may have been underestimated by a factor of up to 2. We see that it is energetically feasible that the IR radiation from N63A, N49, and N49B could be due to collisional heating. In fact the IR luminosity is at a level remarkably close to the predicted value.

N186D is peculiar, with a large "IR excess" that cannot be accounted for by collisional heating. It is clear from a comparison of the X-ray and IR data that the dust heating is of a completely different character in N186D. An additional source of heating, must be important in the energy balance if the very high hot-dust to hot-gas ratio is to be explained.

SNR Evolution
We have identified three SNR's where a substantial fraction of the IR luminosity is most plausibly ascribed to dust which is heated by gas-grain collisions. In these SNR grain cooling exceeds atomic processes by an order of magnitude and we estimate that Λ_{dust} ~ 2x10^{-22} erg cm^3 s^{-1} under the conditions prevailing in these SNR's. For the Galaxy Λ_{dust} ~ 8x10^{-22} erg cm^3 s^{-1} at 6x10^{6}K since the dust-to-gas ratio which is 4 times higher than in the LMC.

Our estimate of Λ_{dust}, for a Galactic dust-to-gas ratio, implies that the dynamics of a Galactic remnant will be affected by cooling at a time-scale

$$t_{dyn} = 9000 \, \varepsilon_{51}^{2/11} \, n_0^{-7/11} \quad \text{yr.}$$

where ε_{51} is the supernova energy in units of 10^{51} erg, and n_0 is the pre-shock gas number density in units of cm^{-3}. Departure from adiabaticity occurs sooner than predicted by calculations which only included atomic processes and we might expect that dust cooling speeds the evolution by a factor of ~3.

In order to establish under what conditions grain cooling modifies SNR evolution it is important to investigate grain destruction. Using the thermal grain sputtering rate calculations of Draine and Salpeter (1979) we find that on entering the shock the ratio of grain lifetime t_g to the time for grains to affect the dynamics is

$$t_g / t_{dyn} \approx 3a \, (\varepsilon_{51} \, n_0^2 \, t)^{-2/11},$$

where a is the grain radius in units of 0.1μm. Thus, if $\varepsilon_{51}=1$, the lifetime of a 0.1μm grain always exceeds t_{dyn} so long as $n_0 < 20$cm^{-3}. Accordingly, if the ISM were homogeneous with a mean density of ~ 1 cm^{-3} then the SNR lifetime would be very short, and the SNR would hardly have time to relax to Sedov expansion before radiative effects modified the dynamics. Any structure in the ISM profoundly changes this conclusion.

Consider a two phase model for the ISM consisting of diffuse clouds with $n=20$cm^{-3} and an intercloud medium of $n=0.1$ cm^{-3} (c.f. Spitzer 1978). The grain lifetime in the intercloud medium is 3×10^5 yr and exceeds even the time for atomic cooling to affect the dynamics. In the low density medium, which supports the X-ray emitting blast wave, $t_{dyn}=4\times10^4$ yr. By this time the shock temperature has dropped to 1×10^6 K, and dust and atomic cooling rates are approximately equal. Consequently, the SNR remains adiabatic while dust cooling dominates, and although dust cooling will speed shell formation after t_{dyn}, dust and atomic cooling will be equally important in this process.

The shocks driven into the clouds will not be significantly modified by grain cooling since the grain lifetime is short at high density. This conclusion is supported by IR spectroscopy of the SNR IC443 where $\sim 30\%$ of the iron bearing grains are destroyed in shocks propagating into clouds with $n_0\sim10$-20 cm^{-3} (Wright et al. in this volume, and Graham, Wright & Longmore, 1987)

These observations were obtained as part of the UK Science and Engineering Research Council collaboration in the IRAS project, which was developed and operated by the Netherlands Agency for Aerospace Programs, the US National Aeronautics and Space Administration, and the UK Science and Engineering Research Council. We are particularly indebted to the software support provided by Drs Abolins and Fairclough as part of the Rutherford-Appleton Laboratories Starlink IRAS data reduction project. M. F. B. acknowledges the receipt of an SERC Advanced Fellowship, and J. R. G. is supported by a Royal Commission for the Exhibition of 1851 Fellowship, and a NATO/SERC Fellowship. This work was supported in part by US Department of Energy Contract DE-AC03-76SF00098

References

Draine, B. T., & Salpeter, E. E. 1979, *Ap. J.*, **231**, 77.

Gatley, I., Becklin, E. E., Werner, M. W., & Wynn-Williams, C. G. 1977, *Ap. J.*, **216**, 277.

Graham, J. R., Evans, A., Albinson, J. S., Bode, M. F., & Meikle, W.P.S. 1987, *Ap. J.*, **319**, 000

Graham, J. R., Wright, G. S., & Longmore, A. J., 1987, *Ap. J.*, **313**, 847.

Harvey, P. M., Hoffmann, W. F., & Campbell, M. F.1979 *Ap. J.*, **227**, 114.

Harvey, P. M., Campbell, M. F., & Hoffmann, W. F. 1979 *Ap. J.*, **228**, 445.

Harvey, P. M., Thronson, H. A., & Gatley, I. 1979, *Ap. J.*, **231**, 115.

Lequeux, J., Peimbert, M., Rayo, J. F., Serrano, A., & Torres-Peimbert, S. 1979, *Astr. Ap.* **80**, 55.

Mathewson, D. S., Ford, V. L., Dopita, M. A., Tuohy, I. R., Long, K. S., & Helfand, D. J. 1983, *Ap. J.Suppl Ser.*, **51**, 345.

Ostriker J. P., & Silk, J. 1973, *Ap. J. Lett*, **184**, L113.

Raymond, J. C., Cox, D. P., & Smith, B. W. 1976, *Ap. J.*, **204**, 290.

Spitzer, L. 1978, *Physical Processes in the Interstellar Medium*, Wiley, New York.

POSSIBILITIES FOR OBSERVATIONS WITH THE INFRARED SPACE
OBSERVATORY OF EMISSION FROM SHOCK–HEATED DUST IN SNRs

J. Svestka

Prague Observatory, Prague, Czechoslovakia *

Abstract: The possibilities for observing infrared emission from shock-heated dust in SNRs with the future Infrared Space Observatory (ISO) are illustrated with calculations of the ISOPHOT-P and ISOPHOT-C flux densities and integration times for radiation from six selected SNRs in eight wavelength bands between 4μm and 180μm.

The flux densities of infrared radiation in eight wavelength bands (4, 25, 35, 50, 75, 105, 140, and 180 μm) from four selected galactic SNRs (Tycho, Kepler, Cas A and G292.0+1.8) and two extragalactic SNRs. (1E0102.2-7219 in the SMC and an SNR in NGC4449) are estimated using results derived from IRAS observations (Braun 1985) and calculations of infrared emission from shock-heated dust (Draine 1981). The integration times are determined from the expected sensitivities of the multiband-multiaperture photopolarimeter (ISOPHOT–P) and the far infrared camera (ISOPHOT–C) (see Tables 1 and 2). The data for the extragalactic SNRs were obtained from Inoue et al. 1983, Raymond 1984, Blair et al. 1983 and Blair et al. 1984. The range of acceptable individual integration times was taken to be between 2 seconds and 1 hour, the total integration time up to 24 hours, and the minimum required signal–to–noise ratio equal to 10.

With these restrictions it should be possible to map Cas A in all eight bands, Kepler in seven bands between 4μm and 140μm, Tycho in six bands between 4μm and 105μm, and G292.0+1.8 in five bands between 25μm and 105μm. Similarly, observations of 1E0102.2-7219 in five bands between 25μm and 105μm and of the SNR in NGC4449 in the 25μm band should be possible. Finally, observations can be made with ISOPHOT–P with different apertures in the 4μm and 25μm bands for Cas A and Kepler, and in the 25μm band only for the other SNRs. Total observing time would be about 15.5 hours. The detailed results are given in Table 3.

* A significant part of these calculations was made during the author's stay at the Max-Planck-Institut für Kernphysik, Heidelberg, F. R. G.

Table 1   ISOPHOT-P (Multiband-Multiaperture Photopolarimeter)

Wavelength range (μm)	3 ... 30
Total number of spectral bands	10
Central wavelength (μm)	4, 6.5, 10, 16, 25, others TBD
Spectral resolution	2.5, 2.5, 2.5, 2.5, 2.5, others TBD
Total number of apertures	15
Field of view (arc sec)	5, 8, 12, 20, 30, 40, 60, 80, 110, 150, 180, others TBD
Polarization measurements	3 grid polarizers with 0^0, 60^0, 120^0
Min. detectable flux[1] (mJy)	
Photometry	
4 μm[2]	0.18
25 μm[3]	5.0
Polarimetry	
4 μm[2]	0.44
25 μm[3]	12.4

Table 2   ISOPHOT-C (Far Infrared Camera)

	Array I	Array II	Array III
Wavelength range (μm)	30 ... 60	60 ... 120	120 ... 200
Pixels	4 x 4	3 x 3	2 x 2
Broad bands, Central wavelength (μm)	45	90	160
Broad bands, $\lambda/\Delta\lambda$	2.5	2.5	2.5
Narrow bands, Central wavelength (μm)	35 , 50	75 , 105	140 , 180
Narrow bands, $\lambda/\Delta\lambda$	4	4	4
Min. detectable flux[1] (mJy)			
Photometry	21	32	140
Polarimetry	52	80	350
Polarisation Measurements	3 grid polarizers with 0^0, 60^0, 120^0		

[1] Integration time 100 s; S/N = 10; broadband filter
[2] NEP = $5 \cdot 10^{-18}$ W·Hz$^{-1/2}$
[3] NEP = $3 \cdot 10^{-17}$ W·Hz$^{-1/2}$

Table 3

TYCHO:	4μ	25μ	35μ	50μ	75μ	105μ
Flux (Jy):	0.003	23	63	56	25	9
Time (s) :	225	2	2	3	9	64

5x5 matrix with spacing 2'

25μ:	180"	80"	40"	20"	12"	8"	5"
Time (s):	2	2	2	20	150	760	3600

KEPLER:	4μ	25μ	35μ	50μ	75μ	105μ	140μ
Flux (Jy):	0.0015	11	15	14	6	2	0.5
Time (s) :	23	2	2	2	4	33	2000

2x2 matrix with spacing 2'

4μ:	180"	80"
Time (s):	23	590

25μ:	180"	80"	40"	20"	12"	8"	5"
Time (s):	2	2	2	3	17	85	560

Cas A:	4μ	25μ	35μ	50μ	75μ	105μ	140μ	180μ
Flux (Jy):	0.015	190	170	150	80	26	7	1
Time (s) :	2	2	2	2	2	2	52	2540

3x3 matrix with spacing 2'

4μ:	180"	80"	40"
Time (s):	2	30	480

25μ:	180"	80"	40"	20"	12"	8"	5"
Time (s):	2	2	2	2	2	2	10

G 292.0 +1.8:	25μ	35μ	50μ	75μ	105μ
Flux (Jy):	15	56	50	25	12
Time (s) :	2	5	6	18	75

6x6 matrix with spacing 2'

4μ:	180"
Time (s):	1050

25μ:	180"	80"	40"	20"	12"	8"
Time (s):	2	2	6	95	730	3600

Table 3 (cont'd)

1E0102.2-7219:	25μ	35μ	50μ	75μ	105μ
Flux (mJy):	120	150	140	60	20
Time (s) :	2	2	3	30	260

25μ:	20"	12"	8"	5"
Time (s):	2	5	23	150

in NGC 4449:	25μ
Flux (mJy):	3
Time (s):	280

References

Braun, R.,1985.*Ph.D. Thesis,* , Univ. of Leiden.

Draine, B. T.,1981. *Astrophys. J.,* **245**, 880.

Inoue, H., Koyama, K., and Tanaka, Y.,1983.*Supernova Remnants and their X-ray Emission,* , (ed. J. Danziger and P. Gorenstein) Reidel, Dordrecht p. 535.

Raymond, J. C.,1984. *Ann. Rev. Astr. Astrophys.,* **22**, 75.

Blair, W. P., Krishner, R. P., and Winkler, P. F.,1983. *Astrophys. J.,* **272**, 84.

Blair, W. P., Raymond, J. C., Fesen, R. A., and Gull, T. R.,1984. *Astrophys. J.,* **279**, 708.

INFRARED SPECTROSCOPY OF SUPERNOVA REMNANTS

A.F.M. Moorwood[1], E. Oliva[2] and I.J. Danziger[1]

[1] European Southern Observatory,
   D-8046 Garching bei München
[2] Arcetri Observatory, I-50125 Firenze

Abstract: Spectra of several galactic and LMC supernova remnants have been obtained at R ~ 1500 in the 1-2.5μm region with the cooled grating/array spectrometer, IRSPEC, on the ESO 3.6m telescope. The brightest lines observed are from [FeII]. In RCW 103, which exhibits the highest surface brightness, H (Brγ at 2.17μm) and H2 (1-0 S(1) at 2.12μm) lines have also been detected. A good correlation is found between [FeII](1.64μm) and Hβ implying similar ionization structures and Fe abundances in a wide variety of remnants. The actual [FeII](1.64μm)/Hβ ratio of ~0.2 also implies a high depletion factor (≥0.9) for Fe. Electron densities, extinctions and [FeII] luminosities derived from these data are presented and discussed and attention drawn to a potentially interesting discrepancy between H(Brγ) and Hβ fluxes.

Introduction: Transitions of [FeII] falling in the atmospheric windows between 1μm and 5μm are of great interest for determining Fe abundances, electron densities and the extinction in shock excited regions, particularly relatively high density regions where visible lines might either not be excited or obscured by dust. In the case of supernova remnants, the feasibility of such observations was first demonstrated by Seward et al. (1983) who measured strong [FeII](1.644μm) emission in MSH 15-52 using a filter spectrometer. This line has also been found to be bright in IC 443 (Graham, Wright and Longmore, 1987). With the availability of the cooled grating/array spectrometer, IRSPEC, on the ESO 3.6m telescope (Moorwood et al., 1986) we have pursued observations of this and other [FeII] lines, plus the H (Brγ at 2.165μm) and H_2 (1-0 S(1) at 2.12μm) lines in some cases, on a sample of supernova remnants including Puppis A, Kepler and RCW 103 in the Galaxy and N63A, N49 and N103B in the Large Magellanic Cloud.

Observations: These have been made since November 1985 using the IRSPEC spectrometer which yields R ≃ 1500 with a 6" × 6" entrance aperture at the ESO 3.6m telescope. The H band, 1.5-1.8μm, spectrum of RCW ·103 shown in Fig. 1 demonstrates the prominence of the [FeII](1.644μm) line but also contains several other [FeII] lines of astrophysical interest. In particular, the relative intensity of the

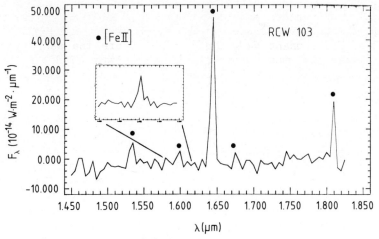

Fig. 1.   H band spectrum of RCW 103.

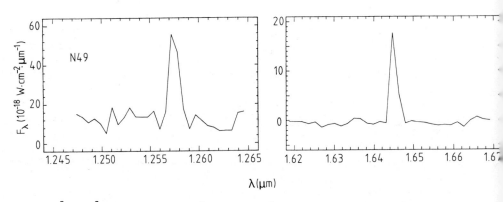

Fig. 2.   [FeII] lines at 1.256μm and 1.644μm in the LMC remnant N49.

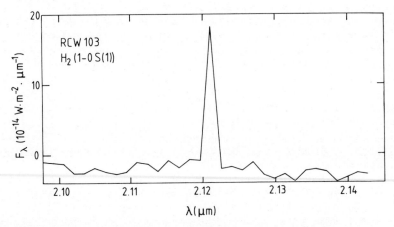

Fig. 3.   Peak of H_2 (1-0 S(1)) line emission in RCW 103.

[FeII](1.599µm) line is density sensitive and the insert spectrum shows a higher s/n ratio measurement of this line made with a longer integration time. Fig. 2 shows the [FeII](1.644µm) and [FeII](1.256µm) lines in the LMC remnant N49. These lines share the same upper level and their ratio is thus a sensitive extinction indicator. In the absence of extinction, the 1.256µm line is the stronger by ≃ 30% but IRSPEC itself is relatively less sensitive at this wavelength than at 1.64µm. This latter line has been detected in all the remnants observed to date and RCW 103 in the Galaxy and N63A and N49 in the LMC have been partially mapped in order to better estimate the total luminosities in this line. Of all the remnants, RCW 103 exhibits the highest surface brightness and, in addition to the several [FeII] lines detected, is the only case where H (Brγ at 2.165µm) and H_2 (1-0 S(1) at 2.12µm) emission (fig. 3) has also been detected.

Results and Conclusions:

- Fe/H. For the remnants Kepler, RCW 103, N63A and N103B for which suitable optical data are available we find an extremely good correlation between the [FeII](1.64µm) and Hβ surface brightnesses. The Crab nebula (observed with a filter spectrometer at the Italian infrared TIRGO Observatory), MSH 15-52 and IC443 also fall on the same relation. This implies that neither the Fe abundance nor the ionization structure exhibits much variation from remnant to remnant. The mean [FeII](1.64µm)/Hβ ratio is measured to be 0.2 or about an order of magnitude smaller than predicted by the shock models of McKee, Chernoff and Hollenbach (1984) for Fe depletion factors ≃ 0.5 which are in accord with the shock destruction of grains predicted by Seab and Shull (1983). From the observed ratio we obtain $Fe^+/H^+ ≃ 10^{-5}$ or $Fe/H < 2.10^{-6}$ with $Fe^+ = Fe$ and $H^+ ≤ 0.2\ H$ computed from a simple ionization structure model. The data appear therefore to indicate a much higher Fe depletion factor $\gtrsim 0.9$ and thus less shock destruction of grains than expected.

    In RCW 103 we can also derive the Fe/H relative abundance using our measurement of H (Brγ) on the [FeII](1.64µm) peak. The result is a factor ≃ 4× larger and the ratio Brγ/Hβ the same factor smaller than expected applying standard recombination theory. In this specific case, the discrepancy might be attributed to possible lack of spatial coincidence between the Brγ and Hβ observations. In N49 however we also observe a Brγ upper limit which is too low relative to Hβ and Graham et al. (1987) have noted the same discrepancy in IC443. The fact that Brγ is relatively fainter than expected makes it difficult to account for by extinction and the normal Balmer decrements tend to exclude collisional population effects. As a next step in understanding this problem therefore we are planning more extensive Brγ observations on regions with well measured Hβ fluxes.

- Ne. The following electron densities have been derived from the [FeII](1.60/1.64) line ratios: Kepler (≥ 2.10⁴ cm⁻³), RCW 103 (3-5.10³), N49 (3-9.10³), N63A (4-11.10³). These densities are systematically higher (about a factor of 2 taking the lower limits) than derived from the [SII] line ratios in Leibowitz and Danziger (1983) and Danziger and Leibowitz (1985). As the [FeII] ratio has a higher critical density this could result from a weighting towards higher density regions. Such an effect could of course also be attributed to errors in the transition probabilities and/or collision strengths.

- Av. The [FeII](1.26/1.64) ratios observed on the relatively unobscured LMC remnants N63A and N49 are close to the expected ratio of 1.3 (Nussbaumer and Storey, private communication). In the quite highly obscured galactic remnant RCW 103 the ratio is only 0.8 implying $A_v \simeq 6$ mag. adopting $A_\lambda = A_\alpha (\lambda/0.656)^{-1.85}$ (Landini et al., 1984), $A\alpha/A\beta = 0.65$ and $A_v/A\beta = 0.86$. As this is rather a steep reddening curve which probably overestimates A_v, the infrared value can be considered to be in reasonable agreement with the $A_v \simeq 4.7$ mag. derived by Leibowitz and Danziger (1983) from the Balmer decrement.

- [FeII](1.64μm) Luminosities. Partial maps of RCW 103, N63A and N49 yield lower limits of 200, 200 and 700 $L_\odot$ respectively for the total luminosities in this line.

- H2. A region containing the bright, southern, optical filament in RCW 103 (see Leibowitz and Danziger, 1983 for a photograph) has been partially mapped in the H_2 (1-0 S(1)) line as well as [FeII] (1.64μm). The H_2 emission extends over several arcminutes along the filament but is displaced $\sim 30"$ from the [FeII] emission in the direction away from the centre of the remnant. Its peak surface brightness (fig. 3) reaches $\sim 30\%$ of that in [FeII] (1.64μm) whereas it is only 6% on the [FeII] peak itself.

Acknowledgements: We are grateful to H. Nussbaumer for providing us with [FeII] transition probabilities in advance of publication.

References:

Danziger, I.J., and Leibowitz, E.M.: 1985, Mon. Not. R. astr. Soc., **216**, 365.
Graham, J.R., Wright, G.S., and Longmore, A.J.: 1987, Astrophys. J., **313**, 847.
Landini, M., Natta, A., Oliva, E., Salinari, P., Moorwood, A.F.M.: 1984, Astron. Astrophys., **134**, 284.
Leibowitz, E.M., and Danziger, I.J.: 1983, Mon. Not. R. astr. Soc., **204**, 273.
McKee, C.F., Chernoff, D.F., and Hollenbach, D.J.: 1984, Galactic and Extragalactic Infrared Spectroscopy (M.F. Kessler and J.D. Phillips, eds., D. Reidel Publ. Co.), p. 103.
Moorwood, A.F.M., Biereichel, P., Finger, G., Lizon, J.-L., Meyer, M., Nees, W., and Paureau, J.: 1986, The Messenger, **44**, 19.
Seab, G., and Shull, J.M.: 1983, Astrophys. J., **275**, 652.
Seward, F.D., Harnden, F.R. Jr., Murdin, P., and Clark, D.H.: 1983, Astrophys. J., **267**, 698.

INFRARED SPECTROSCOPY AND MAPPING OF IC443

Gillian S. Wright[1,2], James R. Graham[3], & Andrew J. Longmore[2]

[1]United Kingdom Infrared Telescope, [2]Royal Observatory, Edinburgh,
[3]Lawrence Berkeley Laboratory, University of California

ABSTRACT

We present infrared (IR) spectroscopy and imaging of the supernova remnant IC443. [FeII] 1.644μm, Brackett γ and H$_2$ 1-0 S(1) were detected. The most striking feature of the spectra is the high [FeII] 1.644μm/Brackett γ ratio, which is typically ~30. We argue that this ratio is due to shock excitation. Two 5'x3' maps of the remnant in the 1.644μm [FeII] line are used to study the line excitation. The IR [FeII] line is a sensitive probe for regions of shocked gas which is especially useful where the extinction is high, or the shocks are too slow to excite optical lines.

INTRODUCTION

Optical spectroscopy is a well established tool for the study of fast shocks (~100 km/s). However, a low excitation IR line which suffers less extinction than optical lines, and falls in one of the near IR windows would be a valuable shock diagnostic. The forbidden IR lines arising from the low lying levels of Fe II satisfy these critera.

The SNR IC443 is an excellent object for developing IR spectroscopy as a shock diagnostic since the blast wave is interacting with interstellar material of a variety of densities at different positions around the SNR shell. This paper summarises the first stage of our study of IC443 in which the 1.644μm [Fe II] line is reported in fast optically bright shocks (see also Graham, Wright & Longmore 1987).

OBSERVATIONS

The data were obtained at UKIRT using the circular variable filter (R ~ 120) in the common user IR photometer UKT9. [FeII] emission line maps of the NE rim and the E extremity of the SNR were made with the CVF centered on 1.644mm, using a 20" aperture sampled on a 15" grid. Since the remnant is large (1°) we chopped 3' off the edge of the remnant but did not beamswitch. Figs. 1 & 2 show the continuum subtracted maps.

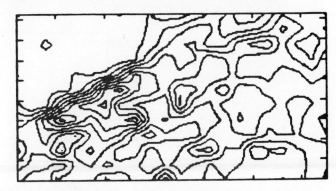

Figure 1
An [FeII]1.644μm image of the NE section of IC443. The contours start at 6σ above the noise, and go up in units of 6σ linearly. The map is 3'x5.5'. The sharp edge of the remnant can clearly be seen, and a number of bright knots are evident.

Spectra were measured on the optically bright NE rim of the remnant (posn. A and 1) and at the E edge (posn. 31). The spectra were obtained in a 20" aperture using a chop of 1.5-3'. Fig. 3 shows the spectrum at position 1.

LINE IDENTIFICATION

At positions A and 1 to the NE we detected lines at 1.644μm and 2.165μm; identified as [FeII]1.6440μm and Brγ (λ=2.1655μm). At position 31 we detected the [FeII] line and a line at 2.121μm which is the 1-0 S(1) transition of molecular hydrogen.

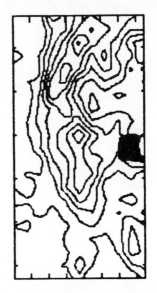

Figure 2
An [FeII] 1.644μm image of the eastern edge of IC443. The edge of the remnant is clearly visible. The map dimensions are 2.75'x5.25'. The contours are the same as in Fig. 1.

IR IMAGES

The [FeII]1.644μm line maps of the optically emitting shocks in IC443 (Fig. 1), demonstrate the ubiquitous nature of this emission. There is a broad correlation between the [FeII] and Hα (Fesen & Kirshner 1980) emission, although close inspection reveals several knots which are much more prominent in the [FeII] map, indicating large variation of the ratio [FeII]/Hα on small spatial scales. Our map of the eastern section (15'SE of the northern map) shows a much higher mean [FeII]/Hα ratio, indicating a significant large scale gradient across the remnant. Comparison of our maps with images in lines other than Hα (Hester, 1987 and private communication) indicates even greater line ratio variations. For example, [OIII]5007Å and [FeII] are both bright in the region covered by Figure 1, but the [OIII] emission associated with Figure 2 is virtually undetectable even   though [FeII] is still almost as bright as it is in Fig. 1a suggesting that [OIII]0.5007μm/[FeII]1.644μm may be a sensitive shock speedometer.

Figure 3
An IR spectrum of IC443 on the NE part of the SNR. The wavelengths of [FeII] 1.644μm and Brγ are indicated. The wavelength scale is not continuous.

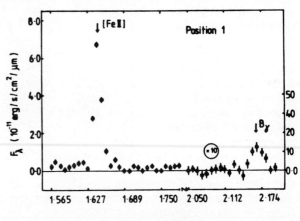

EXCITATION OF [FeII]

The most prominent feature of the IR spectra is the 1.644μm [Fe II] line. The table gives the ratio [FeII]1.644μm/Brγ, (dereddened with $E_{(B-V)} = 0.992$)

IR Line ratios

Position	[FeII]1.644μm/Bγ
A	26.9±2.4
1	41.9±4.9
31	>71 (3 σ)

These ratios are remarkably large. In Orion [FeII]1.644μm/Bγ = 0.057, several orders of magnitude smaller than the measured values in IC443. Clearly the [FeII] line is much brighter relative to hydrogen recombination lines in this SNR shock than in an HII region. This could be due to more efficient excitation of Fe II relative to hydrogen recombination, or due to reduced depletion of Fe in grains, or most likely a combination of both.

The shock models of McKee *et al.* (1984) can be used to predict the ratio [FeII]1.644μm/Bγ. For a preshock density of 10 cm^{-3}, the ratio [FeII]1.644μm/Bγ is 32 and 69 for shock velocities of 40 and 100 km/s respectively. Thus the observed [FeII]1.644μm/Bγ ratio is comparable to the calculated values for these conditions. Optical [FeII] lines of moderately low excitation are also present in the spectrum of IC443. However no lines of multiplets which might feed the upper state of the 1.644μm line (a^{4}D), such as a^{4}D-b^{4}F or a^{4}D-b^{2}P have been detected. These facts confirm that the excited states of FeII are populated collisionally.

The [FeII] lines which have been detected arise from upper states with a large range of excitation temperature (8,000-25,000 K), so the line ratios should be sensitive to temperature. The mean temperature derived from the IR/optical [Fe II] line ratios, 5,100±1,000 K, is significantly lower than that inferred by Fesen & Kirshner (1980) from [O III] (25,000K), [N II], [SII], or [O II] (11,000K). The [Fe II] lines must be excited downstream of the N, O, and S lines where the gas is cooler and the ionization lower. There is only a small scatter in the derived temperatures (~ 1,600K) as expected for collisional population. However, there is a significant trend for the temperature derived from different [Fe II] lines to correlate with the excitation energy of the upper level of the line, indicating a range of temperature in the Fe II ionization zone. The temperature derived from some of the lower excitation [Fe II] lines indicates that these lines are excited where hydrogen has recombined (T < 5000K). In the shock models of McKee *et al.* (1984) some emission from Fe II is excited in the H recombination zone , but substantial emission is excited in predominantly neutral gas.where ionization of Fe II is maintained by UV photons penetrating far downstream (~ 10^{15} cm) from the hot postshock gas where they were emitted.

GRAIN DESTRUCTION

Iron is one of the more highly depleted elements in the interstellar medium, and so predicted [FeII] intensities are especially sensitive to modification of elemental depletions by dust grain destruction. If we assume that the shock model calculated for n=10cm^{-3} and v=100km/s is appropriate for position 1(optical spectroscopy of this filament indicates n = 10 - 20 cm^{-3}, v = 65 - 100 km/s) then the [FeII]/Brγ ratio requires that ~ 30% of the iron bearing grains entering the shock must have been destroyed.

THE MOLECULAR HYDROGEN DETECTION

Our detection of [Fe II] and H$_2$ in the same aperture at position 31 is noteworthy because Fe II and H$_2$ cannot coexist in substantial quantities. As the observed emission arises from the very edge of the remnant it seems unlikely that the [Fe II] and H$_2$ lines arise from two distant regions along the same line of sight. The emitting regions are probably separated by a distance less than or comparable to the projected diameter of the aperture on IC443 (0.14 pc). It is possible that the observed emission arises from the SNR shock propagating into two adjacent regions of different density. Alternatively molecules must form downstream of the ionized gas where the [Fe II] lines are excited.

This work was carried out at UKIRT which is operated by the Royal Observatory Edinburgh on behalf of the U.K. Science and Engineering Research Council. J.R.G is supported by fellowships awarded by the Royal Commission for the Exhibition of 1851, and the U.K. Science and Engineering Research Council. This work was also supported in part by the US Department of Energy under contract No. DE-AC 03-76SF00098.

REFERENCES
Fesen, R. A., and Kirshner R. P. 1980, *Ap. J.*, **242**, 1023.
Graham, J. R., Wright, G. S., & Longmore, A. J. 1987, *Ap. J.*, **313**, 847.
Hester, J. J., 1987, *B.A.A.S.*, **18**, No. 4, 30.11.
McKee, C. F., Chernoff, D. F., and Hollenbach , D. J. 1984, *Galactic and Extragalactic Infrared Spectroscopy*, p103. Eds. M.F. Kessler & J.P. Phillips. D. Reidel, Dordrecht.

IC 443: THE INTERACTION OF A SNR WITH A MOLECULAR CLOUD

Michael G. Burton
NASA Ames MS 245-6
Moffett Field, California 94035 USA

Abstract: Observations are presented of shocked line
emission from H_2, CO and HCO+ molecules in the SNR
IC 443. IC 443 is the most luminous galactic H_2 emission
line source yet discovered. The implications for physical
processes in shocked molecular gas are discussed.

Introduction: The SNR IC 443 is a laboratory for the study of the
interaction of a shock wave with a molecular cloud. The expanding
shell of the SNR is interacting with ionised, atomic and molecular gas
at different positions within the remnant. Near its SE edge it is
running into a dense molecular cloud, exciting molecular hydrogen line
emission, as first observed by Treffers (1979). Shock-excited line
emission has also been observed in CO, OH, HCN, HCO+ and CS (eg. White
et al., 1987), as well as high-velocity 21cm atomic hydrogen emission
(eg. Braun & Strom, 1986). This paper reports detailed observations of
the molecular line emission from H_2, CO and HCO+ in IC 443. The
molecular hydrogen data were obtained at the UKIRT and the CO and HCO+
data with the Nobeyama 45m radio telescope. A more complete discussion
of some of these observations can be found in Burton (1986), Burton
(1987) and Burton et al. (1987).

Molecular Hydrogen Observations: Figure 1 presents a map of the 1-0
S(1) H_2 line, at 2.12 μm, in IC 443. It is incomplete as the NE and
central portions of Figure 1 have not yet been observed. The H_2 line
emission is seen to come from several bright peaks distributed along a
sinuous ridge which forms a nearly complete ring, at least 20 pc long,
and lying between the two optical lobes of the source. There are over
20 resolved peaks, and the two brightest, of comparable peak flux, are
located diametrically opposite each other. The total H_2 line
luminosity, allowing for two magnitudes of extinction for the S(1) line
and a distance to the source of 1500 pc, is estimated to be about 2000
$L_\odot$, making IC 443 the most luminous and extended galactic H_2 source

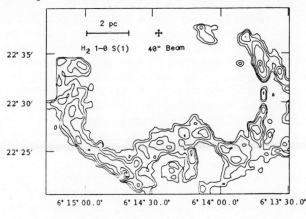

Figure 1.
1-0 S(1) line
emission in
IC 443

yet discovered. The width of the emission ridge is unresolved in
Figure 1. A 5" resolution, fully-sampled map of the peak emission
region is shown in Figure 2. Although line emission is detected from
all over this region, the majority of it comes from the emission ridge;
the width of the ridge remains unresolved even at this higher
resolution, corresponding to a spatial scale of <0.03 pc.
     Figure 3 shows a spectrum in the K window of the emission from the
H_2 emission peak. The emission consists entirely of H_2 lines, with
negligible continuum. No Br γ recombination radiation is seen, which
limits the degree of ionisation of the gas. The spectrum is typical of
shock-excited sources. Measurements of the line profiles at several
locations (not shown) show typical FWHM's of 20-30 km/s. The peak
velocity of the emission, however, varies considerably with position
(by up to 60 km/s in the positions measured).

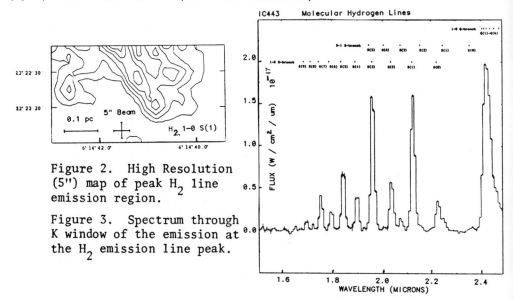

Figure 2. High Resolution
(5") map of peak H_2 line
emission region.

Figure 3. Spectrum through
K window of the emission at
the H_2 emission line peak.

CO and HCO+ Observations: Partial maps of CO, HCO+ and HCN (1-0) line
emission (White et al. 1987) and a complete map of the 21-cm HI
emission (Braun & Strom, 1986) have been published. In the regions of
overlap with the H_2 map the morphology of the emission from all these
species is very similar. In particular the location of the emission
ridge, and the peaks along it, are coincident to the 40" resolution of
the H_2 map. In Figures 4 & 5 are presented CO and HCO+ (1-0) maps of
a portion of the H_2 emission ridge. (The velocity range is -20 to
-10 km/s for each map, and the offsets are from a (0,0) point of
$6^h 14^m 43^s$, 22 23' 00" (1950).) The resolution for these maps is
higher than the H_2 map, and the emission is clearly highly
structured. The emission ridges for the two maps are coincident with
the H_2 emission ridge. The location of most of the CO and HCO+
emission peaks are identical; however two of the peaks are located a
beamwidth apart. Since the lines were observed simultaneously this
offset is probably real; however observations of higher excitation
lines are required to determine whether this is due to differences in

excitation conditions or chemical abundances between the peaks.

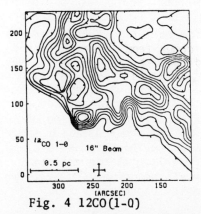

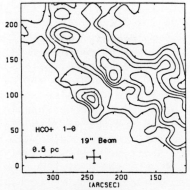

Fig. 4 12CO(1-0)   Fig. 5 HCO+(1-0)

The high velocity resolution attainable for these lines allows a detailed investigation of the shock structure within the source. Figures 6 & 7 present profiles of the CO and HCO+ lines, taken with 0.6 km/s resolution, at the brightest H_2 emission peaks (the top profile is from the bright peak at the NW end of the ridge, and subsequent profiles are from positions moving along the H_2 ridge towards its NE end; the profiles in lines 7 & 8 are from the region of the H_2 emission peak). The profiles are complex, and vary considerably from location to location. At each position, however, the CO and HCO+ profiles are very similar (apart from the rest-velocity spike). Particularly noteworthy are the differences between the two brightest H_2 peaks even though they have comparable H_2 fluxes (compare profiles in lines 1 & 2 with 7 & 8). For profile 5 there is no rest velocity component, the emission starting at -40 km/s and extending to -90 km/s with the HCO+ intensity exceeding the CO intensity.

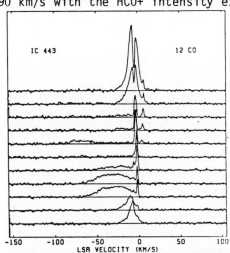

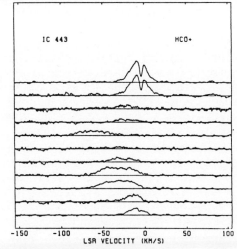

Fig 6. CO profiles at H_2 peaks.   Fig. 7 HCO+ profiles at H_2 peaks.

Discussion: The data have presented clear evidence for the presence of a shock, driven by the expanding gas of a supernova remnant, within a molecular cloud. In this section three aspects of the molecular shock in IC 443 are briefly described; further details can be found in Burton et al. (1987). They are the origin of the shocked gas, the cooling of the hot molecular gas and a model for the source.

The similar morphology of the shocked H_2 and high-velocity HI are suggestive of a partially dissociative shock. The shock leaves behind it accelerated molecular and atomic gas, but little ionised gas. The degree of dissociation varies from point to point, and can be determined from a comparison of the column densities of shocked CO and high-velocity HI. Another possibility is that along each line of sight are contained regions of dissociated atomic gas and regions of shocked molecular gas, the molecular gas presumably remaining in dense clumps where the shock velocity is lowest. A comparison of the emission velocities with those of the fine structure lines of OI and SiII, whose emission is more sensitive to higher velocity shocks, may help discriminate between the models.

Nearly all the emission from IC 443 at optical and near-IR wavelengths is line emission. The H_2 data can be used to investigate the relative importance of H_2 line cooling in the shocked molecular gas. For the H_2 emission peak, estimating the mechanical luminosity delivered per unit area to the shock front (from the X-ray pressure driving the expansion) and equating this to the energy radiated away through H_2 lines, suggests that approximately 80% of the energy input to this region is in fact radiated away through H_2 lines. Since the degree of dissociation appears to be large, a significant fraction of the mechanical energy must also go into dissociating the molecules. We suggest, therefore, that molecular hydrogen line radiation and dissociation provide major, if not the dominant, coolants for the gas.

The ring-like appearance of the H_2 emission ridge, situated between the optical emission lobes, suggests a model for the source. We assume that the quiescent molecular cloud running NW-SE across the source, and roughly bisecting the optical shell, actually contains the SNR. A possible history for IC 443 is that the SN exploded within the remnant of a molecular disk left over from the formation of the star which exploded. The expanding shock wave is running into this disk, compressing and heating the gas, and forming a shell on its inside surface. The shocked molecular and atomic lines are emitted from this shell, and are observed as a thin ring. Above and below the disk the expansion is less impeded, and the blast wave has broken out of the molecular disk and is shocking lower density neutral gas, producing the $H\alpha$ emission of the optical lobes.

Acknowledgements: These results are from a collaboration with Tom Geballe, Peter Brand, Adrian Webster and Tetsuo Hasegawa. I am also extremely grateful to the staff at the UKIRT and Nobeyama observatories for their invaluable assistance in obtaining the observations. This work was done while the author held a National Research Council - NASA Ames Research Associateship.

Braun, R. and Strom, R.G., 1986, Astr. Ap. 164, 193
Burton, M.G., 1986, Ph.D. Dissertation, University of Edinburgh
Burton, M.G., 1987, submitted to QJRAS
Burton, M.G. et al., 1987, submitted to MNRAS
Treffers, R.R., 1979, Ap. J. Lett., 233, L17
White, G., 1987, Astr. Ap. 173, L17

ANALYSIS OF MULTIWAVELENGTH OBSERVATIONS OF IC443

M.L. McCollough[1,2] and S.L. Mufson[2]

[1]Physics Department, Oklahoma State University
[2]Astronomy Department, Indiana University

Abstract: High resolution observations at radio, infrared, optical and X-ray wavelengths have been made (Mufson et al. 1986). The infrared spectrum determined from the IRAS observations has been fitted with models which include shock heated dust, infrared line emission, and radiatively heated dust emission. Numerical simulations of a supernova expanding into a uniform medium, which describe the X-ray and large scale radio emission, are presented.

Introduction: Recent observations have shown that IC443 is expanding into a complex region of the ISM. The appearance and nature of emission at various wavelengths is very strongly governed by the components of the ISM with which the SNR is interacting. Details of the observations and their analysis are presented in Mufson et al. (1986).

IR Spectrum: Integrated and color corrected infrared flux densities for IC443 in the four IRAS bands are shown in Table 1. In modeling these infrared data three major sources of emission are considered: shock-heated dust (BB), IR line emission (lines) and radiation-heated dust (RHD). The method used to fit the data and the results of calculations for $I_{H\beta}= 2.85 \times 10^{-4}$ erg/cm^{2}s sr are given in Mufson et al. (1986). Suggestions have been made (R. Braun, private communication) that the average H$_\beta$ intensity in the IC443 region used in these fits may be substantially lower. We have therefore investigated the effect of using an average H$_\beta$ intensity which is a factor of 10 lower. The results of these calculations are shown in Table 2. From both sets of fits it is apparent that IR

Table 1. Measured Flux Densities

IRAS Band	S_ν
12μ	58 ± 17 Jy
25μ	90 ± 27 Jy
60μ	1330 ± 266 Jy
100μ	1810 ± 360 Jy

Table 2(a). Percentage of flux in the four IRAS bandpasses.
$N_H = 10^{21}$ cm^{-2}

IRAS Band	BB(%)	Lines(%)	RHD(%)
12μ	*	*	*
25μ	65	10	25
60μ	91	4	5
100μ	77	0	23

T = 57 K   * not well fitted by the model
$\tau = 1.4 \times 10^{-6}$ (n = o), $\tau = 1.9 \times 10^{-8}$ (n = 1)
$I_{H\beta} = 2.85 \times 10^{-5}$ ergs/cm^2 s sr

Table 2(b) Percentage of flux in the four IRAS bandpasses. N_H
allowed to vary

IRAS Band	BB(%)	Lines(%)	RHD(%)
12μ	*	*	*
25μ	37	9	54
60μ	76	4	20
100μ	44	0	56

T = 53 K   * not well fitted by the model
$\tau = 1.2 \times 10^{-8}$ (n = 1), $N_H = .3\text{-}3 \times 20^{21}$ cm^{-2}
$I_{H\beta} = 2.85 \times 10^{-5}$ ergs/cm^2 s sr

Table 3.   SN Model Parameters

$E_o = 3 \times 10^{50}$ ergs

n_o (NE) = 0.11 cm^{-3}

n_o (SW) = 0.01 cm^{-3}

$M_{ejecta} = 8M_\odot$

$B_{ISM} = 15\mu G$

$T_{ISM} = 10^4 K$

line emission and radiatively heated dust play an important role in
the IR spectrum of IC443. It should be noted that in the case of
lower line intensities the 12μ flux density is not well fitted by the
three components alone. But as noted by Dwek (1986), the presence
of small dust grains in the shock-heated dust leads to an
enhancement of the short wavelength flux density. This effect can
account for the missing 12μ flux density.

<u>Models of X-Ray and Large Scale Radio Emission</u>: In order to get an
approximation to the X-ray and radio emission, 1-D numerical models
of the structure of a SNR expanding into a uniform density ISM have
been made. The hydrodynamics code used is a version of SOLA-STAR
(Cloutman 1980) modified to handle transient fluid flows in the ISM
(Wolff 1986). To this code magnetic field and cosmic rays have been
added. The parameters used in the code are shown in Table 3. A
density difference of a factor of 10 is needed to explain the
difference between the size of IC443 in the NE to its size in the
SW.

The radio emission is taken to be synchrotron emission from the
swept up magnetic field and cosmic rays. For an initial field of
3μG with equipartition in energy between the field and cosmic rays,
the radio flux density from the models falls two orders of magnitude
short of what is observed. An enhancement of a factor of 5 in the B
field is required to reproduce the observed flux density. In Fig. 1
we show a plot of synchrotron emission as a function of radius for
the NE region.

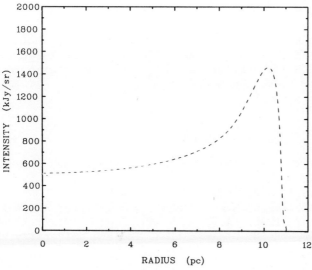

Fig. 1: Synchrotron emission versus radius for the NE region.

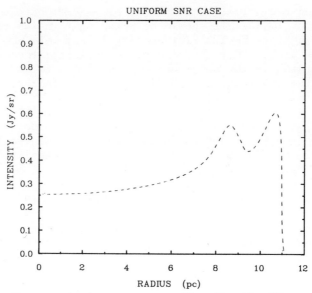

Fig. 2: X-ray emission versus radius for the NE region.

The X-ray emission from a uniform case arises from two sources, the
SN ejecta and the swept-up ISM.   In younger SNR models with ejecta
of several solar masses, the reverse shock traversing the ejecta can
be a strong source of X-rays.   Later when the SNR has swept-up many
times the mass of the ejecta, the model approximates a Sedov
solution as it should.   In fitting our models to IC443 we find it
has not yet evolved to the Sedov phase.   In this stage of its
evolution, IC443 does not appear as a limb-brightened source.   Fig.
2 shows the X-ray emission as a function of radius for the NE.   This
clumpy double peaked picture is more consistant with what is
observed in the NE of IC443.   However, care must be taken in
interpreting IC443 with uniform models, since the region in which
IC443 is evolving is very nonuniform.

References

Cloutman, L.D. 1980, Los Alamos Scientific Laboratory Informal
    Report. LA-8452-MS.
Dwek, E. 1986. Ap.J. 302, 363.
Mufson, S.L., McCollough, M.L., Dickel, J.R., Petre, R., White, R.,
    and Chevalier, R. 1986.   A.J. 92, 1349.
Wolff, M.J., 1986.   Ph.D. Dissertation, Indiana University.

DIFFERENTIAL IMAGING OF [Fe X] in IC 443

Larry Brown, Bruce E. Woodgate and Robert Petre
NASA/Goddard Space Flight Center
Greenbelt, MD 20771

Abstract: This paper presents images of two areas of the supernova remnant IC443 showing emission from the [Fe X] 6374Å red coronal line taken with an emission line differential imaging camera. The areas are in the vicinity of strong soft X-ray emission as observed with the Einstein Observatory. The [Fe X] emission is patchy on the scale of seconds of arc. For the highest emission regions we find an electron density of approximately 100 cm^{-3} and gas pressures of 10^8cm^{-3}K. No correlation is found between the X-ray and [Fe X] knots, and the results support a clumpy, multi-temperature region where the [Fe X] knots are balanced between collapse and evaporation.

Introduction

The middle-aged supernova remnant IC443 is interacting with a large dense interstellar cloudy region, as shown by optical, radio, infra-red and X-ray observations. The X-rays show a hot gas ($T \simeq 1.8 \times 10^7$K) in the interior of the remnant, suggesting a blast wave when the gas was heated to 1100 km/s (Petre et al., 1987). The highest radial velocities observed for individual optical filaments is 220 km/s (Losinskaya, 1979). Evidence for intermediate temperature regions was obtained by detection of the [Fe X] 6374Å red coronal line (Woodgate et al., 1979) which requires a temperature of 1.2×10^6K. We present observations in [Fe X] of the maximum soft X-ray region to investigate the relationship between the two temperature regions.

Observations

The two regions shown in Figure 1 were observed using the differential camera at the NASA/Goddard Space Flight Center 36-inch Telescope. On-band and off-band images through 6Å FWHM filters were observed using the 100x100 pixel photon-counting imager with an S-20 photocathode, and the difference image obtained. The fluxes obtained are shown in Table 1 for Region A. Figure 1 shows the positions of these knots detected in [Fe X] above 3σ compared to the Einstein Observatory HRI X-ray contours (shown in white). The field-of-view is 5 arcmin 50 arcsec. The 3σ surface brightness contour level is 1.5×10^{-6}erg cm^{-2}s^{-1}sr^{-1}.

For comparison with these [Fe X] observations, the Einstein HRI original data was remapped and binned to 10 arcsec to obtain

TABLE 1

Summary of Observed [Fe X] Data

Position RA sec	Dec "	Surface Brightness Max (10^{-6} ergs cm^{-2} s^{-1} sr^{-1})		Area# 3sd 2sd arcsec^2		Statistical Significance of Max(sd)
Region A	06:14:00 +22:50:00	[1950.0]				
E1 +8.0	+ 16	2.8	±.5	221	453'	5.2
SE +5.2	- 47	2.6	±.5	147	294	5.2
E2 +5.0	- 9	1.8	±.5	122	221f	3.6
S" -1.1	-103	3.1	±.6	49	147	5.0
W -3.1	+ 9	1.5	±.5	208	281f	3.1
NW -3.6	+ 93	2.5	±.4	282	355'	6.1
WS -7.4	- 16	1.8	±.5	134	208	3.5
SW -8.7	-100	2.7	±.5	98	172f	5.6
NWC -8.7	+114	2.9	±.4	208	417'f	6.9
WN -9.7	+ 54	3.1	±.5	245	380	5.9

Total positive values: 0.143±.024
        values above 3sd: 0.038±.013

Notes: ' Source may extend further.   f Filamentary.
        " Probably not a source.   x Extends around star.
        # Area covered by values above 3sd or 2sd.

TABLE 2

Summary of [Fe X] Calculations

Position	Surface Brightness Max(10^{-6}) (ergs cm^{-2} s^{-1} sr^{-1}) Obser. Absor. Cor.'		Depth Pc" (±.02)	n_e cm^{-3}	P_e (10^8) (cm^{-3} K)
Region A	06:14:00 +22:50:00	[1950.0]			
E1	2.8±.5	35.	0.13	83.	1.0
SE	2.6±.5	33.	0.10	89.	1.1
E2	1.9±.5	24.	0.09	82.	1.0
S	3.1±.6	39.	0.07	120.	1.4
W	1.6±.5	20.	0.11	67.	0.8
NW	2.5±.4	31.	0.12	80.	1.0
WS	1.9±.5	24.	0.09	81.	1.0
SW	2.7±.5	34.	0.08	100.	1.2
NWC	2.9±.4	36.	0.10	95.	1.1
WN	3.1±.5	39.	0.12	92.	1.1

Note: ' Corrected for interstellar absorption by factor 12.5
       " Based on width at half of peak standard deviation.

accurate positions of the brightest points. No overlaps between
the [Fe X] and X-ray bright points were found with a positional
accuracy of 10 arcsec, although the [Fe X] knots are found in the
region of generally high X-ray emission on the smoothed contour
maps.

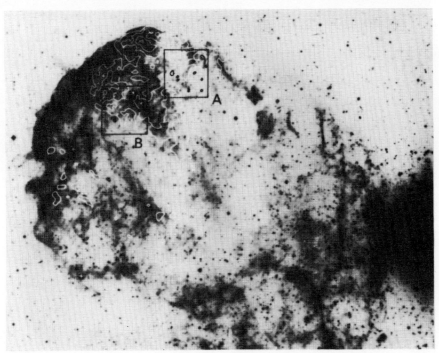

Figure 1.   [Fe X] knots with X-ray contours and optical image.

Discussion

Petre et al. (1987) give N(H) = 6.5 x 10^{21}cm-2 for the
interstellar gas absorption from the Einstein Solid State
Spectrometer data. A normal gas to dust ratio would predict an
absorption factor due to dust of a factor 12.5 for the [Fe X]
line. Applying this correction, and assuming spherical knots, we
estimate the diameters and electron densities in the knots. These
are shown in Table 2.

We calculate the lifetime of these knots:

The lifetime against radiative cooling is given by

t_{cool} = 3kT/2E n_e

where the total emissivity E' = 1.5 x 10-22 erg cm^3s-1.

For the average electron density in Table 2, t_{cool} = 600
yrs. If the knots are embedded in the less dense, hotter plasma as

suggested by the soft X-ray observations, conductive heating will occur. The time for knot evaporation (Cowie and McKee, 1977) is given by

$$t_{evap} = 3.3 \times 10^{20} \, n_e R^2 \, T_L^{-2.5} g \text{ years,}$$

or $\quad t_{evap} = 240 \, n_e R^2$ years $\qquad$ with R in pc, n_e in cm^{-3}

$\qquad t_{evap} = 200$ years for the average knot in Table 2.

Figure 2 shows the [Fe X] knots in the R, n_e plane with the theoretical equilibrium curve between conductive heating and radiative cooling. The position of the observed points is consistent with equilibrium within the large uncertainties in the interstellar absorption correction and the parameters of the theoretical estimates.

Figure 2.

Size and density of
[Fe X] knots compared to
radiative cooling and
conductive heating rates.

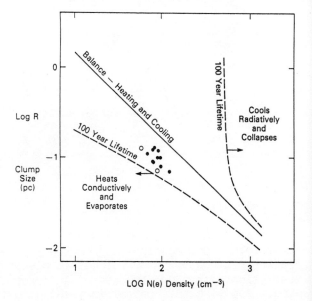

References
Cowie, L. L., and McKee, C. F. 1977, Astrophys. J. 215, 213.
Losinskaya, T. A. 1979, Astr. Astrophys. 71, 29.
Petre, R. Szymkowiak, A. E., and Seward, F. D. 1987, Astrophys. J., (submitted).
Woodgate, B. E., Lucke, R. L., and Socker, D. G. 1979, Astrophys. J. 229, L119.

FABRY–PEROT OBSERVATIONS OF [FeX] IN THE CYGNUS LOOP AND IC443[*]

J. Ballet[1], J. Caplan[2], R. Rothenflug[1], A. Soutoul[1]

1-Service d'astrophysique, CEN SACLAY
91191 GIF sur YVETTE Cedex, FRANCE
2-Observatoire de Marseille
13248 MARSEILLE Cedex 4, FRANCE

Abstract: We present the first results of an observational program of SNRs in the coronal lines of [FeX] and [FeXIV] using the Fabry-Perot spectrophotometer of the Observatoire de Marseille. These support previously published brightnesses.

I INTRODUCTION

The coronal lines of iron present the best opportunity of studying the hot gas inside supernova remnants (SNRs) directly from the ground. Moreover, X-ray observations inform us of the average conditions (temperature and density) along the line of sight whereas the iron lines highlight specific ionization stages or (in an equilibrium situation) specific temperatures. The [FeX] line at 6374 A arises in significantly cooler media (1 10^6K) than the X-ray gas. Its prominence suggests that the density is locally enhanced inside the hot phase (assuming pressure equilibrium).

The first technique used to measure the intensity of these lines (e.g. Woodgate et al.,1979) involved a narrow-band filter. The continuum around the line was estimated and subtracted using nearby regions of the sky (outside the remnant) and of the spectrum. More recently, Teske and Kirshner (1985) obtained images using a CCD detector (and a narrow-band filter). Both groups detected emission clearly associated with the SNRs. Those techniques eliminate the night sky emission very well, but may be affected by the source's continuum or [OI] 6364 A line (in the wing of the [FeX] filter).

We have observed the Cygnus Loop and IC443 with an 80 cm telescope, using relatively large diaphragms (up to 5.2'). A region around the [FeX] wavelength was isolated with a 9 A FWHM interference filter and scanned with a Fabry-Perot interferometer. We adopted a rather low Fabry-Perot resolution (3 to 4 A FWHM) so as to have a high sensitivity for broad lines. This scanning approach is safer (the line can be 'seen') but less efficient than integrating over the whole filter bandpass. Ideally, the line width and the radial velocity can also be determined. We find that a line near 6374 A, which we attribute to [FeX], appears in many places along the filaments. Its intensity roughly matches those found by previous investigators.

II OBSERVATIONAL PROCEDURE

Absolute calibration of our instrument was done once or twice nightly on the stars 58 Aql and 121 Tau. The atmospheric transmission was monitored on the SNR Hα line. We did not attempt to subtract a background spectrum because of the variability of the night sky lines, but simply checked that the 6374 A feature did not show up outside of the SNR. One recording (50 scans) lasted about 15 min, well within the stability limit of the instrument against temperature variations. Successive observations of the

[*]Observations obtained at the Observatoire de Haute Provence (France).

same field were summed whenever statistically compatible.

A typical spectrum is the sum of a small constant (photomultiplier dark noise), a filter-shaped continuum, and one, two or even three lines ([OI] 6364 A, [FeX] around 6374 A and OH 6379 A). The [FeX] line (at most 10% of the background) can be recovered only by fitting the continuum and the lines to the data. The shape of the lines (including [FeX]) was taken to be the Fabry-Perot response broadened to the width of the strongest [OI] lines in the filaments (about 75 km/s).

We estimate the uncertainties with the standard method of computing χ^2 as a function of the intensity of the iron line, leaving all other parameters free. The large number of parameters and their correlations with the iron line result in uncertainties larger than those inferred from the count rates and Poisson statistics. The uncertainty of the surface brightness (and the detection limit) is around 10^{-7} erg cm^{-2} s^{-1} sr^{-1} on a typical field. We should be able to lower this in the near future by closer monitoring of the slow wavelength drifts of the filter and the Fabry-Perot system.

Note that, because these observations of faint emission require a filter bandwidth which is smaller than the free spectral range of the Fabry-Perot, our methods differ from those used previously with this instrument (Caplan and Deharveng, 1985).

III The CYGNUS LOOP

Three areas of the Cygnus Loop were previously observed in [FeX]. Lucke et al. (1980) reported observations across the filaments in the north-east and west parts of the Loop. The surface brightness in their fields was at most 1.4 10^{-7} erg cm^{-2} s^{-1} sr^{-1}. Teske and Kirshner (1985) obtained a CCD image of a field in the east where the X-ray emission (Ku et al., 1984) is maximum. They derived much brighter structures (up to 10 10^{-7} erg cm^{-2} s^{-1} sr^{-1}) on much smaller scales (less than 1′).

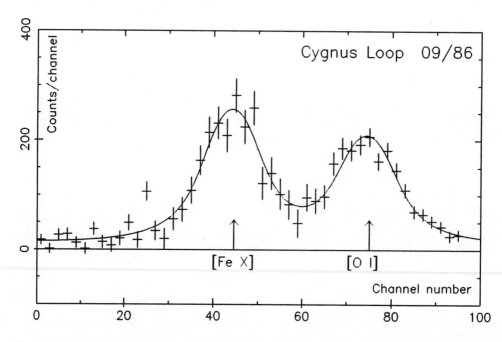

The above figure shows the spectrum (continuum subtracted) obtained in 15 min on a field (5.2′ in diameter) including the bright part of Teske and Kirshner's field. Wavelength increases to the left; the dispersion was 3 channels/A and the instrumental width 4 A. The [FeX] line (on the left) is not Doppler shifted and does not appear broader than the [OI] line. This is a common feature of all the fields where [FeX] is above the detection limit. Our data do not require the existence of velocities greater than $\pm$ 75 km/s. The surface brightness of the [FeX] line in this particular field – our brightest – is 4.5 $\pm$ 1.2 10^{-7} erg cm^{-2}s^{-1}sr^{-1} (uncorrected for interstellar extinction). We reduced the diaphragm diameter to 3.7′: the surface brightness did not increase significantly (4.8 $\pm$ 1.8). The line was also detected (barely above the 90% confidence level) at the peak emission location reported by Lucke et al., at levels compatible with theirs. Precise comparison between their rectangular and our circular fields is difficult.

We observed the north-east part of the Loop during ten nights around the new moon in early September and early November 1986. The [OI] line in the wing of the interference filter usually dominates the spectrum, particularly in November because the lower temperature shifted the filter to the blue. It is much stronger than the [OI] night sky line in most of our fields. The continuum around the filaments is about equal to the sky background, indicating a potential problem in non-spectroscopic studies. We did not subtract the faint stars in our fields, but they are well below this intensity (equivalent to an 8.5 mag star). In 10 (out of 40) pointings along the filaments, the [FeX] line was above the 90% confidence threshold.

IV IC443

The bright north-east filaments of IC443 were observed by Woodgate et al. (1979). They report a peak brightness in [FeX] of 1.3 10^{-7} erg cm^{-2}s^{-1}sr^{-1} somewhat inside the filaments, close to the maximum X-ray emission (Petre et al., 1983). Gensheimer and Teske (1987) obtained a CCD image (in the line) of the same area. As in the Cygnus Loop, their data indicate that locally brighter features (up to 5 10^{-7} erg cm^{-2}s^{-1}sr^{-1}) appear when a better spatial resolution is achieved.

We observed IC443 in early November 1986. Unfortunately, because of the cold, the filter let through much more [OI], which was often ten times stronger in our spectra than the λ6374 feature. Therefore the [FeX] emission at around 2 10^{-7} erg cm^{-2}s^{-1}sr^{-1} that we found along the filaments must be confirmed under better conditions. On a warmer night, and far from the filaments ([OI] much fainter) we measured 2.9 $\pm$ 1.3 10^{-7} erg cm^{-2}s^{-1}sr^{-1} in a 5.2′ circular field near 1(iv) of Woodgate et al.

In late January 1987 we used a new filter, whose bandpass was shifted 3 A redward. We reduced the diaphragm to 2.8′, and explored the bright fields of Woodgate et al. At only one location (see next figure) did we detect the line at the 90% confidence level (2.3 $\pm$ 0.9 10^{-7} erg cm^{-2}s^{-1}sr^{-1}), but the upper bounds in the other fields were not very tight (3 10^{-7}).

We have also, for this same region, observed the spectrum at the wavelength of [FeXIV] (5303 A). Beside the night sky line at 5299 A (OH) we find a feature, with a typical intensity of 1 $\pm$ 0.5 10^{-7} erg cm^{-2}s^{-1}sr^{-1} in 5.2′ fields, which we believe to be [FeXIV] redshifted to 5304–5 A. This result is still very preliminary.

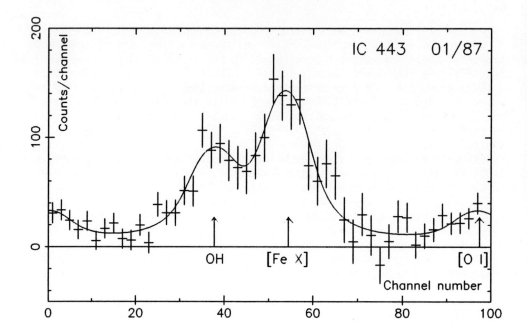

The above figure is the spectrum obtained in 45 min on a 2.8′ field near 1(iii) of Woodgate et al. Wavelength increases to the left; 6374 A is at channel 55. The dispersion was 3.5 channels/A and the instrumental width 3 A. The transmission of this new filter was much lower in the [OI] line (at channel 97, and channel 0 in the next order), but higher in the 6379 A line at channel 37 (probably OH) that did not appear in the former spectra (section III).

V CONCLUSION

We have adapted the Fabry-Perot spectrophotometry technique to the coronal [FeX] λ6374 line. The present sensitivity of the method is 1 10^{-7} erg cm^{-2} s^{-1} sr^{-1}. The line was positively measured in numerous fields in the Cygnus Loop and a few fields in IC443. Differences in size and shape between our diaphragms and those used for narrow-band filter photometry preclude a direct comparison, but no strong discrepancy exists. We note however that the very strong continuum in the filaments (locally as high as the night sky) must be allowed for in non-spectroscopic techniques.

There is no evidence in our data for lines broader than 2 A; the radial velocity dispersion from field to field is of this order. This pleads for [FeX] being attached to slow material ('clouds').

References
Caplan,J.,and Deharveng,L. 1985,Astr.Ap.Suppl.,62,63
Gensheimer,P.,and Teske,R.G. 1987,preprint
Ku,W.H.-M.,Kahn,S.M.,Pisarski,R.,and Long,K.S. 1984,Ap.J.,278,615
Lucke,R.L.,Woodgate,B.E.,Gull,T.R.,and Socker,D.G. 1980,Ap.J.,235,882
Petre,R. et al. 1983,Proc.IAU symp 101,ed.Danziger-Gorenstein(Reidel),289
Teske,R.G.,and Kirshner,R.P. 1985,Ap.J.,292,22
Woodgate,B.E.,Lucke,R.L.,and Socker,D.G. 1979,Ap.J.(Letters),229,L119

NEW OBSERVATIONS OF NON-RADIATIVE SHOCKS IN THE CYGNUS LOOP

J. J. Hester[1] and J. C. Raymond[2]

[1]IPAC, California Institute of Technology
[2]Harvard Smithsonian Center for Astrophysics

ABSTRACT: We present deep Hα and [O III] images and echelle
spectra of non-radiative shocks in the NE Cygnus Loop. The
contrast between the smooth Hα structure and the clumpier
[O III] indicates that portions of the sheet-like front are
beginning to go radiative. The column depth through the
shock seems to be $\lesssim 10^{17.5}$ cm^{-2} for the entire region -- a
remarkable constraint for such a large structure.

INTRODUCTION: The Balmer-line emission from non-radiative shocks
arises in a zone of collisional excitation immediately behind the shock
front. The Hα line profile shows both narrow and broad components with
widths related to the pre- and post-shock temperatures, respectively.
From such a profile, Raymond et al. (1983, hereafter RBFG) derived a
shock velocity of order 200 km s^{-1} for one position. The strength of
forbidden line emission, particularly the presence of [N II] and [S II]
in a spectrum presented by Fesen and Itoh (1985), seems to require some
radiative contribution to the emission. An Hα image presented by
Hester, Raymond, and Danielson (1986, hereafter HRD) shows the two faint
filaments just visible on the POSS red print to be part of a continuous
band of very sharp filaments and diffuse emission.

OBSERVATIONS: Images of a field centered on the two brightest
filaments were obtained through a 15Å FWHM Hα filter and a 30Å FWHM
[O III] λ5007 filter using a new reimaging camera at the 60" telescope
at Palomar Observatory. The camera system employs a 306 mm collimator,
a 58 mm camera lens, and a spherical field lens to put a 16' X 16' field
onto an 800 X 800 pixel TI CCD at 1.2 arcsec pixel^{-1} with an effective
speed of f/1.65. This camera covers a field with about 5 times the area
of the system used by HRD.

The images were calibrated using the spectrophotometry of RBFG.
Hα/Hβ was assumed to be 3 (close to the value of 2.8 reported by Fesen
and Itoh). The Hα image is a single 3000 s exposure with a detection
threshhold (2σ in a 3 X 3 pixel sample) of 3 X 10^{-7} ergs cm^{-2} s^{-1} sr^{-1}
(which would correspond to an emission measure of ~ 3.5 at T = 10^4 K).
The [O III] image is a stack of a 10000 s and a 6000 s exposure, and has
a detection threshhold of 1.5 X 10^{-7} ergs cm^{-2} s^{-1} sr^{-1}. The background
is nonuniform in the [O III] frame due to charge transfer problems
associated with a bright star in the field. Dot density representations
of the images are presented in Figure 1, along with Einstein HRI data.
(The coverage of the X-ray data is incomplete in the northern part of
the field.) An enlargement of an approximately 2' square region just
south of the center of the field is shown in Figure 2. Contours of
[O III] are superposed on a dot-density representation of Hα.

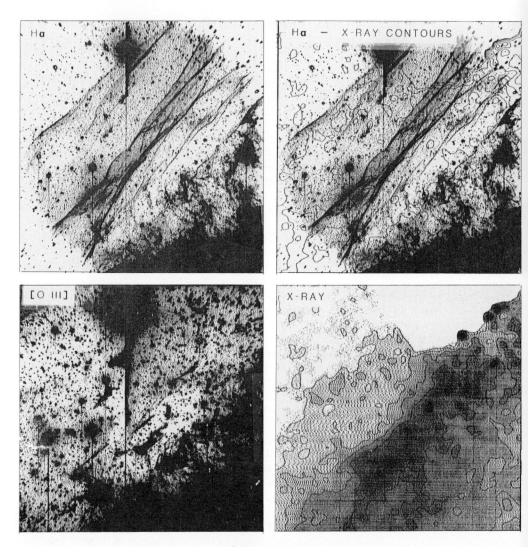

Figure 1.

High resolution spectra were obtained for several positions using the echelle spectrograph on the 4-m telescope at Kitt Peak. The spectrograph was used in a long slit mode, with an optical flat replacing the cross disperser and order separation obtained with a narrow band filter. Figure 3 shows Hα line profiles for the leading and trailing filaments in Figure 2, labelled position 1 and 2, respectively.

RESULTS: The extent, smoothness, and continuity of the Hα emission is the most immediate result. The relative brightness of the diffuse Hα emission in the three regions bounded by lines of prominent filaments can be explained by a single sheet of emission (cf., Hester 1987). From the perspective of the observer the sheet appears to extend from the bright radiative emission to the second line of non-radiative filaments,

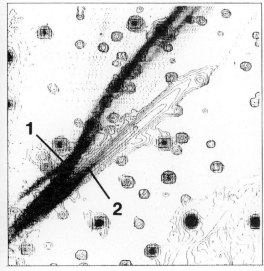

Figure 2

double back and again reach tangency at the line of filaments ~ 1-2' to the SW, and from there extend to the outermost tangency ~ 7' to the NE. Regions of X-ray emission are bounded by lines of optical filaments (cf., Hester and Cox 1986). This includes filaments in the extreme NE (outside of the field of HRD) which bound the X-ray "halo" reported by Ku et al. (1984). The X-ray brightness is roughly correlated with the surface brightness of the Hα filaments that it trails.

The FWHM of the broad component at position 1 is 163 km s^{-1}. The width of the trailing filament (2) is 116 km s^{-1}. All other positions observed had velocities in the 115 – 130 km s^{-1} range. The position with the broadest Hα is also the least pronounced in [O III]. With the exception of position 1, these widths are less than the 160 km s^{-1} width reported by RBFG. A higher signal to noise spectrum of their position was obtained, and yielded a velocity width of 133 km s^{-1}. Observation of a smaller width and evidence below suggesting that some of the [O III] emission is due to a radiative contribution weakens arguments in RBFG favoring Coulomb equilibration behind the shock. In fact, the shock velocity of ~ 120 – 130 km s^{-1} inferred from the present data assuming Coulomb equilibration is uncomfortably small. We prefer a shock velocity of 160 km s^{-1}, consistent with rapid equilibration. For position 1, this gives $v_s \approx 200$ km s^{-1}.

The [O III] emission is dominated by a few bright regions where [O III]/Hβ can exceed 100. These are clearly portions of the shock that are going radiative. Ignoring these, [O III]/Hβ has a fairly uniform value of ~ 0.1 for filaments which are bright enough to be seen in [O III]. The brightest (and therefore best studied) Hα filaments are atypical in that they lie in close proximity to regions of bright [O III]. The optical spectrum of RBFG has [O III]/Hβ ~ 50% higher than average for faint [O III] filaments.

In Figure 2, [O III] strengthens along a filament as Hα weakens. Preliminary models support the interpretation that as progressively more of the [O III] zone forms, the UV from the shock will preionize hydrogen. For a 160 km s^{-1} shock in a medium that was originally 30% neutral, this occurs at a column density of ~ $10^{17} - 10^{17.5}$ cm^{-2}. This corresponds to a time scale of ~ 200 – 700 years, during which time the shock will have travelled ~ .03 – 0.1 pc, or a distance of ~ 10" – 30" on the sky. While very strong magnetic support behind the shock may increase these distances by as much as a factor of ~ 3, the fact remains that over a very short time scale (a few hundred years) the shock newly encountered material with density ~ 1 cm^{-3} over a surface with an area

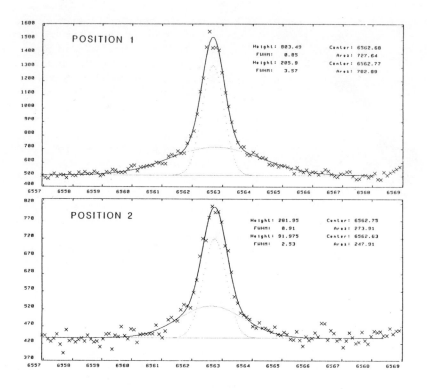

Figure 3

of order 100 pc². Furthermore, this surface has a significant
curvature that fits well with the outline of the rest of the remnant.
    We cautiously suggest that these observations could be most easily
understood if the Cygnus Loop were the result of a cavity explosion.
The presence of bright X-ray emission immediately interior to shocks
that do not seem to be fast enough to produce the X-rays also indicates
that the shock has recently undergone rapid deceleration over a large
area. Finally, the remarkably smooth morphology of the emission
(despite evidence that shock velocities vary significantly in the
region) is easier to understand if the denser more inhomogeneous medium
were encountered only recently.

REFERENCES:

Fesen, R. A., and Itoh, H. 1985, Ap. J., **295**, 43.
Hester, J. J. 1987, *Ap. J.*, **314**, 187.
Hester, J. J., and Cox, D. P. 1986, *Ap. J.*, **300**, 675.
Hester, J. J., Raymond, J. C., and Danielson, G. E. 1986, *Ap. J.*
    (*Letters*), **303**, L17 (HRD).
Ku, W. H.-M., Kahn, S. M., Pisarski, R., and Long, K. S. 1984, *Ap. J.*,
    278, 615.
Raymond, J. C, Blair, W. P., Fesen, R. A., and Gull, T. R. 1983, *Ap. J.*,
    275, 636 (RBFG).

THE EVOLUTION OF OLDER REMNANTS

S. A. E. G. Falle,
Department of Applied Mathematical Studies,
The University,
Leeds LS2 9JT, U.K.

Abstract. An older remnant will be defined as one in which radiative cooling occurs somewhere and has swept up enough mass for the details of the explosion to be less important than the state of the interstellar medium in which the explosion occurred. Without discussing any particular remnant in detail, I will consider how large and small scale density variations in the ambient medium affect the appearance and energetics of such remnants. Finally I will show that radiative instabilities can modify the emission spectrum of radiative shocks in such a way that a naive interpretation of these spectra can be very misleading.

Introduction. In the days when supernova remnants, or at least the theoretical ones, were spherical and expanded into a uniform medium, it was easy to decide what was meant by an older remnant. Remnants had three phases, free expansion, Sedov and radiative (Woltjer 1972) and a remnant was regarded as old if it had entered the radiative phase.

Unfortunately, such a simple picture is no longer adequate even as an idealisation of what happens. We know that the interstellar medium is inhomogeneous on many scales, some of which correspond to the sizes of supernova remnants and we also know that Type II supernovae can significantly modify the interstellar medium in their neighbourhood. All this means that there may not be a simple division of supernova remnant evolution into three phases.

For the purposes of this article, I will call a remnant old if radiative cooling is important somewhere, but will not insist that a significant part of the explosion energy has been radiated away. Observationally this means that a remnant is old if it has a filamentary structure whose emission is characteristic of radiative shocks. Examples are the Cygnus Loop, Shan 147 and Vela. As far as theory is concerned we have to look at the effects of radiative cooling on the dynamics and appearance of the remnant.

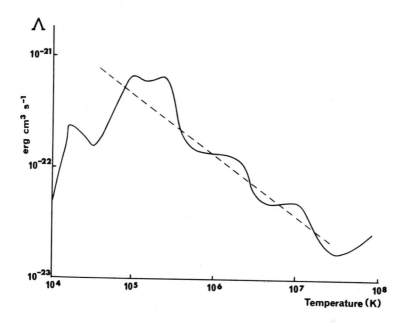

Figure 1. The radiative cooling rate per unit volume for an optically thin plasma. The straight line is the approximation (1) (Kahn 1976).

The radiative cooling rate for an optically thin gas in the relevant temperature range is shown in figure 1. Although this is not the most recent calculation, it has a maximum at about 10^5 K , which is what is important as far as the dynamics is concerned. Note that it does not include the effect of dust cooling which might well dominate above 10^5 K, depending upon the dust to gas ratio (Dwek these proceedings).

One of the nice things about this cooling law is that in the range 5×10^4 K $< T < 5 \times 10^7$ K it is very well approximated by $T^{-1/2}$ power law,

$$\Lambda = A\rho^2 (p/\rho)^{-1/2} \quad (A = 3.9 \times 10^{32} \text{ c.g.s}). \tag{1}$$

Kahn (1976) showed that this assumption makes it possible to calculate the effect of radiative cooling on the overall energetics independently of the details of the dynamics. I will first describe how this works for spherical remnants, and then show how it can be extended to supernovae in plane stratified media.

Spherical Remnant. Suppose that a supernova explosion has energy E_0, ejects mass M_e and occurs in a uniform medium with density ρ_0. Then there will be a Sedov phase provided

$$\left(\frac{E_0}{10^{51}} \right)^{-0.74} \left(\frac{M_e}{M_\odot} \right)^{5/6} \left(\frac{\rho_0}{10^{-24}} \right)^{0.2} < 4 . \tag{2}$$

This is simply the condition that the remnant enters the Sedov phase before radiative cooling becomes important. It is based on Gull's (1973) calculations, which show that it looks like a Sedov solution once it has swept up about 50 M_e, combined with Cox's (1972) estimate of when radiative cooling becomes important.

One would prefer a Sedov phase to exist, because then all the details of the original explosion can be ignored and only the energy E_0 matters. Condition (2) is satisfied for all plausible values of E_0, M_e and ρ_0, but unfortunately it ignores the fact that a Type II supernova can modify its surroundings, either because of its ionizing radiation (Shull, Dyson, Kahn & West 1985), or its stellar wind (Charles, Kahn & McKee 1985). There is also a good deal of observational evidence that this occurs (Braun these proceedings).

I am going to ignore these complications and assume that the original state of the ambient medium is more important than the details of the explosion. Then in a uniform medium radiative cooling becomes important when the post shock temperature is

$$T = T_{sg} = 1.2 \times 10^6 \left(\frac{E_0}{10^{51}} \right)^{0.1} \left(\frac{\rho_0}{10^{-24}} \right)^{0.2} K \tag{3}$$

Cox (1972). For almost all supernova remnants this is in the range for which the approximation (1) is valid.

If we now ignore magnetic fields and assume that all shocks are strong, then as long as $T \geq 5 \times 10^4$ K everywhere, things only depend on E_0, ρ_0 and A. From these we can form a characteristic mass, length and time given by

$$m_c = \frac{(2.02E_0)^{6/7}}{\rho_0^{2/7} A^{3/7}} = 7.3 \times 10^{36} \left[\frac{E_0}{10^{51}} \right]^{6/7} \left[\frac{\rho_0}{10^{-24}} \right]^{-2/7} \text{gm},$$

$$l_c = \frac{(2.02E_0)^{2/7}}{\rho_0^{3/7} A^{1/7}} = 1.9 \times 10^{20} \left[\frac{E_0}{10^{51}} \right]^{2/7} \left[\frac{\rho_0}{10^{-24}} \right]^{-3/7} \text{cm}, \qquad (4)$$

$$t_c = \frac{(2.02E_0)^{3/14}}{\rho_0^{4/7} A^{5/14}} = 1.2 \times 10^{13} \left[\frac{E_0}{10^{51}} \right]^{3/14} \left[\frac{\rho_0}{10^{-24}} \right]^{-4/7} \text{s}.$$

We expect radiative cooling to become important when the swept up mass is about m_c and the radius and age will then be approximately l_c and t_c respectively. Notice that these numbers are about right for the Cygnus Loop and IC443.

If $\gamma = 5/3$ and the cooling rate is given by (1), then the energy equation becomes

$$\frac{dK}{d\tau} = -\frac{2}{3} K^{-1/2}, \qquad (5)$$

where $\tau = t/t_c$ is a dimensionless time and

$$K = \frac{pt_c^2 m_c^{2/3}}{\rho^\gamma l_c^4}$$

is the dimensionless adiabatic constant.

From (5) we can see that the rate of change of K of a fluid element is independent of its dynamics as long as the approximate cooling law holds. The time at which an element of gas cools is then

$$\tau_{cool} = \tau_s + \left[K(\tau_s) \right]^{3/2}. \qquad (6)$$

Here τ_s is the time at which the element passed through the shock and the second term is the time it takes for K to decrease to zero according to equation (5).

In the Sedov phase we have

$$M = \frac{4\pi}{3} \tau_s^{6/5}, \quad K(\tau_s) = \frac{4}{25} \frac{\upsilon(\gamma)}{\tau_s^{6/5}},$$

(7)

$$\upsilon(\gamma) = \frac{2(\gamma - 1)^\gamma}{(\gamma + 1)^{\gamma+1}} = 0.07 \quad \text{for } \gamma = 5/3.$$

Here M is the mass interior to the fluid element in units of m_c. Inserting (7) in (6) gives.

$$\tau_{cool} = \left[\frac{3M}{4\pi} \right]^{5/6} + \left[\frac{4\upsilon}{25} \right]^{3/2} \left[\frac{4\pi}{3M} \right]^{3/2}.$$

(8)

Hence at time τ all the gas for which $\tau > \tau_{cool}$ will have cooled. From figure 2a one can see that cooling first occurs at $\tau = 0.18$ and that its onset is sudden in the sense that most of the mass of the remnant cools at times only slightly later than τ_{cool}. At later times only a small fraction of the mass near the centre remains hot.

Plane Stratified Density Distribution. It is obviously important to see how the ideas of the previous section are modified if the supernova explodes in a non-uniform environment. To get a feel for what happens, let us consider a plane stratified exponential density distribution

$$\rho(z) = \rho_0(0)e^{-z/h},$$

(9)

and assume that the scale height h is large enough for there to be an initial spherical Sedov phase.

Laumbach and Probstein (1969) derived approximate equations for the motion of the shock by assuming that all the energy and mass in each solid angle is conserved and concentrated near the shock. These equations are

$$Z = \frac{r_s}{h} \cos \theta, \quad \frac{dZ}{d\tau'} = U,$$

$$\frac{dU}{d\tau'} = \left[\frac{i - G(Z,\phi)U^2}{F(\phi)\phi} \right], \quad \frac{d\phi}{d\tau'} = \frac{\rho_0(Z)}{\rho_0(0)} Z^2 U,$$

(10)

$$\tau' = t \left[\frac{E_0 |\cos \theta|^5}{4\pi\rho_0(0)h^5} \right]^{1/2}, \quad \begin{matrix} i = 1 & 0 \le \theta < \pi/2 \\ \\ i = -1 & \pi/2 \le \theta < \pi. \end{matrix}$$

Here r_s is the radius of the shock and θ is the angle between the shock and the z axis. Notice that these equations need only be integrated twice, once for each value of i, to get the shock position for all θ and t.

Integration of these equations shows that the velocity of the upward moving shock ($0 \leq \theta < \pi/2$) initially decreases, reaches a minimum at $\tau' = 2.4$ and then increases again. Garlick (1983) and Falle, Garlick and Pidsley (1984) have shown that the acceleration of the rising shock calculated from the approximate solution agrees remarkably well with the results of full numerical calculations. We can therefore use the approximate solution to estimate the effects of cooling.

There is now a dimensionless parameter which we can define to be

$$\beta = \frac{l_c}{h} = \frac{1.9 \times 10^{20}}{h} \left[\frac{E_0}{10^{51}} \right]^{2/7} \left[\frac{\rho_0(0)}{10^{-24}} \right]^{-3/7} . \tag{11}$$

Clearly for $\beta \ll 1$ cooling occurs while the remnant is still spherical, while for $\beta \gg 1$ there will be no cooling in the top part of the remnant since the rising shock will have begun to accelerate long before cooling sets in.

The energy equation is as before except that the entropy behind the shock is now a function of θ as well as time. In fact

$$K_s(\tau',\theta) = \frac{(\beta|\cos \theta|)^3}{8.08\pi} \frac{v(\gamma)U^2}{[\rho_0(Z)/\rho_0(0)]^{\gamma-1}} , \tag{12}$$

where $U(\tau')$ and $Z(\tau')$ are determined from the approximate equations (10). Using our previous approximation for τ_{cool}, we get

$$\tau_{cool} = \frac{(8.08\pi)^{1/2}}{(\beta|\cos \theta|)^{5/2}} \tau'_s + \frac{(\beta|\cos \theta|)^{9/2}}{(8.08\pi)^{3/2}} \left[\frac{v(\gamma)U^2}{[\rho_0(Z)/\rho_0(0)]^{\gamma-1}} \right]^{3/2} \tag{13}$$

As before we really want τ_{cool} as a function of the mass interior to the fluid element. From (10) we have

$$M = \frac{4\pi}{(\beta|\cos \theta|)^3} \frac{1}{\rho_0(0)} \int_0^Z \rho_0(x)x \, dx = \frac{4\pi}{(\beta|\cos \theta|)^3} \phi(Z), \tag{14}$$

where $M/4\pi$ is the mass per unit solid angle interior to the shock.

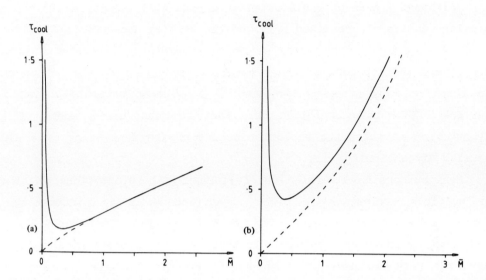

Figure 2. Cooling time (solid line) as a function of the Lagrangean coordinate $\bar{M}$. Mass interior to the shock (dashed line). a) Spherical remnant, b) $\beta|\cos \theta| = 2$. (Falle, Garlick & Pidsley 1984).

Figure 2b shows the result for the upwards moving shock ($\theta < \pi/2$) for $\beta|\cos \theta| = 2$. In contrast to the spherical case, there is now hot gas immediately behind the shock as well as in the interior of the remnant. The amount of hot gas behind the shock depends very sensitively on $\beta|\cos \theta|$ since it is determined by the ratio of the two terms on the right hand side of equation (13). This means that for $\beta|\cos \theta|$ greater than a certain value the gas does not cool behind the upward moving shock, while for smaller values it does.

In fact for $\beta|\cos \theta| > 2.8$, cooling occurs after the rising shock has begun to accelerate and so it is reasonable to suppose that the gas does not cool in a cone with semi-angle θ_0 given by

$$\beta|\cos \theta_0| = 2.8 . \tag{15}$$

The amount of hot gas in this cone can be considerably greater than that produced by a spherical remnant in the same mean density, so these effects ought really to be taken into account when making estimates of the rate at which supernovae inject energy into the interstellar medium.

Although I have only discussed an exponential density distribution, similar results are obtained provided the density decreases sufficiently rapidly with z for the rising shock to accelerate. Furthermore in all such cases the top part of the remnant disconnects from the bottom part so that for many purposes they can be considered separately. Falle & Garlick (1982) have exploited this fact to construct a model of the Cygnus Loop in which the explosion occurs on the dense side of a plane density discontinuity.

Many authors have carried out full numerical calculations of single or multiple explosions in plane stratified media (e.g. Chevalier & Gardner 1974; Tenorio-Tagle, Rozyczka & Yorke 1985; Tomisaka & Ikeuchi 1986; McCray these proceedings). In theory such calculations should give us much more information than the kind of analysis I have just described. Unfortunately there is a major snag, namely that in none of them is the cooling region adequately resolved. The result is that not only is the radiative energy loss incorrect, but the various instabilities are not correctly modelled. This does not mean that these calculations are useless, but it does mean that they must be treated with some caution.

Small Scale Inhomogeneities. In the previous section I looked at the effect of density variations whose scale was of the same order as the size at which the remnant becomes radiative. However, the appearance of remnants like the Cygnus Loop suggests that there are also irregularities on much smaller scales. Indeed McKee & Cowie (1975) have argued that in the Cygnus Loop we only see optical filaments when shocks propagate into small clouds.

The interaction of a plane shock with density inhomogeneities has been looked at by many authors (e.g. Sgro 1975; Chevalier & Theys 1975; Woodward 1976; Nittmann, Falle & Gaskell 1982; Hamilton 1985; Heathcote & Brand 1983). Although we have a rough idea of what happens, at least in the adiabatic case, there are a number of important details which are not clear. Since the propagation of shocks in non-uniform media is discussed by McKee (these proceedings), I will concentrate on some laboratory experiments on adiabatic flow past rigid bodies and suggest how these might help us to understand the much more complicated astrophysical problem.

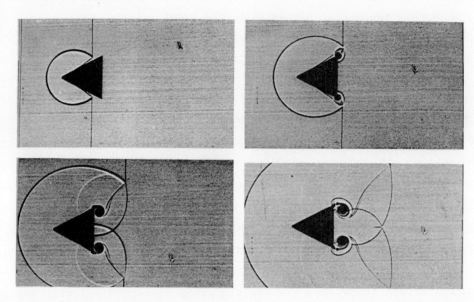

<u>Figure</u> 3. Shadowgraphs of shock diffraction on a rigid wedge (van Dyke 1982, p. 146).

Figure 3 shows a series of shadowgraphs of a shock incident on a rigid wedge. Although this is adiabatic flow and a wedge is perhaps not the most relevant shape, it does show a number of features of interest. Firstly there is a reflected shock which becomes the bow shock at large times provided the flow is supersonic behind the incident shock. There is also a diffracted shock which reflects off the symmetry axis at the rear of the wedge. Initially this is a regular reflection, but as the angle of incidence increases it becomes a Mach reflection with a post shock pressure which is substantially higher than that behind the incident shock. Finally intense vortices are formed at the corners and then move downstream at the local fluid velocity.

What we have to decide is how much of this is relevant to the astrophysical problem. First consider the geometry of the object. Provided the object is not streamlined for supersonic flow, all the features that I have described will be present whatever the shape of the object. For example the flow past a cylinder or sphere is very similar except that two sets of vortices are formed (Bryson & Gross 1961).

Of course supernova remnants do not encounter rigid objects, at least not of significant size. However, from pressure balance we have

$$V_c = V_e \left[\frac{\rho_e}{\rho_c} \right]^{1/2}, \tag{16}$$

where V is a shock velocity and the subscripts e and c refer to the exterior and cloud respectively. So if the cloud is much denser than its surroundings, it deforms slowly compared to the timescale of the exterior flow. The cloud therefore behaves like a rigid body, at least as far as the transient stage of the exterior flow is concerned.

Once the transients in the exterior flow have died away, we get a quasi-steady flow past a slowly deforming body. In principle one could calculate this flow, find the pressure distribution on the cloud and so determine how it deforms. In practice this is very difficult, but we can nevertheless draw some qualitative conclusions.

For almost any shape, the minimum pressure will be at the widest part of the body and the difference between this pressure and the maximum pressure will be of the order of the ram pressure. So, provided the exterior flow is not very subsonic, the cloud will expand sideways at something like its sound speed and will therefore disrupt in about a sound crossing time. That such a sideways expansion occurs can be seen in van Dyke (1982 p.86) which shows a water drop suddenly immersed in an air stream whose ram pressure is considerably higher than that due to surface tension. One consequence of this disruption is that clouds cannot be coherently accelerated to anything like the exterior flow speed (Nittmann, Falle & Gaskell 1982).

From this we can see that laboratory experiments are really very useful as long as the exterior flow is adiabatic and the density contrast is large. In particular we can use them to check the reliability of numerical simulations such as those in Woodward (1976) and Nittmann, Falle & Gaskell (1982). Unfortunately none of the calculations in the astrophysical literature are for cases for which the experiments are relevant. It would be interesting to do a high density contrast adiabatic calculation with a modern numerical scheme just to see how good the results are.

Finally let us consider the influence of cooling on shock-cloud interactions. The first thing is that cooling can amplify small density fluctuations in the undisturbed gas (Chevalier & Theys 1975). This means that inhomogeneities that would be unimportant in the adiabatic case can have observable effects once cooling occurs. Apart from this we might get high velocity dense regions where the shocks intersect at the rear of the cloud (Tenorio-Tagle & Rozyczka 1984). There are also various radiative instabilities which would cause corrugations in the shock fronts so that both the interior and exterior flow is much more complicated than in the adiabatic case.

The trouble is that, although we have some idea of the consequences of cooling, we are unable to come up with reliable quantitative results which can be compared with the observations. Numerical simulations, at least those carried out so far, are too crude to be much use. What is needed is high resolution numerical calculations supported by the sort of analysis described earlier. In this context I think that Whitham's area rule for shock propagation might be very useful (Whitham 1974).

Despite all these theoretical difficulties, we can say something. In the first place the dominant shocks in the cloud will tend to propagate parallel to the primary shock since the lateral shocks are much slower. So if only the cloud shocks are radiative, we will tend to see tangential filaments such as those in the Cygnus Loop. On the other hand if the exterior shocks cool, we should see large scale lacy structures such as those in Shan 147 and Vela.

Radiative Instabilities. We can write the cooling rate shown in figure 1 in the form

$$\Lambda = A\rho^2 \Phi(p/\rho c_*^2) , \qquad (17)$$

where

$$c_*^2 = \frac{kT_*}{\mu m_h} ,$$

(18)

and T_* is a reference temperature. T_* can be chosen to be the temperature at the maximum of Φ ($T_* = 10^5$ K). If we then set $\Phi(T_*) = 1$, we get $A = 2 \times 10^{26}$ c.g.s.

For a spherical remnant the flow is now governed by the parameters E_o, ρ_o, A, and c_* and from these we can form a dimensionless parameter

$$\alpha = \frac{T_{sg}}{T_*} = 10 \left(\frac{E_0}{10^{51}} \right)^{0.11} \left(\frac{\rho_0}{10^{-24}} \right)^{0.22} .$$

(19)

Here T_{sg} is the temperature defined by equation (3). α only affects the evolution of the remnant if there is radiatively cooling gas at temperatures below T_*.

Let us now look at the stability of gas whose cooling rate is given by (17). Suppose that $\Phi(T) \propto T^s$. Then if cooling occurs at constant pressure, the cooling time increases with increasing temperature if s < 2, while for constant density this is true for s < 1. This suggests that there is instability if s < 2 for constant pressure cooling and s < 1 for constant density.

Now the pressure will remain roughly constant if the cooling time $t_{cool} \gg t_{dyn}$ where t_{dyn} is some dynamical timescale. Conversely the density will remain constant if $t_{cool} \ll t_{dyn}$. Suppose that a region of initial size ℓ begins to cool and that the resulting compression is one dimensional. Then

$$\ell(t) \propto \frac{1}{\rho(t)} .$$

The relevant dynamical time is obviously the sound crossing time, so

$$t_{dyn} = \frac{\ell}{c} \propto \frac{1}{\rho T^{1/2}}.$$

On the other hand we have for the cooling time

$$t_{cool} \propto \frac{p}{\rho^2 T^s} \propto \frac{1}{\rho T^{s-1}}.$$

Hence

$$\frac{t_{dyn}}{t_{cool}} \propto T^{s-3/2} , \tag{20}$$

and so increases as the gas cools if $s < 3/2$. If t_{dyn} ever becomes much smaller than t_{cool}, then we expect a large pressure imbalance to occur which will lead to the formation of shocks. A necessary condition for this is $s < 3/2$. For the interstellar cooling law this condition is satisfied for $T > T_*$ and so we expect this kind of instability for spherical remnants if $\alpha > 1$. From (19) we can see that this should happen for almost all such remnants.

Various authors have looked at radiative instabilities. Both Avedisova (1974) and McCray, Stein & Kafatos (1975) carried out a linearised stability analysis with the post shock pressure held fixed. They found that density fluctuations grow if $s < 3$ for perturbations with wavelength much greater than the cooling length, while $s < 2$ is required if the wavelength is much shorter than the cooling length. However, these are isobaric instabilities and do not lead to the formation of additional shocks.

In my numerical calculations of thin shell formation in spherical remnants (Falle 1975, 1981), I found that cooling led to the formation of multiple shocks which caused large variations in the speed of the primary shock. Langer, Chanmugam and Shaviv (1981, 1982) found a similar effect in their calculations of radiative accretion onto white dwarfs.

These results have stimulated a lot of interest in such instabilities. Chevalier & Imamura (1982) used a linearised stability analysis to show that the shock speed will not be constant if $s < 0.8$, even if it is driven by a constant speed piston. To some extent this is confirmed by numerical calculations (Imamura, Wolff & Durisen 1984). Recently Bertschinger (1986) has extended this analysis to two dimensions and shown that in that case instability occurs if $s < 1$.

It has become common practice to deduce the velocity of radiative shocks by comparing the observed optical and UV line ratios with those predicted by steady shock models with various shock speeds (e.g. Raymond et.al. 1980). Unfortunately the above considerations suggest that radiative shocks will not be steady if the shock speed is high enough for cooling to occur in the unstable region of the cooling curve.

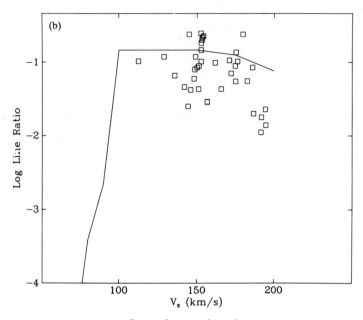

Figure 4. Instantaneous [OIII]5008/[OII]3728 line ratio for an unsteady shock whose mean speed is 175 km s^{-1}. The solid line is the ratio for a steady shock (Innes, Giddings & Falle 1987).

Recently Innes, Falle & Giddings (1987) have shown that, if the detailed atomic physics is included, then radiative shocks will be unsteady if their speed is greater than 130 km s^{-1}. The line ratios then do not correlate with the primary shock speed, nor even with the mean fluid speed, but vary dramatically on the cooling timescale. This effect can be seen in figure 4 which shows the instantaneous [OIII]5008/[OII]3728 line ratio plotted against the instantaneous primary shock speed for a shock driven by a constant speed piston such that the mean shock speed is 175 km s^{-1}. The variations in shock speed were induced by a single sinusoidal density perturbation upstream. The perturbation had an amplitude of 50% of the upstream density and a wavelength 1.4 times the thickness of the cooling region.

Conclusions. I have discussed some of the effects that radiative cooling and density inhomogeneities can have on the evolution of supernova remnants. As far as the overall energetics is concerned, it is clear that the Kahn (1976) and Laumbach & Probstein (1969) approximations give us reasonable estimates of the efficiency with which supernovae inject energy into the interstellar medium.

Interactions with small scale inhomogeneities are more difficult to deal with, but we can use laboratory experiments, numerical simulations and perhaps Whitham's area rule (Whitham 1974) to deduce how clouds of various sizes and densities affect the appearance of remnants. We clearly need a reliable quantitative theory of such interactions in order to make the proper use of the detailed observations that are now possible.

Finally I have indicated that radiative instabilities must exist in radiative remnants and that these make it very difficult to interpret the spectra of radiative shocks. They may also be responsible for at least some of the complex structure seen in old remnants.

References.

Avedisova, V.S. 1974. Sov. Astron., 18, 283.

Bertschinger, E. 1986. Astrophys. J., 304, 154.

Bryson, A.E. & Gross, R.W.F. 1961. J. Fluid Mech., 10, 1

Charles, P.A, Kahn, S.M. & McKee, C.F. 1985. Astrophys. J., 295, 456.

Chevalier, R.A. & Gardner, J. 1974. Astrophys. J., 192, 457.

Chevalier, R.A. & Imamura, J.N. 1982. Astrophys. J., 261, 543.

Chevalier, R.A. & Theys, J.C. 1975. Astrophys. J., 195, 53.

Cox, D.P. 1972. Astrophys. J., 178, 159.

Falle, S.A.E.G. 1975. Mon. Not. R. astr. Soc., 172, 55.

     "   1981. Mon. Not. R. astr. Soc., 195, 1011.

Falle, S.A.E.G. & Garlick, A.R. 1982. Mon. Not. R. astr. Soc., 201, 635.

Falle, S.A.E.G., Garlick, A.R. & Pidsley, P.H. 1984. Mon. Not. R. astr. Soc., 208, 925.

Fesen, R.A., Blair, W.P. & Kirshner, R.P. 1982. Astrophys. J., 262, 171.

Garlick, A.R. 1983. J. Comput. Phys.,52,427.

Gull, S.F. 1973. Mon. Not. R. astr. Soc.,161,47.

Hamilton, A.J.S. 1985. Astrophys. J.,291,523.

Heathcote, S.R. & Brand, P.W.J.L. 1983. Mon. Not. R. astr. Soc.,203,67.

Imamura, J.N., Wolff, M.T. & Durisen, R.H. 1984. Astrophys. J.,276,667.

Innes, D.E., Giddings, J.R. & Falle, S.A.E.G. 1987. Mon. Not. R. astr. Soc.,226,67.

Kahn, F.D. 1976. Astr. & Astrophys.,50,145.

Langer, S.H., Chanmugam, G. & Shaviv, G. 1981. Astrophys. J.,245,L23.

              "   1982. Astrophys. J.,258,289.

Laumbach, D.D. & Probstein, R.F. 1969. J. Fluid Mech.,35,53.

McCray, R., Stein, R.F. & Kafatos, M. 1975. Astrophys. J.,196,565.

McKee, C.F. & Cowie, L.L. 1975. Astrophys. J.,195,715.

Nittmann, J., Falle, S.A.E.G. & Gaskell, P.H. 1982. Mon. Not. R. astr. Soc.,201,833.

Raymond, J.C., Black, J.H., Dupree, A.K. & Hartmann, L. 1980. Astrophys. J.,238,881.

Sgro, A.G. 1975. Astrophys. J.,197,621.

Shull, P., Dyson, J.E., Kahn, F.D. & West, K.A. 1985. Mon. Not. R. astr. Soc.,212,799.

Tenorio-Tagle, G. & Rozyczka, M. 1984. Astr. & Astrophys.,137,276.

Tenorio-Tagle, G., Rozyczka, M. & Yorke, H.W. 1985. Astron. & Astrophys.,148,52.

Tomisaka, K. & Ikeuchi, S. 1986. Publ. Astron. Soc. Japan,38,697.

van Dyke, M. 1982. An Album of Fluid Motion, Parabolic Press.

Whitham, G.B. 1974. Linear and Nonlinear Waves, Wiley Interscience, chapter 8.

Woltjer, L. 1972. Ann. Rev. Astron. Astrophys.,10,129.

Woodward, P.R. 1976. Astrophys. J.,207,484

LUMINOSITY OF EVOLVING SUPERNOVA REMNANTS

D. F. Cioffi and C. F. McKee

University of California, Berkeley, California, U.S.A.

Abstract. We present an analytic kinematic model for the evolution of a supernova remnant beginning with the Sedov-Taylor adiabatic stage and continuing through the radiative stage. Using this model, we obtain the luminosity of the radiative shock and the hot interior.

I. Introduction

A strong explosion in a homogeneous uniform medium results in a blast wave whose adiabatic evolution is well-understood in terms of the classic Sedov-Taylor (ST) solution (Sedov 1959; Taylor 1950). We can model the early evolution of spherical supernova remnants (SNRs) with this solution, where the radius R_s grows with time t in a power law:

$$R_s = \left(\frac{\xi E}{\rho_o}\right)^{1/5} t^{2/5}, \tag{1}$$

where E is the energy of the explosion, ρ_o is the ambient density, and the numerical constant ξ is found to be 2.026 for $\gamma = 5/3$ (Ostriker and McKee 1987). As the hot gas begins to radiate, however, the evolution deviates from this solution, and, with less energy available to drive the remnant, the expansion rate slows. If one wishes to calculate analytically the continuous luminosity from an evolving SNR, one must possess both accurate kinematics and a sufficient knowledge of the distribution and thermal development of the hot gas. Although here we outline the methods and obtain the total SNR luminosity, in Cioffi and McKee (1987; [CM]) we obtain the broad-band spectrum and calculate the X-ray emission. These luminosities match those found in hydrodynamical simulations.

II. Kinematics, Cooling, and Radiation

Radiative losses in the SNR first set in near the edge and lead to the formation of a dense shell of gas which is driven into the ambient interstellar medium by the pressure of the hot interior gas – in other words, a *pressure-driven snowplow* (PDS). If the cooling function Λ (erg cm^3 s^{-1}) falls with the square root of the temperature T, then, as first realized by Kahn (1976), the entropy of a shocked parcel of gas is an explicit function of time alone. We can thus determine the time at which an element of gas first cools to zero temperature, t_{sf}. The discontinuity in the shock velocity seen in a numerical simulation (see Figure 1) confirms that a shell forms at this time, but since cooling has affected the evolution prior to t_{sf}, we begin the PDS stage a factor of e sooner at

$$t_{pds} \equiv \frac{t_{sf}}{e} = 1.33 \times 10^4 \frac{E_{51}^{3/14}}{\varsigma_m^{5/14} n_o^{4/7}} \text{ yr.} \tag{2}$$

Here we have used a cooling function $\Lambda = 1.6 \times 10^{-19} \varsigma_m T^{-1/2}$ erg cm^3 s^{-1}, where the metallicity $\varsigma_m = 1$ for cosmic abundances (Cioffi, McKee and Bertschinger 1987; [CMB]; also see Cox 1986).

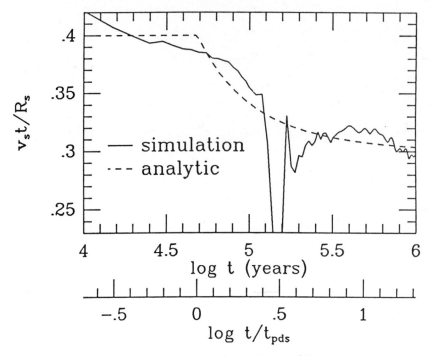

Figure 1. The logarithmic derivative $v_s t / R_s$ versus time.

In Figure 1 we show the logarithmic derivative $v_s t / R_s$ from a hydrodynamical simulation of an SNR expanding into an interstellar medium of hydrogen density $n_o = 0.1 \text{ cm}^{-3}$. The standard PDS power-law solution (McKee and Ostriker 1977) would show a straight horizontal line at $v_s t / R_s = 2/7$, which is discontinuous with the ST solution and, when compared to the hydrodynamical simulation, is too small after the formation of the shell. The SNR retains the "memory" of additional pressure from the ST stage, and cannot relax to a $2/7$ index. We thus choose an *asymptotic* index of $3/10$, and join the PDS solution to the ST solution by means of an "offset" power law (CMB):

$$R_s = R_{pds} \left[\frac{4}{3} \frac{t}{t_{pds}} - \frac{1}{3} \right]^{3/10} , \tag{3}$$

where R_{pds} is obtained from the ST solution at t_{pds}:

$$R_{pds} = 14.0 \frac{E_{51}^{2/7}}{n_o^{3/7} \varsigma_m^{1/7}} \text{ pc.} \tag{4}$$

Figure 1 shows how well the analytic logarithmic derivative agrees with that from the simulation through the transition across the formation of the shell. The luminosity of the radiative shock is $L = \frac{1}{2}\rho_o v_s^3 \left(4\pi R_s^2\right)$ and we find that the product $R_s^2 v_s^3$ almost always agrees with the simulation to within 20% except near t_{sf}, where v_s falls too quickly. CMB show that this fit remains good so long as $t \lesssim 20\, t_{pds}$.

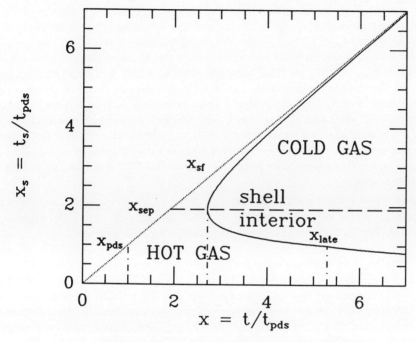

Figure 2. The thermal structure of an SNR.

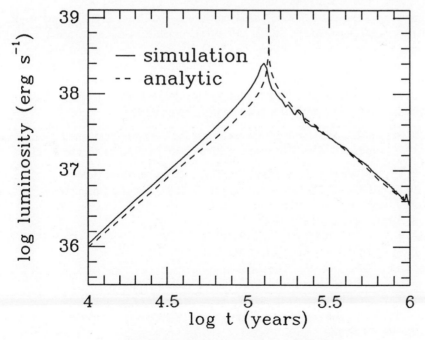

Figure 3. Total luminosity versus time.

We need to integrate through the hot gas to calculate the radiation from the interior. If we again use the $T^{-1/2}$ cooling law, and the solutions (3), we can construct Figure 2, which shows the cooling of an SNR in terms of the normalized times at which a gas element was shocked, $x_s \equiv t_s/t_{pds}$ (CM). At any fixed time $x \equiv t/t_{pds}$, one proceeds vertically along the x_s axis to ascertain the thermal structure of the remnant. At the separation time, $x_{sep} = 1.92$, we "flag" the gas element which will be the first to cool completely. At the shell formation time, $x_{sf} = 2.72$, this element cools to zero temperature and separates the SNR into three zones: i) an extremely narrow, hot region behind the shock; ii) a cold shell and iii) the hot interior, which consists of material shocked prior to x_{sep}. The cold shell grows from both sides as the interior cools and the material behind the now-radiative shock also cools. One other time of interest is $x_{late} = 5.29$; after this time any hot gas remaining in the interior of the SNR was shocked during the remnant's ST expansion.

At a given time t we can sum the emission from all elements which were shocked at prior times t_s, where we again assume a $T^{-1/2}$ cooling law. We fully explain the methods in CM. Figure 3 shows the excellent results from this approach.

III. Summary

Lack of space prevents consideration of two additional contributions to the luminosity (see CM): We can calculate the emission at early times from the reverse-shocked ejecta (e.g., McKee 1974) in a manner similar to that just outlined for the interior gas. Secondly, we note that for a real SNR in interstellar space, dust grains may supply a large luminosity during part of the evolution (e.g., Graham *et al.* 1987). This additional energy loss will shorten the PDS onset time somewhat (Dwek 1981), modifying our results slightly, but not strongly affecting the X-ray emission.

To achieve an accurate ($\lesssim 20\%$) analytic luminosity from SNRs in all stages of evolution, the overall kinematics, $R_s(t)$ and $v_s(t)$, must be very accurate ($\lesssim 5\%$) to obtain the correct radiative shock luminosity $L \sim R_s^2 v_s^3$, and the thermal history of the shocked gas must be calculated. Through understanding the dynamics of SNR expansion and the use of offset power laws, we have obtained a simple expression for accurate kinematics. The assumption of a $T^{-1/2}$ cooling law then allows a determination of the thermal structure and luminosity of a post Sedov-Taylor SNR.

Acknowledgements. We thank Ed Bertschinger for the numerical simulation, details of which can be found in CMB. NSF grant AST 86-15177 supports this research.

Cioffi, D. F., and McKee, C. F. 1987, *Ap. J.*, in preparation (CM).

Cioffi, D. F., McKee, C. F., and Bertschinger, E. 1987, *Ap. J.*, submitted (CMB).

Cox, D. P. 1986, *Ap. J.*, **304**, 771.

Dwek, E. 1981, *Ap. J.*, **247**, 614.

Graham, J. R., *et al.* 1987, *Ap. J.*, **319**, 126.

Kahn, F. D. 1976, *Astr. Ap.*, **145**, 50.

McKee, C. F. 1974, *Ap. J.*, **188**, 335.

McKee, C. F. and Ostriker, J. P. 1977, *Ap. J.*, **218**, 148.

Ostriker, J. P., and McKee, C. F. 1987, *Rev. Mod. Phys*, submitted.

Sedov, L. I. 1959, *Similarity and Dimensional Methods in Mechanics*, (New York: Academic Press).

Taylor, G. I. 1950, *Proc. Royal Soc. London*, **201A**, 159.

STABILITY OF RADIATIVE SHOCKS WITH TIME-DEPENDENT COOLING

T. J. Gaetz
University of Virginia,
Charlottesville, VA 22903

Abstract: The stability of a radiative shock subject to
nonequilibrium cooling is investigated. It is found that
high velocity shocks (> 140 km/s) are subject to
oscillational and condensational instabilities.

Introduction: Although steady radiative shock models have proven
useful in the interpretation of spectra from supernova remnants (e.g.
the models of Cox 1972, Raymond 1979, Shull and McKee 1979), it is
becoming apparent that steady shock models are not able to explain the
combined optical and UV data (Benvenuti et al 1980, Raymond et al
1980, 1981, Fesen et al 1982). The likely cause for the spectral
discrepancies is the presence of unsteady shocks (Raymond 1984, and
references therein.) Radiative shocks cooling via power-law cooling
functions are subject to an oscillatory instability (in which the
shock position relative to the driving "piston" varies with time) if
the power-law exponent is sufficiently small or negative. The
instability was demonstrated analytically in the linear regime
(Chevalier and Imamura 1982) and computationally in the nonlinear
regime (Langer et al 1981; Imamura et al 1984).

In an interstellar shock, the assumption of a power-law cooling
function is not adequate. The recombination timescales for important
species can be comparable to the radiative cooling timescale, and the
cooling function becomes history-dependent for temperatures below
$\sim 3 \times 10^6$ K. In this work we combine an accurate numerical gasdynamics
code (based on the PPM method of Colella and Woodward 1984) with a
detailed treatment of the time-dependent ionization evolution and
radiative cooling problem.

The Models: We examine the evolution of a nonlinear perturbation by
starting with a uniform flow hitting a stationary "wall" and following
the evolution until the shock damps to a steady state or a limit cycle
is reached. The gas upstream of the shock front is assumed to be
preionized; we use the results of Shull and McKee (1979) for shocks
with velocities below 130 km/s. For higher velocity shocks we
estimate the precursor conditions by using a steady state ion
equilibrium for a temperature roughly 0.4 times the postshock
temperature (obtained by extrapolating the trends of Shull and McKee
1979). The ionization equilibrium downstream of the shock is allowed
to relax according to the local thermodynamic conditions until the gas
reaches 10^4 K, at which point we "turn off" the cooling. The precursor
density is 9.4 nuclei cm^{-3}. The atomic rates are extracted from the
Raymond and Smith (1977, 1984) code, and the abundances are from Ross

and Aller (1976). We examined nonlinear perturbations for 130, 150, and 200 km/s shocks.

The 200 km/s shock is strongly unstable in the fundamental mode and the shock structure oscillates periodically (figure 1). The time development of the temperature profile through one cycle is shown in figure 2. During the expansion phase (curve a) the shock temperature is higher than that for a steady shock. The cooling length increases rapidly with temperature, and the resulting overpressure drives the shock far beyond the steady-state position. As the shock reaches its maximum position, the cooling in the interior robs the shock of its pressure support (curve b). A secondary shock forms where the flow hits the cold gas.

As the shock falls in and weakens, the shock temperature falls. A cooling instability occurs behind the shock as an overdense region undergoes runaway cooling and collapse (curves c, d). When the gas in the cooling clump gets cold (10^4K in these models), the clump repressurizes and weak shocks are driven into the adjacent material. In curve d we see hot gas in the primary and secondary shocks separated by cold gas. The pressure in the cold gas is not sufficient to halt the collapse of the structure. In curve e the two hot regions are about to collide; this repressurizes the gas and drives a strong shock back out (curve a again), completing the cycle. The qualitative features of the evolution can be seen in a model with power-law cooling ($\propto \rho^2 T^{-1/2}$) and in a model using an isobaric (but time-independent) cooling function. The models with simplified cooling functions did not exhibit the condensational instability in the collapse phase, however.

Innes *et al* (1987) examine the collision of a steady 200 km/s shock with a sinusoidal density perturbation; their figures show features similar to those in figure 2 above. Innes *et al* argue that the evolution becomes aperiodic. The discrepencies between these results likely arise from the different means of exciting the perturbation together with the short time Innes *et al* were able to run the model. The sinusoidal density perturbation may be more effective in exciting transients which would confuse the interpretation of the initial part of the cycle. In addition, their model may not have been carried far enough to show the cyclic behavior.

The 150 km/s shock is unstable, but to a lesser degree than the 200 km/s shock, and the collapse phase is less violent. The clump formed in the condensational instability cools roughly isobarically so that additional shocks are not produced by the clump formation. The 130 km/s shock is stable; the oscillations die away with time. There are early indications of an overtone mode, but the decaying fundamental mode dominates at late times.

It would appear that the transition to instability lies between 130 and 150 km/s. This is roughly in line with expectations based

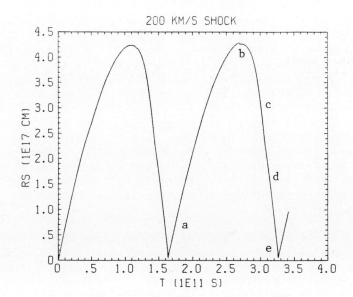

Figure 1. Shock radius (measured from the location of the cold gas)
versus time. The piston velocity is 200 km/s. The flow
is unstable to the fundamental mode.

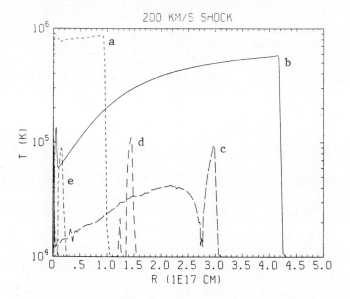

Figure 2. Temperature versus distance from the location of the cold
gas. The piston velocity is 200 km/s.

upon examination of the effective cooling function versus temperature
behind a steady shock. The shocked gas spends most of its time with
temperature comparable to the shock temperature. It is expected that
the stability of small perturbations is determined by the slope of the
effective cooling function near the shock temperature just below the
ionization zone. (Because of the high emissivity in the ionization
zone, the gas cools rapidly through the ionization zone; this zone
usually covers a narrow temperature range, however.) For large
perturbations one must take into account that the slope of the cooling
function behind a steady shock is a function of shock temperature.
Since the slope decreases with increasing shock temperature above 10^5K
(neglecting the high-emissivity ionization zone), there will be a
tendency for large perturbations to be less stable than small
perturbations.

Conclusions: Radiative shocks with velocities above about 140 km/s
are subject to the oscillational instability found earlier in models
cooling by power-law cooling curves. The precise stability limit may
depend on the amplitude of the perturbation. A condensational
instability arises in the collapse phase of the oscillation cycle.
Steady shock models are not an adequate description of high-velocity
radiative shocks.

This work was supported by NSF grant AST 84-13138 to the University of
Virginia and by a grant of time from the Pittsburgh Supercomputing
Center.

References:

Chevalier, R. A. and Imamura, J. N. 1982, *Astrophys. J.* **261**, 543.
Colella, P., and Woodward, P. R. 1984, *J. Comput. Phys.* **54**, 174.
Cox, D. P. 1972, *Astrophys. J.* **178**, 143.
Fesen, R. A., Blair, W. P., and Kirshner, R. P. 1982, *Astrophys. J.*
    262, 171.
Imamura, J. N., Wolff, M. T., and Durisen, R. H. 1984, *Astrophys. J.*
    276, 667.
Innes, D. E., Giddings, J. R., and Falle, S. A. E. G. 1987, *M.N.R.A.S.*
    236, 67.
Langer, S. H., Chanmugam, G., and Shaviv, G. 1981, *Astrophys. J.* **245**,
    L23.
Raymond, J. C. 1979, *Astrophys. J.* **39**, 1.
_____ 1984, *Ann. Rev. Astron. Astrophys.* **22**, 75.
Raymond, J. C., Black, J. H., Dupree, A. K., Hartmann, L, and Wolff,
    R. S. 1980, *Astrophys. J.* **238**, 881.
_____ 1981, *Astrophys. J.* **246**, 100.
Raymond, J. C., and Smith, B. W. 1977, *Astrophys. J. Suppl.* **35**, 419.
_____ 1984, private communication. Ross, J. E., and Aller, L. H.
    1976, *Science* **191**, 1223.
Shull, J. M., and McKee, C. F. 1979, *Astrophys. J.* **227**, 131.

HIGH-RESOLUTION RADIAL VELOCITY MAPPING OF OPTICAL FILAMENTS
IN EVOLVED SUPERNOVA REMNANTS.

H. Greidanus, Sterrewacht Leiden, The Netherlands.
R.G. Strom, Netherlands Foundation for Radio Astronomy.

Abstract: We report on observations of the kinematical
structure of optical filaments in evolved supernova
remnants, using an imaging Fabry-Perot interferometer. The
radial velocity characteristics as seen in [OIII] λ5007
emission in one area in the Cygnus Loop are described, where
four kinematically different components contributing to the
emission can be recognized.

1. Introduction.

We have mapped the radial velocity distribution of the optically
emitting gas in a number of 7'-sized fields in several evolved supernova
remnants, using the Hα and [OIII] λ5007 lines at a resolution of
$1.''2 \times 1.''2 \times 8$ km/s. The observations allow us to study the kinematical
structure of the filaments in two dimensions. This is a considerable
improvement over previous kinematical work on supernova remnants, which
has been limited to one spatial dimension (slit spectra) or a number of
sample points (aperture spectra; conventional Fabry-Perot). Previous work
has, apart from that, mostly been concerned with global expansion
properties. Here, we report briefly on the results for the [OIII] line in
one field in the Cygnus Loop.

2. Observations.

The instrument used was TAURUS, an imaging Fabry-Perot interferometer, on the 2.5 m Isaac Newton telescope of the Roque de los
Muchachos Observatory at La Palma. TAURUS is described in detail in
Atherton et al. 1982; in outline, it works as follows. The ring-modulated
interference pattern, obtained by putting a Fabry-Perot etalon in the
collimated beam, is imaged on a two dimensional detector. This is done
for a large number of consecutive etalon gap sizes, ultimately changing
the gap size by somewhat more than one wavelength. Such a 'scan' produces
a data cube of intensity as a function of two position coordinates (x,y)
and one gap size coordinate (z). With a calibration cube obtained by
observing a Neon lamp in the same way, the gap size coordinate is
transformed to a wavelength coordinate, and thus to radial velocity. (In
the following, all velocities refer to radial velocities.) An
interference filter was used to separate the emission line of interest.
The detector was an Image Photon Counting System (Boksenberg 1972),
characterized by a very low readout noise; its use is indicated by the
necessity to complete a scan as quickly as possible to minimize changes
in atmospheric transparency. The range in velocity Δv is set by the free
spectral range of the etalon, fsr = $\lambda/2G = \Delta v/c$. With a gap size G of
265 μ, $\Delta v = 285$ km/s at $\lambda = 5007$ Å.
        One possible cause of error that should be mentioned here is a
change in alignment between the etalon and the detector, giving rise to a

systematic offset in velocity, with a magnitude varying linearly across the field. However, the alignment was checked during the night, and we do not expect this error to be present to any significant degree.

3. Results and discussion.

Figure 1a shows one of the observed fields, located at $\alpha=20:46:29.1$, $\delta=31°15'07"$ (in the 'carrot', the rich filamentary area in the North-central part of the Cygnus Loop), in total [OIII] $\lambda 5007$ intensity. Other fields in the Cygnus Loop that have been observed show kinematical properties which recur here, and we will limit the presentation to this particular field. Figures 1b and 1c show two position-velocity maps, crosscuts through figure 1a as indicated. In order to bring out the low-level emission, the data have been smoothed to 5" x 5" x 15 km/s. On inspecting the data cube for this field, the emission can be divided into four components, which can also be recognized in these two sample crosscuts: (1) The filaments, showing the strongest emission, narrow in one spatial dimension, generally located at moderate velocities, with (deconvolved) velocity widths of about 45 to 60 km/s; (2) The diffuse gas appearing between the filaments, linking them in a continuous way, going out to more extreme velocities, with a velocity width of the order of 35 km/s; (3) A weak component, very broad in velocity (typically about 150 km/s), present over almost the entire field and enveloping the brighter parts; (4) A weak, smooth component, very narrow in velocity (~15 km/s), appearing at the same central velocity throughout.

The lack of any structure in the velocity of the last component indicates that this is unaccelerated gas; because it is not found in locations where other components are weak (as in the western part of the field), it is probably not unassociated fore- or background material. Photoionized gas ahead of the shock would be a possibility. We do not have an absolute velocity calibration, but putting component 4 at v=0, most filaments appear between v=0 and v=-80 km/s, and only a few are seen at positive velocities up to v=+20 km/s. Following individual, well-defined filaments, the velocity can be seen to change up to 20 km/s over ~1'. Diffuse inter-filament emission is seen up to v=-110 km/s between filaments at negative velocities and up to v=+60 km/s between ones at positive velocities. Apart from some low-level emission associated with component 3, which in places fills the entire spectral range, all emission is well contained within the range.

The kinematical appearance of components 1 and 2 is not inconsistent with the picture of a wrinkled sheet, the velocity characteristics being purely a projection effect, as recently modeled in Hester 1987. On the other hand, translating velocity into density, there must exist a density distribution which, when hit by a shock, gives rise to the observed velocity distribution; at least in the case where velocity is a single-valued function of position, as seen for example in figure 1c at A. This leads to the picture of pre-existing rope-like filaments with not-too-steep radial density profiles. However, a velocity profile which is a continuous, double valued function of position, as seen in figure 1b at B, cannot be interpreted in this way, but is more naturally explained as the edge of an expanding bubble. In any case, the notion that the gas at negative velocities is situated at the front side of the remnant, and the gas at positive velocities at the rear side seems obvious. But there are

Fig. 1a. [OIII] λ5007 emission of a field in the Cygnus Loop, centered at α=20:46:29.1, δ=31°15'07". The grayscale has a logarithmic increment, while the contoursteps are linear.

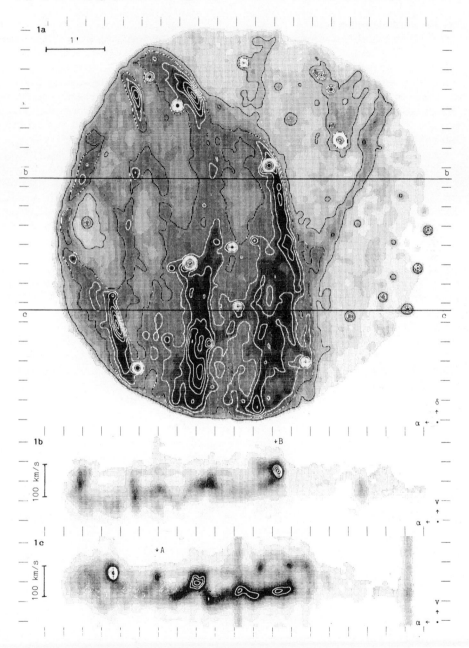

Fig. 1b,c. Position-velocity crosscuts through figure 1a as indicated. Intensity in the 5007 line is plotted as a function of right ascention (this axis is the same as in figure 1a) and radial velocity. The grayscale has a linear increment, as do the contours but with a stepsize 2.5 times larger. Note the stars which appear as vertical bars.

some places where the emission is seen to go continuously from positive
to negative velocities, complicating that idea, as this field is not near
the remnant's edge.

4. Future work.

This contribution reports on the status of our work at the time of
the conference, and is not intended to be the final word. For one thing,
the very limited interpretation given here for the filament/diffuse gas
components 1 and 2, has only considered velocities; a following step will
be to check whether both velocities and intensities are consistent with
the sheet picture or the pre-existing filament picture. Furthermore, the
kinematics of the Hα-emitting gas will have to be analysed and included
in the interpretation.

Observations at sub-arcsecond resolution would be interesting, to
check if the velocity width in components 1 and 2, which is still
considerably above the expected thermal velocity dispersion for oxygen of
12 km/s, would decrease with greater spatial resolution. Finally, to
confirm the photoionized nature of component 4, this type of observation
should be repeated at other wavelengths, to obtain line-ratios for this
component.

References.
Atherton et al. 1982, M.N.R.A.S. <u>201</u>, 661.
Boksenberg 1972, Auxiliary Instrumentation for Large Telescopes,
                 Proc. of ESO/CERN Conf. Geneva, p.205.
Hester 1987, Astroph. J. <u>314</u>, 187.

SUPERSHELLS

Richard McCray
Joint Institute for Laboratory Astrophysics
University of Colorado
Boulder, CO 80309-0440, USA

ABSTRACT: Repeated supernovae from an OB association will, in a few $\times 10^7$ yr, create a cavity of coronal gas in the interstellar medium, with radius > 100 pc, surrounded by a dense expanding shell of cool interstellar gas. Such a cavity will likely burst through the gas layer of a disk galaxy. Such holes and "supershells" have been observed in optical and H I radio emission maps of the Milky Way and other nearby galaxies. The gas swept up in the supershell is likely to become gravitationally unstable, providing a mechanism for propagating star formation that may be particularly effective in irregular galaxies.

1 INTRODUCTION

Cox and Smith (1974) were the first to recognize that the hot interiors of old supernova shells should persist in the interstellar medium (ISM) for timescales $\gtrsim 10^6$ yr, long enough that the hot interiors might interconnect to form a "tunnel" system of coronal ($T \gtrsim 10^6$ K) gas in the ISM with a morphology resembling Swiss cheese. This idea was developed further by McKee and Ostriker (1977), who argued that the supernova rate in the Milky Way is high enough that the porosity of the coronal gas should approach unity. In their model the Swiss cheese falls apart: the substrate is the coronal gas and the H I is found in isolated clouds with dense cores surrounded by warm H I envelopes.

Here I shall describe a model for the interaction of supernova remnants with the ISM that may add an important new ingredient to the McKee-Ostriker model. I depart from their model by assuming that the morphology of the hot gas in the disk of the Milky Way is dominated by clustered type II supernovae from massive Population I stars rather than by randomly distributed type I supernovae from Population II stars. This assumption makes a qualitative difference in the resulting morphology of the hot gas. A cluster of massive stars should, over a few $\times 10^7$ yr, produce many supernova explosions that act in concert to pack the ambient interstellar gas into a giant (radius $R_S \gtrsim 100$ pc) expanding shell called a "supershell." The supershell will likely blast a hole right through the thin H I layer of a disk galaxy, venting a substantial fraction of the supernova energy into the galactic halo. As a result, the porosity of the coronal gas should be less than it would be if the supernovae occurred at random in the disk.

Finally, I shall discuss a most exciting possibility, namely, that supershells may become gravitationally unstable, providing a mechanism for propagating star formation. Although gravitational instability of supershells may provide only a secondary mechanism for star formation in disk galaxies, where spiral density waves dominate, it may be the dominant mechanism in irregular galaxies.

The ideas presented here are discussed and referenced in more detail in papers by Heiles (1987), McCray and Kafatos (1987), and MacLow and McCray (1988).

2 SUPERSHELL DYNAMICS

Consider first the energy delivered to the ISM by a cluster of OB stars. The lifetimes of massive stars are given approximately by $t_{MS} \sim 3 \times 10^7$ yr$[M_*/(10M_\odot)]^{-1.6}$. In a cluster with a "typical" initial mass function, $dN_*/d(logM_*) \sim M_*^{-1.6}$, the increasing numbers of low mass stars fortuitously compensates for the increasing stellar lifetimes so that the mean supernova rate from the cluster should be nearly constant until $t \approx 5 \times 10^7$ yr, the lifetime of the least massive star ($\sim 7M_\odot$, corresponding to spectral type B3) that can explode. A typical newborn OB cluster may have $N_* \sim 20$ stars with $M_* > 7M_\odot$. At first ($t \lesssim 5 \times 10^6$ yr), the ionizing radiation and stellar winds from the most massive O stars will dominate the energy imparted by the cluster to the surrounding ISM; but thereafter, the supernova explosions will dominate, ultimately providing a net energy input ~ 20 times greater.

As a first approximation, we model the impulsive supernova energy input to the ISM by a steady source, with power $P_{SN} \approx 6.3 \times 10^{35}[N_*E_{51}]$ ergs s^{-1}, where N_* is the number of stars in the newborn association with mass $> 7M_\odot$ and E_{51} is the energy of a supernova explosion in units of 10^{51} ergs. Then, if we assume that the ambient medium has uniform atomic density n_0, the supershell expands according to the theory for a stellar wind-driven bubble (Weaver et al. 1977):

$$R_S = 97 \text{ pc } [N_*E_{51}/n_0]^{1/5}t_7^{3/5}, \tag{1}$$

where $t_7 = t/(10^7 \text{ yr})$. Numerical experiments (MacLow and McCray 1988), in which the supernova energy is injected in random impulses, show that equation (1) provides a good approximation to the actual evolution of the supershell after a few supernovae have occurred.

Thus, a "typical" supershell, with, say, $N_*E_{51} = 20$ and $n_0 \approx 1$ cm^{-3} will grow to a radius $R_S \approx 180$ pc within $\sim 10^7$ yr as a result of $\sim$ four supernova events. At that time, the supershell consists of $\sim 7 \times 10^5 M_\odot$ of interstellar gas expanding with velocity $V_S \approx 10$ km s^{-1} and kinetic energy $\approx 10^{51}$ ergs (20% of the net supernova energy). The atomic density of the shell is given by $n_S = n_0(V_S/a_S)^2$, where a_S, the magnetosonic speed in the shell, has a typical value $\sim 1 - 2$ km s^{-1}. Because of its elevated pressure, the gas in the supershell will radiate efficiently and most of its mass will be in the form of cool ($T \lesssim 100$ K) H I or H$_2$. The younger supershells, with $t \lesssim 5 \times 10^6$ yr, may have enough O stars in the cluster to create a thin skin of H II on the inner surface of the supershell that may be visible as a giant optical loop. However, most supershells will be older and larger, so that the O stars will have vanished; these older shells will be visible through H I or CO radio line emission, or perhaps as infrared dust shells.

During the early stages of the evolution of a supershell, approximately 45% of the net supernova energy resides in its interior, providing pressure to drive the expansion. The gas in the interior, which comes from conductive evaporation from the inner surface of the shell, has mass

$$M_i \sim 500 M_\odot [N_*E_{51}]^{0.8} n_0^{-0.06} t_7^{1.2} \qquad (2)$$

and temperature $\sim 10^6$ K. However, after a time

$$t_c \sim 4 \times 10^6 \text{ yr } \zeta^{-1.5} [N_*E_{51}]^{0.3} n_0^{-0.7}, \qquad (3)$$

where ζ is the metallicity in solar system units, the interior radiates at a rate comparable to the average supernova energy input rate, robbing the system of its driving pressure. This pressure never becomes negligible, however: numerical calculations (MacLow and McCray 1988) show that $R_S \propto t^{0.5}$ for $t > t_c$.

A more realistic model for supershell evolution would allow for expansion into an inhomogeneous ISM containing dense clouds embedded in a smoother intercloud medium. In that case the supershell will overtake and entrain clouds as it propagates through the intercloud medium. Clouds punch holes in the shell as the shell overtakes them, but the shell "heals" after it passes the clouds. The main effect of the entrained clouds on the dynamics of the shell is to enhance the radiative cooling of the interior, reducing the value of t_c.

After the first few supernovae, the interior mass of a supershell is sufficient to slow subsequent blast waves to subsonic ($\lesssim 250$ km s^{-1}) velocities before they reach the shell. Thus, our steady-power approximation for the global dynamics of the supershell is reasonable, because these blast waves do not substantially change the interior density profile. In a mature $(t \gtrsim 10^7$ yr) supershell there is a substantial chance of catching one or more supernova blast waves within the interior. If $t \gtrsim t_c$ these blast waves may have entered the radiative phase, and so might be observed as high velocity UV absorption line systems.

It is likely that a supershell will blast a hole through the gas layer of a disk galaxy. The scale height of the H I disk in the Milky Way increases from $z_0 \sim 80$ pc in its inner regions to $z_0 \sim 200$ pc in the solar neighborhood and $z_0 > 500$ pc in its outer reaches. Thus, the radius of a typical supershell in the solar neighborhood becomes comparable to z_0 at a time $t \sim 10^7$ yr.

To understand what happens after this time, MacLow and McCray (1988) have constructed numerical models for the development of super-shells in a stratified medium. [Tomisaka and Ikeuchi (1986) have also cal-culated such models.] In their numerical scheme, MacLow and McCray employ the approximation by Kompaneets (1960, cf. Zeldovich and Raizer 1968) for the dynamics of a thin shell. This approximation, which should be excellent for the problem at hand, consists of assuming that the interior is isobaric and that the surface moves in the direction of its local normal.

Typical results are shown in Figures 1a,b for a model in which the disk gas is assumed to have a density profile $n(z) = n_0(0.8 \exp[-(z/z_0)^2] + 0.2 \exp[-z/3.7z_0])$ (cf. Lockman, Hobbs and Shull 1986). The dimensions of the systems are expressed in units z_0 and times are expressed in units of t_1, the time at which a spherical shell would reach a radius z_0 in a medium of uniform density n_0 (cf. equation 1). Figure 1a shows what happens if the star cluster is located on the midplane of the galaxy, so that the supershell expands symmetrically about this plane. As one would expect, the shell becomes noticeably prolate in the vertical direction for $z \gtrsim z_0$. The shell becomes quite distended and would appear as a ring for $t \gtrsim 5t_1$, at which time $z \approx 3z_0$ and $r \approx 2r_0$. After this time, the expansion in the plane decelerates rapidly, so that the terminal radius of the hole in the plane becomes a measure of the scale height of the gas disk.

Figure 1b shows what happens when the cluster is located at a height $\Delta z = 0.6z_0$ above the midplane. The resulting supershell blows out through only one side of the gas disk. The result should appear as a partial arc.

The most interesting aspect of supershell hydrodynamics is the devel-opment of instabilities. There are three instabilities that we should be concerned about: (1) the dynamical instability of a cold decelerating shell discovered by Vishniac (1983); (2) the Rayleigh-Taylor instability that oc-curs when the shell begins to accelerate; and (3) gravitational instability. In order to investigate the non-linear development of these instabilities, we have begun to model the development of supershells with a supercomputer hydrodynamics code (MacLow, Norman and McCray 1987). Our results do

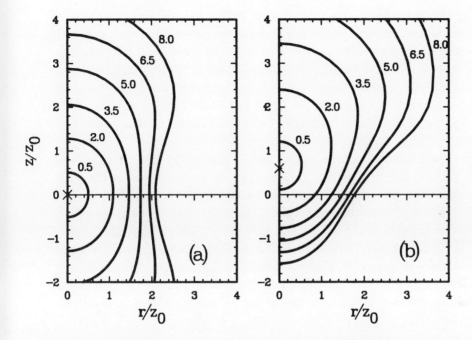

Figure 1: Development of a Supershell in a Stratified Gas Layer. a) Star cluster (marked by cross) located at the galactic midplane; b) Cluster located at a height $0.6z_0$ above the midplane. Curves are labeled by time in units t_1.

show a hint of the Vishniac instability, as do very similar calculations by Ròżyczka (1985), but a detailed comparison of the numerical investigations with the linear theory is still lacking (*cf.* Bertschinger 1986). Our results for the nonlinear development of the Rayleigh-Taylor instability in a stratified disk are very similar to the results of simulations by Ròżyczka and Tenorio-Tagle (1985).

3 OBSERVATIONS OF SUPERSHELLS

Supershells in the Milky Way and other nearby galaxies have been seen through H I 21-cm and H II optical emission line surveys. Heiles (1979, 1984) and Colomb, Poppel, and Heiles (1980) have discussed evidence for giant H I shells in the Milky Way, with radii ranging from ~ 100 pc to $\gtrsim 1$ kpc and kinetic energies ranging from $\sim 10^{50}$ to $\gtrsim 10^{53}$ ergs. The expanding shells have kinematic ages $\sim 10^7 - 10^8$ yr. Although there are a few beautiful examples of full circular arcs, most of the shells are only partial arcs. The complete shells and the largest shells are most often found beyond the solar circle, while many fragments of shells, called "worms" by Heiles (1984), are found in the inner Milky Way. Recently, Brinks and Shane (1984) have discovered ~ 140 holes in velocity-resolved 21-cm emission line maps of M31, with radii $\sim 125 - 300$ pc. Similar H I holes have been found in the Large Magellanic Cloud (Rohlfs *et al.* 1984; Dopita, Mathewson and Ford 1985), and M101 (Allen *et al.* 1978). X-ray emission has been observed from the supershell in Cygnus (Cash 1980) and from the region of Shapley's Constellation III in the LMC (Singh *et al.* 1987; Helfand, Wu, and Wang 1987).

The most prominent optical supershell in the Milky Way is the Gum Nebula, a giant $R \sim 400$ pc H II emission loop in the direction of the Vela supernova remnant (Maran, Brandt, and Stecher 1973). Many similar giant ($R \sim 100 - 200$ pc) and supergiant ($R \gtrsim 200$ pc) H II emission line shells containing clusters of OB stars are seen in the Milky Way (Brand and Zealey 1975; Bochkarev 1985; Fich 1986), the Magellanic Clouds (Westerlund and Mathewson 1966; Davies, Elliott, and Meaburn 1976; Meaburn 1980; Caulet *et al.* 1982; Georgelin *et al.* 1983; Braunsfurth and Feitzinger 1983), and in galaxies of the local group (e.g., Courtès 1977; Courtès, *et al.* 1987).

These observations of giant shells, supershells, and H I holes are all consistent with the theory described above. For $t_7 \lesssim 1$ the ionizing radiation from the OB association will create an H II emission line region on the inner rim of the supershell. For $1 \lesssim t_7 \lesssim 5$ the H II emission will have vanished along with the bright O stars, but the H I shell will continue to grow according to equation (1) as a result of the supernova explosions of the B stars. Thus, roughly 20% of the H I shells should have an associated H II shell, and these younger systems should be somewhat smaller and more rapidly expanding than the older systems. The radii, ages, and kinetic energies of the expanding H I shells are consistent with the theory if they are created by OB associations with $10 \lesssim N_* \lesssim 1000$. The older supershells would be less likely to contain a recognizable cluster, because the remaining B stars are fainter and would have dispersed significantly.

The observation of H I holes in other galaxies without obvious shells surrounding them could be explained if the shells were predominantly H_2; this hypothesis would imply that expanding rings of CO emission should be seen around these holes. Note also that the kinetic energies of the supershells inferred from their H I masses and expansion velocities could be substantial underestimates if most of the mass of the shell is H_2.

The multiple supernova interpretation of the H I supershells requires that the shells contain their internal pressure for $t_7 > 1$. This interpretation seems to require that the supershells are developing in a fairly homogeneous ISM, in order that they remain coherent until they reach radii $R_S \gtrsim 100$

pc. If the intercloud medium is low density ($n_0 \lesssim 10^{-2}$ cm^{-3}) coronal gas, as in the McKee-Ostriker (1977) model, the blast wave from even a single supernova explosion will propagate right out of the galactic disk, and coherent supershells will not form.

The radii of the supershells in spiral galaxies tend to increase with galactocentric radius for two reasons. First, since the interstellar gas density is lower in the outer parts of galaxies, a given net supernova energy can produce a larger shell. Second, as we have shown, a shell will burst through the galactic disk when $R_S \gtrsim 1.3z_0$, and z_0 increases with galactocentric radius. In the inner parts of spiral galaxies the supershells will burst through the disk, leaving "holes" with radius comparable to the disk thickness. The H I "worms" seen by Heiles (1984) in the inner Milky Way are probably the limb-brightened rims of supershells that have burst through the disk. The partial arcs might be fragments of supershells that have burst through only one side of the galactic disk.

Although supershells are seen in disk galaxies, the most spectacular examples are found in the Large Magellanic Cloud. There are two reasons why irregular galaxies are especially favorable environments for supershell formation. First, the interstellar gas in an irregular galaxy has large scale height and low density as a result of the low mass of the galaxy, and possibly also because the gas layer has been disturbed by tidal interactions with neighboring galaxies. Second, irregular galaxies tend to have lower metallicity than the giant spirals, so that radiative cooling is less important (cf. equation 3). Thus, supershells can grow to larger radii in irregular galaxies before they lose their interior pressure as a result of "blowout" or radiative cooling.

A major theme of this conference has been the growing recognition that many supernova remnants are not propagating into any kind of "standard" ISM, which may be fairly represented by uniform gas. Indeed, very interesting evidence has been presented that in many cases the ISM has been drastically modified before the supernova occurred, either by the progenitor star (*cf.* Lozinskaya 1987; Braun 1987), or by previous supernovae. In the latter case supernova remnants would be hard to detect because of their rapid expansion and low surface brightness, but this is now becoming possible (Chu and Kennicutt 1987). The supernovae that give rise to pulsars come from massive stars that usually occur in clusters, so most of these supernovae would occur within a supershell and expand to large radius without becoming radiative. Therefore, it is not surprising that many of the Crab-like SNR's are "Crabs without shells" (Becker and Helfand 1987).

4 POROSITY OF THE INTERSTELLAR MEDIUM

It is clear that coronal gas is a major component of the ISM. Ultraviolet absorption line studies (Frisch and York 1983) and soft X-ray emission maps (McCammon *et al.* 1983) show that the solar system itself is located in a hot bubble, and observations of O VI $\lambda\lambda 1032 - 1037$ interstellar absorption lines in the spectra of more distant stars can be interpreted as the conductive interfaces between interstellar clouds and more distant bubbles of coronal gas (Cowie *et al.* 1979). Two major questions remain: 1) What is the porosity of the coronal gas; and 2) what is the morphology of the cool clouds? The answers to these questions are intimately related to the interaction of supernova remnants with the ISM.

In their theory for the ISM, McKee and Ostriker (1977) begin with an estimate of the porosity of coronal gas created by supernova remnants, given by

$$Q = (\frac{r_{SN}}{A_{disk}H_{SN}})(\frac{4\pi}{3}R_{SNR}^3)(\frac{R_{SNR}}{v_{rms}}), \tag{4}$$

where r_{SN} is the galactic rate of supernova explosions, A_{disk} is the area of the galactic disk, H_{SN} is the thickness of the supernova distribution, R_{SNR} is the radius of a supernova shell after it stops growing, and R_{SNR}/v_{rms} is the time for interstellar gas moving at random velocity v_{rms} to fill in a supernova shell. With reasonable estimates of the galactic supernova rate, McKee and Ostriker find that $Q \sim 1$ if the supernovae are uncorrelated in space and time and the ISM is initially uniform. They construct a model consisting of cool spherical clouds ($T \sim 100$ K, $n \sim 10^2$ cm^{-3}) with warm ($T \sim 10^4$ K, $n \sim 0.4$ cm^{-3}) envelopes embedded in a substrate of coronal ($T \sim 10^6$, $n_0 \sim 10^{-2.5}$ cm^{-3}) gas that is produced by thermal evaporation and removed by blast waves. They find a stationary state in which the coronal gas is heated by the blast waves and cools by radiation.

The factors in equation (4) are very uncertain (Heiles 1987; Shull 1987). The supernova rate may range from $r_{SN} \sim (20$ yr$)^{-1} - (100$ yr$)^{-1}$ and the thickness of the supernova distribution may range from $H_{SN} \sim 100$ pc for type II supernovae to $H_{SN} \sim 1$ kpc for type I supernovae. The greatest uncertainty comes from the value of R_{SNR}, which is given approximately by (Shull 1987)

$$R_{SNR} \approx 50 \text{ pc } E_{51}^{11/35} n_0^{-13/35}. \tag{5}$$

Since n_0 changes from $n_0 \sim 1$ cm^{-3} for $Q \ll 1$ to $n_0 \lesssim 10^{-2}$ cm^{-3} for $Q \gtrsim 1$, equation (4) is nonlinear and predicts a sudden phase change when $Q \gtrsim 0.2$ (the filling factor of coronal gas is given by $f \approx Q/(1 + Q)$). As a result, in the Milky Way (with $A_{disk} = 830$ kpc^2 and $v_{rms} = 10$ km s^{-1}), Q can range from $Q \sim 0.03$ for $r_{SN} = (100$ yr$)^{-1}, H_{SN} = 1$ kpc, and $n_0 = 1$ cm^{-3} to $Q \gg 1$ for $r_{SN} = (20$ yr$)^{-1}, H_{SN} = 100$ pc, and $n_0 < 0.1$ cm^{-3}

If the supernovae in the gas disk are predominantly correlated type II (Population I) events, with $N_* \sim 20 - 1000$ (Heiles 1987), the resulting supershells will most likely blast holes through the galactic disk and one should replace equation (4) by its two-dimensional analogue:

$$Q = (\frac{r_{SN}}{N_* A_{disk}})(\pi R_{SS}^2)(\frac{R_{SS}}{v_{rms}}), \tag{6}$$

where $R_{SS} \sim 1.8 z_0$ is the radius of a supershell when it bursts through the disk. For example, we find $Q \sim 1$ for $r_{SN} = (100$ yr$)^{-1}, N_* = 20$, and $z_0 = 100$ pc. Note the qualitative differences here: in equation (5) Q decreases with N_* and z_0 because an increasing fraction of the net supernova energy is vented to the galactic corona; furthermore, Q is not sensitive to n_0 because the maximum radius of the supershell is determined predominantly by z_0, not n_0.

Of course, equations (4) and (5) provide only rough global estimates of the porosity of a disk galaxy. In reality, all the parameters must be functions of galactocentric radius, and the rates and scale heights of type I and type II supernovae are different functions. If we knew these functions, it would be straightforward to construct more detailed versions of equations (4) and (5), but, given our present ignorance, such an exercise may not be warranted.

Direct observations of the H I distribution in the Milky Way and other nearby galaxies may be a more fruitful approach to understanding the factors determining the porosity of the gas. Observations of the Milky Way show that the warm H I is more pervasive and smooth than predicted by the McKee-Ostriker model (Liszt 1983; Lockman, Hobbs and Shull 1986; Kulkarni and Heiles 1987), suggesting that the model requires some qualitative modification (Shull 1987; Cowie 1987). Another important clue is provided by the beautiful H I image of M31 observed by Brinks and Shane (1984), which has a huge (radius ~ 5 kpc) hole in its center. Perhaps type I supernovae from the galactic bulge predominate in this inner region, so that the ISM there has a morphology in accord with the McKee-Ostriker model, while supershells due to type II supernovae control the morphology of the H I ring seen beyond 5 kpc.

A related aspect of the McKee-Ostriker model that needs further investigation is their assumption that the H I clouds are roughly spherical. If the clouds are made by blast waves, one would expect them to have a sheet-like or sponge-like morphology, because pressure tends to make things flat, not round. (The clouds should tend to become round only when self-gravity becomes important.) Indeed, the 21-cm emission maps of the Milky Way (Colomb, Poppel and Heiles 1980) and the *IRAS* images of infrared cirrus (Low *et al.* 1984) certainly give one the impression that the clouds have sheet-like geometry — *cirrus*, not *cumulus*! If so, blast waves cannot easily circumvent and entrain the clouds.

5   PROPAGATING STAR FORMATION

Since multiple supernovae can pack some $10^6 M_\odot$ of interstellar gas into a supershell, it is natural to ask whether such an expanding shell might become gravitationally unstable. The relevant instability criterion was first derived from a simple energy principle by Ostriker and Cowie (1981) and was verified by Vishniac (1983) with a linear perturbation analysis. When applied to a supershell (cf. equation 1) this criterion gives

$$t_1 \approx 3.2 \times 10^7 \text{ yr } [N_* E_{51}]^{-1/8} n_0^{-1/2} a_S^{5/8} \tag{7}$$

and

$$R_1 \approx 200 \text{ pc } [N_* E_{51}]^{1/8} n_0^{-1/2} a_S^{3/8}, \tag{8}$$

for time and the radius at which gravitational instability first ensues, where a_S is the magnetosonic speed (in km s^{-1}) in the supershell. The typical mass of a gravitationally unstable fragment is given by

$$M_1 \approx 5 \times 10^4 M_\odot [N_* E_{51}]^{-1/8} n_0^{-1/2} a_S^{29/8}. \tag{9}$$

Thus, we see that if $a_S \sim 1$ km s^{-1}, a supershell is likely to break up into gravitationally bound fragments with masses comparable to giant molecular clouds at about the same time that it bursts through the galactic disk.

    The main uncertainty in the theory comes from the interstellar magnetic field, which may inhibit gravitational collapse. If the field is negligible, the sound speed in the cool H I or H_2 shell will have a value $a_S < 1$ km s^{-1}. On the other hand, if the ambient interstellar magnetic field has a typical value $B_0 \sim 1 \mu G (n_0/1 \text{ cm}^{-3})^{1/2}$ (cf. Troland and Heiles 1986), the pressure in the shell will be dominated by the compressed interstellar magnetic field and the magnetosonic speed in the shell will have the typical value $a_S \sim (V_S V_A)^{1/2}$, where $V_A = (B_0^2/4\pi\rho_0)^{1/2} \sim 2$ km s^{-1} is the Alfven speed in the ambient ISM. Thus, it appears that the typical magnetic field in the Milky Way will suppress propagating star formation induced by supershells.

    In disk galaxies, supershells may induce secondary star formation in some instances, but it is clear that the dominant mechanism is provided by the spiral density waves. However, gravitational instability of supershells may be the dominant mechanism for star formation in irregular galaxies. There are three possible reasons. First, as mentioned in §3, supershells can and do grow to greater size in irregulars because of their thicker gas distribution and lower metallicity. Second, because irregular galaxies lack well-organized spiral arms, interstellar gas will continue to accumulate in them until some other mechanism triggers gravitational instability. Indeed, Magellanic irregulars do tend to have higher (~ 1) gas/stellar mass ratios than spirals (~ 0.1). Finally, one may speculate that Magellanic Irregulars do not have an effective interstellar dynamo, and so magnetic suppression of star formation might be less effective.

There is fragmentary evidence that supershell-induced star formation is occurring in the Milky Way and other nearby galaxies (cf. Elmegreen 1985a,b). The most obvious example is Shapley's Constellation III in the LMC, a great arc of bright blue stars stretching some 600 pc (Westerlund and Mathewson 1966). Another example is the distribution of OB associations around the periphery of the Loop IV supershell in the LMC (Dopita, Mathewson, and Ford 1985). The most spectacular example that I have seen is the giant (diameter ~ 500 pc) ring of OB stars pointed out by Bothun (1986) in the Magellanic Irregular NGC 4449.

Finally, under propitious circumstances, supershell-induced star formation might continue for more than one generation, causing a huge excursion in the star formation rate of an irregular galaxy (cf. Gerola, Seiden, and Schulman 1980). Thus, supershells may provide a mechanism for "starburst" activity in galaxies.

This work was partially supported by NASA Grant NAGW-766 under the NASA Astrophysical Theory Program.

REFERENCES

Allen, R. J., van der Hulst, J. M., Goss, W. M., and Huchtmeier, W. 1978, *Astr. Ap., 64*, 359.

Becker, R. H., and Helfand, D. J., these proceedings.

Bertschinger, E. 1986, *Ap. J., 304*, 154.

Bochkarev, N. G. 1985, *Sov. Astr. Letters, 10*, 76.

Bothun, G. D. 1986, *Astr. J., 91*, 507.

Brand, P. W. J. L., and Zealey, W. J. 1975, *Astr. Ap., 38*, 363.

Braun, R. 1987, these proceedings.

Braunsfurth, E., and Feitzinger, J. V. 1983, *Astr. Ap., 127*, 113.

Brinks, E., and Shane, W. W. 1984, *Astr. Ap. Suppl., 55*, 179.

Cash, W., *et al.* 1980, *Ap. J. Letters,* **238**, L71.

Caulet, A., Deharveng, L., Georgelin, Y. M., and Georgelin, Y. P. 1982, *Astr. Ap.,* **110**, 185.

Chu, Y.-H., and Kennicutt, R. C. 1987, these proceedings.

Colomb, F. R., Poppel, W. G. L., and Heiles, C. 1980, *Astr. Ap. Suppl.,* **40**, 47.

Courtès, G. 1977, in *Topics in Interstellar Matter,* ed. H. van Woerden (Dordrecht: Reidel), p. 209.

Courtès, G., Petit, H., Sivan, J.-P., Dodonov, S., and Petit, M. 1987, *Astr. Ap.,* **174**, 28.

Cowie, L. L. 1987, in *Interstellar Processes,* eds. D. Hollenbach and H. Thronson (Dordrecht: Reidel), in press.

Cowie, L. L., Jenkins, E. B., Songaila, A., and York D. G. 1979, *Ap. J.* **232**, 467.

Cox, D. P., and Smith, B. W. 1974, *Ap. J. Letters,* **189**, L105.

Davies, R. D., Elliott, K. H., and Meaburn, J. 1976, *Mem. R. Astr. Soc.,* **81**, 819.

Dopita, M. A., Mathewson, D. S., and Ford, V. L. 1985, *Ap. J.,* **297**, 599.

Elmegreen, B. G. 1985a, in *Birth and Infancy of Stars,* eds. R. Lucas, A. Omont and R. Stora, (Amsterdam: Elsevier) p. 215.

Elmegreen, B. G. 1985b, in *Birth and Evolution of Massive Stars and Stellar Collapse,* eds. W. Boland and H. van Woerden (Dordrecht: Reidel), p. 227.

Fich, M. 1986, *Ap. J.* **303**, 465.

Frisch, P. C., and York, D. G. 1983, *Ap. J. Letters,* **271**, L59.

Georgelin, Y. M., Georgelin, Y. P., Laval, A., Monnet, G., and Rosado, M. 1983, *Astr. Ap. Suppl.,* **54**, 459.

Gerola, H., Seiden, P. E., and Schulman, L. S. 1980, *Ap. J.,* **242**, 517.

Heiles, C. 1979, *Ap. J.,* **229**, 533.

Heiles, C. 1984, *Ap. J. Suppl.,* **55**, 585.

Heiles, C. 1987, *Ap. J.,* **315**, 555.

Helfand, D. J., Wu, X., and Wang, Q. 1987, this conference, not published.

Kompaneets, A. S. 1960, *Soviet Phys. Dokl.,* **5**, 46.

Kulkarni, S. R., and Heiles, C. 1987, in *Galactic and Extragalactic Radio Astronomy,* eds. K. I. Kellerman and G. L. Verschuur (New York: Springer–Verlag), in press.

Liszt, H. S. 1983, *Ap. J.,* **275**, 163.

Lockman, F. J., Hobbs, L. M., and Shull, J. M. 1986, *Ap. J.,* **301**, 380.

Low, F. J., *et al.* 1984, *Ap. J. Letters,* **278**, L19.

Lozinskaya, T. A. 1987, these proceedings.

MacLow, M. M., and McCray, R. 1988, *Ap. J.,* in press.

MacLow, M. M., Norman, M. L., and McCray, R. 1988, these proceedings.

Maran, S. P., Brandt, J. C., and Stecher, T. P. 1973, *The Gum Nebula and Related Problems*, NASA SP-332.

McCammon, D., Burrows, D. N., Sanders, W. T., and Kraushaar, W. L. 1983, *Ap. J.,* **269**, 107.

McCray, R., and Kafatos, M. 1987, *Ap. J.,* **317**, 190.

McKee, C. F., and Ostriker, J. P. 1977, *Ap. J.,* **218**, 148.

Meaburn, J. 1980, *M.N.R.A.S.,* **192**, 365.

Ostriker, J. P., and Cowie, L. L. 1981, *Ap. J. Letters,* **243**, L127.

Rohlfs, K., Kreitschmann, J., Siegman, B. C., and Feitzinger, J. V. 1984, *Astr. Ap.,* **137**, 343.

Ròżyczka, M. 1985, *Astr. Ap.,* **143**, 59.

Ròżyczka, M., and Tenorio-Tagle, G. 1985, *Astr. Ap.,* **147**, 209.

Shull, J. M. 1987, in *Interstellar Processes*, eds. D. Hollenbach and H. Thronson (Dordrecht: Reidel), in press.

Singh, K. P., Nousek, J. A., Burrows, D. N., and Garmire, G. P. 1987, *Ap. J.,* **313**, 185.

Tomisaka, K., and Ikeuchi, S. 1986, *Publ. Astr. Soc. Japan* **38**, 697.

Troland, T. H., and Heiles, C. 1986, *Ap. J.,* **301**, 339.

Vishniac, E. T. 1983, *Ap. J.,* **274**, 152.

Weaver, R., Castor, J., McCray, R., Shapiro, P., and Moore, R. 1977, *Ap. J.,* **218**, 377; errata, **220**, 742.

Westerlund, B. E., and Mathewson, D. S. 1966, *M.N.R.A.S.,* **131**, 371.

Zeldovich, Ya. B., and Raizer, Yu. P. 1968, *Elements of Gasdynamics and the Classical Theory of Shock Waves* (New York: Academic Press).

NUMERICAL MODELS OF SUPERSHELL DYNAMICS

Mordecai-Mark Mac Low, Michael L. Norman[*] & Richard A. McCray
Joint Institute for Laboratory Astrophysics, University of
Colorado and National Bureau of Standards, Boulder, CO
80309-0440

Abstract: Superbubbles play an important role in determining
the state of the ISM in both spiral and irregular galaxies.
We are modeling supershell dynamics in both homogeneous and
stratified atmospheres using ZEUS, a 2-D hydrocode. We find
that when a superbubble blows out of a Gaussian atmosphere,
the cold, dense shell is not greatly accelerated. In
addition, we believe that we observe the Vishniac
overstability in radiative, decelerating shells.

Introduction: It is becoming increasingly apparent that the location of
most Type II SNe in or near their parent OB associations has important
consequences for the state of the ISM. The repeated SNe occurring in a
typical OB association, along with the stellar winds from the hotter
members of the association, will form a large, hot cavity surrounded by a
thin, dense shell of swept up ISM.

In a disk galaxy, with an H I layer only a few hundred parsecs
thick, such a superbubble can blow a hole completely through the disk,
producing structures similar to the "worms" observed by Heiles (1979,
1984) in the Milky Way H I layer. In an irregular galaxy superbubbles
can grow to great sizes. An example of this is the X-ray superbubble
recently discovered in the LMC with radius 500-700 pc, identified with
Shapley III (Singh et al. 1987). Superbubble dynamics are of theoretical
importance as well. In disk galaxies, they provide a means to vent SN
energy to the halo (Ikeuchi 1987), and to accelerate mass out of the
disk, while in irregulars, supershells are probably significant sites of
star formation when they go gravitationally unstable (McCray and Kafatos
1987). In addition, we hope to learn more about the instabilities that a
radiative, decelerating shock is subject to (Vishniac 1983, Bertschinger
1986).

Technique: We are using ZEUS, a general purpose two-dimensional hydrodynamics code described by Norman and Winkler (1986), to model superbubbles in homogeneous and stratified atmospheres. This code has a fluid
interface tracker that makes it well suited to modeling the dynamics of a
thin shell.

We approximate the SNe from an OB association as a continuous energy
source (Mac Low and McCray 1987). We model this source by placing mass
with a temperature 10^3 that of the background medium in the center of the
superbubble. This results in the right order of magnitude of mass being

---

[*]National Center for Supercomputing Applications, Champaign, Il, 61820.

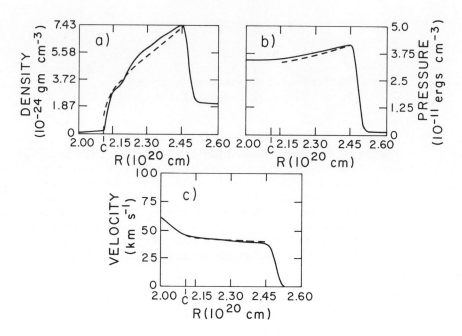

Fig. 1.   Comparison of (a) density, (b) pressure, and (c) velocity profiles of Weaver et al.'s (1977) similarity solution with profiles through an adiabatic supershell modeled with ZEUS. The theoretical position of the contact discontinuity is marked by a C in each plot.

inside the superbubble despite our neglect of the effect of evaporation of mass off the cold shell. We use the cooling function given by MacDonald and Bailey (1981), which is essentially that of Raymond et al. (1976) between 10^5 and 10^7 K. In addition, we include a heating function linearly proportional to density just large enough to hold the background ISM at a steady temperature of 10^4 K.

Weaver et al. (1977) found a similarity solution for the structure of an adiabatic stellar wind bubble which is applicable to the case of an adiabatic supershell. In Fig. 1, we show the close agreement between that solution and profiles through an adiabatic supershell modeled with ZEUS. The theoretically calculated position of the contact discontinuity also agrees very well with the ZEUS result.

Results: In Fig. 2, we compare the density distribution in radiatively cooled and adiabatic supershells. We believe the clumpiness evident in the radiatively cooled model is a manifestation of the Vishniac (1983) overstability of a thin shell. If we are indeed observing this over-stability, then we find that it damps fairly quickly once it becomes nonlinear. Clumps with a factor of 2 density contrast form very quickly in the shell, but they do not condense much further than this.

In Fig. 3 we show the time evolution of a superbubble evolving in a Gaussian atmosphere with a scale height of 100 pc, an atomic number density of 1 cm^{-3} in the plane (we assume 10% He by number) and a temperature

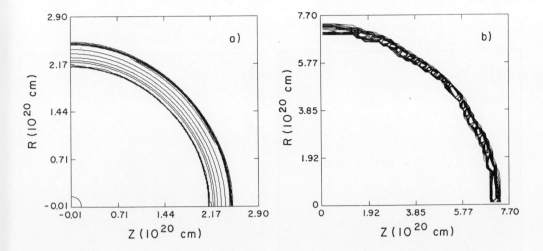

Fig. 2. Density distribution in (a) an adiabatic supershell and (b) a radiatively cooled supershell (at a later time). The contours are linearly spaced from zero to (a) 8.4×10^{-24} cm^{-3} and (b) 2.11×10^{-23} cm^{-3}.

of 10^4 K. The luminosity of this model is equivalent to one SN every 3×10^5 yr, which implies an OB association with ≈ 150 potential SNe. As soon as the supershell begins to accelerate upward Rayleigh-Taylor instabilities set in, disrupting it and allowing the hot interior gas to escape into the low density halo.

The large central spike is a result of our assumption of cylindrical symmetry. In a three-dimensional model, non-axisymmetric modes of the Rayleigh-Taylor instability would be excited, and the central portion of the shell would fragment into several pieces. The key result that the shell fragments are not accelerated by the expanding hot gas would not change, however.

We thank Dr. Mike Van Steenburg, Steve Voels, and CASA for assistance in producing Figure 3. This work was partially supported by NASA grant NAGW-766 under the NASA Astrophysical Theory Program. Computations were performed on the NASA Cray X-MP.

References
Bertschinger, E. 1986, Ap. J., 304, 154.
Heiles, C. 1979, Ap. J., 229, 533.
_____. 1984, Ap. J. Suppl., 55, 585.
Ikeuchi, S. 1987, preprint.
MacDonald, J. and Bailey, M. E. 1981, M.N.R.A.S., 197, 995.
Mac Low, M.-M., and McCray, R. A. 1987, Ap. J., submitted.
Mac Low, M.-M., Norman, M. L., and McCray, R. A. 1987, in preparation.
McCray, R. A. and Kafatos, M. C. 1987, Ap. J., 317, 190.
Norman, M. L., and Winkler, K.-H. A. 1986, in Astrophysical Radiation Hydrodynamics, eds. K-H. A. Winkler and M. L. Norman (Dordrecht: Reidel).

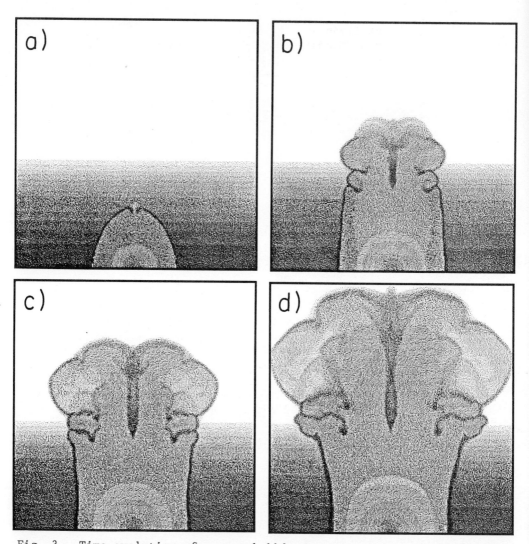

Fig. 3. Time evolution of a superbubble in a Gaussian atmosphere with parameters given in the text. Density (on a log scale ranging from 10^{-6} to 10 cm^{-3}) is shown at times of (a) 3.5 Myr, (b) 4.8 Myr, (c) 5.4 Myr, (d) 6.1 Myr.

Raymond, J. C., Cox, D. P., and Smith, B. W. 1976, Ap. J., 204, 290.
Singh, K. P., Nousek, J. A., Burrows, D. N., and Garmire, G. P. 1987, Ap. J., 313, 185.
Vishnaic, E. T. 1983, Ap. J., 274, 152.
Weaver, R., McCray, R. A., Castor, J., Shapiro, P., and Moore, R. 1977, Ap. J., 218, 377; erratum, 220, 742.

A NONTHERMAL SUPERBUBBLE IN THE IRREGULAR GALAXY IC 10

Evan D. Skillman
Netherlands Foundation for Radio Astronomy
Dwingeloo, The Netherlands

Abstract: New high resolution radio continuum images of
the nearby irregular galaxy IC 10 have revealed a large
(> 250 pc) nonthermal source. The source is roughly
circular with a spectral index of ~ -0.5, and is most
likely a very large supernova remnant. Its large size
suggests that it is the result of several supernovae,
and may be related to the supershells observed in our
own and other galaxies.

Introduction

IC 10 is a nearby irregular galaxy. Because it lies at a very low
galactic latitude (-3°), its distance is very uncertain. Estimates of
its distance range from as near as 1 Mpc (Roberts 1962) to as far as 3
Mpc (Bottinelli et al. 1972). Neutral hydrogen studies of IC 10 have
revealed that it has an extended HI envelope which shows a velocity
gradient of opposite sign to the central condensation (Cohen 1979).
Klein and Grave (1986) have studied the radio continuum emission from
IC 10 and have found a global spectral index of -0.33 between 1 and 10
GHz. Because the thermal free-free emission from HII regions normally
dominates the radio continuum emission from irregular galaxies in the
frequency range of 1 to 10 GHz, it was decided to obtain high
resolution radio observations of IC 10 to search for the source(s) of
the nonthermal component.

New Observations

Radio continuum observations were made at λ6 cm with the
Westerbork Synthesis Radio Telescope (WSRT) and at λ21 cm with the NRAO
Very Large Array. Care was taken to insure that the uv coverages of
the two observations were properly matched, and images at an identical
resolution of 5" (FWHM) were produced. Figure 1 shows the λ21 cm image
of IC 10. Several HII regions were identified by their correspondence
with known Hα emission and thermal (flat) radio spectra, but an
additional extended nonthermal source was also discovered. This large
nonthermal source is nearly circular and is not limb-brightened. It is
confused on one side where it is adjacent to a collection of HII
regions. If one assumes that the major deviations from radial symmetry
are due totally to confusion with the HII regions, a total flux of 39
mJy at λ21 cm and a spectral index of about -0.5 are derived. The
diameter of 0.8' corresponds to a size of 230 pc for the minimum
distance estimate of 1 Mpc. This represents a truly outstanding size
for a single supernova remnant, and makes it more likely that this
source is a result of several supernovae. This source will therefore
be referred to as a superbubble.

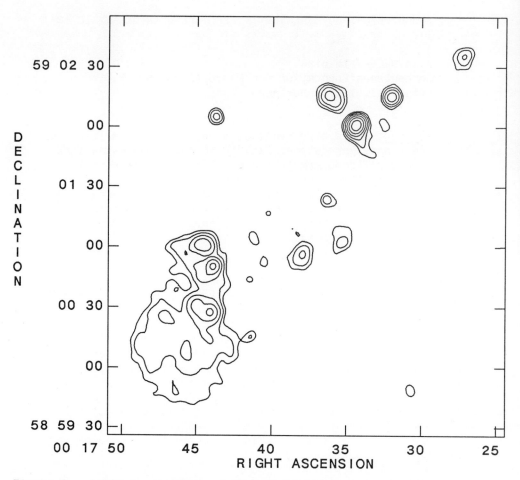

Figure 1:   A λ21 cm continuum image of IC 10 at a resolution of 5".
The contour levels are set at 0.2, 0.4, 0.8, 1.6, 3.2, 6.4, and 12.8
mJy/beam.   The r.m.s. noise in the map is 0.05 mJy/beam.   The
superbubble is the large source in the lower left.   All but one of the
other sources are coincident with Hα emission, have thermal radio
spectra, and are therefore probably HII regions.

    In an effort to study the environment of the superbubble, the
radio continuum images were compared with new synthesis HI observations
of IC 10 made at Westerbork (Shostak and Skillman 1987).   The
superbubble is centered on the most massive HI concentration in IC 10.
The total HI mass of this complex is 8 x 10^6 $(D/Mpc)^2$ solar masses and
it is roughly four times larger in diameter than the superbubble.   HI
position-velocity diagrams were studied in an attempt to try to detect
an interaction between the superbubble and the neutral hydrogen.
Unfortunately, the velocity field in the area of the superbubble is
very complex, and it is difficult to discern discrepant motions.
However, there are no obvious signs of bulk motions of the neutral gas
due to the interaction with the bubble.

Interpretation

The minimum size of 230 pc lies on the border between the ranges
of sizes of the HI shells and supershells discovered in our galaxy by
Heiles (1976, 1979). This size is also at the large end of the
distribution of sizes of the giant shells observed in the Magellanic
Clouds (Meaburn 1980), although taking a larger distance estimate would
make it comparable to the supergiant shells. For our galaxy,
Bruhweiler et al. (1980) have shown that shells can be the natural
result of the evolution of an OB association, while Heiles (1984)
argues that exceptional conditions would be required to give rise to a
supershell. Meaburn (1980) came to essentially the same conclusion
concerning the giant and supergiant shells in the Magellanic Clouds,
although Dopita et al. (1985) have shown how one supergiant shell can
be interpreted as the result of self-propagating star formation.
McCray and Kafatos (1987) have also discussed the creation of a
supershell through self-propagating star formation.

The superbubble in IC 10 may represent evidence of a link between
the giant and supergiant stage. The nonthermal emission makes it clear
that supernovae play an important role, and the large size and
luminosity essentially rule out a single supernova. Thus we appear to
have caught a giant bubble being built by multiple supernovae. The
bubble is most likely in the pressure driven phase identified by McCray
and Kafatos (1987). The HII regions adjacent to the bubble may be
causally connected via self-propagating star formation. Considering
that the size of the HI complex that the bubble is located in is 1
(D/Mpc) kpc, this star formation event could easily propagate into a
structure 1 kpc in size, leaving behind a supergiant shell.

Acknowledgements

The NRAO is operated by Associated Universities Inc., under
contract with the National Science Foundation. The WSRT is operated by
the NFRA, which is financially supported by the Netherlands
Organization for the Advancement of Pure Research (ZWO).

References
Bottinelli, L., Gouguenheim, L., Heidmann, J.:1972, A&A 18, 121
Bruhweiler, F.C., Gull, T.R., Kafatos, M., Sofia, S.:1980, ApJ 238, L27
Cohen, R.J.:1979, MNRAS 187, 839
Dopita, M.A., Mathewson, D.S., Ford, V.L.:1985, ApJ 297, 599
Heiles, C.:1976, ApJ 208, L137
_____.:1979, ApJ 229, 533
_____.:1984, ApJ Suppl 55, 585
Klein, U., Grave, R.: 1986, A&A 161, 155
McCray, R., Kafatos, M.:1987, ApJ 317, 190
Meaburn, J.:1980, MNRAS 192, 365
Roberts, M.S.:1962, AJ 67, 431
Shostak, G.S., Skillman, E.D.:1987, in prep

SUPERSHELLS WITH INDUCED LARGE-SCALE SUPERNOVAE FORMATION

I.G. Kolesnik and S.A. Silich
Main Astronomical Observatory
Academy of Sciences of the Ukrainian SSR,
Kiev, Goloseevo, USSR

Abstract: We propose a model of expanding supershells regulated by star formation which is induced in HI super- clouds by similar shells from a previous evolutionary stage. Compression of pre-existing cloudlets by such shells triggers the formation of massive stars which then explode at the end of their evolution as supernovae. Efficiency of induced supernovae formation must be less than 1% to fit the observational properties of supershells.

Introduction: Recent observations of our own and nearby galaxies have revealed three types of large-scale structures with a characteristic spatial scale of about 1 kpc and mass ~10 $M_\odot$. These are: neutral hydrogen superclouds[6], star complexes[5], HI "holes" and expanding supershells of neutral and ionized hydrogen[2,7,11]. Several mechanisms for the origin of these structures have been proposed[1,3,13,15]. The coincidence of the typical sizes and masses of all of the above objects suggests that superclouds, star complexes, HI "holes" and expanding supershells are just different evolutionary stages of an overall process of the formation and evolution of huge self-gravitating galactic structures. Our main attention is directed to the dynamics of supershells driven by the collective interaction of supernovae with HI superclouds.

The Model: An initial supercloud (Figure 1a), a two-phase gravitationally-bound system, is likely to be composed of cold dense cloudlets embedded in a substrate of low-density warm interstellar gas. Conditions are created in the central parts of superclouds for the formation of giant molecular clouds. Thus giant molecular clouds appear to be only the cores of these larger fundamental structures. Star formation starts in these cores of superclouds. The combined pressure of stellar winds and supernova explosions in an initial OB-association will generate a shell expanding with a velocity[8,15] $u_{sh} \propto R^{-2/3}$. Unusual OB-associations containing more than 10^3 stars are required to produce the largest expanding shells according to the theory of Kafatos and McCray[8]. However, OB-associations usually contain no more than 100 massive stars. The expansion velocities of shells vary within a narrow velocity range as their sizes change from ~0.1 kpc up to ~1.5 kpc[7]. Similarly there is a clear tendency for HI "holes" to be larger for greater expansion velocities[2]. Thus, in studying the dependence of expansion velocities on the radii of supershells, one gains the impression that additional volume sources of energy input must appear during the evolution of a shell. We suggest that collisions of a supershell with the cloudlets randomly

distributed in the supercloud will trigger first the formation of
massive stars and then a wave of supernova explosions moving after the
shell (Figure 1b, 1c).

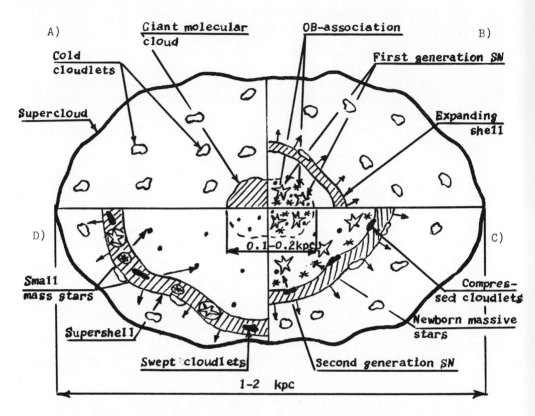

Fig. 1. A scenario of the evolution of a supershell.

Dynamics of the Supershells: Three evolutionary stages are evident for
the different dominant mechanisms of energy input into the cavity.
1. The energy input into the cavity is dominated by the supernova
    explosions in the initial OB-association. Evolution of the shell
    is described by the models of Kafatos and McCray[8], and Tomisaka et
    al.[15].
2. The energy input into the cavity is dominated by newborn supernova
    explosions. This starts immediately after the first newborn
    supernova explosion and ends when the supershell mass increases
    enough to sweep the small cold cloudlets out of the cavity.
3. All massive stars inside the supershell have exploded as
    supernovae. Energy pumping into the cavity stops (Figure 1d).
    The equation of shell motion[10] is:

$$\frac{d^2 u}{dt^2} + (7+3\delta)\frac{u}{R}\frac{du}{dt} + \frac{9\delta u^3}{R^2} + \frac{2(2+3\delta)\pi G \rho_0}{3} = \frac{9(\delta-1)\,\mathcal{E}_0(t)}{4\pi \rho_0 R^4}, \quad (1)$$

where u and R are the shell velocity and radius, ρ_0 is the gas
density in the supercloud corona, and G is the gravitational constant.

Numerical solutions of equation (1) for the expansion of the supershell
are given in Figure 2.

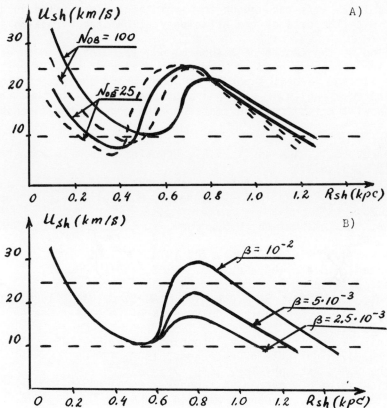

Figure 2. Dependence of the expansion velocity on the supershell
radius   A) for different numbers of stars in the initial
OB-association, N_{OB}, and gas number density, n_0, in the
supercloud corona. Solid curves correspond to $n_0 = 0.5$ cm^{-3}, dashed
lines to $n_0 = 1$ cm^{-3} B) for different efficiencies of induced star
formation β.

Conclusions:
1. Supershell structures are the natural by-products of the sweeping
up of neutral gas from the central parts of superclouds by supernova
explosions at a late stage in the evolution of star complexes. To
understand the dynamics of the largest supershells it is necessary to
take into account the fact that expanding shells can induce star
formation in pre-existing cold dense cloudlets formed as a result of
thermal instabilities. Explosions of massive short-lived ($\sim 10^7$ yr.)
newborn stars then give an additional volume energy supply to the
expansion of the supershell.
2. Gas pressure inside the cavity increases after the second generation
supernova explosions begin, so that a sufficiently large ($R \sim 0.3-0.5$
kpc) shell can expand at ever-increasing speed.

3. The efficiency of induced star formation has to be less than one
percent to produce supershells with radii ~1 kpc and expansion
velocities 10-25 km/s throughout the evolution. This value is in good
agreement with the data of the efficiency of OB star formation in the
Galaxy.
4. Starting at the centre of a supercloud, star formation spreads like
an infection through a hundred parsecs in a time of 10^7-10^8 yr.
Therefore for the largest supershells young stars and associations will
be distributed along their rims and a gradient of age in the stellar
population will exist from the center to the edge of the supershell.

References:
(1) Blinnikov, S.I., Imshennik, V.S., Utrobin, V.P., Pis'ma Astron.
     Zh. 1982, 8, 671.
(2) Brinks, E., Bajaja, E., Astron. Astrophys. 1986, 169, 14.
(3) Bruhweiler, F.C., Gull, T.R., Kafatos, M., Sofia, S., Astrophys. J.
     1980, 238, L27.
(4) Dopita, M.A., Mathewson, D.S., Ford, V.L. Astrophys. J. 1985,
     297, 599.
(5) Efremov, Yu.N. Pis'ma Astron. Zh. 1978, 4, 125.
(6) Elmegreen, B.G., Elmegreen, D.M. MNRAS, 1983, 203, 31.
(7) Heiles, C. Astrophys. J. 1979, 229, 533.
(8) Kafatos, M., McCray, R. Proceedings IAU Symp. N116, 1986, 411.
(9) Kolesnik, I.G., Kinematics and Physics of Celestial Bodies,
     1986, 2, 3.
(10)Kolesnik, I.G., Silich, S.A. Preprint ITP-87-59E, Kiev, 1987.
(11)Meaburn, J. Highlights of Astronomy, 1982, 6, 665.
(12)Silich, S.A., Astrophysics, 1985, 22, 563.
(13)Tenorio-Tagle, G., Astron. Astrophys., 1980, 88, 61.
(14)Tenorio-Tagle, G., Preprint MPA 244, 1986.
(15)Tomisaka, K., Habe, A., Ikeuchi, S., Astrophys. Space Sci.,
     1981, 78, 273.

MOLECULAR CLOUDS ASSOCIATED WITH H I SHELLS

B-C. Koo and C. Heiles
Astronomy Department, University of California, Berkeley,
Berkeley, California U.S.A.

Abstract: We report the detection of molecular clouds associated with three giant galactic H I shells.

Introduction: The interstellar H I is concentrated into shells and
filaments with diameters ranging up to more than a hundred degrees,
which are believed to be formed by stellar winds and/or supernovae,
or by the collision of high velocity clouds with the galactic disk
(Heiles 1984 and references therein). In any case, the shell represents the gas piled up by the interstellar shock waves. This cool and
relatively dense region will be one of the good birth places for molecular clouds and therefore stars.

In this paper, we report the detection of molecular clouds in
three giant H I shells (Heiles 1984): GS 090-28-17, GS 135+29+4, and
GS 193-32+4. We carried out CO observations using the 6 m telescopes
of the Hat Creek Observatory during the scattered period from 1986 to
1987. Initial observations were made along the line segments crossing
the local maximum of H I emission features in well-isolated portions
of each shell. For GS 090-28-17 and GS 135+29+4, small areas ($\sim 1° \times$
$1°$) around the initial detection have been mapped with a rms noise
level of 0.3 K. Also, the ^{13}CO (J=1–0) line was observed for several
points which have shown strong ^{12}CO emission.

H I Shells and Molecular Clouds: The large scale structures of the H
I shells are exhibited by their H I distributions in Figure 1, where
the positions of the detected molecular clouds are also shown by triangles. The H I maps are obtained by integrating the H I survey data of
Heiles and Habing (1974) over the velocity interval of the H I shells.
The resulting H I distributions are represented by gray scale maps superposed on the contour maps; areas with larger column densities are
blacker.

Two shells, GS 090-28-17 and GS 135+29+4, show almost complete
loop structures in the H I distribution. In GS 135+29+4, the velocity
at the peak intensity increases from 0 to 7 km s^{-1} as we move from both
ends of the loop towards (l,b)=(127,39). This is the velocity gradient
expected for an expanding loop with ~ 7 km s^{-1} when we are looking at
parts of it moving away from us. The large shell GS 190-32-4, which is
also known as the Eridanus loop, is one of the best examples that show
the velocity structure of expanding shell (Heiles 1976). Around our
observed position in GS 190-32-4, we see complex filamentary structures. Some of the filaments are parts of another large shell GS 120-
30-8, which has diameter of 100 degrees.

The velocities of the detected molecular clouds agree with H I shell velocities in all cases: CO velocity ranges from -17 to -13 km s^{-1} in GS 090-28-17, 0 to 3 km s^{-1} in GS 135+29+4, and -7 to -4 km s^{-1} and 2 to 5 km s^{-1} in GS 193-32+4. (Two clouds with velocities different from each other by 10 km s^{-1} were detected in GS 193-32+4.) The coincidence of velocities and positions between molecular clouds and H I shells strongly suggests that they are physically associated with each other. In GS 190-32+4, however, the complex filamentary structure prevents us from associating molecular clouds with H I shells.

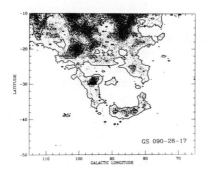
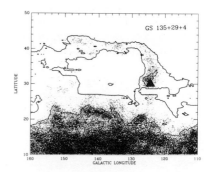

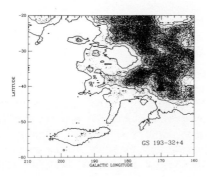

Figure 1. Large scale structures of three giant H I shells are exhibited by their H I distributions. Also, the positions of the detected molecular clouds are shown by triangles. The H I distributions are obtained by integrating over velocities from -22 to -13 km s^{-1} for GS 090-28-17, -4 to 13 km s^{-1} for GS 135+29+4, and -6 to 5 km s^{-1} for GS 193-32-4. The contour level increases from 30 to 110 K km s^{-1} by increments of 20 K km s^{-1} for GS 090-28-17 , 100 to 400 K km s^{-1} by 100 K km s^{-1} for GS 135+29+4, and 100 to 300 K km s^{-1} by 50 K km s^{-1} for GS 193-32+4. Areas with larger column densities are blacker.

Figure 2 shows integrated CO contour maps of the molecular clouds detected in GS 090-28-17 and GS 135+29+4. The beam spacing is $\sim 6'$, which is three times the beam size. The molecular clouds in GS 090-28-17 have sizes ranging from a few minutes of arc to a few tens of minutes of arc. The molecular clouds in GS 135+29+4 are bigger than $1°$ and the observations did not cover the whole cloud. The observed ratio $T_A^*(\ ^{12}CO)/T_A^*(\ ^{13}CO)$ ranges from 5 to 15 and implies that CO is optically thick.

Molecular clouds in GS 090-28-17: The mapping is almost complete for two molecular clouds in this shell, so that their physical parameters can be derived. This shell is seen over the velocity range from -22 km s^{-1} to -13 km s^{-1} and does not show any velocity structure. The kinematic distance to the shell is 3.8 kpc (Heiles 1984). It is unlikely, however, that this is the actual distance to the clouds because then the molecular clouds are located at 1.9 kpc below the galactic plane and have masses greater than $10^5 M_\odot$! There is an impression that the polarization vectors of stars in this area are lying parallel to the shell in the map of Mathewson and Ford (1970), in which case the shell is very close to us, i.e., less than 300 pc. Interstellar Ca II absorption lines had been measured for some of the stars in this area (see Habing 1969). HD 212097 ($l = 87°$, $b = -24°$, distance = 80 pc) has no Ca II lines corresponding to the H I velocities of the shell. On the other hand, HD 215733 ($l = 85°$, $b = -36°$, distance = 2.1 kpc) has a Ca II line at -17.4 km s^{-1}. Another piece of information on the distance may be obtained by assuming that the observed H I mass ($\sim$area$\times$column density) cannot be greater than the original mass in the volume inside the shell. Using a mean H I column density of 5×10^{19} cm^{-2}, we get a lower limit of $\sim 10(n_o/1$ cm$^{-3})$ pc, where n_o is the initial density in this

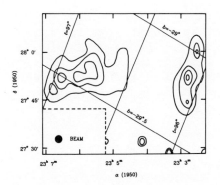
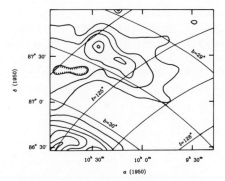

Figure 2. Maps of CO emission detected in two galactic H I shells, GS 090-28-17 and GS 135+29+4. The contour of velocity-integrated antenna temperature increases from 1 K km s^{-1} by increments of 1 K km s^{-1}.

region. Therefore, in summary, the clouds associated with the H I shell are probably located between 80 pc and 300 pc away from us. If we adopt a distance of 200 pc, the sizes ($\sqrt{\text{area}/\pi}$) of two molecular clouds are 0.6 pc and 1.0 pc, respectively. The masses are 80 $M_\odot$ and 340 $M_\odot$ estimated from the ratio $N_{H_2}/\int T_R\,dv = 0.5\times10^{20}$ cm^{-2} (K km s^{-1})$^{-1}$ obtained by de Vries et al. (1987) for the high latitude molecular clouds in Ursa Major.

Since they are physically associated with the H I shell, it is natural to consider them to be formed in the shell. (If they are pre-existing clouds, it is difficult to explain the coincidence of CO velocity with H I velocity.) There are at least two possible mechanisms for the formation of clouds in the shell: First mechanism is of course the gravitational instability. McCray and Kafatos (1987) developed a scenario such that clouds are formed inside H I shells by the onset of gravitational instability. However, unless the observed clouds are at a distance greater than ~2 kpc, they are not gravitationally bound (we used velocity width of 1 km s^{-1}). Alternatively, and more likely, the clouds could be formed by thermal instability in the radiative cooling region behind the shock front (e.g., McCray, et al. 1975). In this case, since the growth rate depends on $\lambda/c\tau_c$ (Schwarz et al. 1972) where λ is the wavelength of perturbation, c is the sound speed of the gas, and τ_c is the radiative cooling time, the clouds with different sizes and different masses will be formed depending upon the wavelength of the initial perturbations and the age of the shell.

References

de Vries, H. W., Heithausen, A., and Thaddeus, P. 1987, Ap. J., in press.

Habing, H. J. 1969, Bull. Astr. Inst. Netherlands, 20, 177.

Heiles, C. 1976, Ap. J. (Letters), 208, L137.

Heiles, C. 1984, Ap. J. Suppl., 55, 585.

Heiles, C. and Habing, H. J. 1974, Astr. Ap. Suppl., 14, 1.

Mathewson, D. S. and Ford, V. L. 1970, Mem.R.A.S., 74, 139.

McCray, R. and Kafatos M. C. 1987, Ap. J., 317, 190.

McCray, R., Stein, R. F., and Kafatos M. C. 1975, Ap. J., 196, 565.

Schwarz, J., McCray, R., and Stein, R. F. 1972, Ap. J., 175, 673.

COLLECTIVE EFFECTS OF SUPERNOVA EXPLOSIONS IN A STARBURST NUCLEUS

Kohji Tomisaka and Satoru Ikeuchi
Tokyo Astronomical Observatory, University of Tokyo,
Mitaka, Tokyo 181, Japan

Abstract: The collective effects of supernova (SN) explosions in the nucleus of a starburst galaxy are studied. It is shown that a large wind bubble with the size of a few hundred parsecs in the nucleus of a starburst galaxy will expand for SN explosion rates of 0.1 per year or greater. The bubble gradually elongates due to density stratification. Finally, the shell breaks near the top and the hot matter flows through the cylindrical shell up to 1-2 kpc above the disk plane. We will discuss the X-ray emission in the halos of galaxies such as M82 and NGC 253 and the distribution of molecular gas in such galaxies in relation to the starburst phenomenon.

Introduction: There has been increasing interest in star formation bursts in many galaxies. We summarize here the observational results of well-known examples of starburst nuclei in the nearby galaxies M82 and NGC 253.

Disk Component: The disk component of a starburst nucleus is observed as a disk or a torus with radius $\sim 200pc$ in a map at near and far IR bands. Recently, very young supernova remnants (SNRs) were found in M82 (Kronberg *et al.* 1985) and NGC 253 (Turner and Ho 1985). These objects trace the IR disk. CO observations,which give us an overall distribution of molecular matter, have shown that H_2 gas forms a "200-pc ring" and a CO ridge or spur-like structure extends perpendicularly from the galactic plane towards the halo (Olofsson and Rydbeck 1984, Nakai *et al.* 1987, Lo *et al.* 1987). The CO ridge surrounds both the IR disk and the radio SNRs cylindrically up to 500 pc from the plane of the disk.

Extended Component: As well as the CO ridge, an extended component is seen in X-ray emission, extending perpendicularly to the disk up to 2-3 kpc in M82 (Watson *et al.* 1984, Kronberg *et al.* 1985) and 1.2 kpc in NGC 253 (Fabbiano and Trinchieri 1984). The extended X-ray emission strongly suggests that the upper halo region is shock-heated by a burst of SN explosions in the disk region. The spur-like structure seems to consist of a gas which resided in the disk and was pushed aside by the X-ray emitting hot gas, as proposed by Unger *et al.* (1984). On the outflow around the starburst nucleus, Chevalier and Clegg (1985) have analytically studied the spherically symmetric wind. In the present paper, by full two-dimensional numerical hydrodynamics we study the evolution of wind bubbles formed by frequent SN explosions after the starburst in the nuclear regions of galaxies.

Model: We assume that star formation and a subsequent SN burst begin abruptly in the interstellar medium which was originally settled in a static equilibrium around a nucleus.
Initial Condition: We assume that the gravitational field is dominated by the stellar component whose density obeys King's distribution. The gravitational force is balanced by the centrifugal force and the pressure gradient. Here, we assume that the rotation is uniform in the z-direction and on the disk plane the centrifugal force is balanced by a part (e^2; $e = 0.9$) of the gravitational force. To fit the rotation curve along the major axis of M82, the parameters are determined as follows: King's core radius $r_c \approx 350pc$, peak rotation velocity $v_{\phi max} \approx 60 km s^{-1}$, and central stellar density $\rho_c \approx 1.4 \times 10^{-22}(v_{\phi max}/60 km\ s^{-1})^2(r_c/350pc)^{-2}e^{-2}\ g\ cm^{-3}$. Further, with the assumption that the gas is entirely isothermal, we take the three-dimensional random velocity of clouds as the isothermal sound speed $c_s = 30 km\ s^{-1}$ (Olofsson and Rydbeck 1984).

SN Explosions: From the distribution of compact nonthermal radio sources, we suppose that the SN burst occurs homogeneously within a disk of diameter $d \approx 300 pc$ and thickness $h \approx 50 pc$. We simulate the SN burst by continuous deposition of energy $E_{SN} \times r_{SN}$ and mass $M_{ej} \times r_{SN}$ into the above disk, where E_{SN} is the energy ejected by a SN, M_{ej} is the mass ejected by a SN, and r_{SN} is the SN rate.

Results: We calculated six cases as summarized in Table 1. We will consider only two cases in the present paper due to space restrictions (see Tomisaka and Ikeuchi 1987 for complete results). Parameters in this simulation are n_0 (density at the center of the disk), r_{SN}, and M_{ej}. The explosion energy is fixed as $E_{SN} = 10^{51} erg$.

Case A: In Fig.1, we illustrate the time evolution of the gas flow. At stage (a) ($t = 10^6 yr$), the cooled dense shell ($n > 30 cm^{-3}$ and $T < 10^4 K$) is formed almost spherically, within which a hot rarefied cavity ($T \geq 3 \times 10^7 K$) spreads.

Table 1. The adopted parameters and the calculated X–ray luminosity.

	n_0 (cm^{-3})	r_{SN} (yr^{-1})	M_{ej} ($M_\odot$)	L_{HRI} (ergs^{-1})	Age (Myr)
A	20	0.1	30	2×10^{40}	6
B	20	0.1	10	2×10^{40}	4
C	100	0.1	30	3×10^{40}	10
D	20	0.01	30	1×10^{39}	9
E	20	0.005	30	4×10^{39}	11
F	100	0.02	30	6×10^{39}	9

The bubble gradually begins to elongate in the z-direction as seen in Fig. 1b ($t = 2 \times 10^6 yr$). In Fig. 1c ($t = 4 \times 10^6 yr$), we can see that the outer shock front expands to $\sim 1 kpc$ in the z-direction. Gas flowing out from the energy injection region soon becomes isotropic, and turns its direction upward at the inward-facing shock. As Rózyszka and Tenorio-Tagle (1985) have pointed out, a part of the shell of the bubble which expands into the region with the large density gradient and is accelerated ($\partial \ln z / \partial \ln t < 1$) breaks due to a Rayleigh-Taylor instability. The instability grows near the z-axis and at age $t \approx 4 \times 10^6 yr$ the shell is broken. Fig. 1d shows the structure at $t = 6.3 \times 10^6 yr$. It is found that a cylindrical wall with a radius $r \approx 600 - 700 pc$ is formed below $z \lesssim 1.6 kpc$, through which hot gas is flowing upwards. Above $z \gtrsim 1.6 kpc$ the dense wall is not formed and the hot gas pushes directly on the ambient matter. Such cylindrical walls with upward flowing hot gas, resembling chimneys, are also seen in superbubbles from sequential SN explosions in OB associations (Tomisaka and Ikeuchi 1986).

Effect of Supernova Rate: We can see the typical case of a low SN rate in Case E where r_{SN} is taken as 5×10^{-3} SNe yr^{-1}. It is found that the hot gas does not flow out to the halo $z \sim 1 kpc$ in $10^7 yr$. We show the structure at an age of $t = 10^7 yr$ in Fig. 2. Although the bubble elongates a little in the z-direction, the dense shell entirely surrounds the hot cavity and the hot gas is completely confined within the wall. In the parameter space of (n_o, r_{SN}), the region where the hot gas halo is formed in $t \sim 10^7 yr$ is shown in Fig. 3. For both Cases D and F, the liberated energy of SNe is efficiently radiated away from the dense wall and the hot gas within the cavity, hindered by the wall, cannot escape to the halo.

Discussion: Fig. 4 shows the relation between the expected distribution of column density of neutral gas and that of soft X-ray intensity in the Einstein HRI band for the situation depicted in Fig. 1d. The ridge of neutral gas is located just outside the X-ray emitting region. This agrees with the relative distribution in M82 of H_2 (derived by Nakai *et al.* 1987) and of X-ray (Watson *et al.* 1984). These show that the observed gas distribution, in which the hot gas is surrounded cylindrically by the cooled gas, is well explained by this model. The starburst nucleus forms a galactic-scale bipolar flow in consequence of the active SN explosions. In Table 1, the X-ray luminosity obtained by the simulation is summarized. It is shown that the luminosity is $\sim 2 \times 10^{40} erg \ s^{-1}$ in the HRI band in Cases A-C, but in Cases D-F the luminosity is only $\sim 10^{39} erg \ s^{-1}$.

This means that a SN rate higher than $\sim 0.1 \ yr^{-1}$ or a mechanical luminosity more than $\sim 3 \times 10^{42} erg \ s^{-1}$ is necessary to explain the observed X-ray luminosity. In Fig.5 we illustrate the profile of the X-ray intensity along the z-axis for Cases A and C. We also plot the observed intensity in M82 and NGC 253. Observed I_{HRI} decreases from $z = 500pc$ to $z = 2kpc$ by only 1/10; this fact fits well with our model calculation, although the expected surface brightness is smaller than that observed.

This work was supported in part by a Grant-in-Aid for Scientific Research from the Ministry of Education, Science, and Culture (61790072) in fiscal 1986 and 1987.

References

Chevalier, R. & Clegg, A. W. 1985, *Nature*, **317**, 44.

Fabbiano, G. & Trinchieri, G. 1984, *Ap.J.*, **286**, 497.

Kronberg, P. P., Bierman, P. & Schwab, F. R. 1985, *Ap.J.*, **291**, 693.

Lo, K. Y. et al. 1987, *Ap.J.*, **312**, 574.

Nakai N. et al. 1987 submitted to *Pub. Astr. Soc. Japan.*

Olofsson, H. & Rydbeck, G. 1984, *Astr. Ap.*, **136**, 17.

Rózyszka, M. & Tenorio–Tagle, G. 1985, *Astr. Ap.*, **147**, 209.

Tomisaka, K. & Ikeuchi, S. 1986, *Pub. Astr. Soc. Japan*, **38**, 697.

Tomisaka, K. & Ikeuchi, S. 1987, submitted to *Ap.J.*

Turner, J. L. & Ho, P. T. P. 1985, *Ap.J.(Letters)*, **299**, L77.

Unger, S. W. et al. 1984 *M.N.R.A.S.*, **211**, 783.

Watson, M. G., Stanger, V. & Griffiths, R. E. 1984, *Ap.J.*, **286**, 144.

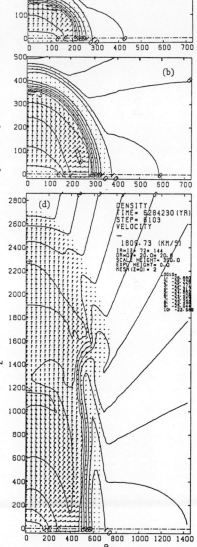

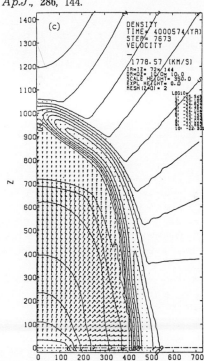

Fig.1: The evolution of the flow in Case A. The flows at (a) $t=10^6$ yr, (b) $t=2\times10^6$ yr, (c) $t=4\times10^6$ yr, and (d) $t=6\times10^6$ yr are illustrated.

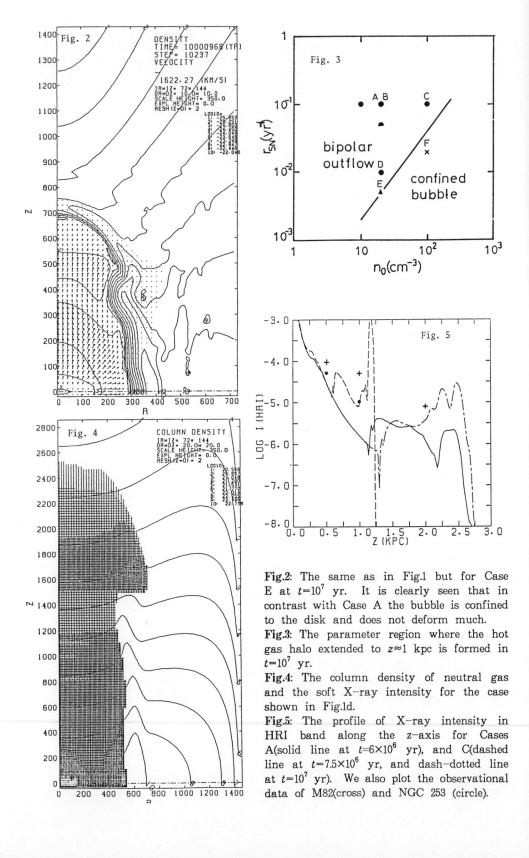

Fig.2: The same as in Fig.1 but for Case E at $t=10^7$ yr. It is clearly seen that in contrast with Case A the bubble is confined to the disk and does not deform much.

Fig.3: The parameter region where the hot gas halo extended to $z\approx 1$ kpc is formed in $t=10^7$ yr.

Fig.4: The column density of neutral gas and the soft X-ray intensity for the case shown in Fig.1d.

Fig.5: The profile of X-ray intensity in HRI band along the z-axis for Cases A(solid line at $t=6\times 10^6$ yr), and C(dashed line at $t=7.5\times 10^6$ yr, and dash-dotted line at $t=10^7$ yr). We also plot the observational data of M82(cross) and NGC 253 (circle).

THE DISTRIBUTION AND BIRTHRATE OF GALACTIC SNRs

Xinji Wu* and D.A. Leahy
Department of Physics, Univ. of Calgary, Canada

Abstract: In this paper we find that the spatial distribution of SNRs shows a peak in the range R = 4-6 kpc; the total number of SNRs in the Galaxy is 749±104; and the birthrate of SNRs is one in every 18±3 years.

Introduction: The distribution of supernova remnants (SNRs) in our Galaxy has been investigated by Ilovaisky and Lequeux (1972) and Kodaira (1974). Since their investigations only gave the distribution for SNRs which have diameters D<30 pc, it is not possible to estimate the total number and the birthrate of SNRs from their work.

We improve the method of Kodaira (1974) and adopt the catalogue of SNRs given by Li (1985). It contains 155 SNRs of which there are 7 without values of angular diameter and flux density. So our statistical sample has 148 SNRs.

1.   Discussion of Observational Selection Effects

Ilovaisky and Lequeux (1972) suggested that three selection effects must be considered: (1) limiting angular diameter; (2) limiting surface brightness; (3) limiting flux density. To attempt to correct for these selection effects, they limited their sample to sources with angular diameters larger than 2 minutes of arc; distances to the Sun d<6.3 kpc; D<30 pc; and flux densities larger than 10 Jy at 1 GHz.

We do not adopt these criteria since to do so would remove many SNRs, more than 50%, from our sample. Kodaira (1974) used an alternative approach in which the incomplete counting is regarded as caused by confusion with noise. The degree of the confusion was to be empirically determined as a function of distance from the observer, independent of the direction. However, he still chose only SNRs with D<30 pc.

We improve the method of Kodaira to correct for the selection effects. We need not remove those SNRs with D<30 pc and S<10 Jy from our sample. Thus we obtain a larger statistical sample.

2.   Statistical Method and Determination of the Surface Density

Because most of the SNRs are distributed in a flat system, the galactic plane, we consider only the surface density of SNRs. We assume that the surface density is symmetric around the galactic center (i.e. is a function of distance to the galactic centre (R) only). Secondly, we assume that the observations of SNRs are complete within 2 kpc of the Sun. In this region the selection effects are not a factor.

*Permanent address: Geophysics Department, Peking University, Beijing, China.

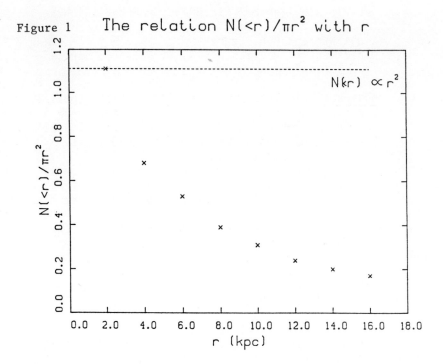

Figure 1   The relation $N(<r)/\pi r^2$ with r

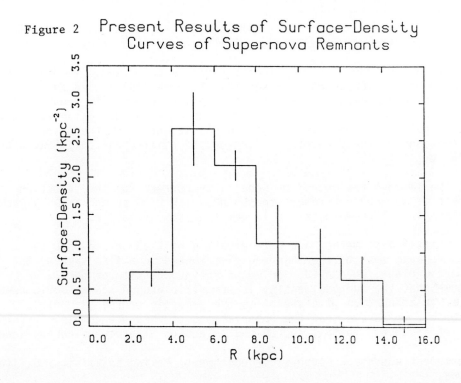

Figure 2   Present Results of Surface-Density
Curves of Supernova Remnants

We first obtain the apparent surface density of SNRs. We adopt the distances given by Li (1985). For the remainder we use the Σ_{1GHz} - D relations given by Li and Wheeler (1984):

$$\Sigma_{1GHz} = 3.1 \times 10^{-17} D_{pc}^{-2.4} \text{ W m}^{-2} \text{ Hz}^{-1} \text{ sterad}^{-1} \text{ for crablike SNRs}$$

$$\Sigma_{1GHz} = 1 \times 10^{-16} D^{-2.77} \text{ W m}^{-2} \text{ Hz}^{-1} \text{ sterad}^{-1} \text{ for shell SNRs}$$

If SNRs were uniformly distributed throughout the galactic disk we should expect $N(<r)$, the number of SNRs with distance less than r, to increase as r^2 and expect $N(<r)/\pi r^2$ to equal a constant independent of r. The values of $N(<r)/\pi r^2$ are not constant with distance r. The result shown in Figure 1 largely reflects the fact that the incomplete counting due to observational selection becomes more serious with increasing distance from the Sun.

Now we consider the derivation of correction factors for the effects of incomplete counting. According to our second assumption the observations of SNRs are complete within 2 kpc of the Sun, so the true surface density is equal to the apparent surface density in this distance range. According to the first assumption, the surface density of SNRs is symmetric around the galactic center. So the surface density of SNRs must be the same within each ring of R to R + ΔR. We use these two assumptions to calculate the correction factors for each counting range. Figure 2 shows the distribution of surface density corrected for observational effects with ±1σ error bars. This distribution shows a peak in the range of R = 4-6 kpc.

3. The Estimation of Total Number and Birthrate of SNRs
     The total number of SNRs in the Galaxy obtained by an integration over the galactic radial distribution of Figure 2 is 749±104.

In order to derive the birthrate of SNRs, we must know the average lifetime of SNRs. Using the formula for shock radius versus time from the adiabatic phase in the standard Sedov model and considering the effect of the three component medium on $R_s(t)$ without cloud evaporation (Cox 1979), we have an age versus diameter relation:

$$t = (\frac{D}{22.4pc})^{2.5} (\frac{n_h f_h}{\varepsilon})^{0.5} \times 10^4 \text{ years} \qquad (1)$$

For calculating ages we choose f_h (filling factor) = 0.75, ε (blast energy in units of 0.75×10^{51} ergs) = 2/3 and n_h (hot intercloud medium density) = 0.005 cm^{-3}.

However young SNRs expanding in the ISM will not have swept up enough mass to have reached the self-similar phase of evolution. For these remnants the expansion is approximately linear with age (Gull 1973). The transition between linear and self-similar phases occurs when the swept up mass from the hot medium is of order one solar mass.

This gives a transition radius of $R_0 = 11.2(n_h/0.005)^{-1/3}$ pc. For an adopted value of n_h, matching $R = V_{ej}t$ to formula (1) at 11.2 pc gives an ejection velocity of 14700 km s^{-1}, in good agreement with SN models (10,000 to 20,000 km s^{-1}). Thus, for remnants less than 11.2 pc in radius we calculate age from the linear formula.

The average age for our SNR sample is 6.8×10^3 years. As a check on the method, the calculated ages for historical supernova remnants are compared with true ages in Table 1. We see that the ages agree very well with the true ages for SN 1604 and SN 1572, and reasonably well for SN 1188 and SN 185. Exceptions are: Cas A, located in a higher density region of ISM; SN 1006, at large distance from the galactic plane (lower density); SN 1054 (the Crab SNR), powered by a central pulsar. This agreement, with explained exceptions, indicates that the formula for age we have chosen gives correct age estimates.

Table 1 Historical SNRs

	Cas A	SN1604	SN1572	SN1188	SN1054	SN1006	SN185
True age (years)	~325	383	415	806	933	981	1902
Calculated age (years)	65	295	398	389	90	4470	2820

Generally, the average lifetime is twice the average age. Using the result for the total number in the Galaxy, the birthrate of SNRs is one in every 18 ± 3 years. This birthrate is higher than most previous estimates which range from one in every 50 to 150 year (Caswell, 1970; Clark and Caswell, 1976), but near the result (one in every 22 yr) of Srinivasan and Dwarakanath (1982). This result is also consistent with the SN birthrate (one in every 11 yr to 30 yr) derived by Clark and Stephenson (1977), Katgert and Oort (1967), and Tamman (1977). This result also agrees with results for pulsar birthrates.

Caswell, J.L. 1970. Astr. Astrophys. 7, 59.
Clark, D.H. and Caswell, J.L. 1976. Mon. Not. R. Astr. Soc., 174, 267.
Clark, D.H. and Stephenson, F.R., 1977. Mon. Not R. Astr. Soc., 179, 267.
Cox, D.P. 1979. Astrophys. J. 234, 863.
Gull, S.F. 1973. Mon. Not. R. Astr. Soc, 161, 47.
Li, Z.W. 1985. Proceeding of Beijing Normal Univ. Vol. 4, p. 65.
Li, Z.W. and Wheeler, J.C. 1984. Bulletin A.A.S., 24, 234.
Ilovaisky, S. and Lequeux, J. 1972. Astr. Astrophys., 18, 169.
Katgert, P. and Oort, J.H. 1967, Bull. AStr. Inst. Netherl., 19, 239.
Kodaira, K. 1974, Publ. Astr. Soc. Japan, 26, 255.
Srinivasan, G. and Dwarakanath, K.S. 1982. J. Astrophys. Astr. 3, 351.
Tamman, G.A., 1977. Ann. N.Y. Acad. Sci., 302, 61.

A STATISTICAL STUDY OF THE CORRELATION OF GALACTIC SUPERNOVA REMNANTS AND SPIRAL ARMS

Zongwei Li (Beijing Normal University)
J. Craig Wheeler, Frank N. Bash, William H. Jefferys
The University of Texas at Austin
Austin, Texas 78712

Abstract: A statistical study of the correlation of Galactic supernova remnants with spiral arms and the disk is presented. SNR apparently have a larger radial scale length than disk stars. We estimate that only about 10 percent of the Galactic SNR have been detected.

1. Introduction

Based on statistical studies of external galaxies, our Galaxy is expected to produce Type I and Type II supernovae in closely equal number. There are as of this writing 162 known galactic SNRs. The basis for our investigation is the observation that SN II (and SN Ib) are tightly correlated with spiral galaxies. SN I (strictly speaking SN Ia) do not correlate with spiral arms, but are roughly of old disk population. We seek a method by which we can determine a correlation of SNR with spiral arms. For this preliminary exercise we use only information on the position of the SNR in Galactic coordinates, rather than uncertain distance estimates. As a test case, we compared the angular distribution of SNR and giant H II regions which are presumed to define the location of the spiral arms. To the eye there did seem to be a correlation.

2. Data base

The work of Milne and Downes and Clark and Caswell resulted in a catalogue of 125 SNR (Milne 1979). Van den Bergh (1983) presented a catalogue of 135 SNR in the Galaxy. Green (1984) gives a list of 145 SNRs. We have added some new SNRs, so the total number of SNRs is augmented to 153. In order to compare the distribution of giant H II regions and SNR, we used the catalogue of Georgelin and Georgelin (1976) and the list of H II regions in Blitz et al. (1982).

3. Monte Carlo simulation model

We have developed a quantitative approach to investigate the questions of the correlation of SNR with arm or disk populations. We use the observed angular distribution of SNR and giant H II regions to form a cumulative distribution with respect to galactic longitude. Two observed distributions can be compared and Kolmogorov-Smirnov statistics used to determine the probability that the two samples are not drawn from the same distribution. In addition, we have constructed Monte Carlo models in which sample objects are distributed in the spiral arms and galactic disk in a prescribed fashion. The correlation of these models with observed distributions can be used in conjunction with K-S statistics in an affirmative fashion to establish a figure of merit for the goodness of fit. The parameters of the Monte Carlo models, such as the opening angle of the spiral arms and the fraction of objects in the disk versus arm population can be varied to obtain the best fit to the observed distribution. We also study the effect of

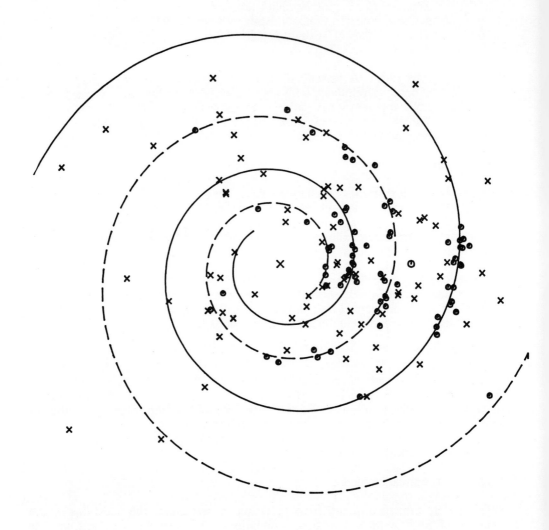

Figure 1 - The radial distribution is given for a Monte Carlo model containing 152 sample SNR with SH = 7.0 kpc, SE = 5.0 kpc, D_O = 3.0 kpc and i = -8°. Half the points are in the exponential disk, half distributed along the assumed two-armed logarithmic spiral. The cumulative angle distribution of such a model distribution gives a reasonable representation of that corresponding to the observed SNR.

selection effects, if the surface brightness of SNR falls below a threshold for detectability (as $1/r^2$). We calculated the following distributions:

(a) The distribution of SN in an exponential galactic disk with coordinates given by the radial distance from the galactic center, R, and the angle with respect to the line of centers to the Sun, θ. A selection effect function was used to bias against distant supernovae.

(b) For the distribution of SNR along the spiral arms, we first assumed that the SNR were uniformly distributed per unit length along the spiral arms, then calculated the corresponding R and θ. We also required a decrease in probability density along the arms according to the exponential law used for the disk population. We also spread the points in a Gaussian distribution laterally to the arms so that they were not infinitesimally thin. A selection effect function was added in the same fashion as for the disk models.

In order to reduce numerical fluctuations each Monte Carlo model is run to produce 500 sample points. Each such model is then run 20 times and the results are averaged to produce the angular distribution to be compared with the observational data.

4. Kolmogorov-Smirnov Two-Sample Test of SNR, Giant H II regions, and Monte Carlo models

The Kolmogorov-Smirnov two-sample test (K-S Test) is a test of whether two independent samples have been drawn from the same population. The two-tailed test is sensitive to any kind of difference in distributions from which the two samples were drawn.

Table 1 gives the probability that the sample of 150 SNR is similar to the Monte Carlo models. The parameters are as follows: SH = 7.0 is the radial scale length in the disk, SE = 5.0 is the scale length of the $1/r^2$ selection effect, D_O = 3.0 is the radius of the selection free circle around the solar neighborhood, f is the fraction of the SNR in the spiral arms, and i is the opening angle of the model two-armed spiral. The probability of positive correlation is maximum for i ~ -8°, characteristic of spiral arms, and for f ~ 0.5, as expected. Figure 1 shows a representative radial distribution for a Monte Carlo model

Table 1
150 SNR vs. Models

$$SH = 7.0 \quad SE = 5.0 \quad D_O = 3.0$$

f	i=-8.9	-8.45	-8.2	-8.0	-7.8	-7.6	-7.4
0.0	0.64	0.64	0.64	0.64	0.64	0.64	0.64
0.1	0.69	0.73	0.73	0.73	0.73	0.69	0.65
0.2	0.73	0.76	0.76	0.78	0.75	0.73	0.69
0.3	0.78	0.86	0.86	0.85	0.84	0.75	0.70
0.4	0.76	0.84	0.86	0.90	0.87	0.80	0.73
0.5	0.69	0.78	0.84	0.90	0.89	0.82	0.70
0.6	0.55	0.69	0.75	0.84	0.91	0.76	0.73
0.7	0.45	0.60	0.69	0.80	0.82	0.65	0.62
0.8	0.38	0.54	0.61	0.74	0.73	0.49	0.46
0.9	0.30	0.47	0.59	0.70	0.63	0.40	0.35
1.0	0.24	0.45	0.49	0.65	0.50	0.29	0.25

with these parameters.

The K-S test shows that the observed samples of SNR and giant H II regions are not, in fact, highly correlated. The H II regions do seem to have a radial scale length ~ 3.5 kpc in the solar neighborhood. For the SNR, however, the probability of positive correlation decreases significantly for a radial scale length of 3.5 kpc, typical of the stellar disk. This suggests that the SNR in the sample are not distributed radially with the same scale length as the stellar disk and that SNR are not concentrated in an inner ring. More effort is underway to examine the impact of selection effects on this tentative conclusion. The Monte Carlo models that reproduce the observed distribution moderately well suggest that the Galaxy contains $\geq$ 1200 SNR, and that only ~ 10 percent have thus been detected.

References

Blitz, L. et al. 1982, *Ap. J.*, **49**, 183.
Georgelin, Y. M. and Georgelin, Y. P. 1976, *Astron. & Ap.*, **49**, 57.
Green, D. A. 1984, *M.N.R.A.S.*, **209**, 449
Milne, D. K. 1979, *Aust. J. Phys.*, **32**, 83.
van den Bergh, S. 1983, *I.A.U. Symposium, No. 101*.

THE FREQUENCY OF SUPERNOVAE IN EXTERNAL GALAXIES

E. Cappellaro and M. Turatto
Asiago Astrophysical Observatory, Italy

Abstract: A new determination of the SN-frequency in
different morphological types of galaxies, obtained from
the material of the Asiago SN-Search, is presented. From
these values we derive a mean interval between SN events in
our Galaxy consistent with values determined in different
ways.

Introduction. Since 1959 the Asiago Astrophysical Observatory has been
carrying out a systematic search for supernovae in about 70 fields by
using 50/40 and 92/67 Schmidt telescopes. The more than 5000 plates
devoted, up to now, to this aim constitute a unique batch of
homogeneous material, accessible in the near future to the whole
astronomical community. This plate archive is the base on which a new
determination of the frequency of supernovae has been recently obtained
(Cappellaro and Turatto, 1987), the main results of which are here
briefly summarized.

The frequency of supernovae. The method of computation of the
frequency is based on the knowledge of "control time" (Barbon, 1968;
Zwicky, 1942).

    The sample of objects used for this investigation consists of all
the galaxies in RC2 with a morphological classification (de Vaucouleurs
et al., 1976) which are included in the fields of the search and which
have known radial velocities. The sample of supernovae is formed by
all the objects present on the plates of the search in the sample
galaxies. The total sample consists of 736 galaxies and 51 SNe. The
unclassified supernovae have been assigned to Type I and II, following
the ratio of occurrence in the corresponding galaxy type from Table 4
of Barbon et al. (1984).

    In this way we overcome the main limitations of the previous
investigations. No assumption has been made on the time-coverage of
the galaxies, each object being treated individually as to distance,
observation time distribution, luminosity, inclination and internal
absorption. Longer coverage of the fields and larger samples allowed
us to compute the frequency in a finer morphological subdivision and to
determine the frequency for Types I and II supernovae independently.

    Table I shows the main results. In the first four rows the
numbers of galaxies and supernovae of different types are listed. Rows
5 and 6 report the total control times, Ct, for different galaxy types,
separately for SNI and II. The control times for SNII are smaller than
for SNI because of their lower luminosity. In rows 7, 8, and 9 the

Table 1.

Galaxy type	E	S0	Sa	Sab	Sb	Sbc	Sc	Scd	Sd	Sm-Im
No.GAL	125	181	81	34	70	51	65	41	24	64
No.SNI	3	6	3	0	2	2	6	2	2	0
No.SNII	0	0	0	0	1	2	7	3	0	1
No.UNCLASS	0	1	0	1	2	4	2	1	0	0
C.t.SNI(yr)	990.9	1558.5	851.5	361.5	753.3	601.0	671.5	608.2	373.4	894.7
C.t.SNII(yr)	806.6	1293.4	744.6	296.5	658.2	533.0	593.9	552.6	338.3	827.4
ν SNI (*)	0.30	0.45	0.35 (<0.13)	0.14	0.44	0.83	1.03	0.41	0.54 (<0.30)	(<0.11)
ν SNII				0.17	0.25	0.56	1.36	0.63		0.12
ν All	0.30 ±0.19	0.45 ±0.18	0.35 ±0.21	0.31 ±0.30	0.70 ±0.31	1.39 ±0.49	2.39 ±0.61	1.04 ±0.42	0.54 ±0.39	0.12 ±0.11
ν SNI (***)	0.09	0.25	0.27 (<0.12)	0.07	0.19	0.39	0.71	0.50	0.79 (<0.48)	(<0.58)
ν SNII				0.11	0.16	0.31	1.17	0.81		0.64
ν All	0.09 ±0.07	0.25 ±0.12	0.27 ±0.17	0.19 ±0.17	0.35 ±0.17	0.69 ±0.24	1.88 ±0.49	1.31 ±0.52	0.79 ±0.61	0.64 ±0.60

(*) $yr^{-1} \times 100$.
(**) SNu (1 SNu = 1 SN per 100 yr per $10^{10} L_{OB}$).

frequencies per average galaxy, ν, are shown. Errors (row 10) are quoted only for the total frequencies and are computed on the hypothesis that the occurrence of supernovae in galaxies is a random event. The last three rows report the frequencies of supernovae per luminosity unit. 25% of the galaxies (8% of Sne) have been rejected, since no data on the magnitude were found in the literature.

Our main conclusions are the following:

- The frequency of supernovae increases from early- to late-type galaxies and it is peaked at Sc's.

- SO galaxies have a SNI rate 3 times higher than Ellipticals, in contrast with the results of Tammann (1982).

- The frequency of SNI in Sc and Scd galaxies is half that of SNII, being smaller than in Tammann (1982), Oemler and Tinsley (1979). This is partially due to their neglect of the different discovery chance for SNI and SNII.

- The SNI rate increases from E to Sd galaxies. This may be related to the different progenitor population of the SNI subclasses, SNI-a and SNI-b (Gaskell et al., 1986). No SNI-b belongs to our sample, but many SNI are without sub-type classification. On the hypothesis that SNI-b occur only in spiral galaxies, the relative ratio SNI-b/SNI-a $\approx$ 0.6, as observed in the last three years; i.e. since this subclass has been fully recognized, the numbers are consistent with the increasing rate of SNI in late spirals being due only to SNI-b.

The Galaxy. In order to compare the expected SN rate in the Galaxy with values deduced in different ways, it is crucial to establish the Galaxy's morphological type and luminosity.

Table 2.

	Predicted	SNR	Pulsar	Historical SNe	Others	Tammann (1982) (external evid.)
T_{SNI} (yr)	88 $^{+94}_{-30}$				100 [7]	45
T_{SNII} (yr)	109 $^{+204}_{-43}$		30÷120 [4]		120 [8]	55
T_{all} (yr)	49 $^{+34}_{-14}$	80±30 [1] 23±7 [2] 30 [3]		45 $^{+66}_{-16}$ 17 $^{+12}_{-5}$		25

(1) Lerche,I.:1981,Astroph.and Space Sci. 74,273; (2) Smirnov,M.A., Sakhibov,F.Kh.:1984,Soviet Astron.28,137; (3) Mills,B.Y.:1983, "Supernova Remnants and their X-ray Emission",IAU Symp.No.101,pag.551, eds.I.J.Danziger and P.Gorenstein,P.,Reidel,Dordrecht; (4) Lyne,A.G., Manchester,R.N.,Taylor,J.H.:1985,M.N.R.A.S.213,613; (5) Van den Bergh, S.:1983,P.A.S.P.95,388; (6) Tammann,G.A.:1982,"Supernovae: a Survey of Current Research",eds.M.J.Rees and R.J.Stoneham,P.,Reidel,Dordrecht; (7) Woosley,S.E.,Weaver,T.A.:1986,Ann.Rev.Astron.Astrophys. 24,205; (8) Heiles,C.:1987,Ap.J. 315,555.

If we consider the Galaxy a Sb-Sbc type with luminosity $L=3.9 \times 10^{10}$ $L_\odot$ (Tammann, 1982), we have a SN rate of 1.13 SNu for SNI, 0.92 SNu for SNII, 2.03 SNu for all SNe. In Table 2 the mean SN intervals in our Galaxy (with the derived internal errors) are compared with other recent determinations. It appears that the mean interval between two SN events, obtained from external galaxies, is twice the analogous value of Tammann (1982), and it is consistent, within the large errors, with independent estimations.

References

Barbon, R.: 1968, Astron. J. 73, 1016.

Barbon, R., Cappellaro, E., Ciatti, F., Turatto, M., Kowal,C.T.: 1984, Astron. Astrophys. Suppl. Ser. 58, 735.

Cappellaro, E., Turatto, M.: 1987, Astron. Astrophys. submitted.

de Vaucouleurs, G., de Vaucouleurs, A., Corwin, H.G.: 1976, Second Reference Catalogue of Bright Galaxies, Univ. Texas Press, Austin (RC2).

Gaskell, C.M., Cappellaro, E., Dinerstein, H.L., Garnett, D.R., Harkness, R.P., Wheeler, J.C.: 1986, Astrophys.J. 306, L77.

Oemler, A., Tinsley, B.M.: 1979, Astron. J. 84, 985.

Tammann, G.A.: 1982, Supernovae: A Survey of Current Research,Ed. Rees, M.J., Stoneham, R.J., D. Reidel, Dordrecht.

Zwicky, F.: 1942, Astrophys. J. 96, 8.

STAR FORMATION, SUPERNOVAE AND THE STRUCTURE OF DISK GALAXIES

Michael A. Dopita
Mt. Stromlo and Siding Spring Observatories
The Australian National University
Private Bag, Woden P.O., ACT 2606, Australia

Abstract: A physical model for bi-modal star formation and the structure of the interstellar medium and the self-regulating evolution of disk galaxies is presented. Stars heavier than about one solar mass are produced as a result of collisions of molecular clouds or in cloud crushing events whereas low-mass stars are produced at a steady rate in dense molecular clouds and the T-Tauri winds resulting maintain the support of these clouds against rapid collapse and fragmentation. Supernova explosions and stellar winds associated with the massive stars maintain the phase structure, and the scale height of the gas. The collective effects of these energetic processes may create a hole in the disk gas, and allow a galactic wind of metal-enriched gas to develop.

1. Phase Structure of the ISM and the Collective Effects of Supernovae.

The energetic processes (winds, ionising radiation and supernova explosions) associated with the young, massive stars exercise the fundamental control of the phase properties, pressure and distribution of the interstellar medium (ISM) in the plane of disk galaxies. This is the basis of the multi-phase models (Field,Goldsmith and Habing 1969; Cox and Smith 1974; McKee and Ostriker 1977; Cox 1979,1980), although these differ in the details of how such a multi-phase medium is set up and maintained.

The local interstellar medium is perhaps not the best place to start to try to understand the phase structure of the ISM. We lie within a region of extensive recent star formation defined by the stars of the Gould's Belt, which extends over some 700pc, and encompasses the Sco-Cen, Taurus and Orion associations (Lindblad and Westin, 1985). Thus, the phase structure derived locally may not be generally applicable.

The Magellanic Clouds, on the other hand, offer a convenient laboratory in which to study the phase structure of the ISM, and evidence for cloud, inter-cloud and coronal components may be found. It has become increasingly clear that Type II supernovae interact with a highly modified ISM. However, Type I supernovae will tend to occur well separated in time and space from their star formation region, and are therefore much more useful in probing "normal" samples of the ISM. From a variety of lines of evidence, Tuohy et $al.$ (1982) have shown that the Balmer dominated SNR are most likely to be the remnants of Type I SNR. For these, the Hα data suggest that the SNR is interacting with a phase of the ISM which has a density of about 0.1 cm^{-3}, whereas the X-ray data gives 0.3 cm^{-3} for the same LMC remnants. This phase can be identified with an intercloud medium. The cloud medium, on the other hand has a density of about 10-30 cm^{-3} (Dopita, 1979; Wilson 1983), and is probably in stochastic pressure balance with the intercloud.

The supernova rate in the LMC is about 1 per 200 years. This is insufficient to maintain a coronal medium with a large filling factor globally. However, in regions of local enhancement of the star formation rate, the collective effects of supernova explosions can be sufficient to strip out the disk HI, creating a bubble of coronal gas which eventually finds its way into an extended hot halo. This is graphically illustrated by the case of Shapley Constellation III (Dopita, Mathewson and Ford, 1985). This region of very extensive star formation over the past 2×10^7 year has produced a hole in the disk HI almost 2kpc across, and energetic processes such as stellar winds and SN explosions have pushed out a shell of HI at a velocity of 36 km.s^{-1}. Such HI will take about 2×10^8 years to return to the plane. Furthermore, once the disk HI is swept out, coronal gas produced by subsequent SN events is free to escape into a hot coronal medium which can pressurise the disk gas. Such coronal gas is undoubtedly present and has been observed directly in soft X-rays in Constellation III (see Helfand, Wu and Wang 1987, this conference, not published).

For the Galaxy, it has become clear that a hot corona of shock-heated gas first suggested by Spitzer (1956) does indeed exist. The presence of this gas is evident locally in the soft X-ray observations (Tanaka and Bleeker 1977; McCammon et al. 1983; Jakobsen and Kahn 1986), and in observations of OVI absorption (Jenkins 1978). As it cools, it gives rise to absorption in highly ionised species such as N V, C IV and Si IV which can be observed with the IUE Satellite (Savage and de Boer 1978,82; York et al. 1982; Pettini et al. 1982; de Boer and Savage 1983). From this work, it is evident that this gas has a (local) scale height for cooling of about 3-4 kpc, and is denser and more confined to the disk towards the inner parts of the Galaxy.

If the gravitational binding energy of this hot gas is less than its thermal energy minus the radiative losses as it streams out into the halo, then a galactic wind, rather than a steady state "galactic fountain" will result. The conditions under which this will occur were discussed in an elegant paper by Chevalier and Oegerle (1979). In our local solar neighbourhood, and in the inner regions of the Galaxy, this condition does not appear to be met. However, the cooling timescale of this material at LMC and SMC abundances is very long, of order $(0.3 - 3) \times 10^9$ years. Even if it cools, the height to which the hot gas rises is sufficiently distant above the plane to be very weakly bound to the system. It is likely that a galactic wind can be driven in this case.

The pattern of elemental abundances (Dopita 1987) strongly suggests that this does indeed occur. The chemical yield of oxygen in the Magellanic Clouds is lower than that of the galaxy by a factor of two to three (e.g. Dufour, 1984). However, data from stellar atmospheres of young disk stars shows that **both** of [Fe/H] and [O/Fe] are lower in the Magellanic Clouds. This is in contradistinction with the Galaxy, for which old disk stars show a **lower** [Fe/H] is coupled with a **higher** [O/Fe] (Tomkin and Lambert, 1984; Tomkin, Sneden and Lambert, 1986; Sneden 1985; Nissen Edvardsson and Gustafsson 1985). Since O is made in massive stars, and Fe in lower mass stars, this difference could be taken to mean that the slopes of the IMFs are different. However, such observational evidence as we have does not support such a conclusion. The alternative hypothesis is that oxygen has been preferentially lost to the system. This could occur through the funnels opened out into the galactic halo by bursts of star formation. These will have filled by the time the Fe-producing Type I events occur, allowing the retention of this element.

2. Star-Formation - Bimodal or Not?

As Silk (1985), pointed out, the essential ingredients of a star formation theory are the initial mass function (IMF), the star formation efficiency and the rate of star formation. Most models of galactic evolution (Audouze and Tinsley 1977; Vader and de Jong 1981) have tended to assume a constant IMF and to reduce the star formation problem to a simple "prescription" of the rate in terms of the local HI gas density (Schmidt 1959), or of HI surface density (Sanduleak 1969; Hamajima and Tosa 1975).

The rôle of the IMF has been receiving increasing attention in recent papers. In our own Galaxy, star formation may well have a bimodal character, with high mass stars being preferentially formed in the vicinity of the spiral arms but low-mass stars being formed throughout the disk (Güsten and Mezger 1983). If the CO-emitting molecular clouds map star-formation regions, then their distribution in the Galaxy appears to offer convincing support of the bimodal hypothesis (Scoville and Good 1987). The CO clouds are clearly divided into two populations which reflect their kinetic temperatures. The warm molecular clouds are clustered, are associated with HII regions and form a spiral arm population. The cold core clouds are distributed throughout the disk. Scoville Sanders and Clemens (1986) argue that, since the star formation efficiency for massive stars appears to decrease as the mass of the parent cloud increases, the formation of these stars must be triggered by an external cause, such as cloud-cloud collisions, rather than internally as in the sequential star formation models.

The apparent segregation of the high-mass and low-mass modes of star formation becomes even more pronounced in starburst regions. Here, several analyses suggest that, in these regions, only the high mass stars are being formed and that the low mass cutoff in the IMF is of order 3 solar masses (Rieke *et al.* 1980,1985; Olaffsson, Bergrall and Ekman 1984; Augarde and Lequeux 1985).

In an inspiring recent paper, Larson (1986) has presented convincing arguments that, provided that the global rate of star formation decreases with time, the IMF is a double peaked function. The division between the "high" and "low" mass sections of the IMF occurs at about one solar mass, so that, from the point of view of galactic chemical evolution, only the high mass mode of star formation is important.

Although, in our own Galaxy, the two modes of star formation appear to be spatially distinct, with the high-mass stars preferentially formed near spiral arms, it is not necessary, or even desirable, to associate this with a density wave trigger. Elmegreen (1985,86) has shown that galaxies of the same Hubble types with and without a density wave have effectively identical star formation rates. The rôle of the density wave is therefore one of spatial ordering of the star formation regions rather than one of enhancement of star formation rate.

3. The High-Mass Mode of Star Formation.

Cloud-cloud collisions, or cloud crushing events generate a dense sheet of shock-compressed material. Let us assume that high-mass star formation results from the development of gravitational instabilities in such a shocked layer (Mouschovias, Shu and Woodward 1974; Elmegreen, 1979,1982; Cowie, 1981; Balbus and Cowie, 1985). Why should this preferentially produce high-mass stars? For an infinite isothermal sheet of surface density $\sigma\ M_{\odot}\ pc^{-2}$, the fastest growing mode of instability (Larson 1985; Field 1985) has a characteristic mass M_C given by

$$M_C = 2.4\ T^2 / \sigma \quad M_{\odot}$$

The typical cloud surface density is of order 100-200 $M_\odot$ pc^{-2}. For low-mass star formation, cloud temperatures are 5-15 K so that M_c lies in the range 0.3-5 solar masses. However, in a shocked sheet, the surface density increases as the mean temperature of the post-shock gas falls towards its equilibrium level. Thus, the most massive modes of instability are triggered first, and fragmentation proceeds to smaller and smaller characteristic Jeans mass. However, this process is terminated when the massive stars reach the main sequence and ionise and break up the shocked layer. The characteristic Jeans mass of fragments is therefore determined by their characteristic temperature, given by the condition that the cooling timescale at this temperature should be comparable to the collapse timescale of the largest fragment.

Cloud-cloud collisions reduce the momentum, and therefore, the velocity dispersion of the gas in the vertical (w-plane). Thus, energetic processes associated with the high-mass mode of star formation must, in the steady state, feed as much momentum into the gas of the ISM as is being lost in cloud-cloud collisions. If $d\sigma_*/dt$ is the surface rate of star formation, and σ_g is the surface density of gas, then:

$$(d\sigma_*/dt) = \beta\, \sigma_g / \tau_{cc} \qquad (3.1)$$

where the cloud-cloud collision timescale is τ_{cc} and where the constant of proportionality β is composed of both a "spontaneous" term and a "stimulated" term which accounts for the fact that a burst of star formation may induce a local overpressure leading to cloud crushing and induced star formation in its vicinity. These processes represent the justification for the model of stochastic self-propagating star formation (Gerola and Seiden 1978; Seiden and Gerola 1979; Feitzinger et al. 1981) which has enjoyed considerable success in reproducing the morphological features of both spiral and irregular disk galaxies. Here we assume that the coefficient of stimulated star formation is linearly related to the spontaneous term, so that β is not too sensitive to the galaxian environment.

Since the cloud-cloud collisions are radiative, the physical parameter which is conserved in the collision is the momentum. In the steady state disk, therefore, the modulus of the sum of the momentum vectors of the individual gas clouds is maintained at a constant value. Thus, in steady-state;

$$\gamma\,(d\sigma_*/dt) = \sigma_g v_g/\tau_{cc} \qquad (3.2)$$

where v_g is the vertical w-velocity dispersion of the gaseous layer and γ is a coupling constant. To the extent that the IMF and the energy yield from the high mass stellar population does not depend on metallicity, γ will be independent of galaxian environment.

Equations (3.1) and (3.2) together imply an observational consequence;

$$v_g = \beta\gamma \qquad (3.3)$$

that is to say that the vertical velocity dispersion of the gas in **all galaxies** will be the same, and independent of the radial coordinate. This is true of all disk galaxies which have so far been observed (van der Kruit and Shostak 1984, van der Kruit 1985; private communication, Meatheringham et al. 1987). The HI velocity dispersion is of order 6-10 km.s^{-1}, and varies little between the arm and interarm regions. However, it is seen to increase in regions of active star formation, such as the 30 Doradus region in the LMC.

If the volume filling factor of molecular clouds in the gaseous disk is f, then the cloud-cloud collision timescale is given approximately by:

$$\tau_{cc} = 2\,z_g\,/v_g\,f^{2/3} \qquad\qquad (3.4)$$

Star formation pressurises the ISM in the plane, and, because the hot gas from supernova explosions can bubble up to form a hot halo to the galaxy, the scale height for pressure variation is long compared with the matter scale height. Thus, in regions where the hot coronal gas generated by supernova explosions is incapable of driving a galactic wind the pressure, P, is simply proportional to the surface rate of star formation ;

$$P = \alpha\,(\,d\sigma_*/dt\,) \qquad\qquad (3.5)$$

We assume that the molecular and atomic gas clouds are in equilibrium with the external pressure, (this is implied by the scaling relationships for individual molecular clouds; Larson, 1981; Chièze 1987), and that they move in the disk potential defined by the gas and the stars with scale height z_*.

This then allows a solution for the net star formation rate per unit area of disk:

$$(d\sigma_*/dt) = \text{const. } z_*^{\,-1/2}.\sigma_t^{1/2}\,\sigma_g \qquad\qquad (3.6)$$

As might have been expected on purely phenomenological grounds, the star formation rate depends primarily on the local surface density of gas. However, there is also a dependence on total surface density, and on the stellar scale height. Since the scale height evolves with time due to stellar diffusion (see below), the z_* term introduces an additional temporal evolution of the star formation rate.

4. The Low-Mass Mode of Star Formation.

It has long been recognised that the lifetime of molecular clouds is at least an order of magnitude longer than their free-fall timescales (Kwan 1979; Blitz and Shu 1980) and that the typical turbulent velocities are highly supersonic. It is clear that an energy source is required to give the required turbulent support. Norman and Silk (1980) and Franco (1983) suggested that the winds from young stellar objects might provide this energetic input. Franco and Cox (1983) were able to derive a stellar birthrate on the assumption that this turbulent input also serves to regulate the rate of low-mass star formation within molecular clouds. Essentially, the structure of a such a cloud at any instant can be regarded as a set of interlocking shells of compressed gas, orbiting each other under their mutual gravitational attraction. Amongst these, just a sufficient number are in a state of collapse under their self-gravity to provide enough new stars for the turbulent support. Thus, in the absence of any external perturbation, the cloud is converted into stars at a nearly constant rate.

Direct observational evidence for such a picture is forthcoming from two separate directions. Firstly, Fukui et al. (1987) have shown that in one cloud, the Orion Southern molecular cloud, the energy input from the CO outflow sources found in an unbiased survey is sufficient to balance the cloud against turbulence dissipation, provided that the timescale over which this operates is an order of magnitude greater than the free-fall timescale. More general evidence to support the Franco and Cox picture comes from the observed mass / radius or velocity dispersion / radius relations. Larson (1981) found from observation that $M \propto R^2$ and that the velocity dispersion, $\Delta v \propto R^{1/2}$. Chièze (1987) has shown that this is exactly what would be expected if the interstellar clouds are close to gravitational instability in a constant pressure environment, and suggests that the sub-condensations may form a gravitational N-body system in a quasi-static virialised condition, which will leave the scaling relationships unchanged for the individual fragments.

Following Franco and Cox (1983), at any instant the volume of the cloud, V_c, should be filled with N_s interacting momentum-conserving shells each of volume V_{int};

$$V_c = N_s V_{int} \tag{4.1}$$

If the interaction timescale is t_{int}, then the rate of star formation per unit volume, S_v, will be given by:

$$S_v = \varepsilon (V_{int} t_{int})^{-1} \tag{4.2}$$

where ε is an efficiency factor. This equation assumes that the cooling timescale in the shock-compressed layer is short compared with t_{int}.

By consideration of the detailed physics of this situation, we arrive at the star formation rate per unit mass, $S_m = S_v / \rho_c$;

$$S_m = const.\rho_c^{\zeta} \tag{4.3}$$

where possible values of the exponent ζ lie in the range $-1/8$ to $+1/4$, respectively. Thus the assumption that the low mass star formation rate is simply proportional to the mass of available gas is, in general, a very adequate assumption.

The rate given by eqn. (3.8) is appropriate for molecular clouds which are in the quasi-equilibrium state. However it must be recognised that clouds are destroyed by stellar winds and supernova explosions in regions of high-mass star formation, and fragments may be blown to large scale height, as has been observed in very active regions of star-formation in the Magellanic Clouds (Caulet et al. 1982; Dopita, Mathewson and Ford 1985) the Galaxy (Heiles 1979, 1984), or in M31 (Brinks 1981). This is the probable origin of the so-called "cirrus" clouds. If molecular clouds reform out of this component, they must do so as a result of coalescent collisions at low relative velocity, followed by radiative shedding of internal turbulent motions. This implies that the steady-state low-mass star-formation of eqn. (3.8) is effective only for a duty factor F given by;

$$F = \tau_{cc} / [\tau_{cc} + \tau_{diss}] \tag{4.4}$$

where τ_{diss} is the timescale of turbulence dissipation. At densities larger than 200 cm^{-3}, the cooling is dominated by molecular species of which by far the most dominant is CO (Goldsmith and Langer, 1978, Hollenbach and McKee,1979), which ensures that the cooling rate declines rapidly toward low abundances. The timescale for dissipation of turbulence is;

$$\tau_{diss} = (E_{therm} + E_{turb} + E_{shear}) / \Lambda \tag{4.5}$$

where Λ is the cooling rate per molecule and the terms, E, represent respectively the energy per molecule in thermal motions, random turbulence and in turbulence due to rotational velocity shear, $(1/r).d(V_{rot}/r)/dr$, across the region over which cloud fragments are accreted i.e., on the local Oort A value. Thus, equations (4.4) and (4.5) show that, in regions of low metal abundance, or in regions of high velocity shear, the duty factor for low-mass star formation becomes low.

5. The Structural Evolution of the Disk.

In order to use the results of the previous two sections in a model of galactic evolution, we require to know how the gaseous and stellar disks evolve with age. Essentially, this resolves itself into three separate problems, the evolution of the gas layer, stellar diffusion and the infall / outflow problem.

The total surface density, $\sigma_T (r)$, in disk galaxies is seen to decline radially outwards according to an exponential law with scale length R_o;

$$\sigma_T (r) = \sigma(0).\exp[-r / R_o]$$

(Freeman 1970, van der Kruit and Searle 1981a,b). Since eqn.(3.3) implies a constant velocity dispersion in the gas layer then the vertical scale height in the gas varies as;

$$z_g = z_g(0).\exp[-r / 2R_o] \tag{5.1}$$

Such a variation is in fair agreement with the observations of the thickness of the HI layer in our Galaxy, assuming a disk scale length of order 4 kpc (Downes and Güsten, 1982).

Stars born in this gas layer will diffuse out of the layer by the dynamical heating which is a result of interactions with gravitational perturbations in the disk, spiral density waves, or giant molecular clouds (Spitzer and Schwarzschild 1951,53; Wielen (1977), Twarog 1980; Vader and de Jong 1981; Lacey 1984 and Villumsen 1985). According to this theory, if stars are born at an intrinsic velocity dispersion $V_*(0)$, then at time t they will have acquired a velocity dispersion $V_*(t)$ given by:

$$V_*(t) = V_*(0) [1 + t / t_{diff}]^{1/3} \tag{5.2}$$

this equation remains valid only for so long as the scattering clouds and the stars can be considered to remain in the same layer. The breakdown of this assumption will lead to a change in the exponent. Weilen (1977) finds empirically that an exponent of 1/2 gives a good fit to the observations, whereas both Lacey (1984) and Villumsen (1985) find that a rather lower exponent in the range 0.25-0.35 is a better fit to the theoretical models. The diffusion or scattering timescale depends on the characteristic mass,M_c, of the scattering centers (molecular clouds or spiral arm density perturbations) and the local density. The Spitzer Schwarzschild formulation gives;

$$t_{diff} = 4 V_*(0)^3 / 3\pi^{3/2}G^2M_c\rho \ln[a] \tag{5.3}$$

where a is an impact parameter. Locally, t_{diff} is of order 5×10^7 years. For the purpose of generality, it is convenient to cast the diffusion time in terms of the diffusion timescale of a disk initially entirely gaseous, $t_{diff}(0)$, which can be related to the initial diffusion timescale for our reference conditions t_r;

$$t_{diff}(0) = \varphi V_g(o)^3 /\rho_g(0) = t_r [\sigma_r / \sigma_T]^2 \tag{5.4}$$

To a first approximation, we can assume that the conversion of matter into stars and remnants proceeds exponentially with time, and that the gas collapses to a thin disk on a timescale which is short in comparison to the current age of the galaxy. These equations can therefore be used to compute the time evolution of the mean stellar velocity dispersion V_*. From many runs with different gas depletion and infall timescales, an analytic fit to V_* is found to be;

$$V_* = V_g [1 + (t / t_r)(\sigma_T / \sigma_r)^2]^m \qquad (5.5)$$

where the exponent, m varies between 0.25 and 0.31, in good agreement with both Lacey (1984) and Villumsen (1985). With the particular value m=0.25, a very interesting result is found. From the above equations it follows that, at any radial position in the galaxy, where the surface density is $\sigma_T(r)$, the scale height of the stars, $z_*(r,t)$, is given by:

$$z_*(r,t) = (V_g^2 / \pi G \sigma_T(r,t)) [1 + (t / t_r)(\sigma_T(r,t) / \sigma_r)^2]^{1/2} \qquad (5.6)$$

Since $t_r \ll t$ at the current time, this simplifies to:

$$z_*(r,t) = z_*(t) = (V_g^2 / \pi G \sigma_r)(t / t_r)^{1/2} \qquad (5.7)$$

Thus we have the very important result that, in a galaxy in which the disk formed at a particular epoch, the **current scale height** of the stars **depends only on the age of the disk**, and is independent of the radial coordinate in the galaxy. The variation in the exponent m from the value of 1/4 is so weak that this result is essentially independent of the actual value of m (see also Lacey and Fall 1983). This result is exactly what is required to explain the results of van der Kruit and Searle 1981a,b;1982, who found that the observed light distribution in edge-on galaxies could be best fitted by a model in which the stellar scale height is constant with radius.

Eqn. (5.7) allows the functional dependence on the rate of high-mass star formation in equation (3.6) to be fully determined. This shows that stellar diffusion, by allowing the gas layer to swell up at late times, has the effect of reducing cloud/cloud collisions and, therefore, the star formation rate.

6. Evolution of the Solar Neighbourhood

The observational material that has been accumulated in our solar neighbourhood over the years is sufficient to place very severe restraints on any model of galactic evolution. The end point of the models is determined by the measured age, metallicity, present-day local gas and stellar content and scale height, estimates of the mass fraction in stellar remnants, the rate of star formation and the gas depletion timescale. The history of the local disk can be inferred by age / metallicity , metallicity / height , stellar dynamics / age and element abundance ratio / metallicity relationships, or by the metallicity distributions of long - lived stars.

There is absolutely no reason to suppose that our solar neighbourhood is in any way peculiar in its properties. It is therefore reasonable to hope that a galactic evolution model which can successfully account for all the locally observed relationships, should also be capable of describing the radial variation of observable parameters in our Galaxy. These include gas content and star formation rates,the gas and stellar scale height, the metallicity distributions of stars and the abundance gradient in the gas.

The local surface density of matter in the region of the sun is about 75 $M_\odot pc^{-2}$ and has a scale height of about 300 pc (Güsten and Mezger 1982; Bahcall 1984a,b). An appreciable fraction of this is in an unseen form, with a surface density of about 30 $M_\odot pc^{-2}$ and a scale height not exceeding 700 pc.The local gas content has been estimated by Güsten and Mezger (1982), using the observations of Burton and Gordon (1978) and, more recently, Lacey and Fall (1985) and Rana and Wilkinson (1986) have extensively reviewed the question, which depends critically on the assumed molecular fraction. Derived values range from 3.0 to 6.5 $M_\odot pc^{-2}$ with a mean of about 5 $M_\odot pc^{-2}$. Thus the gas fraction ranges from 0.04-0.085, with a best guess value of about 0.06.

The local star formation rates are variously computed at between 3 and 8 $M_\odot$ $pc^{-2} Gyr^{-1}$ (Miller and Scalo,1978; Smith,Biermann and Mezger, 1978; Güsten and Mezger ,1982). Thus gas passes through a stellar generation in a timescale of order 1-2 Gyr in the solar neighbourhood. This timescale is oppressively short, a fact that had been noted by Larson, Tinsley and Caldwell (1980), by Rocca-Volmerange, Lequeux and Maucherat-Joubert (1981) in the context of the Magellanic Clouds and, for a sample of other galaxies, by Kennicutt (1983). The problem is therefore not confined to the solar neighbourhood. Even when the return of gas by intermediate mass stars is taken into account, the gas depletion timescale is only 2-8 Gyr, still very short when compared with the disk age of 15 Gyr.

The collapse timescale to the thin disk configuration is taken as a free parameter, but should not be very different from the free-fall timescale in any particular model galaxy. This timescale,τ_{ff}, is given by $\tau_{ff} = 1.65(R_{100} / M_{11})^{1/2}$ Gyr ; where R_{100} is the proto-galactic radius in units of 100 kpc, and M_{11} is the galaxian mass in units of 10^{11} solar masses. For our Galaxy, the mass is of order $M_{11} = 5$ (White and Frenk, 1983; Meatheringham *et al.* 1987). The radius cannot initially have been very much larger than 100 kpc, therefore the free-fall timescale was of order 1 Gyr.

In our model of star-formation, the current gas content of our local region of the Galaxy is determined principally by three efficiency factors; the efficiencies of the high and low mass modes of star formation, and the duty factor for low mass star formation. We express the first of these as an initial gas depletion timescale, τ_{gas}, which is inversely proportional to the initial rate of high mass star formation in a wholly gaseous disk with the reference surface density of 100 $M_\odot .pc^{-2}$. The relative efficiency of low mass star formation, R, is the rate relative to the high mass mode at time t = 0 in these reference conditions, with a duty factor of 1.0. Both these factors are constrained by the observed gas fraction and the current age of the disk. The duty factor depends critically on metallicity and the local turbulent energy content of the coalescing cloudlets. Since the duty factor determines the low-metallicity cut-off of the low mass mode of star formation in the disk, it is very tightly constrained by the observations of the metallicity distribution in the lower main sequence (Pagel and Patchett 1975) .

It turns out that these observational restraints are fairly severe. The power law slope of the IMF, p, is restricted to a very narrow range p $=2.2\pm0.15$ and the sensitivity to changes in τ_{gas} and R is shown in figure 1, from which it can be inferred that the "best guess" parameters are; $\tau_{gas} = 2.2\pm0.4$ Gyr; R = 2.0 ± 1.0 and $\tau_{diss}/\tau_{cc} = 0.2\pm0.1$.

With the parameters set by the above procedure, the model may now be used in a predictive fashion. To give age / metallicity relationships, the variation of metallicity ratios with age, and to predict metallicity / velocity dispersion / scale height relaionships for the local disk stars.

The evolution of the scale height or the velocity dispersion depends, in this model, on both the metallicity / age relationship and the orbital diffusion process. The value of the initial diffusion timescale is determined by the requirement that the initial gas velocity dispersion is 8 km s^{-1}(Stark, 1979); which remains independent of time in our models, and that the current scale height for the disk dwarf stars in the solar neighbourhood is 300 pc (Gilmore and Reid, 1983). This gives an initial diffusion timescale for the reference surface density (100M$_{\odot}$ pc^{-2}) of 1.0±0.3 10^7 years. With this parameter, the theoretical age / metallicity / velocity dispersion relationships are solved, and is given in table 1 with observational determinations by Janes (1979) and Norris (1987). The velocity dispersion of the oldest disk stars is predicted to be 42 km s^{-1} in this model, using an initial gas velocity dispersion taken from Stark (1979) and references therein.

From the discussion in this section, it is clear that the solar neighbourhood presents an adequate number of observational relationships to overdetermine the free parameters in our galactic enrichment model. Although the derived parameters are unique in the sense that they are closely constrained by the observations, it is possible to fit this observational data set almost as well on the basis of a completely different set of assumptions. See, for example, Vader and de Jong (1981). The true test of the model lies in its ability to predict the gross structural properties of disk galaxies, see above, and in its ability to account for the radial dependence of observable physical and chemical parameters of galaxies in the context of a single set of star formation efficiency parameters. This will now be tested in the case of our Galaxy.

TABLE 1: METALLICITY / AGE / DYNAMIC RELATIONSHIPS FOR DISK RED GIANTS

Twarog (1980)		Janes	Norris	Model Results		
Age (Gyr)	[Fe/H]	V_z (km.s^{-1})	V_z (km.s^{-1})	Age (Gyr)	[Fe/H]	V_z(km.s^{-1})
0.0	......	8.0		0.0	......	8.0
0.5	0.18	16.7	15.5	0.5	0.10	8.5
2.0	0.10	13.0	16.2	2.0	0.05	10.3
3.4	0.00	13.9	12.0	3.4	0.00	12.3
5.3	-0.08	16.3	14.1	5.3	-0.09	15.0
7.2	-0.17	18.5	14.4	7.2	-0.21	17.8
8.6	-0.26	21.6	17.2	8.6	-0.30	21.2
9.7	-0.35	23.1	18.2	9.7	-0.41	24.7
10.8	-0.46	26.0	21.3	10.8	-0.58	29.2
11.6	-0.57	35.4	33.0	11.6	-0.68	32.1

7. Global Evolution of the Galaxy.

The hypothesis of bimodal star formation has already been the subject of evolutionary models (Güsten and Mezger, 1983; Larson 1986; Wyse and Silk 1987). The major problem here is to avoid ad-hoc assumptions about the relative star formation efficiencies in the high and low mass modes. In the context of the bimodal star-formation model presented here, there remains one obstacle to the application of the solar neighbourhood model to the Galaxy at large. The duty factor for low mass star formation, and by implication the timescale for the dissipation of turbulence, has only been determined locally. The local Oort A value is 15 km.s^{-1}.kpc^{-1}. In the region over which molecular clouds reform, roughly equal to the thickness of the molecular cloud region of the disk, about 140 pc, the shear heating term is of order 2-5 times as large as the thermal term. In fact, there is little difference between models provided that the shear term exceeds the thermal and random terms, so we adopt a value of 4 for the ratio of these terms locally, with a probable error of a factor of two.

The observational material on gas content, star formation rates and oxygen abundance, have been adequately reviewed recently by Lacey and Fall (1985), and to facilitate comparison with their results, these observations as summarized in their figures 2, 3c and 3d are used here. We have supplemented the abundance data they used by the Mezger *et al.* (1979) observations of HII regions near the galactic center.

The results of the closed box model (with a 1.0 Gyr collapse timescale) are shown as a solid line in figure 1. It is clear that the model gives an excellent fit to the run of gas content and to

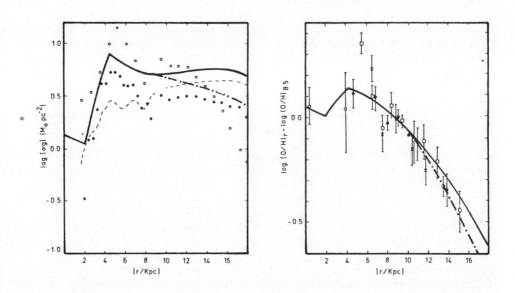

Figure 1: *The run of gas content, left, and oxygen abundance, right, with radius in the Galaxy, adapted from Lacey and Fall (1985). The solid curve is for the model without outflow, and the dot dash curve is for the model in which the gas loss increases linearly with radius outside 9 kpc.*

the oxygen abundance variation between 0 and 10 kpc in the disk. Note that, by contrast with the simple model, the abundance gradient in the gas is quite steep. This is the result of the change in effective yield with radius associated with the partial suppression of the low mass mode of star formation in the inner regions of the galaxy.

The model predicts the presence of a central hole in the gas distribution which corresponds to the changeover from a flat rotation curve to a solid-body rotation curve near the center. The disappearance of the shear term leads to a rapid re-formation timescale for molecular clouds close to the galactic center, and a correspondingly more efficient low mass mode of star formation. As a result, the gas is depleted in a shorter timescale, and a lower gas content and oxygen abundance result. Other galaxies also show this central hole (Wevers 1984), and, in the cases where the resolution is adequate to draw conclusions, this also appears to be associated with the transition from flat to solid-body rotation. Perhaps the clearest example of the phenomenon is M31 (Brinks 1981,1984). Here the transition is particularly abrupt, and corresponds to high precision with the inner edge of the HI-rich disk.

Since the low-mass mode of star formation develops when a critical metallicity is reached in the gas, the abundance gradient shown by the stars will be less than that which applies to the gas. Also, since the rate of star formation passes through a peak in the models, we expect that the average metallicity of the stars will be lower than that of the gas at any radius. Lewis (1986) has determined the abundances and the radial coordinates of some 600 disk giants, and indeed, finds this to be true.

8. Application to External Galaxies.

Dopita (1985), showed that the Donas and Deharveng sample of galaxies gave a good observational correlation between the specific rate of star formation of massive stars, and the gas fraction. In the model presented here, the theoretical tracks represent only a first-order fit to the observed correlation. However, they also demonstrate that scatter on this correlation may be induced by secondary parameters such as disk size (in scalelengths), age, total surface density, and variations in the relative efficiencies of the two modes of star formation caused principally by the mean velocity shear, in closed box models, or by galactic winds, in open models.

Provided that the epoch of galaxy formation is well-defined, at least for the larger spirals, we expect to find a second correlation, between the metallicity of the disk and its surface brightness. This will occur because the relative number of high mass stars ever formed, and the conversion of the matter into stars is more complete in regions of higher surface densities. Such a correlation was found by Wevers (1984), in the form of a correlation between the abundance - sensitive HII region line ratio $\log([OIII] + [OII])/H\beta$, as observed by McCall (1982) and the J-band surface brightness, μ_J. Provided that the mass to light ratio of the stellar population remains constant, μ_J will be proportional to the logarithm of the surface density. In figure 2 is shown the comparison between the observation and model correlations, where the HII region line ratios have been converted to absolute abundance using the calibration of Dopita and Evans (1986). The agreement is very good provided that a surface brightness μ_J of 20.0 corresponds to a surface density of about 1000 $M_\odot \cdot pc^{-2}$. According to the model, scatter in this correlation results

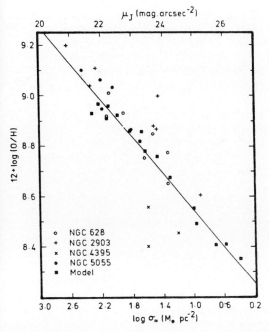

Figure 2: *The correlation between the oxygen abundance and disk surface brightness. This uses the results of McCall (1982) for HII regions, converted to abundance using the calibration of Dopita and Evans (1986), and the surface photometry of Wevers (1984). Also shown are points in the oxygen abundance / surface density plane from a variety of models (filled squares), and the best fit line to the models. The observational points and the models are in good agreement if the four galaxies plotted are approximately coaeval, and if $\mu_J = 20.0$ mag corresponds to a disk surface density of $1000\ M_\odot.pc^{-2}$.*

principally from the local shear in the rotation, and from scatter in the true ages of disk galaxies.

9. Conclusions

The physical model of bimodal star formation developed here has been successfully applied to observations of the solar neighbourhood and the Galaxy in general. We have shown that the observational data in our solar neighbourhood is sufficiently extensive to overdetermine the problem and to restrict all the free parameters of the theory to within very narrow ranges. With these parameters, the theory accounts for the chemical and structural evolution of the solar neighbourhood and can give a good description of the radial variation of gas content, star formation rates and metallicity and the metallicity distribution in the Galaxy.

For external galaxies, the model predicts a constant stellar scale height across the disk, a constant axial velocity dispersion in the gas, and a correlation between the gas fraction and the specific star formation rate. All of these are observed structural parameters of disk galaxies. The model also predicts the observed correlation between the surface density and the metallicity in a given galaxy, or from one galaxy to another, provided that they were born at a common epoch.

In all models, the low-mass mode of star formation, dominant below one solar mass, becomes rapidly more important as the metal abundance exceeds a critical threshold. The early disk history of these galactic models is biased towards the formation of more massive stars. This allows a greater fraction of the total matter, typically of order 30% of the total, to become trapped in dark remnants such as neutron stars, black holes and white dwarfs. This is a similar result to that obtained by Larson (1986), except that we find a rather smaller fraction of matter is trapped in dark remnants.

We predict that galaxies showing a central hole in their HI distribution, or in the rate of high-mass star formation, are galaxies in which these inner sections show solid-body rotation and in which the outer portion of the rotation curve is flat.

Finally, it is tempting to speculate that the star formation processes discussed here in the context of disk galaxies, may be equally applicable to other galaxian environments. For example; the collision of two galaxies would be expected to produce a very intense burst of high mass star formation. The formation of globular clusters in the halos of galaxies could be the result of quiescent low-mass star formation in primordial concentrations of matter over the collapse timescale of the galaxy, when cloud-cloud collisions are unimportant. Elliptical galaxies might result from the development of a multi-phase medium in the halo phase, with star-formation running to completion before disk formation can complete.

References

Audouze, J., and Tinsley, B.M. 1977,*Ann. Rev. Ast. Ap.*, **14**, 43.

Augarde,R. and Lequeux,J. 1985, *Ast. Ap.* (in press)

Bahcall, J.N. 1984a, *Ap. J.*, **276**, 169.

_____ 1984b, *Ap. J.*, **287**, 926.

Balbus, S.A. and Cowie,L.L. 1985, *Ap.J.*, **277**,550.

Blitz, L. and Shu, F.H. 1980, *Ap. J.*, **238**, 148.

Boer, K.S. de, and Savage, B.D. 1983, *Ap. J.*, **265**, 210.

Brinks, E. 1981, *Ast. Ap.*, **95**, L1.

_____. 1984, Ph.D. Thesis, University of Leiden.

Burton, W.B. and Gordon, M.A. 1978, *Ast. Ap.*, **63**, 7.

Caulet, A., Deharveng, L., Georgelin, Y.M. and Georgelin, Y.P.,
        1982, *Ast. Ap.*, **110**, 185.

Chevalier, R.A. and Oegerle, W.R. 1979, *Ap. J.*, **227**, 398.

Chièze, J.-P. 1987, *Ast.Ap.* (in press).

Cowie, L.L. 1981, *Ap. J.*, **245**, 66.

Cox, D.P. 1979, *Ap. J.*, **234**, 863.

_____. 1980, *Ap.J.*, **245**, 534.

Cox, D.P. and Smith, B.W. 1974, *Ap.J. (Lett.)*, **189**, L105.

Dopita, M.A. 1979, *Ap. J. Suppl.*, **40**, 455.

_____. 1985, *Ap.J. (Lett.)*, **295**,L5.

_____. 1987, IAU Symposium #115, *"Star Forming Regions"*, Tokyo,
        Japan 11-15 Nov., eds M.Peimbert and J.Jugaku, Reidel:Dordrecht., p 501.

Dopita, M.A. and Evans, I.N. 1986, *Ap. J.*,

Dopita, M.A., Mathewson, D.S. and Ford, V.L. 1985, *Ap. J.*, **296**,

Donas, J. and Deharveng, J.M. 1984, *Astr. Ap.*, **140**, 325.

Downes, D. and Güsten, R. 1982, *Astron. Gesellschaft*, **57**, 207.

Elmegreen, B.G. 1979, *Ap.J.*, **231**,372.

_____. 1982, *Ap.J.*, **253**,655.

_____. 1985, in*"Birth and Evolution of Massive Stars and
        Stellar Groups"*, eds. W. Boland and H. van Woerden, Dordrecht : Reidel, p 227.

_____. 1986, IAU Symposium #115, *"Star Forming Regions"*, Tokyo,
        Japan 11-15 Nov., eds M.Peimbert and J.Jugaku, Reidel:Dordrecht.

Field, G.B., Goldsmith, D.W., and Habing, H.J. 1969, *Ap.J. (Lett.)*, **155**, L149.

Feitzinger, J.V., Glassgold, A.E., Gerola, H. and Sieden, P.E.
        1981, *Astr. Ap.*, **98**, 371.

Franco, J. 1983, *Ap. J.*, **264**, 508.
Franco, J., and Cox, D.P. 1983, *Ap.J.*, **273**, 243.
Freeman, K.C. 1970, *Ap. J.*, **160**, 811.
Fukui,Y., Sugitani, K., Takabe, H., Iwata, T., Mizuno, A., Ogawa, H. and Kawabata, K. 1987,*Ap. J. (Lett.)* (in press).
Gerola, H. and Seiden, P.E., 1978, *Ap.J.*, **223**,129
Gilmore, G. and Reid, I.N., 1983, *M.N.R.A.S.*, **202**, 1025.
Goldsmith, P.F. and Langer, W.D. 1978, *Ap. J.*, **222**, 881.
Güsten, R., and Mezger, P.G. 1982, *Vistas in Astronomy*, **26**, 159.
Hamajima,K., and Tosa, M. 1975,*Pub. Astr. Soc. Japan*, **27**,561.
Heiles, C. 1979, *Ap.J.*, **229**, 533.
_____. 1984, *Ap. J. Suppl.*, **55**, 585.
Hollenbach, D. and McKee, C.F., 1979, *Ap. J. Suppl.*, **41**,555.
Jakobsen, P., and Kahn, S.M. 1986, *Ap. J.* , **309**, 682.
Janes, K.A. 1979, *Ap. J. Suppl.*, **39**, 135.
Jenkins, E. 1978, *Ap. J.*, **220**, 107.
Kennicutt, R.C. 1983, *Ap. J.*, **272**, 54.
Kennicutt, R.C., and Kent, S.M. 1983, *A. J.*, **88**, 1094.
Kruit, P.C. van der and Shostak, G.S. 1984, *Astr. Ap.*, **134**, 258.
Kruit, P.C. van der and Searle, L. 1981a, *Astr. Ap.*, **95**, 105.
_____. 1981b, *Astr. Ap.*, **95**, 116.
_____. 1982, *Astr. Ap.*, **110**, 61.
Kwan, J. 1979, *Ap.J.*, **229**, 567.
Lacey, C.G. 1984, Ph.D. thesis, University of Cambridge.
Lacey, C.G. and Fall, S.M. 1983, *M.N.R.A.S.*, **204**, 791.
_____. 1985, *Ap. J.*, **290**, 154.
Larson, R.B. 1972, *Nature Phys. Sci.* **236**, 7.
_____. 1976, *M.N.R.A.S.*, **176**, 31.
_____. 1981, *M.N.R.A.S.*, **194**, 809.
_____. 1985, *M.N.R.A.S.*, **214**, 379.
_____. 1986, *M.N.R.A.S.*, **218**, 409.
Larson, R.B., Tinsley, B.M. and Caldwell, C.N. 1980, *Ap.J.*, **237**, 692.
Lewis, J.R. 1986, Thesis, Australian National University.
Lindblad,P.O. and Westin, T.N.G. 1985, in"*Birth and Evolution of Massive Stars and Stellar Groups*", eds. W. Boland and H. van Woerden, Dordrecht : Reidel, p 33.
McCall , M.L. 1982, Ph.D. Thesis, University of Texas.
McCammon, D., Brunner, A.N., Coleman, P.L., and Kraushaar, W.L. 1971,*Ap. J. (Lett.)* , **168**, L33.
McKee, C.F. and Ostriker, J.P. 1977, *Ap.J.*, **218**, 148.
Meatheringham, S., Dopita, M.A., Ford, H.C. and Webster, B.L. 1987, *Ap. J.* (in press).
Mezger, P.G., Pankonin, V.,Schmid-Burgk, J., Thum, C., and Wink, J. 1979, *Ast. Ap.*, **80**, L3.
Miller, G.E. and Scalo, J.M. 1979, *Ap. J. Suppl.*, **41**, 513.
Mouschovias, T. Ch., Shu, F.H. and Woodward,P. 1974, *Astr.Ap.*, **33**, 73.
Nissen, P.E., Edvardsson and Gustafsson, b. 1985, in ESO Workshop No. 21 on "*Production and Distribution of C,N,O Elements*", Ed. I.J. Danziger, F. Matteucci and K. Kjär, p.131.
Norman, C. and Silk, J. 1980, *Ap.J.*, **238**, 158.
Norris, J. 1987, *Ap. J. (Lett.)*, in press.
Olaffsson, K., Bergrall, N. and Ekman, A., 1984, *Ast. Ap.*, **137**, 327.
Pagel, B.E.J. and Patchett, B.E. 1975, *M.N.R.A.S.*, **172**, 13.

Pettini, M., Benvenuti, P., Blades, J.C., Boggess, A., Boksenberg, A., Grewing, M., Holm, A., King, D.L., Panagia, N., Penston, M.V., Savage, B.D., Wamsteker, W., and Wu, C. -C. 1982, *M.N.R.A.S., 199*, 409.

Rana, N.C., and Wilkinson, D.A. 1986, *M.N.R.A.S., 218*, 497.

Reike, G.H., Catri, R.M., Black, J.H., Kailey, W.F., Mc Alary, C.W., Lebofsky, M.J. and Elston,R. 1980, *Ap. J., 290*, 116.

Rieke,G.H., Lebofsky,M.J., Thomson, R.I., Low,F.J. and Tokunaga,A.T. 1985, *Ap. J., 238*, 24.

Rocca-Volmerange, B., Lequeux, J. and Maucherat-Joubert, M. 1981, *Ast.Ap., 104*, 177.

Savage D.B. and de Boer, K.S. 1979, *Ap. J. (Lett.), 230*, L77.

_____ . 1981, *Ap. J. , 243*, 460.

Schmidt, M. 1959, *Ap. J., 129*,243.

Scoville, N.Z. and Good, J.C. 1987, *Ap. J.* (in press).

Scoville, N.Z., Sanders, D.B. and Clemens, D.P. 1986, *Ap. J. (Lett.), 310*, L77.

Seiden, P.E. and Gerola, H. 1979, *Ap. J., 233*, 56.

Silk,J. 1987, IAU Symposium #115,*"Star Forming Regions"*, Tokyo, Japan 11-15 Nov.1985   eds M.Peimbert and J.Jugaku, Reidel:Dordrecht.

Smith, L.H., Biermann, P. and Mezger, P.G. 1978, *Ast. Ap., 66*, 65.

Sneden, C. 1985, in ESO Workshop No. 21 on*"Production and Distribution of C,N,O Elements"*, Ed. I.J. Danziger, F. Matteucci and K. Kjär, p.1.

Spitzer,L.and Schwarzschild ,M. 1951, *Ap. J., 114*,106.

_____ . 1953, *Ap. J.,118*, 106.

Spitzer, L 1956, *Ap. J., 124*, 20.

Stark, A.A. 1979, Ph.D. Thesis, University of Princeton.

Talbot, R.F. 1980, *Ap. J., 235*, 821.

Tanaka, Y. and Bleeker, J.A.M., 1977, *Space Sci. Rev., 20*, 815.

Tomkin, J. and Lambert, D. L. 1984, *Ap. J., 279*, 220.

Tomkin, J., Sneden, C. and Lambert, D. L. 1986, *Ap. J., 302*, 415.

Tuohy, I.R., Dopita, M.A., Mathewson, D.S., Long, K.S., and Helfand, D.J., 1982, *Ap. J., 261*, 473.

Twarog, B.A. 1980, *Ap. J., 242*, 242.

Vader,J.P., and de Jong, T. 1981, *Ast. Ap., 100*, 124.

Villumsen, J.V. 1985, *Ap. J., 290*, 75.

Wevers, B.M.H.R. 1984, Thesis, University of Groningen

White, S.D. and Frenk, C.S. 1983, in*"Kinematics, Dynamics and Structure of the Milky Way"*, Ed. W.L. Shuter (Dordrecht:Reidel), p343.

Wielen, R. 1977, *Astr. Ap., 60*, 263.

Wilson, I.R., 1983. Ph.D. Thesis, Australian National University

Wyse, R.F.G. and Silk, J. 1987,in Proceedings of the 2nd. IRAS Conference, *"Star Formation in Galaxies"*, Ed. N. Scoville (in Press).

York, D.G., Blades, J.C., Cowie, L.L., Morton, D.C., Songaila, A. and Wu, C.-C. 1982,*Ap. J., 255*, 467.

GRAVITATIONAL AND DYNAMICAL INSTABILITIES OF A DECELERATING PLANE-PARALLEL SLAB OF FINITE THICKNESS

G. Mark Voit
Center for Astrophysics and Space Astronomy, and
Joint Institute for Laboratory Astrophysics
University of Colorado
Boulder, CO 80309 (USA)

Abstract: In order to explore how supernova blast waves might catalyze star formation, we investigate the stability of a slab of decelerating gas of finite thickness. We examine the early work in the field by Elmegreen and Lada and Elmegreen and Elmegreen and demonstrate that it is flawed. Contrary to their claims, blast waves can indeed accelerate the rate of star formation in the interstellar medium. Also, we demonstrate that in an incompressible fluid, the symmetric and antisymmetric modes in the case of zero acceleration transform continuously into Rayleigh-Taylor and gravity-wave modes as acceleration grows more important.

Shock fronts in the interstellar medium may be generated by several common mechanisms: supernovae, strong stellar winds, ionization fronts, to name a few. As these shocks sweep through the interstellar medium they compress the ambient gas into shells of considerably higher density. In increasing the density of the gas, the shocks decrease the characteristic gravitational collapse time, $t_{ff} \sim \sqrt{1/G\rho}$. However, in these compressed shells, Jeans-collapse theory is no longer valid because the shell thicknesses of interest are much smaller than the Jeans lengths of the compressed gas. There has been considerable debate in the literature regarding whether or not shocks can accelerate the process of star formation by enabling gravitational collapse to proceed more quickly. We intend in this paper to illuminate some of the issues in that debate by demonstrating that certain claims in the literature are conceptually ill-grounded, and by presenting a toy model of our own to illustrate the relevant physical phenomena.

We concern ourselves in this paper primarily with the instabilities of a plane-parallel slab of finite thickness, bounded by contact discontinuities on both sides, where the pressures on the two sides are, in general, dissimilar. Elmegreen and Lada (1977) examined the critical stability of an isothermal, self-gravitating, decelerating slab of gas bounded by contact discontinuities on both surfaces. They found that no instabilities arise in the slab until its age is of order t_{ff} in the unshocked medium (t_{norm}), i.e. shocks do not accelerate star formation. Elmegreen and Elmegreen (1978) calculated dispersion relations for the modes of a slab of isothermal, self-gravitating, stationary gas bounded by contact discontinuities on both surfaces. They found that the slab becomes unstable to deformational instabilities on timescales much shorter than t_{norm}, but that instabilities to gravitational collapse do not arise until the age of the slab is of order t_{norm}, in agreement with Elmegreen and Lada (1977). Vishniac (1983) found dispersion relations for the gravitationally unstable modes of an infinitesimally thin shell of gas having a polytropic equation of state and bounded on one side by shock jump conditions and on the other by a contact discontinuity. His results show that gravitational collapse of the shell can

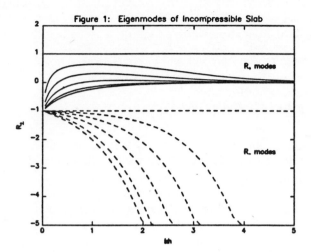

Figure 1 illustrates how the eigenmodes of an incompressible slab change with changing $\mathcal{P}$ and differ at differing dimensionless wavenumber kh. The quantity $R \equiv -\eta/\zeta$, where η and ζ are the amplitudes of the perturbative displacements of the top and bottom surfaces, characterizes a given mode. The solid lines give the R_+ modes for $\beta = 2$ ($p_2 = 2p_1$) and $\mathcal{P} = 0.0, 0.2, 0.5,$ 1.0, 2.0, 3.0; the dashed lines give the R_- modes for the same values of $\mathcal{P}$. The quantities $R_\pm$ decrease monotonically with increasing $\mathcal{P}$. Note that for $\mathcal{P} = 0$, the R_+ modes is symmetric $(R_+ = 1)$, and the R_- mode is antisymmetric $(R_- = -1)$. When $\mathcal{P} \neq 0$, at large kh (small wavelength), $R_+ \to 0$, and $R_- \to -\infty$, that is, the surfaces decouple into separate waves on top and bottom. As $\mathcal{P}$ grows, decoupling occurs at longer and longer wavelengths (smaller kh).

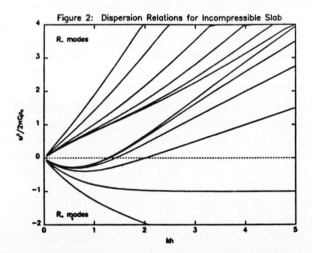

Figure 2 displays the dispersion relations of R_+ and R_- modes – dimensionless frequency squared ($\omega^2/2\pi G\rho_0$) versus dimensionless wavenumber (kh). The relations are plotted for $\beta = 2$ and $\mathcal{P} = 0.0, 0.2, 0.5, 1.0, 2.0, 3.0$. For R_- modes, $\omega^2 > 0$ always, the modes are always stable, and ω^2 always increases with increasing $\mathcal{P}$. The R_+ modes go unstable ($\omega^2 < 0$) for at least some kh, no matter what the value of $\mathcal{P}$, and ω^2 decreases with increasing $\mathcal{P}$. As $\mathcal{P}$ becomes large, the R_+ modes become Rayleigh-Taylor modes, and the R_- modes become gravity waves. That this physics should emerge is not surprising.

occur at times considerably smaller than t_{norm} in the unshocked medium – that shocks <u>can</u> accelerate the rate of star formation in the ISM.

Elmegreen and Lada (1977) begin with an isothermal, plane-parallel slab bounded by two contact discontinuities at pressures p_1 and p_2. The $\hat{z}$ direction points perpendicular to the slab and the $\hat{x}$ direction along it. They then introduce a density perturbation

$$\rho_1 = \rho_0 e^{ikx} \theta(z).$$

To this system they apply boundary conditions

$$\theta(z_1) = \theta(z_2) = 0,$$

where $z_1 \equiv z(p_1)$ and $z_2 \equiv z(p_2)$ are defined in the unperturbed system. In an isothermal slab the pressure perturbation, P_1, obeys $P_1 = (const.)\rho_1$, so

$$P_1(z_1) = P_1(z_2) = 0,$$

and

$$P(z_1) = P_0(z_1) = p_1$$
$$P(z_2) = P_0(z_2) = p_2.$$

If the boundary of the perturbed slab is displaced by some $\eta(x,t) \ll |z_1 - z_2|$, then

$$P(z_1 + \eta) = P(z_1) = p_1.$$

In this system, dP/dz is monotonic and nonzero at the boundary; therefore, $\eta = 0$ by necessity, and the boundaries must be rigid. Elmegreen and Lada find an instability criterion valid only for a small and nonphysical subset of the possible modes – those for which the boundaries of the slab remain fixed.

Elmegreen and Elmegreen (1978) consider an isothermal, plane-parallel slab of gas bounded by two contact discontinuities at the same pressure; thus, the slab is symmetric about its midplane. They assume all perturbed quantities take the form $\hat{f}(x,z,t) = f(z)e^{i(kx+\omega t)}$. Allowing the top and bottom surfaces to ripple either symmetrically or antisymmetrically about the midplane according to the displacement $\hat{\eta}(x,t) = \eta e^{i(kx+\omega t)}$, they calculate numerically the dispersion relations, $\omega(k)$, for the symmetric and antisymmetric modes. They find that the antisymmetric modes are always stable and that the symmetric modes are unstable to wavelengths longer than of order the thickness of the slab. However, the symmetric modes merely deform the slab, causing it to quilt – they do not collapse gravitationally. Elmegreen and Elmegreen then solve for the time when gravitational instability does set in by calculating when the gas contained within one characteristic wavelength of the symmetric mode will collapse. They fail to recognize that collapse over much longer wavelengths will occur much earlier.

In order to explore the stability of a decelerating plane-parallel slab of gas bounded by two contact discontinuities at differing pressures, p_1 and $p_2 = \beta p_1$, we consider the case of the incompressible slab, which we can solve analytically. Assuming perturbed quantities take the form $\hat{f}(x,z,t) = f(z)e^{i(kx+\omega t)}$, we allow the displacements of the two surfaces to ripple according to $\hat{\eta}(x,t) = \eta e^{i(kx+\omega t)}$ and $\hat{\zeta}(x,t) = \zeta e^{i(kx+\omega t)}$. Since we introduce two degrees of freedom, namely η and ζ, into the problem, we obtain the two dispersion relations

$$\omega_{\pm}^2 = 2\pi G \rho_0 \left[(R_{\pm} e^{-kh} - 1) \right.$$
$$\left. + \frac{kh}{e^{kh} - e^{-kh}} \left\{ (e^{kh} - 2R_{\pm} + e^{-kh}) - \frac{(\beta - 1)}{2} \mathcal{P} (e^{kh} + 2R_{\pm} + e^{-kh}) \right\} \right]$$

$$R_{\pm} = \frac{U(e^{kh} + e^{-kh}) \mp \sqrt{U^2(e^{kh} - e^{-kh})^2 + 4V^2}}{2(V - U)}$$

$$U = kh(\beta - 1)\mathcal{P}$$

$$V = 1 - e^{-2kh} - 2kh,$$

where ρ_0 is the density of the slab, h is the slab thickness, σ is the surface density of the slab, $R \equiv -\eta/\zeta$ describes the eigenmodes, and $\mathcal{P} \equiv p_1/\pi G \sigma^2$ is a dimensionless quantity which describes the relative importance of self-gravity in determining the structure of the slab. We shall refer to the two eigenmodes as $R_{\pm}$ modes. We plot $R_{\pm}$ in Fig. 1 and $\omega_{\pm}^2$ in Fig. 2. When $\beta = 1$, our dispersion relations are qualitatively identical to those that Elmegreen and Elmegreen find for an isothermal slab bounded by equal pressures. We believe this correspondence justifies our admittedly nonphysical assumption of incompressibility.

In summary, we have shown that Elmegreen and Lada (1977) and Elmegreen and Elmegreen (1978) made conceptual errors in their treatments of the instabilities of isothermal slabs of gas. They concluded incorrectly that shocks do not accelerate star formation. A correct approach to the shock induced gravitational instability, albeit in the thin shell approximation, is given by Vishniac (1983). We have also solved for the stability of a slab of incompressible gas bounded by differing pressures. The dispersion relations we find demonstrate that the symmetric and antisymmetric modes of Elmegreen and Elmegreen (1978) are related to Rayleigh-Taylor and gravity waves, respectively.

REFERENCES

Elmegreen, B. G., and Elmegreen, D. M. 1978, *Ap. J.*, **220**, 105.
Elmegreen, B. G., and Lada, C. J. 1977, *Ap. J.*, **214**, 725.
Vishniac, E. T. 1983, *Ap. J.*, **274**, 152.

SELF-REGULATING STAR FORMATION IN ISOLATED GALAXIES

Antonio Parravano

Universidad de los Andes, Fac. Ciencias, Dept. Fisica,
Grupo de Astrofisica, Merida 5101, VENEZUELA

Abstract

Recent investigation (Parravano, 1987) shows that the diffuse phases of the ISM condense mainly by the transition from warm gas to small cool clouds (WG → SC). In this work we introduce the new hypothesis that the star formation rate (SFR) in isolated galaxies is self-regulated in such a way that it maintains Pmax close to the ISM gas pressure. Here Pmax is the gas pressure at the marginal state of stability for the transition WG → SC. This hypothesis leads to a relation between global galactic parameters which appears to be applicable to various morphological groups of isolated galaxies.

The ISM Gas Model

As is shown in Figure 3 of Parravano (1987) the thermochemical equilibrium curve is strongly dependent on the Fuv parameter.* The Fuv parameter (Jura, 1976; Spitzer, 1978; Parravano, 1987) may be written as:

$$ Fuv = \left(\frac{Ye/0.15}{\rho s/3 \text{ g cm}^{-3}} \right) \left(\frac{Qa}{a/2 \; 10^{-6} \text{ cm}} \right) \left(\frac{160 \text{ Md}}{\text{Mh}} \right) \left(\frac{U\lambda}{7.10^{17} \text{ erg cm}^{-3} \text{ Å}^{-1}} \right) \quad (1) $$

where Fuv = 1 for the mean Galaxy conditions. The curve "log (Pmax) vs log (Fuv)" has a nearly linear dependence with a slope of 1.35. Thus, when Pmax is near the gas pressure, an increase (decrease) of the non-ionizing UV radiation tends to inhibit (further) cloud formation and consequently star formation.

For the ISM gas pressure a dependence $P \propto M^{\alpha} R^{-\beta}$ is assumed. Here $\alpha=2$ and $\beta=4$ if the gas pressure is maintained by stellar energy input (Dopita, 1985) or $\alpha=2.5$ and $\beta=4.5$, if hydrostatic equilibrium is assumed.

Hypothesis

In isolated galaxies, the critical pressure Pmax, above which the transition WG → SC occurs, is close to the ISM gas pressure P.

*See last page for a glossary of terms and a list of approximations.

If the hypothesis is correct (P≈Pmax) and the assumed approximations are not far off, isolated galaxies should follow the relation

$$\log \ Mh = A \ + P1 \ \log(Mwd \ Lb \ d1/d2) + P2 \ \log \ d1 \qquad (2)$$

where P1=0.40 and P2=0.39 if the gas pressure is maintained by stellar energy input.   If hydrostatic equilibrium is assumed P1=0.35 and P2=0.47.

Results

To test the validity of equation (2) we used the Sc and Sd galaxies of the sample of Chini et al. (1986), the irregular galaxies of the sample of Hunter et al. (1986) and M82 and the blue compact galaxy IZW18 of the sample of Gondhalekar et al. (1986).   Figure 1 shows the relation log(Mh) vs 0.35 log (Mwd Lb d1/d2) + 0.47 log (d1) for these galaxies.   A least-squares fit to the sample points gives a straight line with a correlation coefficient of 0.97 as well as the expected slope.   The typical error bars are also shown in Figure 1 with the assumption that all the parameters have a relative error of 0.2.

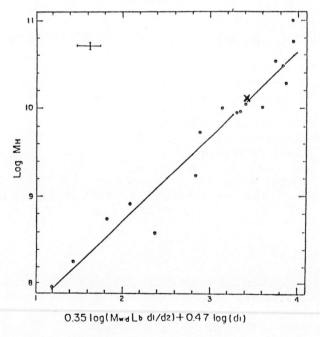

0.35 log(Mwd Lb d1/d2)+0.47 log(d1)

Fig. 1:   The relation log(Mh) vs 0.35 log (Mwd Lb d1/d2) + 0.47 log(d1). The Galaxy is plotted as x.

Discussion

It is necessary to improve the multiple approximation used to derive equation (2) in order to reach definitive conclusions; however many general conclusions may be given:
- For the assumed ISM gas model SFR is self-regulated in such a way that it holds Pmax close to the ISM gas pressure P.
- From all the parameters present in the model the mean energy density U_λ is the one which can vary most rapidly and can transport information most efficiently. Thus U_λ may be considered as the physical quantity that regulates star formation.
- Dust effects are important in determining the SFR in isolated galaxies.
- Blue Compact Galaxies have been described as galaxies that show intense bursts of star formation at the present epoch (Gondhalekar et al. 1986), but here it is shown that the high UV luminosity is due to the low dust content. The sample member NGC 1569 is the one which had the major deviation from the model. This was possibly due to a burst of star formation.
- Interaction between galaxies may produce perturbations in all the model parameters. If the interaction perturbs galaxy parameters to points above (below) the curve in Figure 1 the SFR tends to increase (decrease).

A model including the ISM inhomogeneities and the time lag between the WG $\rightarrow$ SC transition and star formation is necessary to reach definitive conclusions.

In a non-explicit manner, the galactic rate of supernovae plays an important role in this model. Many of the parameters used here are sensitive to the supernova distribution in space and time (i.e. the energy density of non-ionizing UV radiation, the ISM gas pressure and the WG filling factor). Additionally, it is well known that supernova shell fragmentation is an important source of neutral clouds.

Acknowledgements

The author is very grateful to the Centro Cientifico IBM de Venezuela, to VIASA, to the CDCH of the Universidad de los Andes and to the local organizing committee of the I.A.U. Colloquium 101 for financial assistance.

Glossary

WG   - ISM warm gas
SC   - ISM small cool clouds
Pmax - Gas pressure at the marginal state of stability for the
       transition WG $\rightarrow$ SC due to thermal instabilities
Ye   - Mean efficiency for photoelectric emission
ρs - Solid density of grains
Qa   - Grains efficiency factor for absorption
a   - Mean radius of grains

Mg   - Total gas mass
MH   - Total hydrogen mass
M*   - Total stellar system mass
Md   - Total dust mass
Mwd - Total warm dust mass (i.e. T~50 K)
U_λ - Mean non-ionizing UV density energy
Lb   - Blue luminosity
d1   - Major galaxy diameter
d2   - Minor galaxy diameter

Approximations

i)   U = Cte.Lb/(d1d2)

ii) Cte, Mwd/Md, MH/Mg and Mg/M* are the same for all the galaxies
     in the sample

References

Chini, R., Kreysa, E., Krugel, E., and Mezger, P.G., 1986,
     Astr. Ap., 166, L8
Dopita, M., 1985, Ap. J., (Letters), 295, L5
Gondhalekar, P., Morgan, D.H., Dopita, M., and Ellis, R.S., 1986,
     Mon. Not. R. Astr. Soc., 219, 505
Hunter, D.A., Gillett, F.C., Gallagher, J.S., Rice, W.L. and Low, F.J.,
     1986, Ap. J., 303, 171
Jura, M., 1976, Ap. J., 204, 12
Parravano, A., 1987, Astr. Ap., 172, 280
Spitzer, L., 1978, Physical Processes in the Interstellar Medium,
     Wiley, New York

SUPERNOVAE BLAST WAVES IN PROTO-DWARF GALAXIES

Alberto Noriego-Crespo
Canadian Institute for Theoretical Astrophysics

Peter Bodenheimer
Lick Observatory

Abstract: Gas mass loss in proto-dwarf galaxies can be efficiently driven out by blast waves created by the first generation of supernovae. There is, however, a threshold set by the total gravitational potential beyond which gas mass loss does not occur. This limit is in agreement with the one predicted by some Cold Dark Matter senarios.

I. Introduction: Supernovae explosions could explain the chemical and dynamical evolution of dwarf galaxies (Vader 1986, Wyse and Silk 1985; hereafter WS); and could provide an important clue to several outstanding cosmological questions such as the the origin of Lyman absorber clouds (Shields *et al.* 1987), a mechanism for bias galaxy formation and, therefore, a mechanism to differenciate between dwarf and normal galaxies (Dekel and Silk 1986). The mass loss process is expected to be strongly constrained by the overall gravitational potential of the galaxy; WS have shown that the condition for gas removal for a galaxy after one free-fall time and in terms of the one dimensional velocity dispersion σ_{1D}, is given by:

$$\sigma_{1D} \leq 0.58 \left(n_{0\chi}^3\right)^{1/22} E^{4/11} G^{3/22} m_p^{1/22} M_{sn}^{-3/11} \qquad (1).$$

where $n = n_{tot}/\chi = n_0/1\text{cm}^{-3}$ is the gas density; E is the supernova energy; M_{sn} is the mass in forming stars required to create one supernova; G is the gravitational constant and m_p the proton mass. For $\chi = 10$, $E = 10^{51}$ erg and $M_{sn} = 100 M_\odot$, $\sigma_{1D} \sim 57 n_0^{1/22}$ km s^{-1}, or in terms of the three dimensional velocity dispersion, $\sigma_{3D} \lesssim 104 n_0^{1/22}$ km s^{-1}, or a total mass of $\sim 10^9$ $M_\odot$. In these calculations we are interested in how the gas mass loss process depends on the total system mass, the amount of gas and the initial gas metallicity.

The effect of the first generation of stars, as a main mechanism to drive the gas mass loss in the presence of non- luminous matter (NLM), has been recently addressed by Dekel and Silk (1986), and it is in this context that we have calculated models of blast waves created by early supernovae in proto-dwarf galactic systems (see Noriega-Crespo and Bodenheimer 1987).

II. Models with NLM: Dwarf galaxies could be embedded in massive non-luminous halos (Aaronson 1986, Kormendy 1987). The luminous and non-luminous gravitational potentials were modeled by King distribution (King 1966). The gas dynamics in these potentials was followed by means of a one dimensional hydrodynamical code. The ratio between the luminous mass and NLM was chosen to be approximately 1/9 or 1/30; the ratio of their respective core radii was chosen to be $\sim 1/12$. The supernovae rate is assumed to be dominated by Type II supernovae. The number of massive stars $(M > 20 M_\odot)$ was initially determined from a Salpeter initial mass function, but essentially it was considered a free parameter. Since the King models

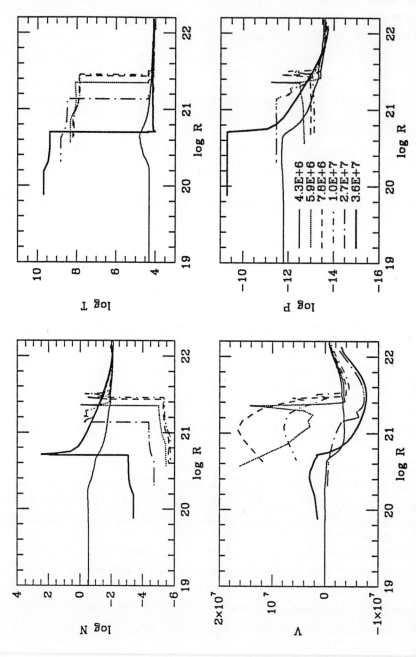

Fig. 1 The evolution of Case 3. Density in cm^{-3}, temperature in K, velocity in cm s^{-1} and pressure in dyne cm^{-2} are plotted as function of radius (in cm) at different times (in yrs). The initial time corresponds to the structure left by the HII region, at $\sim 4.3 \times 10^6$ yrs (thin solid line). The gas driven by the supernovae explosions moves outwards up to $\sim 2.7 \times 10^7$ yrs. After this time the evolution is dominated by the infalling gas; at $\sim 3.6 \times 10^7$ yrs, most of the gas is collapsing at large velocities (thick solid line).

are characterized by a tidal radius, gas that flows beyond this radius has been removed from the model. The fraction of gas has been assumed to be comparable to the total mass in stars (see Table 1). All the cases have an ionizing flux of 2×10^{53} photons s^{-1}, an initial gas temperature of 3×10^3 K, a 5 kpc tidal radius, and a stellar component of 10^8 $M_\odot$.

Computed models are shown on Table 1. The models are in a "non-equilibrium" state, in the sense that the residual gas is initially assumed to be infalling although star formation has already begun. The photoinization of the gas was carried on for a time t_{sn} and then turned off. At this moment the supernovae energy was introduced.

Table 1. Supernovae in Proto-dwarf Galaxies with NLM

(1) Case	(2) ρ_0 g cm^{-3}	(3) Z $Z_\odot$	(4) M_{nlm} $M_\odot$	(5) M_{gas} $M_\odot$	(6) E_{sn} erg	(7) E_b erg	(8) t_{sn} yrs	(9) Mass Loss
1	1.3×10^{-26}	10^{-2}	9.0×10^8	$1. \times 10^8$	$2. \times 10^{55}$	7.9×10^{53}	4.9×10^6	yes
2	1.3×10^{-26}	1	9.0×10^8	$1. \times 10^8$	$2. \times 10^{55}$	8.0×10^{53}	4.8×10^6	no
3	1.3×10^{-26}	10^{-2}	3.0×10^9	$1. \times 10^8$	$2. \times 10^{55}$	2.5×10^{54}	4.3×10^6	no
4	4.2×10^{-27}	10^{-2}	3.0×10^9	3.2×10^7	$2. \times 10^{55}$	7.8×10^{53}	4.2×10^6	no
5	4.2×10^{-27}	1	3.0×10^9	3.2×10^7	$2. \times 10^{55}$	8.0×10^{53}	3.4×10^6	no

Column (1) gives the number of the case, (2) the central gas density in c.g.s units, (3) the gas metallicity, with Z = 1 for a solar abundance, (4) the total non-luminous mass of the model in solar masses, (5) the mass of gas in solar masses, (6) the supernovae energy in ergs,(7) the binding energy of the gas in the gravitational potential of the model in ergs,(8) the time in years, at which the supernovae energy was inserted in the model.

There is just one case from Table 1, Case 1, in which mass loss occurred because of a single blast wave. The escape velocity or, equivalently, the one dimensional velocity dispersion at the stellar core radius (1 kpc) for this case was $\sim$ 60 km s^{-1}. The binding energy of the gas (E_b) was $\sim$ 26 times smaller than the supernovae energy (E_{sn}). The mass gas loss process begins at $\sim 10^7$ yrs after the initial blast, and continues fairly rapidly. At $\sim 3.2 \times 10^7$ yrs, approximately 70% of the gas has been removed, and at $\sim 7. \times 10^7$ yrs $\sim$ 98% of the primeval gas has left the system. Case 1 behaves as expression (1) predicts. The importance of the initial gas metallicity is illustrated in Case 2, where for identical parameters as in Case 1 except for higher Z, gas mass loss did not occur in $\sim 10^8$ yrs because of increased energy loss by cooling.

Models more massive than Case 1 were calculated (Table 1), with total masses of $\sim 3.0 \times 10^9$ $M_\odot$, or an escape velocity of $\sim$ 100 km s^{-1}. There was no gas mass loss in any of the cases during $\sim 4. \times 10^7$ yrs. In all cases the gas has stalled, turned around and collapsed. It could be argued, however, that the gas could bounce back from this collapse; this seems unlikely since at the last time calculated, the gravitational binding energy of the gas was several times larger than the kinetic and thermal energies; and energy losses by radiative cooling have become more important. The evolution of Case 3 is shown in Fig. 1. The supernovae energy was equivalent to 2×10^4 supernovae explosions (10^{51} erg per supernova). Since most of the mass of gas is accreted with velocities of ~ 30 km s^{-1}, just the innermost regions ($r \leq 1$ kpc) have expanded during the supernovae evolution.The maximum radius reached by the blast wave was less than 1 kpc.

III. Discussion: The idealized models that have been calculated are in agreement with expression (1) as long as the initial gas metallicity is low (Case 1). High metallicity increases the effect of radiative cooling and reduces the supernovae effect (Case 2 and 5). In models with total masses higher than 3.0×10^9 $M_\odot$, or escape velocity ~ 100 km s^{-1} for our King models, there was no gas mass loss. The models have concentrated in the first burst of star formation, and have assumed implicitly that star formation has been effective enough to transform at least 50 % of the gas into stars. For cases below the threshold where gas mass loss occurred,this seemed quite reasonable. For cases above the threshold, however, even though $\sim 90\%$ of the gas has been transformed into stars, and all the SN available from the IMF were used, there was no mass loss.

References.

Aaronson, M., 1986, in *Stellar Populations*, ed by C.A. Norman, A. Renzini, and M. Tosi. Cambridge University Press.

Dekel, A., and Silk, J., 1986, Ap. J., **303**, 39.

King, I., 1966, Astron. J. **71**, 64.

Kormendy, J., 1987, in *Nearly Normal Galaxies: From the Planck to the Present*, ed. S.M. Faber. New York: Spring-Verlag, in press.

Noriega-Crespo, A., and Bodenheimer, P., 1987, in preparation.

Shields, G., Silk, J., Wyse, R. F.G., 1987, preprint.

Vader, J.P., 1986, Ap.J., **305**, 669.

Wyse, R.F.G., and Silk, J., 1985, Ap. J. Lett., **296**, L1. (WS).

HIGHLIGHTS OF I.A.U. COLLOQUIUM 101:
EXCERPTS FROM A PANEL DISCUSSION

Chairman:
Sidney van den Bergh
Panellists:
Craig Wheeler, Rob Fesen, Richard Strom, David Helfand, Carl Heiles

Editors' Note: At the conclusion of IAU Colloquium 101 a panel discussion was held. The panellists were asked to give their opinions on significant new developments presented at the Colloquium and to outline questions to be answered in the future. The editors have summarized the discussion.

SIDNEY VAN DEN BERGH: I would like to take a few minutes to give my own impressions of our meeting. A good reference point is the I.A.U. Symposium 101 ("Supernova Remnants and their X-ray Emission") which took place in Venice five years ago. In comparing this meeting with the Venice meeting, it strikes me that the theoreticians now seem to be more realistic; they are no longer talking about a smooth interstellar medium - they have lumps in the medium, and lumps in the supernova remnants. There was a real conundrum in Venice that has now been resolved. At that time there were two kinds of supernovae - Type I and Type II - but there were three kinds of supernova remnants - objects like Tycho, ones like Cas A and objects like the Crab. Now we realize that there are three kinds of supernovae and we seem to agree which kinds of supernovae can produce which kinds of remnants. Another big difference was that five years ago we didn't have SN1987a - this has added a great deal of interest to the present meeting. There are new problems that were brought up at this meeting that were not discussed in Venice. For instance, why is it that many plerions, including the Crab Nebula, don't have a high velocity shell around them?

CRAIG WHEELER: As a person interested in hydrodynamics and the evolution of supernovae I come to a meeting such as this asking "What can we learn from supernova remnants about the kinds of stars that exploded?". At this meeting we have seen some progress on this subject, in particular from the modelling of X-ray spectra (*e.g.* Smith and Jones, Itoh *et al.*, and Hamilton and Fesen). I would also like to know the energy of a supernova for comparison with the predictions of models - carbon deflagration versus core collapse, and a prompt explosion core collapse versus neutrino pumping. I have learned at this conference that there is no answer yet to this second question. I would also like to know the mass and composition of the ejecta, and I would like to know the frequency of supernovae.

As Dick McCray mentioned, the different types of supernova precursors, depending upon the the amount of mass they shed, where they are, and what they do to the ISM, may, from their winds, make different bubbles and super-bubbles. We are after that global problem - the connection between star formation, winds and bubbles, supernovae and supernova remnants, shocks in the interstellar medium, and once again, star formation. From the point of view of stellar evolution, I think that we are making some progress because we are asking different kinds of questions. I was particularly interested in the questions of supernovae and supernova remnants in or near HII regions (Chu and Kennicutt) and molecular clouds (Dubner and Arnal; Burton; Fürst *et al.*).

ROB FESEN: The main theme I've discerned from this conference concerns the importance of the circumstellar material in affecting the local environment and evolution of young supernova remnants, particularly those of Types Ib and II. Lozinskaya showed what a massive star could do to its local interstellar medium - the UV radiation, the stellar winds and wind blown bubbles, and the ejection of slow shells of material. She showed some fine examples of multiple shell structures and one example of a pre-supernova candidate. Rob Braun also showed a perfect example of a wind blown bubble in the cavity next to IC443. Shelled and unshelled Crab-like remnants are, I think, just different phases in the evolution of the same type of object. A "naked Crab" would be one that comes from a massive star progenitor where the shock wave is still moving through the stellar wind cavity. A "shelled Crab" is an older version where the shock wave is now moving through the circumstellar material. I had thought it was going to be impossible to distinguish between remnants of Type I and Type II supernovae, but if Type I's come from white dwarfs, they won't churn up the interstellar medium. This may explain why objects like SN1006 and Tycho form very uniform, smooth shells in the X-ray and radio, and Balmer dominated emission in the optical. Precursors of Type II's really do affect their local ISM and the remnant's early evolution is affected strongly by the mass loss of the progenitor star; Cas A, 3C58 and Kepler are perfect examples (I would call Kepler a Type II rather than Ib because there is evidence in the X-rays and optical for considerable mass around the remnant).

Long ago, in Peter Shull's observations of the LMC remnants N49 and N63A, he found very high velocity features along filaments near the peripheries of those objects. This may be the same phenomenon that Blair, Chu and Kennicutt are seeing in the high surface brightness remnants in M33 where they have observed extremely high expansion velocities, 200 to 300 km/s in what seems to be a fairly large remnant. These might be explained by the theoretical work of Norman *et al.* which suggests a blast wave interacting with very clumpy circumstellar material. On the other hand the optical images and spectra presented partially here by Hester and Raymond, and the high resolution spectra of Greidanus and Strom for old objects like the Cygnus Loop present a completely different picture; it is the interstellar, not circumstellar medium and it is fairly uniform on the large scale. You can't look at the pictures of Hester and Raymond and think there are a lot of clumps; there really are non-radiative sheets there. I wonder whether we will be able to say anything specific about any one region in the interstellar medium - about its density, about the shock velocity and abundances, or whether we will only be able to discern general trends.

Eli Dwek stressed the importance of dust as a density diagnostic in high temperature supernova remnants. On the other hand, Arendt has looked at IRAS data for a large number of supernova remnants in our galaxy and found only 25 to 30 percent with a fair amount of detectable infrared emission. Is this just a sensitivity problem in removing the complex galactic background or is it really a problem of what sort of medium the large remnants have moved into? Maybe when a supernova remnant encounters a dark cloud you see the infrared emission, but if it does not, like SN1006, then you simply don't see the infrared emission. Three more things. I was really impressed with (i) the large number of supernova remnants discovered optically by Long *et al.* in a small region of M33, (ii) the work on Puppis A of Winkler *et al.* showing that the ejection blobs of remnants last at least 3600 years, which is very surprising, and (iii) the work of Ballet *et al.*, Jerius and Teske, and Brown *et al.* on the coronal line emission which they try to match spatially with the radiative or X-ray gas.

I have a list of questions for the future which the people in the optical, UV, and infrared will have to answer. (i) Where is the Crab's outer shock? Why can't we confirm or deny its existence? (ii) If Cas A really was a Wolf-Rayet star, where is its wind blown bubble? The star should have produced a big bubble, and we don't see much evidence for one. (iii) Where are all the other O stars around these young remnants? (iv) Cas A simply cannot be the youngest remnant in our galaxy - it is 300 years old - where are all the other ones? (v) Where is the blast wave in Cas A - is it ahead of the knots or have the knots punched through? (vi) Finally there is the problem of CTB80. It is not a Crab and it is not a shell, it is not a regular remnant, but it certainly is a supernova remnant and it has an X-ray point source, but it is not clear how it fits into the whole picture.

TATJANA LOZINSKAYA: I would like to say something about the fast shell around the Crab nebula. We can speculate that we are beginning to see a shell-like radio source around the Crab nebula because there are some hints that the spectrum is steeper and the polarization smaller on the periphery. If so, then the remnant is not "Crab-like", but a composite shell-Crab showing the very beginning of a radio synchrotron shell. If I am right (I don't insist) then the question arises - where does the turbulent amplification of the magnetic field take place? If it is between ejected and swept-up gas then the Crab is peculiar because of the small expansion but if it is between the pulsar driven plasma and the ejecta then we might expect to see the fast shell.

RICHARD STROM: I thought I would begin by saying just what radio emission tells us. The one great advantage of these long wavelengths is the lack of extinction or, as radio astronomers might say, absorption. The second thing about radio emission is that it is a good tracer of shock fronts and indicates the presence of magnetic fields. The final thing is that radio photons, being low energy quanta, are plentiful and that is why radio surveys, such as the one described by the Bonn group (Reich *et al.*) still provide the best way of finding new galactic supernova remnants. You might ask, should we be looking for more? My answer is yes. Further searches should turn up more bizarre and rare types and it may help us to correct biases in catalogs. They may also reveal some young supernova remnants, ones that are just turning on. If one supernova explodes in the galaxy every 50 years or so, there must be some out there which are now brightening. In Roger Chevalier's review we heard about the models of radio emission from young remnants - it seems to me that the radio light curves will be useful as probes of circumstellar conditions.

There has been some very beautiful work describing radio line emission, primarily HI and millimetre molecular lines. I think we are just about to see a great expansion in the millimetre work, with several telescopes that have just come on line. We have seen some lovely examples of emission from IC443 - one wonders why people haven't been looking there to discover new species because it seems to be so rich in molecules.

At Symposium 101 five years ago we felt that the radio $\Sigma - D$ relationship, certainly as a distance indicator, was finished, but I see it has tentatively come back this time, in Colloquium 101, in Elly Berkhuijsen's talk, and in Jim Caswell's.

I've read many papers in which the first statement is that magnetic fields are dynamically unimportant so we can ignore them. In Don Cox's talk, we heard of the importance of the magnetic pressure in the interstellar medium. I was glad to hear in Roger Blandford's talk about the modelling of shocks in which magnetic fields, quasi-parallel and quasi-perpendicular, are included. Radio observations can help to locate the shock fronts - we should also attempt to understand more about the role of the magnetic fields. Some questions. Should we carry out more surveys or should we try to understand the remnants that we already have? Are all extended non-thermal galactic radio sources supernova remnants? And when do supernova remnants, as distinct from the supernovae themselves turn on as radio sources?

DAVID HELFLAND: I can answer one of the questions. We know that all non-thermal radio sources in the galaxy are *not* supernova remnants.

CARL HEILES: Do we know that all supernovae form remnants that are observable in the radio?

RICHARD STROM: We don't know.

CRAIG WHEELER: There were several poster papers that tried in various ways to estimate how many supernova remnants of the kind that are observed actually exist in the galaxy - the numbers are between 700 and 1200. It seems to me we should keep hunting for these things.

DAVID HELFAND: I would like to thank the previous speakers for not leaving me much time because there isn't much to say. We haven't learned anything new about the X-ray emission of supernova remnants in some time - or at least we don't have any new data to talk about. Only twelve out of the seventy contributed papers were in any way related to X-rays, which is simply a statement that there is a need for the capability for making more X-ray observations, and I think that further observations *are* going to be useful. Cox and McKee and McCray and Dopita convinced everyone that to understand how supernovae interact with the interstellar medium we probably have to go to larger scales than supernova remnants themselves. We probably will find many more old ones and they are likely to tell us more again about the interstellar medium. Going to larger scales can include X-ray emission; there was a fascinating poster on the X-ray emission of the Large Magellanic Cloud (Helfand, Wu and Wang, not published) which might lead us to test some of the global models of the interstellar medium which we heard a significant amount about. On the other hand, if we are talking about Craig Wheeler's question "what can the supernova remnant emission tell us about the supernova?" then our hopes must be pegged on a more detailed study of the half dozen very young remnants that we know already.

A comment to what Rob Fesen said about the naked Crabs and clothed Crabs. It should be noted that the second most X-ray luminous Crab-like object out of the seventeen or eighteen that are known (G29.7-0.3) is inside a shell and without a doubt the X-ray emission tracks the current energy output of the neutron star since the particle lifetimes are only a year or so. That would suggest it is probably not just an age effect that distinguishes between the two - it probably has more to do with the progenitor and perhaps the surrounding medium.

Now, to come to the future and what needs to be done in X-ray astronomy. There are two or three planned missions as most of you know, which I would say don't match very well the crucial questions which we need to answer for the study of supernova remnants. I think what we need for this subject is spatially resolved moderate and high resolution X-ray spectroscopy and unfortunately none of the missions planned before the end of the century really delivers that. We don't need 1 arcsecond resolution for spectroscopy, we need 1 arcminute resolution and the reason is very simple - there just aren't enough photons for any reasonable size detector to use 1 arcsecond resolution. That is the only way we are really going to determine the key things about supernova remnants which X-rays can tell us: the temperatures, the densities, and the compositions of the gas and maybe even the kinematics given high enough resolution. Therefore I cannot see an optimistic outlook for the future of X-ray observations of supernova remnants.

CARL HEILES: Q stands for questions, and I have four of them here; Q also stands for the fraction of the volume of the interstellar medium that is filled by the hot ionized medium of McKee, Ostriker, Cox and Smith. The four questions relate to Q. Question 1: What fractions of supernovae are of Type Ia and of Type Ib? This is important because SNe Type I were thought to be randomly distributed in space and time and would have produced individual bubbles that were uncorrelated. The problem in the past has been that the derived rates of Type I combined with the energies which we believed they had, meant that the Q value due to these remnants alone was so large that many observers didn't believe it. We now have two sub-types of Type I's and only the Ia's will produce these uncorrelated bubbles. Since the frequency of Type Ia's is smaller, so is the value of Q. The Q for these uncorrelated supernovae also depends on the explosion energy and we have heard several values for that energy here this week.

Question 2 has to do with the other type of supernova - the type that are correlated in space and time. They are correlated in two ways. First, they are correlated with each other and second, they are correlated with molecular clouds - something that has been made very clear at this meeting, in a more general sense by the paper of Lozinskaya, and also in a number of individual papers of which the most spectacular one was by Fukui and Tatematsu. When one of these clusters of supernovae goes off, instead of blowing a spherical bubble in the interstellar medium, there is enough energy to punch through the plane and you have to consider the fraction of the area of the plane filled with hot bubbles. We have seen calculations by Tomisaka and Ikeuchi, and by MacLow et al. of this effect, which occurs only if the energy released by the combined effects of a cluster is enough to actually break through the plane. That depends on (a) the energy per supernova, and (b) the number of stars per cluster that will go supernova. In order to avoid an embarrassingly large Q value for these correlated supernovae we can put many more stars per cluster so that we reduce the number of clusters that go off and thereby reduce the number of holes in the plane.

Another way to accomplish this might be through the association with molecular clouds that we have heard about. This leads me to my next question. Question 3: I would like to know if a typical molecular cloud in the galaxy or a typical Giant Molecular Cloud is large enough to act as a significant condensation centre to soak up an appreciable fraction of the energy which would otherwise go into producing these holes.

The last point has to do with the $5\,\mu G$ average magnetic field in the galactic plane that Don Cox derived by considering the pressure produced by the weight of overlying layers. Question 4: How does this large field affect the evolution of individual remnants and, in particular, of super-shells? Does it soak up the energy in the way that Cox suggests? I find this very interesting. First, it would reduce the volume filling factor of the shells and super-shells. Wouldn't it be interesting if Cox is right - much of this supernova energy may simply go into Alfvén waves and heat up the surrounding Warm Neutral Medium! It would solve several problems at the same time. I think that this is a very interesting suggestion which needs some theoretical attention.

MIKE SHULL: Ellen Zweibel and I have worked out a theoretical treatment of hydromagnetic wave heating. The source is hydromagnetic waves from planetary nebulae and supernovae and the damping provides the heat. In this way you can provide enough energy to heat the Warm Neutral Medium. Of course there are large uncertainties in this theory. The more substantial comment I have for you is on your Question 3 - how do massive molecular clouds affect the evolution of supernova remnants? Some years ago there were two papers on supernovae in molecular clouds, one by Craig Wheeler and one by myself, in which we concluded these would be very strong infrared sources.

CARL HEILES: That is an extremely good point and I think your work is very important. The fact that supernovae of these types are correlated with molecular clouds means that some fraction will go off inside molecular clouds and those molecular clouds will essentially soak up all the energy. Here I was referring to the case where you have supernovae sitting outside the molecular cloud - that is a problem related to the one you treated. I would like to see a good treatment of this.

REFERENCES TO PAPERS IN THIS VOLUME:

Arendt, R. G., page 379.

Ballet, J., Caplan, J., Rothenflug, R., and Soutoul, A., page 411.

Berkhuijsen, E. M., page 285.

Blair W. P., Chu Y.-H., and Kennicutt, R. C., page 193.

Blandford, R. D., page 309.

Braun, R., page 227.

Brown, L., Woodgate, B. E., and Petre, R., page 407.

Burton, M. G., page 399.

Caswell, J. L., page 269.

Chevalier, R. A., page 31.

Chu, Y.-H., and Kennicutt, R. C., page 201.

Cox, D. P.,page 73.

Dopita, M. A., page 493.

Dubner, G. M., and Arnal, E. M., page 249.

Dwek, E., page 363.

Fukui, Y., and Tatematsu, K., page 261.

Fürst, E., Reich, W., Hummel, E., and Sofue, Y., page 253.

Greidanus, H., and Strom, R. G., page 443.

Hamilton, A. J. S., and Fesen, R. A., page 59.

Hester, J. J., and Raymond, J. C., page 415.

Itoh, H., Masai, K., and Nomoto, K., page 149.

Jerius, D., and Teske, R. G., page 145.

Long, K. S., Blair, W. P., Kirshner, R. P., and Winkler, P. F., page 197.

Lozinskaya, T. A., page 95.

MacLow, M.-M., Norman, M. L., and McCray, R. A., page 461.

McCray, R., page 447.

McKee, C. F., page 205.

Norman, M. L., Dickel, J. R., Livio, M., and Chu Y.-H., page 223.

Reich, W., Fürst, E., Reich, P., and Junkes, N., page 293.

Smith, B. W., and Jones, E. M., page 133.

Tomisaka, K., and Ikeuchi, S., page 477.

Winkler, P. F., Tuttle, J. H., Kirshner, R. P., and Irwin, M. J., page 65.

INDEX OF AUTHORS

All authors of each paper are listed in this index. The page number opposite the principal author of a paper is in boldface.

INDEX OF ASTRONOMICAL OBJECTS

SUBJECT INDEX